T0383067

Nanopharmaceuticals in Regenerative Medicine

Nanopharmaceuticals in Regenerative Medicine

Edited by
Harishkumar Madhyastha and
Durgesh Nandini Chauhan

CRC Press
Taylor & Francis Group
Boca Raton London New York

CRC Press is an imprint of the
Taylor & Francis Group, an **informa** business

First edition published 2022
by CRC Press
2 Park Square, Milton Park, Abingdon, Oxon, OX14 4RN

and by CRC Press
6000 Broken Sound Parkway NW, Suite 300, Boca Raton, FL 33487-2742

CRC Press is an imprint of Informa UK Limited

British Library Cataloguing-in-Publication Data
A catalogue record for this book is available from the British Library

ISBN: 978-0-367-72113-8 (hbk)
ISBN: 978-0-367-72116-9 (pbk)
ISBN: 978-1-003-15350-4 (ebk)

DOI: 10.1201/9781003153504

Typeset in Times
by SPi Technologies India Pvt Ltd (Straive)

Contents

Foreword

Nanopharmaceutics is a branch of nanobiotechnology with vast applications in diagnostics, regenerative medicine, and drug development in current science of medicine. Within a short span of two decades, the subject has expanded into a promising arena for clinical and translational medicine. The biomedical scientists show immense interest in nanomaterials due to their extraordinary surface to volume area, tunable optical emission, unique electrical, and magnetic behaviour, which particularly helps in drug discovery research. The hybridisation of nanotechnology and tissue regeneration will open a new path of innovation and will have potential application to treat incurable diseases. The book '*Nanopharmaceuticals in regenerative medicine*' is an informative compilation of nanomedicine, combining description of pharmaceutical formulations and their mechanisms of action. The book provides the comprehensive bundle of information and accurate scientific information on nanopharmaceutical use in regenerative medicine and would be epochal to the scientific community, especially clinicians and pharmacists.

I applaud the editors, Dr. Harishkumar Madhyastha who has been my colleague for many years at University of Miyazaki and Smt. Durgesh Nandini Chauhan for the excellent compilation of chapters contributed by well-known scientists and academicians from different countries. All 18 chapters are different from each other in content, but share a single objective of nanopharmaceutical advancement. The most notable chapters include therapeutic applications, technological innovations, and tissue regeneration. The authors successfully navigate the chapter contents with updated literature. I believe '*Nanopharmaceuticals in regenerative medicine*' will remain a valuable resource for years to come.

Prof. Dr. Tsuyomu Ikenoue, MD. Ph.D.
President
University of Miyazaki
Miyazaki-Japan.

Preface

Trajectory of scientific thoughts is propelling rapidly through good research communications. Research ideas will be broadcasted through good publications which are mainly dispersed by review manuscripts, book chapters, etc. A comprehensive scientific dissertation serves as a satellite stop reference book for budding academicians, scientists, professionals, and technologists. With extensive and annotated knowledge and information, the book is a gateway for knowledge dispersion and escalation, community curation, and finally betterment of society. With the advancement of scientific knowledge, a new paradigm of science, nanobiotechnology, is emerging in the area of biomedical science and regenerative medicine. In regenerative medicine arena, nanotechnologies have wide and high-impact benefits like drug development, diagnostics, and delivery system. This book provides an in-depth knowledge on applied nanobiomedical contents for university graduates, researchers, and technocrats with striking balance between fundamentals and applications for regenerative medicine. The book contains 18 chapters covering a wide range of topics related to chemistry, pharmacy, and material science. The chapters are broadly classified into three categories; potential insights into smart technologies, interpretations of different modes as delivery systems, and tissue engineering and generation aspects. Each chapter includes multidisciplinary approaches and recommendations to use the nanotechnologies for tissue regeneration with meaningful conclusions and attracts new ideas for future development. Chapters 1–5 emphasize the applications of nanoparticles in regenerative therapy. Chapters 6–12 focus on different technological approaches devoted to tissue recalcitrant engineering. Chapters 13–18 elucidate the updates on nanomaterials in the field of tissue regeneration, with special focus on osteoporosis, cancer, and cardiology with a pharmaceutical angle.

Date: 16 April 2021

Dr. Harishkumar Madhyastha
Durgesh Nandini Chauhan

Editors

Dr. Harishkumar Madhyastha Ph.D., FBRSI. Harishkumar Madhyastha, faculty at Department of Applied Physiology University of Miyazaki, Miyazaki, Japan. With two Ph.D. degrees, he ignited his research career as a scientist in *Spirulina* biotechnology at MCRC-Chennai. Later on, he pursued postdoctoral research at Miyazaki University that culminated in a faculty position in the Department of Applied Physiology at the University of Miyazaki from 2006. His current research interests include generation and delivery of nanosized metallic payloads for regenerative diseases application. His academic credentials are credited with more than 80 *Sci-E* indexed papers; *h*-value of *29*, clarivate analytic cumulative impact factor of *204.5* and RG score of *33.76* with *six* international patents. His research has been presented in conferences more than 100 and has been frequently picked up by national and international media. He is also actively involved in many international projects including ongoing Indo-Japan scientific and academic collaborations. He is Fellow of Biotechnology Research Society of India (FBRSI), Fellow of Royal Biological Society-London (FRBS-UK). He is an officially recognised Indo-Japan academic spokesperson of University of Miyazaki and engaged in outreach programs to further strengthen the cohesive relationship between Indian academicians and University of Miyazaki-Japan.

Mrs. Durgesh Nandini Chauhan, M.Pharma, has completed her B.Pharm degree in Pharmacy from the Rajiv Gandhi Proudyogiki Vishwavidyalaya, Bhopal, India and her M.Pharma in pharmaceutics from Uttar Pradesh Technical University, currently known as Dr. A.P.J. Abdul Kalam Technical University, Lucknow in 2006. She is presently working as Assistant Professor in Columbia Institute of Pharmacy, Raipur, Chhattisgarh, India. Mrs. Durgesh Nandini Chauhan has 14 years of academic (teaching) experience from Institutes of India in pharmaceutical sciences. She taught subjects as pharmaceutics, pharmacognosy, traditional concepts of medicinal plants, drug-delivery phytochemistry, cosmetic technology, pharmaceutical engineering, pharmaceutical packaging, quality assurance, dosage form designing and anatomy, and physiology.

She is member of Association of Pharmaceutical Teachers of India (APTI), SILAE: Società Italo-Latinoamericana di Etnomedicina (The Scientific Network on Ethnomedicine, Italy), and so forth. She has written more than 10 publications in national and international journals, 16 book chapters, and has edited 7 books. She is also active as a reviewer for several international scientific journals and active participant in national and international conferences.

Contributors

Aakriti
Medical College
Baroda, Vadodara, Gujarat

Philippe Abdel-Sayed
Musculoskeletal Medicine Department
Lausanne University Hospital, University of Lausanne
Switzerland

Augustus Adams
Department of Chemical and Biomolecular Engineering
North Carolina State University
Raleigh, NC, USA

Eric Allémann
School of Pharmaceutical Sciences and Institute of Pharmaceutical Sciences of Western Switzerland
University of Geneva
Switzerland

Lee Ann Applegate
Musculoskeletal Medicine Department
Lausanne University Hospital, University of Lausanne
Switzerland

Amthul Azeez
Post Graduate & Research Department of Zoology
J.B.A.S. College for Women
Chennai, Tamil Nadu, India

Ajay Bahl
Department of Cardiology
Postgraduate Institute for Medical Education and Research
Chandigarh, India

Manju Bala
Department of Pharmaceutical Sciences
Maharshi Dayanand University
Rohtak, Haryana, India

Neena Bedi
Department of Pharmaceutical Sciences
Guru Nanak Dev University
Amritsar, Punjab, India

Nitu Bhaskar
Material Research Centre, Indian Institute of Science
CV Raman Road
Bengaluru, Karnataka, India

Ishani Bhat
Nitte University Center for Science Education and Research
Nitte (Deemed to be University)
Deralakatte, Mangalore, Karnataka, India

Durgesh Nandini Chauhan
Columbia Institute of Pharmacy
Raipur, Chhattisgarh, India

Pooja A Chawla
Department of Pharmaceutical Chemistry and Analysis
ISF College of Pharmacy
Moga, Punjab, India

Viney Chawla
University Institute of Pharmaceutical Sciences and Research
Baba Farid University of Health Sciences
Faridkot, Punjab, India

Michael A. Daniele
Department of Electrical and Computer Engineering
Joint Department of Biomedical Engineering
North Carolina State University
Raleigh, NC, USA

Anju Dhiman
Department of Pharmaceutical Sciences
Maharshi Dayanand University
Rohtak, Haryana, India

Harish Dureja
Department of Pharmaceutical Sciences
Maharshi Dayanand University
Rohtak, Haryana, India

Abdul Faruk
Department of Pharmaceutical Sciences
HNB Garhwal University (A Central University)
Srinagar-Garhwal, Uttrakhand, India

Munish Garg
Department of Pharmaceutical Sciences
Maharshi Dayanand University
Rohtak, Haryana, India

Mahlegha Ghavami
Department of Medical Biotechnology, Faculty of Medical Sciences
Tarbiat Modares University
Tehran, Iran

Madhyastha Harishkumar
Department of Applied Physiology
Faculty of Medicine
University of Miyazaki
Miyazaki, Japan

Zahra Sadat Hashemi
ATMP Department, Breast Cancer Research Center
Motamed Cancer Institute, ACECR
Tehran, Iran

Nathalie Hirt-Burri
Musculoskeletal Medicine Department
Lausanne University Hospital, University of Lausanne
Switzerland

Jainey Puthenveettil James
Department of Pharmaceutical Chemistry
Nitte (Deemed to be University)
NGSM Institute of Pharmaceutical Sciences (NGSMIPS)
Mangalore, Karnataka, India

Olivier Jordan
School of Pharmaceutical Sciences and Institute of Pharmaceutical Sciences of Western Switzerland
University of Geneva
Switzerland

Anmoldeep Kaur
Department of Pharmaceutical Sciences
Guru Nanak Dev University
Amritsar, Punjab, India

Saeed Khalili
Department of Biology Sciences
Shahid Rajaee Teacher Training University
Tehran, Iran

Madhu Khullar
Department of Experimental Medicine and Biotechnology
Postgraduate Institute for Medical Education and Research
Chandigarh, India

Katie M. Kilgour
Department of Chemical and Biomolecular Engineering
North Carolina State University
Raleigh, NC, USA

Pankaj Kumar
Department of Pharmaceutical Chemistry
Nitte (Deemed to be University)
NGSM Institute of Pharmaceutical Sciences (NGSMIPS)
Mangalore, Karnataka, India

Alexis Laurent
Research Department
LAM Biotechnologies SA
Switzerland

Jin Liu
Key Laboratory of Shanxi Province for Craniofacial Precision Medicine Research
College of Stomatology, Xi'an Jiaotong University
China

Lucky. R
Post Graduate & Research
Department of Zoology
J.B.A.S. College for Women
Chennai, Tamil Nadu, India

Tao Ma
Department of Oncology and Diagnostic Sciences
University of Maryland School of Dentistry
Baltimore, MD, USA

Bangera Sheshappa Mamatha
Nitte University Center for Science Education and Research
Nitte (Deemed to be University)
Deralakatte, Mangalore, Karnataka, India

Maruyama Masugi
Department of Applied Physiology
Faculty of Medicine
University of Miyazaki
Miyazaki, Japan

Stefano Menegatti
Biomanufacturing Training and Education Center
North Carolina State University
Raleigh, NC, USA

Adam C. Midgley
Key Laboratory of Bioactive Materials, Ministry of Education
State Key Laboratory of Medicinal Chemical Biology
College of Life Sciences
Nankai University, Tianjin, China

Anupam Mittal
Department of Translational and Regenerative Medicine
Postgraduate Institute for Medical Education and Research, Chandigarh, India

Zahra Mohammadpour
Biomaterials and Tissue Engineering Department, Breast Cancer Research Center
Motamed Cancer Institute, ACECR
Tehran, Iran

Seyed Morteza Naghib
Nanotechnology Department, School of Advanced Technologies
Iran University of Science and Technology
Tehran, Iran

C.M. Noorjahan
Post Graduate & Research Department of Zoology
J.B.A.S. College for Women
Chennai, Tamil Nadu, India

Alexandre Porcello
School of Pharmaceutical Sciences and Institute of Pharmaceutical Sciences of Western Switzerland
University of Geneva
Switzerland

Priyanka
Department of Pharmaceutical Sciences and Drug Research
Punjabi University
Patiala, Punjab, India

Madhyastha Radha
Department of Applied Physiology
Faculty of Medicine
University of Miyazaki
Miyazaki, Japan

Wassim Raffoul
Musculoskeletal Medicine Department
Lausanne University Hospital, University of Lausanne
Switzerland

Mark A. Reynolds
Department of Advanced Oral Sciences and Therapeutics
University of Maryland School of Dentistry
Baltimore, MD, USA

Yusnita Rifai
Faculty of Pharmacy
Hasanuddin University
Makassar, South Sulawesi, Indonesia

Anthony de Buys Roessingh
Department of Pediatric Surgery
Lausanne University Hospital, University of Lausanne
Switzerland

Bharti Sapra
Department of Pharmaceutical Sciences and Drug Research
Punjabi University
Patiala, Punjab, India

Abraham Schneider
Department of Oncology and Diagnostic Sciences
University of Maryland School of Dentistry
Baltimore, MD, USA
Member, Marlene and Stewart Greenebaum Cancer Center
University of Maryland School of Medicine
Baltimore, MD, USA

Kamal Shah
Institute of Pharmaceutical Research
GLA University
Mathura, UP, India

Gaurav Sharma
Department of Translational and Regenerative Medicine
Postgraduate Institute for Medical Education and Research
Chandigarh, India

Kusha Sharma
Department of Pharmaceutical Sciences
Guru Nanak Dev University
Amritsar, Punjab, India

Maryada Sharma
Department of Otolaryngology
Postgraduate Institute for Medical Education and Research
Chandigarh, India

Gurpreet Singh
Department of Pharmaceutical Sciences
Guru Nanak Dev University
Amritsar, Punjab India

Ramandeep Singh
Department of Cardiology
Postgraduate Institute for Medical Education and Research
Chandigarh, India

Mubeen Sultana. D
Post Graduate & Research Department of Zoology
J.B.A.S. College for Women
Chennai, Tamil Nadu, India

Yifan Tai
Key Laboratory of Bioactive Materials, Ministry of Education
State Key Laboratory of Medicinal Chemical Biology
College of Life Sciences
Nankai University, Tianjin, China

Abhay Tharmatt
Department of Pharmaceutical Sciences
Guru Nanak Dev University
Amritsar, Punjab, India

Brendan L. Turner
Joint Department of Biomedical Engineering
North Carolina State University - University of North Carolina Chapel Hill
Raleigh, NC, USA

Michael D. Weir
Department of Advanced Oral Sciences and Therapeutics
University of Maryland School of Dentistry
Baltimore, MD, USA

Hockin H. K. Xu
Department of Advanced Oral Sciences and Therapeutics
University of Maryland School of Dentistry
Baltimore, MD, USA
Member, Marlene and Stewart Greenebaum Cancer Center
University of Maryland School of Medicine
Baltimore, MD, USA
Center for Stem Cell Biology & Regenerative Medicine
University of Maryland School of Medicine
Baltimore, MD, USA

Nakajima Yuichi
Department of Applied Physiology
Faculty of Medicine
University of Miyazaki
Miyazaki, Japan

Zeqing Zhao
Department of Orthodontics, School of Stomatology
Capital Medical University
Beijing, China

1

An Insight into Advanced Nanoparticles as Multifunctional Biomimetic Systems in Tissue Engineering

Kusha Sharma and Abhay Tharmatt
Guru Nanak Dev University, Amritsar, Punjab, India

Pooja A Chawla
ISF College of Pharmacy, Moga, Punjab, India

Kamal Shah
GLA University Mathura, India

Viney Chawla
University Institute of Pharmaceutical Sciences and Research, Baba Farid University of Health Sciences, Faridkot, Punjab, India

Bharti Sapra
Punjabi University, Patiala, Punjab, India

Neena Bedi
Guru Nanak Dev University, Amritsar, Punjab, India

CONTENTS

DOI: 10.1201/9781003153504-1

Introduction

Tissue engineering (TE) aims to restore living tissues that have been lost or damaged. It was first introduced in the late 20th century. TE holds the potential of engineering damaged tissues and organs with recent technological advancements in stem cell therapy, bioengineering, and nanotechnology (Kim et al., 2013). Since 1997, when Cao et al. displayed the auricular implant grown on the back of a mouse and galvanised the province of biomedical research, the whole-organ engineering has been a proof-of-concept and expanding area of interest in regenerative medicine (Cao et al., 1997). Due to increasing incidences of diabetes, obesity, ageing population, and growing trauma cases, tissue engineering markets have huge potential. In a report, the global tissue engineering market is estimated to grow to approximately 110 billion dollars in the coming years (Tian et al., 2007). Regulatory authorities have recently approved the use of tissue-engineered skin to treat burn wounds and skin ulcers.

The first tissue-engineered skin products that were developed for the treatment of burns include Integra (Integra Lifesciences Corporation, USA), Epicel (Genzyme Biosurgery, USA), and TransCyte (Smith and Nephew, UK). Several products are available for chronic skin ulcer management, such as Apligraf (Organogenesis, USA), Dermagraft (Smith and Nephew, UK and Advanced Tissue Sciences, USA), EpiDex (Euroderm, Germany), Epibase (Laboratories Genévrier, France), and BioSeed-S (BioTissue Technologies, Germany). The tissue-engineered products are also commercially available for cosmetic surgery applications such as BioSeed-M and MelanoSeed (BioTissue Technologies AG, Germany) (Law et al., 2017). Several stem cell engineered products are currently in development and are being tested in clinical and preclinical trials (Wang et al., 2017). Nanoparticles (NP) are solid colloidal particles, and their size range is 10–200 nm. Their small size provides a high surface to volume ratio, one of their most attractive intrinsic properties (Acharya and Sahoo, 2011). Due to their size and surface properties, NPs can be exploited as theranostic agents, drug delivery systems, genetic material, and growth factors. NPs used in biomedicine must be biocompatible, and the stability of nanocarriers in biological media is crucial when formulating nanomedicines (Figure 1.1).

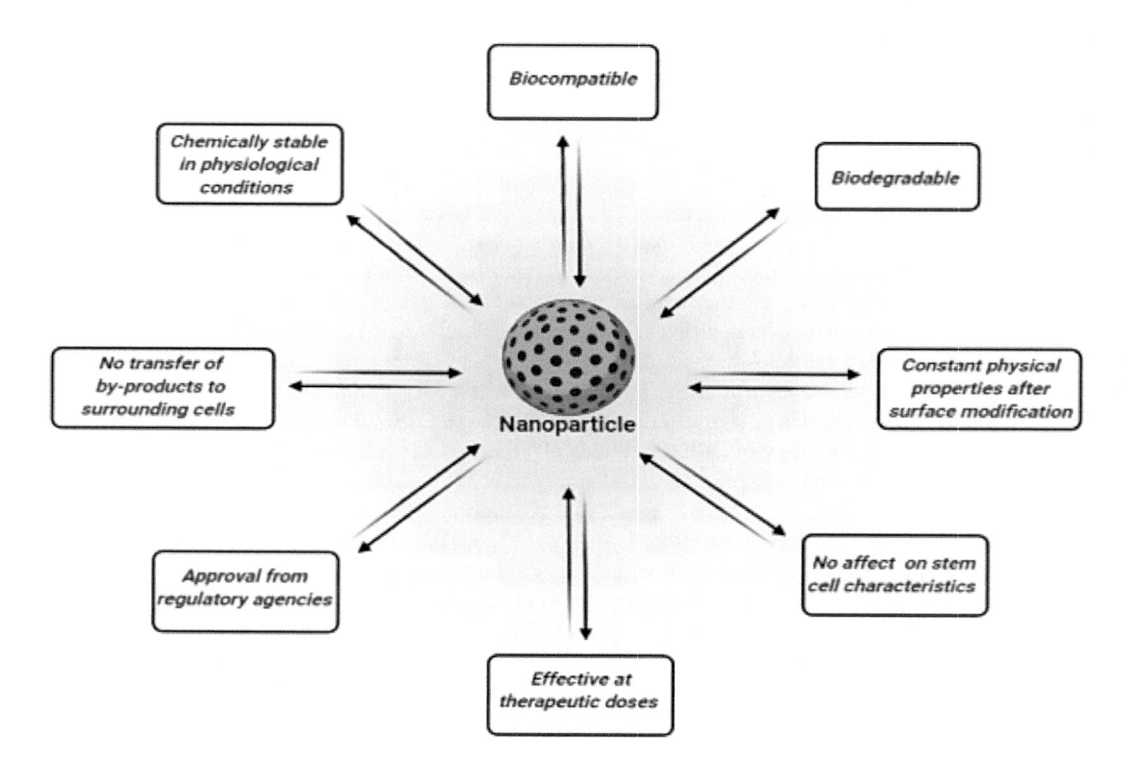

FIGURE 1.1 Desired characteristics of NPs in biomedicine.

The future biomedical applications of NPs are in developing of scaffolds for TE, as carriers of bioactive compounds, cells, and wound dressings. Nanotechnology is being applied in TE to improve the mechanical and biological performances of tissue constructs. The use of nanoparticles is highly advantageous in TE due to their smaller size and large surface area, similar to peptides and small protein molecules. They can smoothly diffuse across cell membranes and facilitate cellular uptake (Green et al., 2008). Furthermore, their sizes and surface characteristics can be easily modified. NPs naturally mimic the nanometre size scale of tissue extracellular matrix (ECM) (Wang et al., 2015). In this chapter, the role of nanoparticles in tissue engineering and regenerative medicine has been discussed.

Over the years, metallic and polymeric NPs have been used as diagnostic and therapeutic agents. The evolution of new ceramic materials for biomedical applications has also grown hastily. Nanoscale ceramics mainly silica (SiO_2) and titanium oxide (TiO_2) are being studied for their significance in regenerative medicine and improving their physical-chemical properties to reduce their cytotoxicity in biological systems.

Metallic Nanoparticles

Iron Oxide Nanoparticles

Iron is a naturally occurring metal in the human body, and the body has evolved to metabolise these particles. Therefore, iron oxide NPs are biocompatible. Due to their superparamagnetic nature resulting in a lack of interparticle attraction, NPs have a minimum risk of accumulation (Chaudhury et al., 2014). In general, magnetite and maghemite crystals with particle sizes 20–150 nm exhibit superparamagnetic order known as superparamagnetic iron oxide NPs (SPIONs). Their physicochemical properties are generally studied using Fourier transform infrared spectroscopy (FT-IR), X-ray photoelectron spectroscopy, transmission electron microscopy (TEM), scanning electron microscopy (SEM), energy-dispersive X-ray analysis, X-ray diffraction, zeta potential, nanoparticle tracking analysis, and dynamic light scattering (Navaei et al., 2016). Due to their size, magnetic properties, and biocompatibility, iron oxide NPs (SPION) have transpired as exceptional contrast agents in magnetic resonance imaging (MRI). Moreover, the MRI signal intensity can be significantly controlled without altering its in vivo stability. Due to their ideal characteristics, SPOINs have been approved as possible substitutes for Gadolinium (III) contrast agents, among the most widely used contrast agents in MRI (Charlton et al., 2016).

The SPION-based MRI imaging probes are accepted as T2-weighted contrast agents. Such contrast agents are applied to label, track, and activate stem cells and other cell types. It is a promising non-invasive imaging technique that provides the opportunity to monitor cell status after transplantation and accelerate progress in regenerative medicine (Cromer Berman et al., 2013) (Figure 1.2).

Two such MRI agents are ferumoxides (Feridex/Endorem) of 120–180 nm particle size and ferucarbotran (Resovist) of about 60 nm particle size. These are clinically approved for liver MRI. Resovist can be administered as a bolus, while Feridex must be administered as a slow infusion and is only used in delayed phase imaging. The iron nanoparticles are opsonised and taken up by phagocytic Kupffer cells of normal RES. In Kupffer cells, following the phagocytosis, the NPs exert significant T2/T2* relaxation effects. The hepatocytes negatively improve T2 or T2*-weighted images leading to enhanced exposure of pathological lesions that hampers the spotting of reticuloendothelial cells. The degree of SPIO uptake and the consecutive level of signal intensity drop is used to identify and characterise lesions (Wang et al., 2015). In a multicentric clinical study, Feridex-enhanced T2-weighted images were reported to reveal more lesions, which were not observed in unenhanced images in 27% of cases. This additional information can change therapy in more than 50% of cases (Sahle et al., 2019).

Gold Nanoparticles

Gold nanoparticles (AuNPs) are colloids consisting of nanometre-sized particles of gold. Due to their strong affinity to gold, conjugation of different ligands such as polypeptides, antibodies, and proteins containing moieties such as amines, phosphines, and thiols is possible (González-Béjar et al., 2016). In biomedicine, colloidal AuNPs are potential candidates due to their non-reactive and non-immunogenic

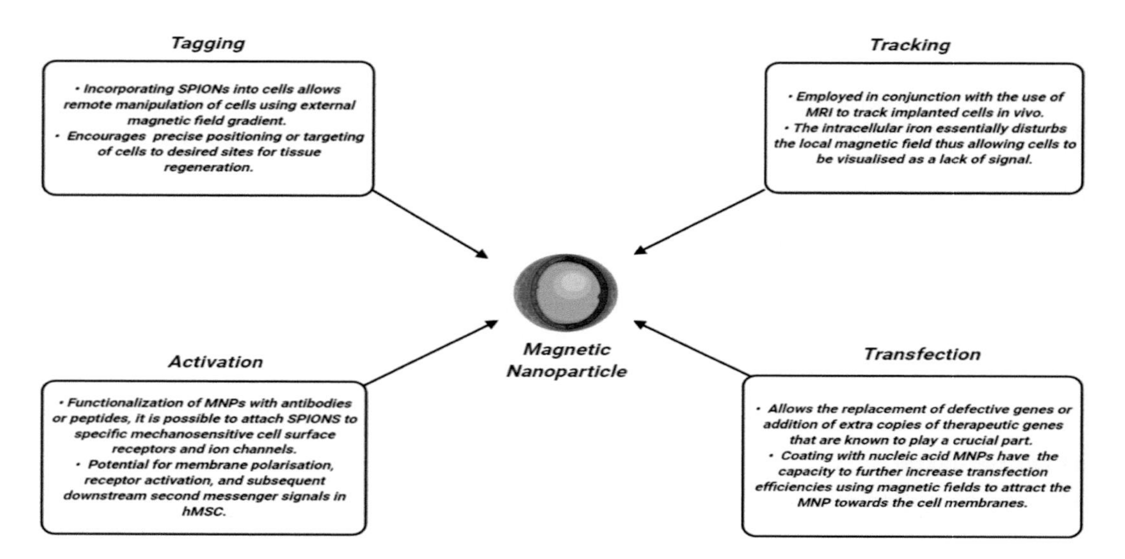

FIGURE 1.2 Applications of magnetic nanoparticles in the field of regenerative medicine.

attributes, biocompatibility, and biodistribution that can be prepared and altered easily. AuNPs can be effectively coupled with other biomolecules without changing their biological function. The gold nanoparticles are promising drug delivery candidates when formulated using compatible tissue reagents (Vacanti, 2006). PEG-functionalised AuNPs, prepared by chemically linking PEG to AuNPs, have been reported to enhance post-SCI recovery by reducing acute phase damage without harmful reactions such as body weight loss and health problems (Paul and Sharma, 2006).

Owing to desired properties, including less expense, non-toxicity, osteogenic differentiation induction, AuNPs are potential candidates as osteogenic agents. Heo et al., 2014 have developed a hybrid hydrogel made of gelatin and AuNPs (Gel-AuNP) as a regeneration strategy for bone tissue repair (Heo et al., 2014). The in vitro and in vivo evaluation of the Gel–AuNP composite hydrogel was effective in bone tissue engineering. The cellular experiments showed that the Gel–AuNP system promoted proliferation, viability, osteogenic differentiation, and significantly high-alkaline phosphate activities in human adipose-derived stem cells, dose dependently. AuNPs were also reported to promote the osteogenic transcriptional profile of mesenchymal stem cells (MSCs) to facilitate the differentiation of MSCs into functional osteoblast cells (Yu et al., 2014). Over the years, AuNPs have emerged as potential X-ray contrast agents due to their biocompatibility, ability to target tumours, high X-ray absorption coefficient and their competence to replace iodine in CT imaging. Since gold is a metal with a high atomic number, it offers better X-ray diminution and is considered ideal for CT imaging (Ahn et al., 2013). Due to their longer circulation time and better contrast, AuNPs are examined as substitutes to gadolinium, barium, or iodine-based agents (Markides et al., 2012).

Silver Nanoparticles

Applications of silver nanoparticles (AgNPs) have been explored in regenerative medicine and widely for wound dressings. Silver ions were found to disrupt the integrity of the bacterial cell by binding to various bacterial enzymes and proteins. Silver can alter critical functions in the bacterial cell, including permeability, enzymatic activity, cellular oxidation and respiratory processes, causing bacterial death (De la Riva et al., 2010). Many studies on the use of AgNPs in tissue engineering scaffolds reported lower infection rates with high biocompatibility. AgNPs have been tested on scaffolds made of diverse materials, including cellulose acetate, polyvinyl alcohol, chitosan-alginate, gelatin and PCL, resulting in strong antimicrobial activity (Ferreira, 2009). A study reported the cytotoxic and osteogenic differentiation effects of AgNPs and $AgNO_3$ in urine-derived stem cells (USCs). The AgNPs were found to induce actin polymerisation, improve cytoskeletal tension, and activate RhoA, which facilitated the osteogenic differentiation of USCs. AgNPs are sufficient for integration into tissue-engineered coatings that use urine-derived stem cells as seed cells (Rahmani Del Bakhshayesh et al., 2018).

Ceramic Nanoparticles

Titanium Oxide Nanoparticles

Research on the possible applications of titanium oxide nanoparticles biomedicine was first carried out owing to its potential role as photosensitising agents that could be utilised in cancer treatment and photodynamic inactivation of antibiotic-resistant bacteria (Sundaramurthi et al., 2014). Owing to its inert nature, TiO_2 scaffolds are being studied for several biomedical applications, including the development of implants in bone tissue engineering. The addition of TiO_2 to the chitosan scaffold was aimed for bone regeneration and was reported to decrease the scaffold's fast degradation. The connection between chitosan and nano-TiO_2 made it extremely porous, making it an effective replacement for bone tissue engineering (Kutsuzawa et al., 2006).

Silica Nanoparticles

Silica nanoparticles (SNPs) received great attention owing to their extensive biomedical applications and their exceptional intrinsic properties such as large surface area, biocompatibility and biodegradability. Due to their extraordinary biocompatible properties, SNPs may be used as ligands to monitor stem cell therapy. Their ability to integrate multiple contrast agents into a single factor can provide clinicians with multiple monitoring controls (Qin et al., 2014). The application of MSNs in tissue engineering allows the development of advanced technologies that enable guidance and tracking of stem cell transplantation to achieve a controlled and targeted stem cell delivery (Jiang et al., 2010). In one of the previous works, 89Zr4+ was immobilised by covalent attachment with p-isothiocyanatobenzyldesferrioxamine (DFO-NCS), a complexing agent to make large-pore MSNs. Quantitative 89Zr4+ labelling was achieved within just a short time due to the high concentration of DFO material on MSNs, and the high signal strength of the nanoparticles was demonstrated by effective PET imaging using a mouse model, indicating that 89Zr4+-DFO MSNs are potential PET imaging agents for long term in vivo imaging (Min et al., 2015). It is important for a tissue engineering scaffold to measure the pore size, distribution and interconnectivity in three dimensions. The characterisation of MSNs is generally done using mercury intrusion porosimetry and, most recently, by X-ray microtomography, which assesses the material dimensions non-destructively. MSNs exhibit tunable pore structures and high surface areas suitable for developing MR-enhancing hybrid materials. The porous nanostructures provide a worthy alternative to integrating many Gd chelates while maintaining all of them open to the media around them (Cao et al., 1997).

Zinc Oxide Nanoparticles

One of the promising applications of zinc oxide (ZnO) nanoparticles for TE owns to the biocompatibility and antibacterial properties of ZnO along with favourable effects arising from the interaction of living cells with ZnO nanoparticulate surfaces (Gritsch et al., 2019). A study by Colon et al., (2006) demonstrated the potential of ZnO and TiO_2 NPs in reducing *S. epidermidis* adhesion and increasing osteoblast functions essential for enhancing the efficiency of orthopaedic implantations. ZnO and TiO_2 nanoparticles also increased the osteoblast adhesion and alkaline phosphatase activity (ALP) compared to their respective microparticles (Colon et al., 2006). Foroutan, 2015 studied the effects of ZnO NP on human MSC differentiation into osteoblasts and reported that ZnO nanoparticles significantly upregulated the gene expressions of osteogenic indicators such as ALP, osteocalcin, and osteopontin (Foroutan, 2014).

Polymeric Nanoparticles

Polymeric NPs are composed of biocompatible and nontoxic biodegradable polymers, including polysaccharides (e.g., alginate, chitosan, dextran, heparin, hyaluronic acid, and pullulan), proteins (e.g., albumin, elastin, gelatin, and silk) and synthetic polymers (e.g., polyamides, polyesters, polyanhydrides, polyacrylates, and polyurethanes) alone or in conjugation in combination with other materials to provide specific functionalities (Mosselhy et al., 2021, Choi et al., 2017). The characterisation of polymeric NPs characterisation is important to ascertain their applicability and evaluate nanotoxicology and exposure assessment. The composition and concentration, surface properties, size, shape, crystallinity, and dispersion state are characterised using several analytical techniques, including chromatography, dynamic light scattering, electrophoresis, electron microscopy, near-infrared spectroscopy, and photon correlation spectroscopy (Zong et al., 2005). The NPs' surface area can be determined by the adsorption of inert gas (such as N_2) by changing the pressure conditions to form a gas monolayer.

Recently, biodegradable polymeric NPs loaded with ciprofloxacin were studied as drug delivery systems aimed for TE to prevent bacterial infection followed by implantation. The prepared poly (DL-lactide-coglycolide) NPs showed predictable, controlled release properties, ensuring a sustained release of the drug at the site of action to improve the therapy. Besides, NPs were more efficient and safer against *S. aureus* and *P. aeruginosa*, indicating that NPs loaded with antibiotics can be effectively used as carriers to deliver antibiotics (Gaspar et al., 2018). In another study, the potential of biodegradable polymer–DNA nanoparticles was demonstrated using poly (β-amino esters) to produce modified stem cells that can efficiently express angiogenic factors. Nanoparticles made of biodegradable polymers can carry foreign DNA into a cell. Nanoparticles loaded with a gene responsible for vascular endothelial growth factor (VEGF) production successfully transformed human mesenchymal stem cells (hMSCs) and human embryonic stem cell-derived cells (hESdCs), which resulted in increased hVEGF, cell viability, and promoted tissue vascularisation upon engraftment into target cells (Dakal et al., 2016).

Applications of Nanotechnology in Different Fields of TE

Stem Cell Engineering

NPs have many favourable properties, including large surface area, cell membrane activity, and increased intracellular uptake, making them suitable carriers for the delivery of genes and growth factors into stem cells. Some potential applications of NPs in stem cell research include non-invasive tracking of transplanted stem cells and intracellular delivery of small drug molecules, nucleic acid products (DNA, RNA, miRNA, and siRNA), and protein peptides. NPs also facilitate the observation of stem cell differentiation and the physiological condition of stem cells. Usually, all of these applications involve the cellular uptake of NPs. Owing to their nanoscale diameter, NPs can enter the stem cells. In most situations, the charged nanoparticles facilitate their cellular uptake (Grenho et al., 2015).

Several clinical trials involving the use of stem cells are being conducted to treat severe wounds (*via* tissue regeneration), cardiovascular, tissue/organ graft-related diseases (Ferreira, 2009). Continuous tracking of stem cells is critical to study the therapeutic effects or adverse events at the graft site. Magnetic resonance imaging is the best available non-invasive technique for monitoring the transplanted stem cells.

SPIONs have shown their potential for utilisation in MRI. SPIONs respond to external magnetic field gradients, and there is no residual magnetisation after their removal, which reduces the risk of aggregation. SPION labelling is used to observe the transplanted neural stem cells in rat brains after seven weeks of transplantation, providing information about stem cell migration in the nervous system (Cruz-Maya et al., 2019). Non-invasive detection tools that can monitor stem cell differentiation in situ are highly needed to study the degree of stem cell differentiation. NPs can be employed as biosensors on stem cells by covalent or non-covalent tagging of some immobilised biological substrate molecules onto the NP surface (Dias et al., 2017). Based on surface-enhanced Raman spectroscopy (SERS) findings, a study reported that 3D graphene oxide loaded in AuNP efficiently detected neural stem cell differentiation.

The substrate-laden NPs facilitated the effective detection of a single MSC differentiation stage by using electrochemical and electrical techniques (Kitsara et al., 2017).

Gene delivery can be a powerful strategy to study stem cell biology and advance tissue engineering therapies. Nanoparticle-based delivery systems could play an important role in gene transfection in stem cell-centred regenerative medicine. NPs immobilise the vector at the regeneration site, minimise damage, and provide a wide area for cell membrane interaction for receptor-mediated endocytosis (Huang et al., 2020). A recent study developed biodegradable polymeric NPs for non-viral gene transfer to human embryonic stem cells (hESC), which showed four times higher gene delivery efficacy than a leading transfection agent, Lipofectamine 2000. The nanoparticles were reported to have minimal toxicity on hESC morphology or showed any non-specific differentiation (Kutsuzawa et al., 2006).

NPs have a promising role in stem cell engineering. However, NPs can also show adverse effects or cytotoxicity and might alter the self-renewal and differentiation abilities of the stem cells. Impending research should carefully assess the risk to benefit ratio involved with the use of nanoparticles. For example, several studies have not reported the cytotoxic or adverse effects associated with the use of AuNPs. NPs can exert unpredictable molecular responses after a breakdown, which should be considered seriously before the stem cell therapies.

Ocular Regeneration

Several degenerative ocular diseases require the advancement of non-invasive regenerative techniques, including acute macular degeneration (AMD), cataracts, diabetic retinopathy, glaucoma, and other congenital retinal defects such as Leber amaurosis, retinitis pigmentosa and X-linked retinoschisis (San Thian et al., 2006). The current treatment options include laser surgery, photodynamic therapy, vitrectomy, and delivery of angiostatic steroids. Some current therapies fail to address the challenge, whereas others are associated with severe adverse effects (Langer, 2000). So, increasing efforts have been undertaken in the last decade to apply nanotechnology-based systems for the regeneration of lost or damaged eye tissues. Several biocompatible materials are being designed into nanoscaffolds that have a significant role in treating retinal and ocular regeneration (Kaul and Ventikos, 2015). The nanofibrous membranes can imitate natural substrate for retinal pigment epithelium to a great extent, leading to the engineering of in vivo human retinal pigment epithelium monolayer while maintaining its biofunctional characteristics. For corneal tissue regeneration, the scaffolds should be optimised to have sufficient porosity along with mechanical and optical properties (Nukavarapu et al., 2008, Rodriguez-Contreras et al., 2018). A variety of natural and synthetic polymers were used for the fabrication of nanoscaffolds, including a mixture of polyvinyl acetate with collagen type I, collagen with PLGA, PCL with PGS, collagen type I with chondroitin sulphate, and silk fibroin with poly (L-lactic acid-co-ε-caprolactone) (Cruz-Maya et al., 2019, Pellegrini et al., 1997). Nanoscaffolds are currently being developed for lens regeneration, which is generally injected into the lens capsule following the removal of its content (Alaminos et al., 2006).

Nanoparticles in Dental Regeneration

TE is considered an optimal approach for repairing craniofacial bone defects and cartilage destruction resulting from congenital disorders, cancer, infection, and trauma. Due to their morphological similarity to bone tissue and their exceptional surface properties such as surface chemistry, energy, topography, and wettability, nanophase ceramics are widely used as bone replacements, filling and coating, and materials (Webster et al., 2000, Webster et al., 2001, Webster et al., 2005). Nanohydroxyapatite (HA) has been reported to enhance osteoblast adhesion and function and improve mineralisation, indicating their role as a suitable candidate for clinical use. Several studies suggested nanosized HA to have better bioactivity than the micro-sized HA (Nukavarapu et al., 2008, Sato et al., 2006). The same theory applies to other nanoceramics, particularly aluminium, titanium, and zinc oxide (Sarao et al., 2014). A previous study reported that osteoblast adhesion was improved by 146% with 23 nm zinc oxide and 200% with 32 nm of titanium nanoparticle than their respective micro-sized counterparts (Webster et al., 2005).

The clinical use of NanOss (Angstrom Medica Inc), a bone void filler that is the first nanotechnological medical device, was approved by the US Food and Drug Administration (USFDA) in 2005. It consists of calcium orthophosphate NPs that resemble the composition and performance of human bones. Currently, it is being used to treat various general bone injuries (Paul and Sharma, 2006). Ostim is another commercial injectable formulation (synthetic nano-HA in water suspension). It is used to treat alveolar ridge augmentation, metaphyseal fracture, and osteotomies (Smeets et al., 2008). Nanotechnology plays an important role construction of biofunctionalised biomaterials for periodontal tissue repair. Tissue bioengineering with genetically designed NP, a biomimetic of mineralised tissues, leads to the manufacturing of in vitro teeth tissues (Yuan et al., 2011). In a previous attempt, Chen et al., 2005 developed HA nanorod surface with monolayer surfactants, which gave them exact surface characteristics for self-assembling enamel into a prism-like structure at the water-air interface. The size of synthesised HA nanorods was similar to human enamel crystals. Biomimetic approaches are being used for developing nanomaterials to remineralise enamel lesions. Researchers have anticipated that that nano-HA could encourage remineralisation of bone tissue through adhesion (Chen et al., 2005). Some studies also suggest that nano-HA may be employed to deliver the calcium to the mouth to facilitate enamel remineralisation. By acting as suitable biomimetic tools, NPs could improve dental tissue rejuvenation. Several nanomaterials are being engineered to address dental issues such as periodontal defects, bone degeneration, brittle teeth, and lost teeth (Huang et al., 2009, Huang et al., 2019).

Nanoparticles in Skeletal Regeneration

Current options to treat bone defects include replacing lost bone with allogeneic or similar bone grafts. Nevertheless, bone grafting is restricted due to high failure rates, donor site morbidity, risks of surgical complications, and limited availability of free bone tissue to be harvested (Wang and Yeung, 2017). Biomaterials under development could simulate the natural extracellular matrix of the bone structure. These biomaterials are being engineered with the application of nanotechnology (Yao et al., 2019). Because the ECM structure has nanoscale properties, nanomaterials are studied as biomimetics for the ECM structure (Barnes et al., 2007).

Owing to its ability to link cells and ECM formation, collagen is a widely studied material for developing a scaffold to back the bone growth (Liu et al., 2018). In a study, SEM and micro-computed tomography images were used to assess the encouraging role of poly-lactic acid (PLA)/collagen-I nanofibrous scaffold in osteogenic differentiation of hMSCs. Chitosan-based biomaterials promoted bone regeneration owing to their biocompatibility and low immune response. In a recent study, chitosan/HA electrospun scaffolds were reported to encourage cell proliferation and mineral deposition and provide osteoinductivity (Huang et al., 2020).

In addition to mixing polymers to supply 3D scaffolds, NPs have been recently investigated to promote regeneration by delivering bioactive molecules (e.g., growth factors) to support bone formation (Gritsch et al., 2019). De la Riva et al. (2010) developed an abrushite–chitosan system to control the release kinetics and localisation of VEGF and PDGF at the bone defect site, thereby promoting bone healing. The abrushite–chitosan system also protects the growth factors from systemic exposure (De la Riva et al., 2010). Owing to its inert nature, TiO_2 scaffolds are being studied for several biomedical applications, including the development of implants in bone tissue engineering. The addition of TiO_2 in chitosan scaffold aimed for bone regeneration was reported to decrease the fast degradation of the scaffold, and the interaction between chitosan and nano-TiO_2 made it highly porous, rendering it an efficient alternative for bone tissue engineering (Ikono et al., 2019).

Biomimetic Nanoscaffolds for Nerve Regeneration

Currently, nerve autografts and allografts are principally used in nerve regeneration; however, their application is associated with problems such as donor site morbidity, loss of function or formation of painful neuromas at the donor site, a limited amount of donor nerves, and the risk of immunorejection (Fathi-Achachelouei et al., 2019). Thus, research is being conducted to develop artificial nerve conduits to enhance nerve regeneration techniques. Various synthetic materials such as polycarbonate, poly(lactic acid), poly(lactic-co-glycolic acid) (PLGA) and silicon are widely being studied for fabricating nerve conduits as chemical changes to these polymers allow changes in the porosity, biocompatibility, geometric configuration, mechanical strength and degradation (Nectow et al., 2012).

Natural ECM contains collagen fibres of nanometre diameter; efforts are being made to fabricate ideal ECM scaffolds for nerve regeneration using electrospinning. The application of electrospun nerve fibres in nerve regeneration has been reported by many researchers (Liu et al., 2011, Yu et al., 2014). In a study, poly(L-lactide-co-glycolide) polymer fibres were used to fabricate the nerve conduits and implanted into the sciatic nerve of the rat. The study reported no inflammatory response, and rats showed successful nerve regeneration during one month of study. The developed scaffolds were reported to be flexible, permeable, and with no inflammation symptoms (Lee et al., 2006). In a study, cultures of neuronal stem cells (NSCs) with porous polymeric nanofibrous scaffold have been attempted. The nanostructured PLLA scaffold was fabricated using a biodegradable poly (l-lactic acid) (PLLA) and mimic a natural extracellular matrix. The NSCs have shown differentiation on the nanostructured scaffold in cell cultural tests indicating its potential use as a carrier in nerve TE (Yang et al., 2004).

Nanoparticles in Cardiac Regeneration

Currently, both synthetic and natural hydrogels are being used widely in cardiac TE, including poly (N-isopropyl acrylamide, poly-L-lactide, alginate, gelatine, and poly (ethylene glycol) (PEG). The native myocardium achieves heart contraction, allowing conductivity between cardiomyocytes, promoting the dissemination of the action potential. However, in traditional scaffolding materials, conductivity is limited due to the many large pores within the structures (Kamaleddin and Medicine, 2017). Different scaffolding materials have been integrated into NPs to improve the electrical conductivity and imitate the nanofibrous structures of the native myocardium ECM (Kitsara et al., 2017).

In an investigation, a UV-cross-linkable gold (GNR)-incorporated gelatin methacrylate (GelMA) hybrid was fabricated for its utilisation in cardiac TE (Cromer Berman et al., 2011). The incorporated nanomaterial was reported to promote the hydrogel matrix's mechanical stiffness and electrical conductivity, and the hybrid hydrogels seeded cardiomyocytes represented excellent viability, cell retention, and metabolic activity. Similarly, carbon-based nanomaterials, such as carbon nanotubes (CNTs), have enhanced conductivity-related characteristics. For instance, a study reported the synthesis of cardiac patches by planting neonatal rat cardiomyocytes onto CNT-incorporated photo-cross-linkable gelatin methacrylate hydrogels. The electrically conductive and nanofibrous complexes formed by CNT were reported to improve cardiac cell adhesion and organisation and provide outstanding mechanical integrity and improved electrophysiological functions.

Microscale scaffolds cannot effectively replicate the ECM atmosphere, so nanoscale scaffolds such as nanofibres are being studied widely to develop heart tissue constructs. Both synthetic and natural polymers (e.g., collagen, protein, and fibrinogen), synthetic polymers [e.g., poly(lactide), poly(glycolide)], and their copolymers [(PLGA), poly(ε-caprolactone), poly(ethylene-co-vinyl alcohol)] are being spun into nanofibres using electrospinning (Cao et al., 1997). In a study, nanofibres were fabricated using biodegradable non-woven poly(lactide)- and poly(glycolide)-based (PLGA) platforms for cardiac TE applications (Zong et al., 2005). The results suggested that the nanofibrous construction of the matrix permitted the cardiomyocytes to use external hints for isotropic or anisotropic growth. The cardiomyocytes also interacted with the surrounding nanofibrous network to shape their growth to follow the scaffold-prescribed direction. Similarly, a nanofibrous assembly comprised of poly(l-lactic acid) and polyaniline, designed for cardiac TE, was described to show better biocompatibility and a significant outcome on cell differentiation while improved cellular distribution, arrangement, and cell–cell interactions (Wang et al., 2017).

Nanoparticles in Skin Regeneration

Acute and chronic wounds are one of the major causes of morbidity and mortality, affecting many populations across the world, leading to increased research in skin regeneration. However, the development of composite epidermal and dermal layers that can efficiently mimic in vivo skin, greater reviving ability, and negligible scarring is a huge challenge in this field (Singer and Boyce, 2017). Nanomaterials have an imperative role in skin regeneration owing to their high surface-area-to mass ratio promoting cell–matrix interactions. The main advantage of NPs is their competence to carry, protect, and control the release of drugs. Drug-associated side effects could be reduced due to their ability to deliver directly in a controlled manner, thus reducing the drug dose to a wound bed (Rahmani Del Bakhshayesh et al., 2018). Chronic wounds resulting from burns and injuries are at great risk of microbial infections. The increased incidence of multi-drug resistant microbial (e.g., methicillin-resistant *Staphylococcus aureus*) infections requires novel delivery systems that incorporate antimicrobial components in the scaffold for effective treatments (Mosselhy et al., 2021). Currently, the applicability of antibacterial agent-loaded NPs is widely assessed in skin tissue engineering. AgNPs naturally possess antibacterial and antifungal properties and are incorporated in various antibiotic therapies. The application of AgNPs in treating burn and diabetic wounds was investigated by Tian et al., 2007. It was found that NPs have also accelerated the rate of healing in animal studies along with showing antimicrobial effect. In another study, metallic nanosilver particles-collagen/chitosan hybrid scaffold was reported to act as a bactericidal, anti-inflammatory agent and promote wound healing by regulating macrophage activation and fibroblast migration (Mo et al., 2017).

Severe skin wounds require engineered biomaterials to cover and facilitate tissue revival. An ideal tissue-engineered skin scaffold should act like the native tissue to function and secure the wound bed from blood/fluid loss, infection, and promote exudate discharge. Various systems are being studied as dermal scaffolds, i.e., gels, nanofibres, microfibres, films, and membranes (Sundaramurthi et al., 2012). Nanofibres are ideal candidates for tissue engineering due to their skin-like mechanical properties, simulating the local extracellular matrix, high porosity, and large surface area. The nanofibres were electrospun into nanometre diameters (Figure 1.3).

A wide range of natural and synthetic polymers were electrospun and assessed for their efficiency to support skin regeneration (Table 1.1).

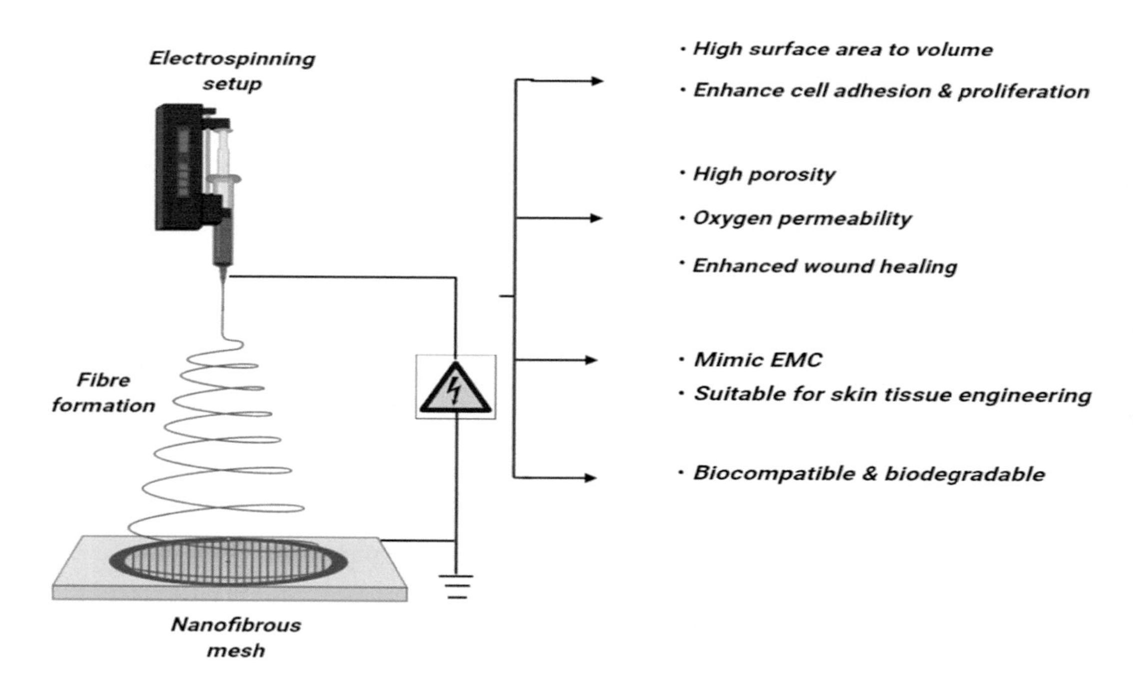

FIGURE 1.3 Electrospun nanofibres for skin regeneration.

TABLE 1.1

List of Studies Conducted on Various Types of Polymeric Nanofibres in the Field of Skin Regeneration

Polymer	Diameter	Active Constituent	Result	Reference
Gelatin and Poly(L-lactic acid)-co-poly-(e-caprolactone) (PLLCL)	Less than 700 nm	Penicillin– 146 streptomycin solution	The nanofibres showed sustained release	(Jung et al., 2013)
Polycaprolactone (PCL)	382 nm ± 140 nm	Potassium phthalimide salt and 2-chlorocyclopentanone	Enhanced fibroblast growth while maintaining normal cell morphology	(Hough et al., 2006)
Poly (D,L-lactic-co-glycolic acid) (PLGA)	400 nm ± 25 nm	Angiogenin and Curcumin	In vivo angiogenesis and anti-infection properties of composite nanofibres were transplanted into the infected full-thickness burn wound	(Mo et al., 2017)
Polycaprolactone (PCL)	130 nm	Gentamicin sulphate	The encapsulation of drug improved the bioactivity of nanofibres	(Ahn et al., 2018)
Bioactive glass	21.87 ± 0.21 Mpa	BG precursor solution	The nanofibres showed antibacterial activity against *S. aureus and* induced the secretion collagen and vascular endothelial growth factor	(Abdul Khodir et al., 2018)
Gelatin	339 ± 91 nm and 276 ± 88 nm	1,4-butanediol diglycidyl ether (BDDGE)	Non-toxic with capability of tailoring gelatin's mechanical and physical properties and ability to synthesis of new extracellular matrix	(Zhou et al., 2017)
Collagen	310 ± 117 Nm	Tilapia skin	Promoted viability, migration and differentiation of HaCaTs	(Dias et al., 2018)
PLGA	398–135 nm	Bone marrow derived mesenchymal stem cells (BM-MSCs)	Implanted BM-MSCs were found to promote epithelial edge in growth and collagen synthesis	(Pilehvar-Soltanahmadi et al., 2016)
Chitosan	177.6–40.5 nm	Poly(caprolactone) (PCL)	Served as provisional matrix to promote wound closure	(Zielińska et al., 2020)
Cellulose acetate	396.66 ± 0.90 nm	Soya protein	Promoted fibroblast proliferation, migration, infiltration, and integrin β1 expression.	(Markides et al., 2012)

(*Continued*)

TABLE 1.1 (Continued)

Polymer	Diameter	Active Constituent	Result	Reference
Gelatin and Polycaprolactone	417 ± 165 nm	Gel powder of pig skin	Provided high porosity which contributed to an ideal water vapour permeability rate and good mechanical and biological properties	(Liu et al., 2011)
Polycaprolactone (PCL)	0.71 ± 0.33 μm	Spirulina	Enhanced viability and proliferation of mouse fibroblasts	(Ahn et al., 2013)
Poly (3-hexylthiophene) (P3HT)	503 ± 116, 421 ± 128 and 314 ± 101 nm	Phalloidin (fluorescein isothiocyanate)	Showed better cell proliferation along with retaining cell morphology	(Fathi-Achachelouei et al., 2019)
Alignate-PCL	1.35 ± 0.45 μm	Spirulina	Capable of holding large amounts of water, to support the backbone of the nanofibres	(Tan and Marra, 2010)
Chitosan & Gelatin	120 &220 nm	Trifluoroacetic acid and Dichloromethane	Chitosan-gelatin-blend nanofibres were found to enhance tensile strength as compared to gelation nanofibres	(Kamaleddin and Medicine, 2017)
3D silk fibroin	18.56 ± 0.1 μm	Cocoons	Use of 3D nanofibre scaffolds presented a high porosity with controlled full thickness	(Jung et al., 2013)
Silk fibroin	250 nm	Human amniotic membrane	Due to fibroin nanofibre coating, mechanical properties of HAM improved significantly	(Colon et al., 2006)
Peptide	6 mm	Heparin	Nanofibre-treated burn wounds formed well-organised and collagen-rich granulation tissue layers with greater density of newly formed blood vessels	(Qin et al., 2014)
Poly(lactic-co-glycolic acid) (PLGA)		Heparin sodium salt	Stimulated cellular activity by enhancing the bioactivity of the released GF	(Levengood et al., 2017)

TABLE 1.2

Recent Patents Representing Utilisation of Nanotechnology in the Field of Tissue Engineering and Regenerative Medicine

S. No	Patent	Type of Nanoparticles	Summary of Invention	Remarks	Reference
1.	US 6,552,172 B2 (Apr. 22, 2003)	Fibrin nanoparticles	Provides methods for preparing fibrin NPs and microbeads of different sizes with compositions	Provides method for incorporating active agents into a cell, method for isolating stem or progenitor cells from a biological sample and a composition of fibrin particles bound to stem or progenitor cells	(Marx and Gorodetsky, 2003)
2.	US2006/016.5805A1 (Jul. 27, 2006)	Magnetic nanoparticles	Provides method of use of magnetic pole matrices for tissue engineering and targeting systematic therapy for cardiovascular disease using magnetic polymer nanoparticles based gene/drug delivery	The magnetic pole matrices possesses advantages such as distributing the magnetic nanoparticles conjugated with genes or drugs locally and uniformly on the artificial surface which essentially solved the blood vessel blocking problem along with promoting the adhesion of the cells on specific locations	(Steinhoff et al., 2006)
3.	WO2011145963A1 (Nov. 24, 2011)	Non-leaching polymeric nanoparticles	The invention is related to methods of NP preparation particularly for the intracellular delivery of hydrophobic drugs	These nanoparticles can be used to control the activity and differentiation of cells, including stem cells through the intracellular delivery of the hydrophobic drugs	(Ferreira et al., 2011)
4.	US 8,172.997 B2 (May 8, 2012)	Cerium oxide	Provides method of preparation of novel non-agglomerated ultra-fine ceramic oxide nanoparticles by a novel Sol micro emulsion process	Applications of the NPs include benefits for wound healing, treating arthritis and joint diseases, anti-aging and the treating of inflammations	(Seal et al., 2012)
5.	US 2012/0087868 A1 (Apr. 12, 2012)	Mesenchymal stem cells having nanoparticles loaded cells	The invention also provides methods of manufacturing the nanoparticle-loaded cells, and methods for treatment and/or imaging	The interaction of NPs and NP-loaded cells with one or more electromagnetic radiations or magnetic fields provides opportunities for imaging of tissues and/or detection of various diseases, diseased cells, disease states and the like	(Todd and Danilkovitch, 2012)

(*Continued*)

TABLE 1.2 (Continued)

S. No	Patent	Type of Nanoparticles	Summary of Invention	Remarks	Reference
6.	US 2013/0273011 A1 (Oct. 17, 2013)	Stem cell generated nanoparticles (40–100 nm)	Discloses stem cells and derived nanoparticles useful for the treatment of acute radiation syndrome and other inflammatory conditions The endometrial regenerative cells are administered to patients having been exposed to radiation to augment recovery of endogenous hematopoietic stem cells	The invention pertains to stem cell therapeutic activities and treatment of hematopoietic injuries through administration of mesenchymal stem cells that are capable of increasing survival after radiation injury through accelerating hematopoietic recovery. It also pertains to the use of exosomes secreted by said mesenchymal stem cells for achievement of this purpose	(Ichim et al., 2013)
7.	US 8,663,675 B2 (Mar. 4, 2014)	Polymeric nanoparticles	Injectable matrix is used to create a three-dimensional matrix system in an area of the desired tissue or organ	Various embodiments herein relate to an injectable matrix used for regeneration, reconstruction, repair, or replacement of organ or tissue	(Ghoroghchian and Ostertag, 2014)
9.	US 9,109,203 B2 (Aug. 18, 2015)	Magnetic nanoparticles are coated with silica	Silica-coated MNPs were developed using the water-in-oil (w/o) reverse micelle method	The NPs can be utilised for targeting neural stem cells	(Yung et al., 2015)
10.	US 2016/0030600 A1 (Feb. 4, 2016)	Nanoparticle (lipid based and polymer based), has a size from about 100 nm to about 400 nm	This invention relates to nanoparticles for use in the in vivo diagnostics of epicardial derived cells (EPDCs) to nanoparticles for use in the treatment of cardiac injury	Targeted delivery of phagocytable particles to EPDCs can be exploited in the delivery of therapeutic agents to EPDCs, for instance in regenerative therapy approaches for cardiac injury and in the delivery of labelling agents to EPDCs, which is for instance applicable in in vivo imaging of these stem cells	(Schrader, 2016)

11.	US 2016/0067273 A1 (Mar. 10, 2016)	O-gal nanoparticles	Provides compositions and methods for preparation of synthetic O-gal nanoparticles	The present invention gives compositions and methods comprising molecules and nanoparticles that can replace the O-gal nanoparticles generated from natural materials in the induction of repair and regeneration of injuries	(Galili, 2016)
12.	US 9,750,845 B2 (Sep. 5, 2017)	Polymeric nanoparticles	It describes methods and compositions of nanocomposites consisting citric acid copolymers, poly L-lactic acid (PLLA) and polylactic-co-glycolic acid (PLGA)	The present invention describes a composition aimed for strengthen scaffolds for bone tissue engineering applications and, to a lesser extent, for soft tissue regeneration	(Ameer and Webb, 2017)
13.	US 2018/0360768 A1 (Dec. 20, 2018)	Mineral-based nanoparticles	The silicate nanoparticles are synthesised through a process where the precipitate of sodium silicate is mixed with one or more elements and compounds and milled into NPs	NPs comprising silicate that induce human mesenchymal stem cells (hMSCs) into a cartilage lineage through the upregulation of cartilage-specific genes resulting in the transformation of the cell phenotype into that of a chondrocyte, i.e., a cartilage producing cell.	(Gaharwar and Carrow, 2018)

Patents

The pioneer work in the field of TE primarily focused on what materials and structures were needed to be utilised to develop tissue equivalents (Wang et al., 2017). In the coming years, the research was more focused on physiological concepts and mechanisms involved in TE including cell attachment, differentiation and the role of growth and differentiation factors in TE. Next, the research revolved on evolving the knowledge for application of new polymers and different methods for engineering of matrices for creating of new tissues. In the last few decades, advances in the field of nanotechnology have enabled fabrication of biomimetic microenvironment at nanoscale leading to increased efforts in utilisation of engineered NPs as biomimetic nanoscaffolds (Liu et al., 2018, Sundaramurthi et al., 2014). Table 1.2 reviewedsome of the most important patents in this field.

Future Prospects and Limitations

Different biomaterial-based NPs including metals, ceramics, and polymers have demonstrated promising role in TE application. However, there are still many challenges in the field that needs to be addressed. The focus of current research should be on the study of toxicity, carcinogenicity, and teratogenicity of NPs and appropriate evaluation tools need to be developed so as to ensure safety of human health in terms of the use of nanomaterials in regenerative medicine (Yang et al., 2004). NPs are being studied as a promising targeted contrast agent for non-invasive diagnosis; however, so far only iron oxide NPs are being utilised in clinical application (Chen et al., 2005). The reason for this could be the difficulty in achieving desirable pharmacokinetic properties and particle homogeneity with NPs along with the lack of clinical studies related to their elimination, toxicity, and biodegradation. Therefore, there is a dire need for the development of specific guidelines and precautionary measures for applications of NPs in biomedical field.

Summary

Owning to their biocompatibility and surface modification, various types of NPs have been studied to be utilised as a diagnostic and therapeutic agent. NPs for skin have been studied extensively in previous years' regeneration mainly for the treatment of burn victims and for wound healing. At present, several tissue-engineered skins are available for human use. By acting as suitable biomimetic tools, NPs have shown great potential for improving dental, ocular, skeletal, and cardiac tissue regeneration. In this chapter, we have reviewed the use of nanostructures structures in TE in different ways such as by acting as antimicrobial agents, providing constant release of several growth factors, or by acting as contrasting agents for monitoring of stem cells. Although NPs offer a promising role in TE, the lack of teratogenicity, carcinogenicity, and toxicity studies needs to be addressed. Despite the challenges, nanotechnology has potential to stimulate the reconstruction of complex tissue architectures in future.

REFERENCES

Abdul Khodir W., AbdulRazak A., Ng M., Guarino V., Susanti D. et al. 2018. *Journal of Functional Biomaterials 9* (2): 36. doi:10.3390/jfb9020036.

Acharya S., Sahoo S. 2011. *Advanced Drug Delivery Reviews 63* (3): 170–183. doi:10.1016/j.addr.2010.10.008.

Ahn S., Chantre C., Gannon A., Lind Johan U. et al. 2018. *Advanced Healthcare Materials 7* (9): 1701175. doi:10.1002/adhm.201701175.

Ahn S., Jung S., Lee S. 2013. *Molecules 18* (5): 5858–5890. doi:10.3390/molecules18055858.

Alaminos M., Sánchez-Quevedo M., Munoz-Ávila J., Serrano D. et al. 2006.*Investigative Ophthalmology & Amp; Visual Science 47* (8): 3311–3317. doi:10.1167/iovs.05-1647.

Barnes C., Sell S., Boland E., Simpson D.et al. 2007. *Advanced Drug Delivery Reviews 59* (14): 1413–1433. doi:10.1016/j.addr.2007.04.022.

Boccaccini A., Ma P. 2014. *Tissue engineering using ceramics and polymers*, Editor Ed. Elsevier. https://www.elsevier.com/books/tissue-engineering-using-ceramics-andpolymers/boccaccini/978-0-85709-712-5

Cao Y., Vacanti J., Paige K., Upton J. et al. 1997. *Plastic and Reconstructive Surgery 100* (2): 297–302. doi:10.1097/00006534-199708000-00001.

Caspani S., Magalhães R., Araújo J., Sousa C. 2002. *Materials 13* (11): 2586. doi:10.3390/ma13112586.

Charlton J., Pearl V., Denotti A., Lee J.et al. 2016. *Nanomedicine: Nanotechnology, Biology and Medicine 12* (6): 1735–1745. doi:10.1016/j.nano.2016.03.007.

Chaudhury K., Kumar V., Kandasamy J., RoyChoudhury S. 2014. *International Journal of Nanomedicine 9*: 4153. doi:10.2147/IJN.S45332.

Chen H., Clarkson B., Sun K., Mansfield J. 2005. *Journal of Colloid and Interface Science 288* (1): 97–103. doi:10.1016/j.jcis.2005.02.064.

Choi J., Kim M., Chung G., Shin H. 2017. *Biotechnology and Bioprocess Engineering 22* (6): 679–685. doi:10.1007/s12257-017-0329-3.

Colon G., Ward B., Webster T. 2006.*Journal of Biomedical Materials Research Part A 78* (3): 595–604. doi:10.1002/jbm.a.30789.

Cromer Berman S., Walczak P., Bulte J. 2011. *Wiley Interdisciplinary Reviews: Nanomedicine and Nanobiotechnology 3* (4): 343–355. doi:10.1002/wnan.140.

Cromer Berman S., Wang C., Orukari I., Levchenko et al. 2013. *Magnetic Resonance in Medicine 69* (1): 255–262. doi:10.1002/mrm.24216.

Cruz-Maya I, Guarino V, Almaguer-Flores A, Alvarez-Perez M.A., Varesano et al. 2019. *Journal of Biomedical Materials Research. Part A 107*(8): 1803–1813. doi:10.1002/jbm.a.36699.

Dakal T., Kumar A., Majumdar R., Yadav V. 2016. *Frontiers in Microbiology 7*, 1831. doi:10.3389/fmicb.2016.01831.

De la Riva B., Sánchez E., Hernández A., Reyes et al. 2010. *Journal of Controlled Release 143* (1): 45–52. doi:10.1016/j.jconrel.2009.11.026.

Dias J., Baptista-Silva S., De Oliveira C., Sousa et al. 2017. *European Polymer Journal, 95*: 161–173. doi:10.1016/j.eurpolymj.2017.08.015.

Dias J., Baptista-Silva S., Sousa A., Oliveira et al. 2018. *Materials Science and Engineering 93*: 816–827. doi:10.1016/j.msec.2018.08.050.

Fathi-Achachelouei M., Knopf-Marques H., Ribeiro da Silva C.E., Barthès et al. 2019. *Frontiers in Bioengineering and Biotechnology 7*: 113. doi:10.3389/fbioe.2019.00113.

Ferreira L. 2009. *Journal of Cellular Biochemistry 108* (4): 746–752. doi:10.1002/jcb.22303.

Ferreira L., Malva J., Bernardino L., De Sousa T., Silva J., Non-leaching nanoparticle formulation for the intracellular delivery of hydrophobic drugs and its use to modulate cell activity and differentiation.WIPO Patent 2011145963A1, Nov. 24, 2011.

Foroutan T. 2014. *Nanomedicine Journal 1* (5): 308–314. doi:10.7508/NMJ.2015.05.004.

Gaharwar A.K., Carrow J.K. Mineral-based nanoparticles for arthritis treatment. U.S. Patent 2018/0360768 A1, Dec. 20, 2018.

Galili U., Compositions and methods for preparation of synthetic alpha-gal nanoparticles and for their clinical use. U.S. Patent 2016/0067273 A1, Mar.10, 2016.

Gaspar L., Dórea A., Droppa-Almeida D., de Mélo Silva et al. 2018. *Journal of Nanoparticle Research 20* (11): 289. doi:10.1007/s11051-018-4387-z.

Ghoroghchian P.P., Ostertag E., Biodegradable nanoparticles as novel hemoglobin-based oxygen carriers and methods of using the same. U.S. Patent 8663675 B2, Mar. 4, 2014.

González-Béjar M., Francés-Soriano L., Pérez-Prieto J. 2016.*Frontiers in Bioengineering and Biotechnology 4*: 47. doi:10.3389/fbioe.2016.00047.

Green J., Zhou B., Mitalipova M., Beard C. et al. 2008. *Nano Letters 8* (10): 3126–3130. doi:10.1021/nl8012665.

Grenho L., Salgado C., Fernandes M., Monteiro et al. 2015. *Nanotechnology 26* (31): 315101. doi:10.1088/0957-4484/26/31/315101.

Gritsch L., Conoscenti G., La Carrubba V., Nooeaid et al. 2019. *Materials Science and Engineering 94*: 1083–1101. doi:10.1016/j.msec.2018.09.038.

Heo D., Ko W., Bae M., Lee et al. 2014. *Journal of Materials Chemistry B 2* (11): 1584–1593. doi:10.1039/c3tb21246g.

Hough S., Clements I., Welch P., Wiederholt K. 2006. Differentiation of mouse embryonic stem cells after RNA interference-mediated silencing of OCT4 and Nanog, *Stem Cells 24* (6): 1467–1475. doi:10.1634/stemcells.2005-0475.

The global tissue engineering market is expected to reach USD 11.5 billion by 2022. https://www.prnewswire.com/news-releases/the-global-tissue-engineering-market-is-expected-to-reach-usd-11-5-billion-by-2022--300683950.html (Assessed on Jan 15, 2021).
Huang Q., Liu Y., Ouyang Z., Feng Q. 2020. *Bioactive Materials 5* (4): 980–989. doi:10.1016/j.bioactmat.2020.06.018.
Huang S., Gao S., Yu H. 2009. *Biomedical Materials 4* (3): 034104.doi:10.1088/1748-6041/4/3/034104.
Huang S., Wu T., Yu H. 2019. *In Nanobiomaterials in Clinical Dentistry*, 495–515. doi:10.1016/B978-0-12-815886-9.00020-6.
Ichim T., Bogin V., Patel A., Stem cells and stem cell generated nanoparticles for treatment of inflammatory conditions and acute radiation syndrome. U.S. Patent 2013/0273011 A1, Oct. 17, 2013.
Ikono R., Li N., Pratama N., Vibriani A., Yuniarn et al. 2019. *Biotechnology Reports 24*: e00350. doi:10.1016/j.btre.2019.e00350.
Jiang X., Lim S., Mao H., Chew S. 2010. *Experimental Neurology 223* (1): 86–101. doi:10.1016/j.expneurol.2009.09.009.
Jung S., Kim D., Ju J., Shin H. 2013. *In Vitro Cellular & Developmental Biology-Animal 49* (1): 27–33. doi:10.1007/s11626-012-9568-y.
Kamaleddin M. 2017. *Nanomedicine: Nanotechnology, Biology and Medicine 13* (4): 1459–1472. doi:10.1016/j.nano.2017.02.007.
Kaul H., Ventikos Y. 2015. *Tissue Engineering Part B: Reviews 21* (2): 203–217. doi:10.1089/ten.TEB.2014.0285.
Kim T., Lee K., Choi J. 2013. *Biomaterials 34* (34): 8660–8670. doi:10.1016/j.biomaterials.2013.07.101.
Kitsara M., Agbulut O., Kontziampasis D., Chen Y., Menasché et al. 2017. *Actabiomaterialia 48*: 20–40. doi:10.1016/j.actbio.2016.11.014.
Kutsuzawa K., Chowdhury E., Nagaoka M., Maruyama K., Akiyama et al. 2006. *Biochemical and Biophysical Research Communications 350* (3): 514–520. doi:10.1016/j.bbrc.2006.09.081.
Langer R. 2000. *E-biomed: A Journal of Regenerative Medicine 1* (1): 5–6. doi:10.1089/152489000414507.
Law J., Liau L., Saim A., Yang Y., Idrus R. 2017. *Tissue Engineering and Regenerative Medicine 14* (6): 699–718. doi:10.1007/s13770-017-0075-9.
Lee D., Choi B., Park J., Zhu S., Kim et al. 2006. *Journal of Cranio-Maxillofacial Surgery 34* (1): 50–56. doi:10.1016/j.jcms.2005.07.011.
Levengood S., Erickson A., Chang F., Zhang et al. 2017. *Journal of Materials Chemistry B 5* (9): 1822–1833. doi:10.1039/C6TB03223K.
Liu J., Wang C., Wang J., Ruan H., Fan C. 2011. *Journal of Biomedical Materials Research Part A 96* (1): 13–20. doi:10.1002/jbm.a.32946.
Liu S., Mou S., Zhou C., Guo L., Zhong et al. 2018. *ACS Applied Materials &Interfaces 10* (49): 42948–42958. doi:10.1021/acsami.8b11071.
Markides H., Rotherham M., El Haj A. 2012. *Journal of Nanomaterials* 2012. doi:10.1155/2012/614094.
Marx G., Gorodetsky R., Fibrin nanoparticles and uses thereof. U.S. Patent 6,552,172 B2, Apr. 22, 2003.
Min L., Edgar T., Zicheng Z., Yee Y. 2015. Biomaterials for bioprinting, pp. 129–148. In *3D Bioprinting and Nanotechnology in Tissue Engineering and Regenerative Medicine*. Academic press.
Mo Y., Guo R., Zhang Y., Xue W., Cheng et al. 2017. *Tissue Engineering Part A 23* (13–14): 597–608. doi:10.1089/ten.tea.2016.0268.
Mosselhy D.A., Assad M., Sironen T., Elbahri M. 2021. *Nanomaterials 11* (1): 82. doi:10.3390/nano11010082.
Navaei A., Saini H., Christenson W., Sullivan R.T., Ros et al. 2016. *ActaBiomaterialia 41*: 133–146. doi:10.1016/j.actbio.2016.05.027.
Nectow A.R., Marra K.G., Kaplan D. 2012. *Tissue Engineering Part B: Reviews 18*(1): 40–50. doi:10.1089/ten.TEB.2011.0240.
Nukavarapu S.P., Kumbar S.G., Brown J.L., Krogman et al.2008. *Biomacromolecules 9* (7): 1818–1825. doi:10.1021/bm800031.
Paul W., Sharma C.P. 2006. *American Journal of Biochemistry and Biotechnology 2* (2): 41–48.
Pellegrini G., TraversoC.E., Franzi A.T., Zingirian M., Cancedda et al. 1997. *The Lancet 349* (9057): 990–993. doi:10.1016/S0140-6736(96)11188-0.
Pilehvar-Soltanahmadi Y., Akbarzadeh A., Moazzez-Lalaklo N., Zarghami N.J. 2016. *Artificial Cells, Nanomedicine, and Biotechnology 44* (6): 1350–1364. doi:10.3109/21691401.2015.1036999.
Qin H., Zhu C., An Z., Jiang Y., Zhao et al. 2014. *International Journal of Nanomedicine 9*: 2469. doi:10.2147/IJN.S59753.

Rahmani Del Bakhshayesh A., Annabi N., Khalilov R., Akbarzadeh A.et al. 2018. *Artificial Cells, Nanomedicine, and Biotechnology 46* (4): 691–705. doi:10.1080/21691401.2017.1349778.

Rodriguez-Contreras A., Bello D.G., Nanci A.J. 2018. *Applied Surface Science 445*: 255–261. doi:10.1016/j.apsusc.2018.03.150.

San Thian E., Huang J., Best M., Barber H., Brooks et al. 2006. *Biomaterials 27*(13): 2692–2698. doi:10.1016/j.biomaterials.2005.12.019.

Sarao V., Veritti D., Boscia F., Lanzetta J.2014.*The Scientific World Journal* 2014. doi:10.1155/2014/989501.

Sato M., Sambito A., Aslani A., Kalkhoran M., Slamovich et al. 2006. *Biomaterials 27*(11):2358–2369. doi:10.1016/j.biomaterials.2005.10.041.

Schrader J., Targeted delivery of nanoparticles to epicardial derived cells (epdc). U.S. Patents 2016/0030600 A1, Feb. 4, 2016.

Seal S., Cho H., Patil S., Mehta A., Cerium oxide nanoparticle regenerative free radical sensor. U.S. Patent 8172997 B2, May. 8, 2012.

SingerJ., BoyceT. 2017.*JournalofBurn Care & Research38*(3):605–613.doi:10.1097/BCR.0000000000000538.

Smeets R., Jelitte G., Heiland M., Kasaj A., Grosjean et al. 2008. Hydroxylapatit-Knochenersatzmaterial (Ostim®) bei der Sinusbodenelevation. *Schweizer Monatsschrift für Zahnmedizin 118*(3): 203.

Steinhoff G., Steinhoff K., Li W., Ma N., Magnetic pole matrices useful for tissue engineering and treatment of disease. U.S. Patent 2006/016.5805A1, Jul. 27, 2006.

Sundaramurthi D., Krishnan M., Sethuraman J.2014.*Polymer Reviews 54*(2):348–376. doi:10.1080/15583724.2014.881374.

Sundaramurthi D., Vasanthan S., Kuppan P., Krishnan M., Sethuraman et al. 2012. *Biomedical Materials 7*(4): 045005. doi:10.1088/1748-6041/7/4/045005.

Tan H., Marra G.2010. *Materials 3*(3): 1746–1767. doi:10.3390/ma3031746.

Tian J., Wong K., Ho M., Lok N., Yu et al. 2007. *Chem Med Chem: Chemistry Enabling Drug Discovery 2*(1): 129–136. doi:10.1002/cmdc.200600171.

Todd G., Danilkovitch A., Nanoparticle-loaded cells. U.S. Patent0087868 A1, Apr. 12, 2012.

Vacanti C. 2006. *Tissue Engineering 12*(5): 1137–1142. doi:10.1089/ten.2006.12.1137.

Wang L., Wu Y., Hu T., Guo B., Ma et al. 2017. *Actabiomaterialia 59*: 68–81. doi:10.1016/j.actbio.2017.06.036.

Wang W., Yeung W. 2017. *Bioactive Materials 2* (4): 224–247. doi:10.1016/j.bioactmat.2017.05.007.

Wang Z., Wang K., Lu X., Li C., Han et al. 2015. *Advanced Healthcare Materials 4* (6): 927–937. doi:10.1002/adhm.201400684.

Webster J., Ergun C., Doremus H., Siegel W., Bizios et al. 2000. *Journal of Biomedical Materials Research: An Official Journal of The Society for Biomaterials, The Japanese Society for Biomaterials, and The Australian Society for Biomaterials and the Korean Society for Biomaterials 51*(3): 475–483. doi:10.1002/1097-4636(20000905)51:3<475:aid-jbm23>3.0.co;2-9.

Webster J., Ergun C., DoremusH., Siegel W., Bizios et al. 2001. *Biomaterials 22*(11): 1327–1333. doi:10.1016/s0142-9612(00)00285-4.

Webster J., Hellenmeyer L., Price L. 2005. *Biomaterials 26*(9): 953–960. doi:10.1016/j.biomaterials.2004.03.040.

Yang F., Murugan R., Ramakrishna S., Wang X., Ma et al. 2004.*Biomaterials 25*(10): 1891–1900. doi:10.1016/j.biomaterials.2003.08.062.

Yao Q., Zheng W., Lan H., KouL., Xu et al. 2019. *Materials Science and Engineering: C 104*: 109942. doi:10.1016/j.msec.2019.109942.

Yu W., Jiang X., Cai M., Zhao W., Ye et al. 2014. *Nanotechnology 25*(16): 165102. doi:10.1088/0957-4484/25/16/165102.

Yuan Z., Nie H., Wang S., LeeH., Li et al. 2011. *Tissue Engineering Part B: Reviews 17*(5): 373–388. doi:10.1089/ten.TEB.2011.0041.

Yung K.L., Li H.W., Lui N.P., Tsui Y.P., Ho S.L., Chan Y.S., Shum K.Y., Tsang E.S.C. Method of extracting neural stem cells using nanoparticles. U.S. Patent 9109203 B2, Aug. 18, 2015.

Zhou T., Sui B., Mo X., Sun J. 2017. *International Journal of Nanomedicine 12*: 3495. doi:10.2147/IJN.S132459.

Zielińska A., Carreiró F., Oliveira A.M., Neves A., Pires et al. 2020. *Molecules*, *25* (16): 3731. doi:10.3390/molecules25163731.

Zong X., Bien H., ChungY., Yin L., Fang et al. 2005. *Biomaterials*, *26*(26): 5330–5338. doi:10.1016/j.biomaterials.2005.01.052.

2

Two-Dimensional Nanomaterials for Drug Delivery in Regenerative Medicine

Zahra Mohammadpour
Motamed Cancer Institute, ACECR, Tehran, Iran

Seyed Morteza Naghib
Iran University of Science and Technology, Tehran, Iran

CONTENTS

Introduction

Two-dimensional (2D) materials are layered structures, which upon top-down exfoliation, delaminate to ultrathin nanomaterials. While the other two dimensions can be quite large, the thickness of 2D nanomaterials can go down to the uni- or multi-atomic layer. Diverse chemical functionalities, unique physicochemical characteristics, and most importantly, large specific surface area endowed by their high anisotropy have made 2D nanomaterials appealing to various research endeavours, including sensing, catalysis, energy storage, electronics, and biomedicine (Salahandish et al. 2018a; Liu et al. 2015; Choi et al. 2018; Chimene et al. 2015; Samadi et al. 2018; Salahandish et al. 2018b, Salahandish et al. 2018c). In biomedicine, 2D nanomaterials have widely been exploited in the design of nano-bio interfaces for biosensing (Zhu et al. 2015; Luong and Vashist 2017; Cao et al. 2018), drug delivery (Shim et al. 2016; Wang et al. 2017), non-invasive diagnostics (Chen et al. 2018a), tumour therapy/theranostics (Mohammadpour and Majidzadeh-A 2020; Chen et al. 2017b; Yang et al. 2018a), and tissue engineering (Shin et al. 2016; Goenka et al. 2014).

Tissue engineering is a powerful tool for reconstructing damaged organs (Hashemi et al. 2015). Graphene and graphene-like nanomaterials have been great choices for the fabrication of bioactive scaffolds. Stimulation of stem cell differentiation, controlled release of the drug under internal or external stimulus, biodegradability, and comparable mechanical strength and flexibility to organs are among the

DOI: 10.1201/9781003153504-2

several characteristics of 2D nanomaterials that led to remarkable success in tissue regeneration. Several examples are given in the present chapter and other reports (Zhang et al. 2020a; Gaharwar et al. 2019). They demonstrated that advancements in bone, muscle, cardiac, neuron, and cartilage tissue engineering, as well as immunomodulation, have been made by exploitation of 2D nanomaterials in the scaffold design. Furthermore, the inherent antibacterial property of 2D nanomaterials (Zhang et al. 2020d) and the possibility of loading antimicrobial agents with high capacity on the surface of 2D nanomaterials (Salguero et al. 2020) proved their potential for application in wound repair.

Here, we summarised the applications of various 2D nanomaterials as delivery carriers in regenerative medicine. Specifically, repairing different tissues (vein, muscle, bone, skin, cardiac, cartilage, ocular, dental, etc.) and cancer immunotherapy are covered in this chapter. The sections are organised based on the nanomaterial type. Graphene, black phosphorous (BP), layered double hydroxides (LDHs), clays, metal oxides, and metal-organic frameworks (MOFs) are discussed. A separate section showcasing the promise of other 2D nanomaterials (polymeric nanosheets and transition metal dichalcogenides (TMDs)) in tissue engineering has also been considered. Finally, in the conclusion section, challenges and prospects of using 2D nanomaterials in regenerative medicine are outlined.

Graphene for Bone Tissue Engineering

The chemical, physical, and mechanical properties of graphene and their derivatives have endowed them a great potential in a broad spectrum of research fields including bio- and nanomedicine (Naghib 2019; Salahandish et al. 2018a; Kalkhoran et al. 2018; Askari et al. 2019; Mamaghani et al. 2018; Naghib et al. 2020a; Naghib et al. 2020b). Their promise as a vehicle for carrying diverse cargos (drug, genes, nanoparticles, proteins, etc.) has been reported extensively (Gooneh-Farahani et al. 2019; Kalkhoran et al. 2018; Gooneh-Farahani et al. 2020; Askari et al. 2019). Graphene oxide (GO) has been widely employed in regenerative medicine. The oxygenated functional groups at the edge sites and the hydrophobic basal plane interact with drugs, genes, or proteins through hydrogen bonding, electrostatic interactions, or π–π stacking. Such interactions preserve biomolecules (proteins, DNA) from enzymatic degradation. Besides, they guarantee the sustained and continuous release of the cargo over time.

A suitable scaffold for tissue engineering should have good mechanical properties. In bone tissue engineering, sufficient elasticity and stress-bearing capacity hold the platform at the defective area. To this end, graphene-based materials with superb mechanical stability are of great interest. Sedghi et al. blended biocompatible polymers, zinc-curcumin, and GO to fabricate biocompatible nanofibres (Sedghi et al. 2018). They investigated the capacity of the resulting coaxial nanofibres for bone tissue engineering. While GO enhanced the mechanical stability of the scaffold, zinc-curcumin improved cell proliferation, antibacterial activity, and cytocompatibility. Improved tensile strength of the nanofibres in line with the increase in GO concentration originated from the strong interactions between the oxygenated groups of GO with the polymer matrix and the high specific surface area of GO. In another study, Lee et al. designed glycol chitosan-hyaluronic acid (HA) injectable hydrogel and incorporated GO into the hydrogel structure (Lee et al. 2020). Not only the mechanical stability of the scaffold was reinforced by GO but also osteogenesis was induced by it. The authors did not elucidate the mechanism of action of GO in this study.

Several reports have used graphene and its derivatives as a carrier. Graphene-based materials can carry large amounts of drugs or other (bio) molecules, provide sustained release, and preserve bioactivity. In gene delivery, they can condense, protect, deliver, and release the genetic material. In this respect, Zhang et al. covalently attached PEG and PEI to the surface of GO (nGO-PEG-PEI) and used the nanocomposite to immobilise small interfering RNA (siRNA) (Zhang et al. 2017b). The transfection efficiency of the nanovector was so high that the targeted gene, which was responsible for the production of casein kinase-2 interacting protein-1 (Ckip-1) as a negative regulator of bone formation, was silenced. Therefore, the differentiation of MC3T3-E1 cells was dramatically enhanced, evidenced by the increased alkaline phosphatase (ALP) production and extracellular matrix (ECM) mineralisation. They hypothesised that the osteogenic potential of nGO-PEG-PEI might come from the ability of GO to bind with osteogenic enhancing biomolecules through π–π stacking or electrostatic interactions. There are other examples in the literature in which graphene-based materials have served as vehicles to deliver salvianolic acid B (Wang et al. 2020b), methyl vanillate (Jiao et al. 2019), tetracycline hydrochloride (Mahanta et al. 2019), basic fibroblast growth factor (bFGF), bone morphogenetic protein-2 (BMP-2) (Ren et al. 2018), and alendronate (Zeng et al. 2020), all of which promoted osteogenesis and bone repair. Graphene can also modulate the release of cargo from scaffolds. Laurenti et al. designed a bioactive and biocompatible scaffold comprising mesoporous zinc oxide covered with GO (Laurenti et al. 2019). They observed the finely tuned loading/releasing property of gentamicin sulphate and proposed the system for bone tissue engineering. They attributed the controlled release property to the existence of nanopores in the structure of the GO layer. Therefore, it prevented the burst release of the drug and prolonged the delivery process. Also, GO induced the formation of a compact and continuous apatite layer in simulated body fluid.

Novel fabrication strategies of scaffold-based and scaffold-free systems based on graphene for bone regeneration are reported. Li et al. designed a three-dimensional organic–inorganic hybrid material and proposed it as a bone-grafting scaffold (Li et al. 2018b). They prepared peptide nanosheets (PNSs) by self-assembly and covered them on three-dimensional graphene foam (GF-PNSs). Hydroxyapatite was then mineralised on GF-PNSs, after which the final composition of the scaffold was obtained (GF-PNSs-hydroxyapatite). They investigated the viability of L-929 cells in the presence of powders of the ground scaffold and observed 98%cytocompatibility after 24 hours. Although the authors did not conduct further cellular experiments, they suggested that the high porosity, tunable shape, and interconnected structures of the 3D scaffold make it suitable for vascularisation and medium transport. In another study, Zhang et al. introduced a "growth-factor-immobilised cell sheets" approach to fabricate bone-cartilage junction (Zhang et al. 2017d). They synthesised GO-Fe_3O_4 nanocomposites. BMP-2, as osteogenic induction factor, and TGF-β3, as a chondrogenic induction factor, was then immobilised on the nanocomposite surface. Surface modification of Fe_3O_4 NPs with nGO promoted the interaction of growth factors with the nanocomposite and their delivery to dental-pulp stem cells. Enforcement of an external magnetic field aligned the cells into which the nanocomposites were located. A composite tissue of chondrogenic and osteogenic cell sheets was formed and proposed for osteochondral defect regeneration. Figure 2.1 shows a nanocomposite comprising rGO and nanoscaledhydroxyapatite used in bone tissue engineering (Nie et al. 2017).

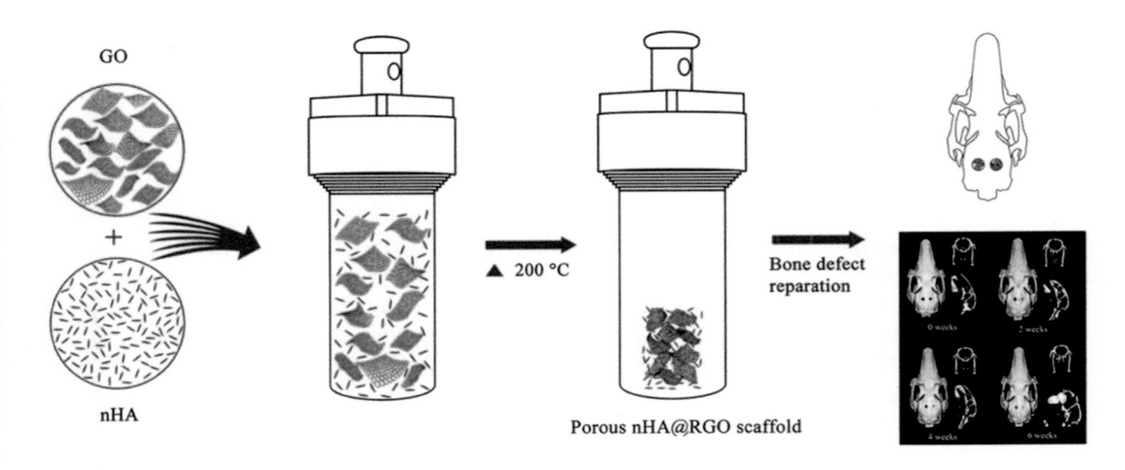

FIGURE 2.1 3D porous scaffold synthesised by self-assembly of rGO and nanoscaled hydroxyapatite composite for bone tissue engineering (Nie et al. 2017).

Graphene for Wound Healing

The integration of bioactive components and graphene-based matrices has been proposed to be promising for wound healing (Mellado et al. 2018; Liu et al. 2017a). Mellado et al. developed a composite aerogel based on GO and poly vinyl alcohol (PVA) into which an extract of grapes rich in proanthocyanidins was incorporated (Mellado et al. 2018). High absorption of blood demonstrated the aerogel's haemostatic property. The cumulative release of the extract did not exceed 20%, probably due to the strong interaction between the aerogel and the extract in the form of covalent bonding. Tanum et al. investigated the sustained and prolonged delivery of nitric oxide as a therapeutic gas from a solid film (Tanum et al. 2019). A molecular source of NO gas was conjugated to the surface of GO by covalent functionalisation. Due to the multilayer nature of the film, the released NO was physically trapped, which made the diffusion length long. The gas was stabilised by hydrogen bonding. Therefore, the unfavourable burst release of NO was prevented, and the release time of the gas was prolonged. The tortuosity of the path caused by the stacked GO sheets was also responsible for tunable oxygen delivery. Jalani et al. showed that the integration of GO with perfluorocarbon emulsions slowed down the release of oxygen and stabilised the emulsion (Jalani et al. 2017). The perfluorocarbons can absorb large amounts of oxygen and are suitable as alternatives to blood for the oxygen supply of tissues.

Delivery of the drug to the wound area is also possible through the stimuli-responsiveness of graphene-based nanosheets. In this respect, Altinbasak et al. embedded rGO into poly acrylic acid (PAA) nanofibre (PAA@rGO) and used the resulting NIR stimuli-responsive drug delivery mats in wound dressing (Altinbasak et al. 2018). The platform was activated by irradiation of NIR light as an external stimulus, and antibiotics (ampicillin and cefepime) were released on demand. The mats functioned reproducibly, and they were reusable. The interaction of antibiotics with the surface of rGO was established through non-covalent interactions. The photothermal effect of rGO was responsible for the active drug release without any sign of inflammation or bacterial foci in the superficial skin-damaged infection model. In another study, Liu demonstrated that the electrical stimulus could trigger the release of lidocaine hydrochloride based on the electrical conductivity of rGO (Liu et al. 2012). Passive release of the payload was prevented as the rGO in the rGO–PVA membrane acted as a physical barrier.

Non-healing diabetic wounds have also been treated by graphene-based natural scaffolds (Chu et al. 2018; Fu et al. 2019). Liu's research group used an acellular dermal matrix (ADM) as a natural scaffold. ADM is biocompatible and retains the basic dermal architecture. As a transplanting carrier for pluripotent mesenchymal stem cells, the ADM-rGO scaffold promoted stem cell adhesion and proliferation. The inclusion of rGO into ADM enhanced mechanical stability. Full-thickness wounds of diabetic mice were repaired. The composite platform also provided robust vascularisation, collagen deposition, and rapid re-epithelialisation of newborn skin.

Graphene for Immunotherapy

Graphene-based nanomaterials have shown excellent prospects in immunotherapy. Gurunathan et al. conducted a mechanistic study on the pathways involved in cellular toxicity and immunomodulatory of GO and vanillin-functionalised GO (V-rGO) (Gurunathan et al. 2019). Molecular evaluations were performed on THP-1 cells, a human acute monocytic leukaemia cell line. The results revealed that both nanomaterials stimulated cancer cell lines to release cytokines and chemokines, including IL1-β, TNF-α, GM-CSF, IL-6, IL-8, and MCP-1. However, the released cytokines were significantly higher in the case of V-rGO than that observed for GO. GO has been employed as a carrier for the delivery of immunostimulatory CpG oligodeoxynucleotides (ODNs) (Zhang et al. 2017a; Tao et al. 2014). Zhang et al. showed that the cellular uptake of CpG ODNs that were loaded on GO-chitosan (GO-CS) nanocomposites was higher than free CpG ODNs and GO/CpG ODNs complexes (Zhang et al. 2017a). Higher levels of interleukin-6 (IL-6) and tumour necrosis factor-α (TNF-α) confirmed the observation. Tao et al. further showed that CpG delivery was controlled by the photothermal effect of GO-PEG-PEI (Tao et al. 2014).

Graphene has also been used as a vaccine delivery platform to activate antigen-presenting dendritic cells (DCs). Sinha et al. functionalised graphene with dextran and used it as an ovalbumin carrier to activate cytotoxic T cells (Sinha et al. 2017). Increased colloidal stability and cell internalisation of the nanocomposite was ascribed to the dextran. At the same time, the high loading of ovalbumin on the surface of the nanocomposite was related to the large specific surface area of GO and its hydrophobic basal plane. Based on the in vivo experiments, they observed that the production of cytotoxic T cells was induced by ovalbumin-loaded graphene nanocomposite, which subsequently resulted in the prevention of tumour growth. Therefore, they proposed the delivery system potentially as a cancer vaccine. The same research group also developed a 3D alginate scaffold into which graphene was embedded (Sinha et al. 2019). They loaded it with ovalbumin, granulocyte-macrophage colony-stimulating factor, and CpG. They offered the system to be promising in cancer vaccination.

The communication among the cells of the immune system is established via clusters of cell surface molecules rather than isolated ligand–receptor pairs. The latter activates the immune response weakly. To mimic receptor nanoclustering for intense activation of the immune system, Loftus et al. covalently attached the nanoscale GO (NGO) to natural killer (NK) cell-activating antibodies (Loftus et al. 2018). This way, the molecular cluster of antibodies was generated. They successfully stimulated and activated NK cells. The assembly of immunomodulatory drugs into clusters via the implementation of NGO as a template led to a significantly higher secretion of INF-γ compared to a solution of the same antibodies applied as individual molecules.

Graphene for Muscle and Neuron Tissue Engineering

Graphene-based nanomaterials have the capacity as functional platforms for culturing muscle cells. Myogenic differentiation of C2C12 cells on GO and rGO was studied by Ku et al. (Ku and Park 2013). A comparative study among graphene derivatives was conducted based on biocompatibility, cell adhesion, proliferation, and differentiation. Both nanosheets were biocompatible and induced myotube formation. However, quantitative analysis of the immunofluorescence data for myosin heavy chain (MHC) and myogenin confirmed that myogenic differentiation was better on the surface of GO. Higher expressions of MyoD, myogenin, Troponin T, and MHC genes for the cells that were grown on the GO surface were also observed. The effect was ascribed to nanotopography and, more importantly, the surface oxygen content of GO that promoted the adsorption of serum proteins. GO has also been integrated into alginate for the encapsulation of muscle cells (Ciriza et al. 2015). At the optimum concentration of GO, the cell viability and metabolic activity of microencapsulated muscle cells increased.

GO-incorporated polyacrylamide hydrogel was patterned by femtosecond laser ablation followed by chemical reduction (Park et al. 2019). The micropatterned conductive hydrogel provided an appropriate topography for myoblast maturation. In addition, the inclusion of rGO in the hydrogel structure enabled electrical stimulation for the differentiation of myoblasts. In another advanced architecture for myoblast differentiation and cell sheet formation, Kim et al. developed an active scaffold that not only provided support for cell proliferation and differentiation but also enabled monitoring the physiological characteristics associated with the cultured cells (Kim et al. 2015). The prepared muscle-on-a-chip scaffold was employed for drug screening purposes. Besides, the differentiated cell layers were transferred in vivo for muscle tissue regeneration.

Lee et al. developed a composite of poly(3,4-ethylene dioxythiophene):poly (styrene sulphonate) (PEDOT:PSS) and GO/rGO and investigated its performance as a biocompatible neural interface (Lee et al. 2019). The nanocomposite electrodes were not toxic to PC12 neural cells. Also, gene expression for the production of GAP-43 and synapsin on the surface of PEDOT:PSS/GO or PEDOT:PSS/rGO was significantly higher than that of PEDOT. Better cellular communication through the conductive matrix of the nanocomposite could be responsible for more increased intracellular signalling. The authors proposed that the graphene-based microelectrode could potentially be employed as an implantable neural electrode. The electrical conductivity of graphene-based nanomaterials is enticing in other aspects of neural tissue regeneration, including the on-demand release of macromolecular therapeutics. In this regard,

Magaz et al. developed a hybrid biocomposite of silkworm fibroin and rGO that was loaded with nerve growth factor-β (NGF-β) (Magaz et al. 2020). Upon application of a pulsatile electrical stimulus, the growth factor was released over a ten-day period. The authors proposed that hybrid biocomposite might be suitable for the design of personalised scaffolds.

Black Phosphorus

Black phosphorous (BP) nanosheets with unique optical properties and biodegradation into harmless compounds have garnered considerable interest in biomedicine. Hou et al. used BP nanosheets for the treatment of acute kidney injury (Hou et al. 2020). Through scavenging the reactive oxygen species (ROSs), BP nanosheets could alleviate the cellular apoptosis associated with oxidative stress. The authors stated that the flake-like morphology, antioxidative property, and minimal cytotoxicity make BP nanosheets intriguing candidates for the treatment of acute kidney injury and other ROS-related diseases. Another well-known property of BP nanosheets is their ability to convert NIR light into heat. This characteristic has been widely used in biomedicine (Mohammadpour and Majidzadeh-A 2020). In a study by Tong et al., highly efficient bone regeneration was induced on an osteoimplant (BPs@poly(lactic-co-glycolic acid) (PLGA)) (Tong et al. 2019). The mild heat generated by the photothermal conversion property of BP nanosheets led to in vitro and in vivo osteogenesis under remote control. The expression of cellular heat shock proteins played an important role. The photothermal conversion of BP nanosheets can also be used to on-demand light-controlled drug delivery for the regeneration of bone (Wang et al. 2018) and fighting bacterial infection (Guo et al. 2020).

BP nanosheets have actively been employed in chemo-photoimmunotherapy (Nguyen et al. 2019; Ou et al. 2020). Ou et al. synthesised BP nanosheets using the "plug-and-play nanorisation" technique (Ou et al. 2020). On the resulting uniform nanosheets, the authors added doxorubicin (D), chitosan–PEG (c), and folic acid (F). They encapsulated the modified nanosheets with siRNA (s) and programmed death-ligand 1 (PL) to realise BP-DcF@sPL. The combinatorial therapeutic function of nanoplatforms was used against colorectal cancer. Upon NIR irradiation, a considerably higher drug release from BP-DcF@sPL was observed; thus, higher cytotoxicity to HCT116, HT29, and MC-38 cells was achieved compared to controls. Besides, around 20% higher maturation of DC cells and 37% higher activation of T cells were found. A comprehensive study of the parameters associated with the activation of the immune system against the tumour exhibited a collectively activated immune response, the ultimate effect of which was prolonging the survival period of the treated mice group. Figure 2.2 represents that cell proliferation and osteogenesis are promoted on 3D printed scaffolds based on 2D BP and GO. (Liu et al. 2019a).

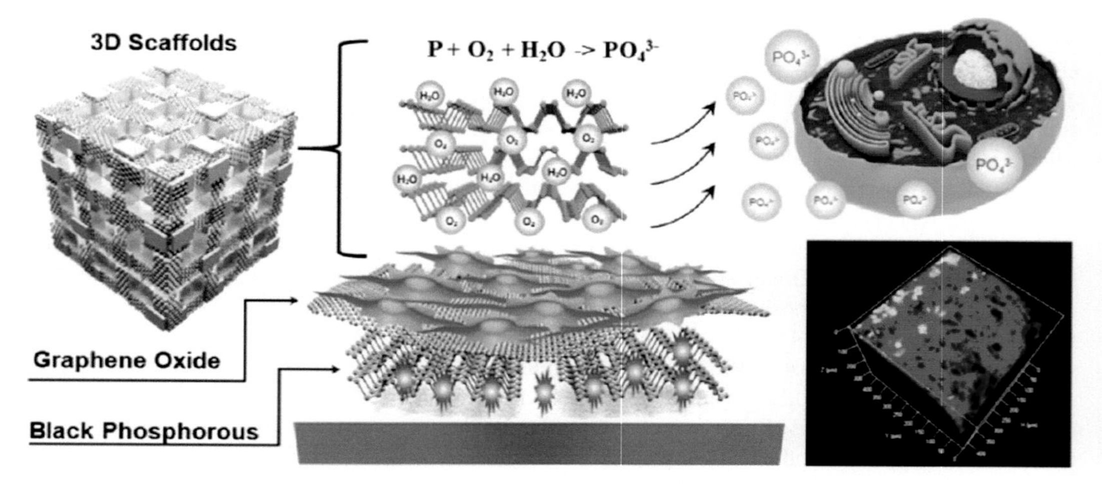

FIGURE 2.2 Schematic representation of the 3D printed scaffold comprising 2D BP and GO (Liu et al. 2019a). Copyright (2019) American Chemical Society.

Layered Double Hydroxides

Layered doubled hydroxides (LDHs) are a class of anionic clays that consist of positively charged metal hydroxide layers and interlayer anionic molecules for charge balance. Their structure is close to brucite. Incorporation of various metal cations into the layers and the interlayer anions' exchangeability give rise to a flexible composition and a multitude of properties for these laminated structures. The expansive space between the layers of LDHs has made them hosts for various (bio) molecules, drugs, nanoparticles, etc. LDH nanoparticles have shown in vivo biocompatibility (Figueiredo et al. 2018). In regenerative medicine, LDHs have been actively employed for wound healing, immunotherapy, bone regeneration, drug delivery, etc. Isoniazid, an anti-tuberculosis drug, was sustainably released from MgAl LDHs (Saifullah et al. 2014). Saifullah et al. found that the nanodelivery system provided higher biocompatibility than free isoniazid (Saifullah et al. 2014). The possibility to intercalate poorly water-soluble drugs is another feature of LDH nanoparticles. As an example, Gao et al. loaded ibuprofen and ketoprofen into LDHs and mixed them with polycaprolactone (PCL) to fabricate organic–inorganic nanohybrids (Gao et al. 2017). While the drug release from the LDH itself was relatively fast during the first four hours, the release rate from the nanohybrids was much slower. Due to such characteristics, the authors proposed nanofibres as appropriate candidates for implantable drug delivery systems.

Zhang et al. prepared MnNi LDH hierarchical structures and used them for wound healing (Zhang et al. 2020d). The intrinsic oxidase-like characteristic of MnNi LDHs, which led to the production of superoxide radicals in the presence of oxygen, accounted for the antibacterial property (both gram-negative and gram-positive bacteria). Furthermore, MnNi LDHs showed promising anti-infective activities to wounds that were infected by *S. aureus* and suppression of the bacterial growth at the wound site; therefore, a faster healing process was observed for the mice group that were treated with MnNi LDHs. As mentioned earlier, LDHs can be the hosts of various drugs (Munhoz et al. 2019; Salguero et al. 2020). Salguero et al. intercalated ciprofloxacin (Cip) into the interlayer structure of ZnAl LDH and fabricated a composite film by the inclusion of LDH-Cip into a hyaluronan matrix (Salguero et al. 2020). Based on the in vitro antibacterial investigations, the authors proposed that the film might be appropriate for topical drug delivery to skin wounds.

In immunotherapy, LDHs act as carriers to deliver multi-antigen epitopes. Zhang et al. built efficient peptide-based cancer vaccines, based on LDH-BSA nanoparticles, with the capacity to efficiently deliver three different antigens as well as an immunostimulant to DCs, and induce strong T-cell responses (Zhang et al. 2018). To this end, they functionalised the LDH nanosheets with BSA to enhance the colloidal stability in various biological media. They assembled tumour antigens (Trp2, a melanoma tumour-associated antigen; M27, MHC I restricted; M30, MHC II restricted) as well as CpG into the interlayer space of LDH-BSA and used the nanoformulation for the prevention of melanoma growth. Results indicated that the peptide-loaded LDH-BSA was uptaken by cells and escaped into the cytoplasm. The effect of this was to activate T cells strongly and inhibit tumour growth. They ascribed the immunomodulatory characteristic of the developed nanovaccine to the high stability of the modified LDH dispersions. The same research group co-delivered Trp2 and indoleamine 2,3-dioxygenase siRNA (siIDO) to DCs, which activated T cells against melanoma tumours (Zhang et al. 2017c). The inclusion of siRNA and the peptide into the LDH structure protected them against enzymatic degradation. In another study, Yang et al. loaded miR155 into MgAl-LDH (LDH@155) and used it to reverse the immunosuppression tumour microenvironment (Yang et al. 2019). They also employed classical "priming + boosting" vaccination strategy and employed four combinations of vaccination routes (IV + IV, IV + SC, SC + IV, and SC + SC) for tumour immunotherapy (Zhang et al. 2020b). IV and SC stand for intravenous and subcutaneous, respectively. The nanovaccines were constructed by loading antigen and CpG into LDH nanoparticles. They found that "IV-priming + SC-boosting" was the most efficient vaccination strategy for preventing melanoma tumour growth at its early stages.

The high loading capacity of LDH nanosheets and efficient delivery of therapeutic molecules to defected sites have promoted researchers to employ LDH-based nanocarriers for bone regeneration. Yasaei et al. intercalated simvastatin, a stimulator for bone generation, into ZnAl LDHs (Yasaei et al. 2019). Sustained-release behaviour of the drug was reported. Thus, it was introduced as a potential drug

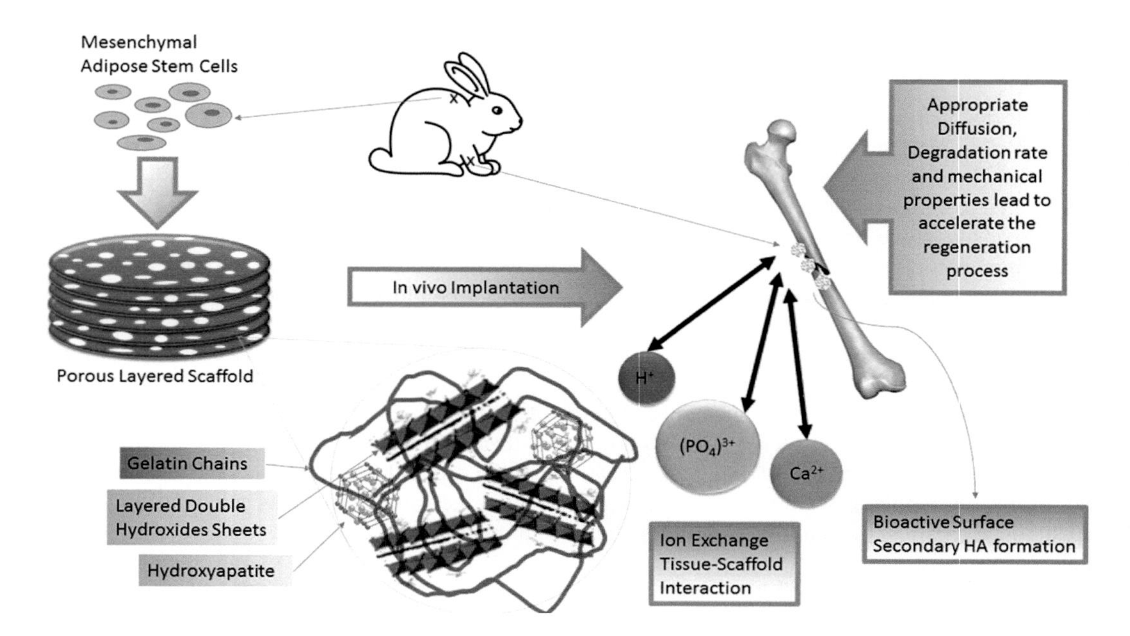

FIGURE 2.3 Schematic representation of the LDH-based scaffold applied in bone tissue engineering (Fayyazbakhsh et al. 2017).

carrier for bone repair. Chen et al. fabricated LDH-chitosan interconnected porous structures and loaded them with PFTα (Chen et al. 2017a). The therapeutic bone scaffolds released PFTα and induced osteogenesis of mesenchymal stem cells (in vitro) and bone repair (in vivo). The latter was evaluated 12 weeks post-implantation. In one step further, the chemical composition of the LDH itself and its capacity for high loading and release of drug molecules constituted a theranostic platform for both X-ray diagnosis and bone repair, respectively (Kim et al. 2016). Kim et al., who prepared risedronate-loaded ZnAl LDHs (RS-LDHs), conducted the study. A hybrid of PLGA and RS-LDH was attached to a clinically approved bone plate and placed on the area of the cranial bone defect in a rat model. As demonstrated by in vivo μCT analysis, after eight and 12 weeks post-implantation, the volume of new bone formed at the defective site was significantly higher for RS-LDH than that of the control groups. Apart from being effective in attenuating the X-ray radiation (Kim et al. 2016), the metal cations inserted into the skeleton of LDH nanomaterials took part in the complex ligation processes with cell surface receptors (Kang et al. 2017). Kang et al. prepared MgFe LDHs into which adenosine molecules were sandwiched (2017). Upon dissolution of the LDH structure and co-release of adenosine and Mg^{2+} cation (ligation activator), the adenosine A2b receptor was synergistically activated and osteogenic differentiation and bone repair was promoted in vivo. An LDH-hydroxyapatite/gelatin bone tissue engineering scaffold was fabricated, characterised, and studied in vivo as shown in Figure 2.3 (Fayyazbakhsh et al. 2017).

Clays for Wound Healing

Clays are natural minerals that compose negatively charged layers into which positively charged ions are intercalated. They have shown massive potential in biomedicine. High loading capacity, long-term controlled release, cost-efficiency, and eco-friendly are amongst the various advantages that motivate researchers to employ clay minerals in different biomedical applications. In wound healing, clays have been the host for intercalation of various antimicrobial agents such as chlorhexidine (Ambrogi et al. 2017), ciprofloxacin (Kevadiya et al. 2014), rifampicin (Morariu et al. 2020), and carvacrol (Tenci et al. 2017). Sun et al. synthesised nanoplatforms based on montmorillonite (MMT), capable of drug delivery for simultaneous targeting of bacteria and fungi (Sun et al. 2019). They intercalated quaternised chitosan

(QCS), 5-fluorocytosine (5-FC), and metal copper ions into the interlayer space of MMT (QCS/MMT/5-FCCu). Bacterial viability tests, fluorescence imaging, and SEM analysis confirmed the high inhibitory effect of QCS/MMT/5-FCCu against Gram-positive and Gram-negative bacteria as well as fungi. Besides, the wounds of animal models that were treated with QCS/MMT/5-FCCu recovered with barely any inflammatory reaction, which was demonstrated by H&E staining. Interestingly, the cytotoxicity tests revealed that unlike the free form of the antimicrobial agent (5-FCCu), the ones that were intercalated into the MMT structure showed negligible toxicity to L929 cells. Examination of major organs by H&E staining further confirmed the biosafety of QCS/MMT/5-FCCu. MMT, as solid support, has also been used for the storage and release of nitric oxide as a therapeutic gas for antimicrobial, antithrombotic, and wound healing applications (Fernandes et al. 2013, 2016).

Kiaee et al. reported the stimuli-responsive drug delivery for wound healing applications (Kiaee et al. 2018). Clay-based nanoformulations were used to achieve the purpose. They designed an active drug delivery platform. The electronic wound dressing device could modulate the pH, which in turn induced the topical drug delivery to chronic wounds. In the electronic circuit, the anode was coated with a hydrogel layer (pH-sensitive poly(ethylene glycol)-diacrylate/Laponite) into which drug-loaded chitosan nanoparticles were embedded. Electronic manipulation of the circuit resulted in pH elevation. The chitosan nanoparticles under basic conditions dehydrated and released the drug. Turning off the DC voltage reverted the pH to acidic values, and drug release stopped. Cytocompatibility and wound closure capability of the flexible wound healing patch were confirmed through cell proliferation tests and in vitro scratch wound assay.

Clays for Bone Tissue Engineering

Clay-based biomaterials have presented numerous opportunities for constructing bioactive cell-laden scaffolds for efficient repair of bone defects. Besides, due to their high loading capacity, clays are excellent choices for loading bone-forming factors, including BMP-2. Clinical studies have displayed the ability of BMP-2 in eliciting bone formation. To exert the high efficiency of this therapeutic protein, precise and prolonged delivery is of high importance. Laponite, a layered nanosilicate-based platform, has attracted a myriad of interests in bone reformation (Prabha et al. 2019). Apart from its high loading capacity, previous studies have shown that the leakage of silicate and lithium ions from the Laponite nanostructures enhances cell spreading on substrates, upregulates the expression of osteogenic genes, and induces the production of mineralised matrix (Xavier et al. 2015). Driven by such properties, Zhang et al. explored the co-delivery of nanosilicates and growth factors from a 3D hydrogel scaffold for osteoinduction (Zhang et al. 2020e). They synthesised hydrogels consisting of HA and dextran through Schiff base chemistry. They introduced Laponite@BMP-2 complexes into the hydrogel matrix. The synergistic effect of simultaneous delivery of Laponite nanoplatelets and BMP-2 was evaluated. The proliferation rates of stem cells incubated with Laponite and Laponite@BMP-2 complexes were higher than that of BMP-2 alone. Also, the ALP relative activity of the cells that were incubated with Laponite@BMP-2 complexes was markedly upregulated compared with Laponite or BMP-2, which was ascribed to the synergistic effects of both the released ions from the Laponite structure and the sustained release of BMP-2. Evaluation of the protein release kinetics showed that bare hydrogel completely released BMP-2 during the first week, the mechanism of which was probably dominated by diffusion. In case of Laponite-containing hydrogels, the release time prolonged to four weeks. In the next step, for evaluating the in vivo osteogenic capacity of the modified hydrogel, stem cells were loaded into hydrogel and injected into the defective region of a rat calvarial defect model. After eight weeks post-operation, the μ-CT scan revealed that the defective site that was cured by hydrogel/Laponite@BMP-2 was almost completely covered by the new bone. The results of this study revealed that pragmatic challenges associated with direct administration of BMP-2 including fast clearance and very short half-life can be avoided by employing bioactive clay-based scaffolds. In a similar study, Laponite in combination with gelatin and alginate were used for the fabrication of an injectable hydrogel, the composition of which was similar to ECM (Liu et al. 2020). Hydrogels were loaded with stem cells and injected into a bone defective rat model. A significant accelerated bone healing was achieved.

Orrefo's group conducted active research towards advanced biofabrication approaches. To this end, they developed cell compatible bioinks by the inclusion of Laponite into various biomaterials to achieve bioactive cell-laden constructs for skeletal tissue regeneration (Cidonio et al. 2019a; Cidonio et al. 2019b; 2020). In one report, gelatin methacryloyl (GelMA) was mixed with Laponite and turned into a cell supportive structure by extrusion-based 3D bioprinting (Cidonio et al. 2019b). They observed higher drug localisation in the porous structure of Laponite-GelMA rather than GelMA. The characteristic was attributed to Laponite. Furthermore, the addition of Laponite up to an optimised concentration improved the quality of the bioprinted scaffold. Cell viability and proliferation were assessed by encapsulation of human bone marrow stromal cells into the 3D platform. Unlike the GelMA controls, cells that were encapsulated into Laponite-GelMA consistently proliferated over 21 days. Osteogenic differentiation of the stem cells localised in the Laponite-GelMA scaffold was visualised by alizarin red staining. Interestingly, even in the absence of full osteogenic media (without dexamethasone), mineralisation areas appeared. Laponite-GelMA hydrogel discs that were loaded with vascular endothelial growth factor (VEGF) were implanted in a chick CAM model. Compared with other controls, extensive integration and vascularisation was observed for Laponite-GelMA-VEGF. Increased drug loading and retention in the Laponite structure accounted for better ex vivo performance of the scaffold.

Metal Oxides and Metal-Organic Framework Nanosheets

Analogous to other 2D nanomaterials, metal oxide nanosheets possess large specific surface areas and the ability to protect surface adsorbed biomolecules from enzymatic degradation. Manganese dioxide, in particular, has intriguing features that have made it unique in biomedical applications. High biodegradability is one of its well-known characteristics. Also, MRI activity of the released Mn^{2+} upon the nanosheets' degradation and photothermal conversion efficiency have made MnO_2 nanosheets special for cancer therapy. Taking advantage of the biodegradability of MnO_2 nanosheets, Yang et al. developed a bioscaffold for neural stem cell transplantation (Yang et al. 2018b). Enhanced stem cell differentiation on the MnO_2-based nanoscaffolds was attributed to the high binding affinity of MnO_2 nanosheets towards ECM proteins such as laminin. The calculation of the binding energies between the two parties through density functional theory proved the favourable interaction between ECM proteins and MnO_2nanosheets, which supports preferable cell adhesion on the surface of these nanosheets. The MnO_2-based nanoscaffold was then fabricated by self-assembly of atomically thin MnO_2 nanosheets, ECM proteins, and therapeutic drugs. Immunostaining data showed a significant neuronal stem cell differentiation and neurite outgrowth compared to the control substrate. As an essential aspect of transplanted bioscaffolds, the biodegradability of the MnO_2-based structure was investigated. It was revealed that the biodegradability was tunable by changing the number of the assembled layer. Unlike GO as a negative control, MnO_2 nanosheets were degraded without any external trigger. In vivo, stem cell transplantation was performed using the biodegradable scaffold into the animal model that suffered from spinal cord injury. Rapid biodegradation and differentiation of stem cells into mature neurons were also detected in vivo. In another study, Wang et al. exploited the enzyme mimetic property of MnO_2 nanosheets for cancer therapy and wound healing purposes (Wang et al. 2020a). They consolidated MnO_2 nanosheets into chitosan-based injectable hydrogels. They proposed that the ability of MnO_2 nanosheets to decompose H_2O_2 as the most abundant endogenous ROS to oxygen could disappear the hypoxic and oxidative cellular microenvironment of wounds, which is a factor that postpones wound healing. As a result, inflammation was reduced, and fibroblast proliferation accelerated.

There are some immunosuppressive pathways, the result of which is escaping tumours from being recognised by the immune system. The antibodies that block programmed cell death protein 1 (PD1) and its ligand (PDL1) can restore the immune responsivity towards tumours. The therapeutic strategy is called checkpoint blockade immunotherapy (CBI). However, it has been reported that radical therapy as an immunostimulatory treatment is also immunogenic. In radical therapy, the tumour cells are killed by ROSs that are generated either by an external stimulus such as ultrasonic waves (sonodynamic therapy) and photons (photodynamic therapy) or by endogenous tumour microenvironments (chemodynamic therapy). The latter proceeds through metal ion-mediated decomposition of hydrogen peroxide and ROS generation. Lin's research team showed that a combination of CBI and chemo/photodynamic therapy could synergistically restore the immune responsivity towards tumours (Ni et al. 2019; Ni et al. 2020a; Ni et al. 2020b). To this end, they employed Cu^{2+} containing MOF nanosheets (Cu-TBP) (Ni et al. 2019). Diverse chemical composition and porous structure have made MOFs useful for loading different cargos. Besides, the inclusion of redox-active metal ions and radiosensitisers as ligands in the structure of MOFs provides efficient means for conducting chemo and photodynamic therapies. Cu-TBP MOFs were efficiently delivered into the cells. The released Cu^{2+} metal cations and TBP photosensitisers took part in ROS production through chemo and photodynamic pathways and regressed the local tumour. The Cu-TBP was injected inside the tumour, and the anti-PD-L1 antibody was given to the animal model every three days. Not only the primary tumours were eradicated but also the distant tumours of one-third of mice disappeared due to the anti-tumour immunity induced by the synergistic therapy. The treated mice were rechallenged with new tumour cells after a month. Interestingly, they remained tumour free due to the induced immune memory. The same group employed nanoscaled MOFs (Hf-DBP) to reverse the immunosuppression of tumours by co-delivery of imiquimod and αCD47 (Ni et al. 2020a). Imiquimod was loaded inside the pores of the MOF, and αCD47 adsorbed on its surface. Imiquimod is a hydrophobic drug molecule that activates the toll-like receptor-7 (TLR-7) pathway. αCD47 is an antibody that blocks the CD47 checkpoint signalling molecule on the surface of tumour cells. The TLR-7 pathway activated inflammatory macrophages (M1), which subsequently took part in cytokine secretion, phagocytosis, and antigen presentation to naïve T cells inside the lymph nodes. The activated T cells then invaded the tumour cells, of which the PDL1 proteins were blocked by αPDL1. The T cells recognised and killed the cancer cells (Figure 2.4). This way the macrophage therapy was synergised with CBI, which eradicated both the primary and distal tumours of the animal models.

Other Nanosheets

Several research groups have displayed the promise of polymeric nanosheets in regenerative medicine. Wang et al. loaded PCL nanosheets with vasoactive intestinal peptides (VIPs) and studied the modified nanosheets' practical applicability for wound healing (Wang et al. 2016). They grafted dopamine (DA) on the surface of PCL nanosheets, followed by the addition of VIPs as the bioactive functional units. After acetone addition, the VIP-loaded microspheres were generated, in situ, on PCL nanosheets' surface. Incorporating VIPs as an angiogenesis agent into the polymer matrix and its controlled delivery at the wound site showed enhanced efficiency of wound repair. The histological results of the repaired wound of animal models with PCL-DA-VIP nanosheets revealed that the granulation tissue was thicker than that observed for PCL-DA. Furthermore, the generation of vascular tissue was confirmed by measuring the angiogenic markers. Western blotting confirmed that the amount of CD31 was significantly higher in the mice group treated with PCL-DA-VIP than the PCL-DA group. In another study, Liu et al. explored PLGA nanosheets' usability as siRNA delivery in dental transplantation (Liu et al. 2017b). They assembled antagomiR204-modified gold nanoparticles on the surface of PLGA nanosheets and investigated their dental osseointegration effect on diabetic rats. The released biomolecule inhibited the expression of miR204 in bone mesenchymal stem cells, thus, increased the osteogenic capacity. The efficiency of the modified nanosheets was ascribed to the stable attachment of PLGA sheets on titanium surface after in vivo transplantation and its slow degradability, which endowed gradual release of the modified gold nanoparticles.

Polymer-based nanosheets have also been employed as injectable materials for minimally invasive therapeutics. Low rigidity and shape conformity are two factors that determine their favourable functionality as injectable implants. In this regard, Yamagishi et al. developed a self-expandable, injectable, and magnetic field guidable polymeric platform (Yamagishi et al. 2019). They proposed it as an advanced medical device to deliver cells, sensors, drugs, or engineered tissues to the target sites. They embedded magnetic nanoparticles into polyurethane-based shape-memory polymers. An additional layer of PLGA was interfaced with the modified nanosheets to make them suitable carriers of cellular or molecular components. Due to the incorporation of magnetic nanoparticles, the sheet could be guided under an external magnetic field. In another report, Fujie et al. demonstrated syringe injectable cell-laden polymeric

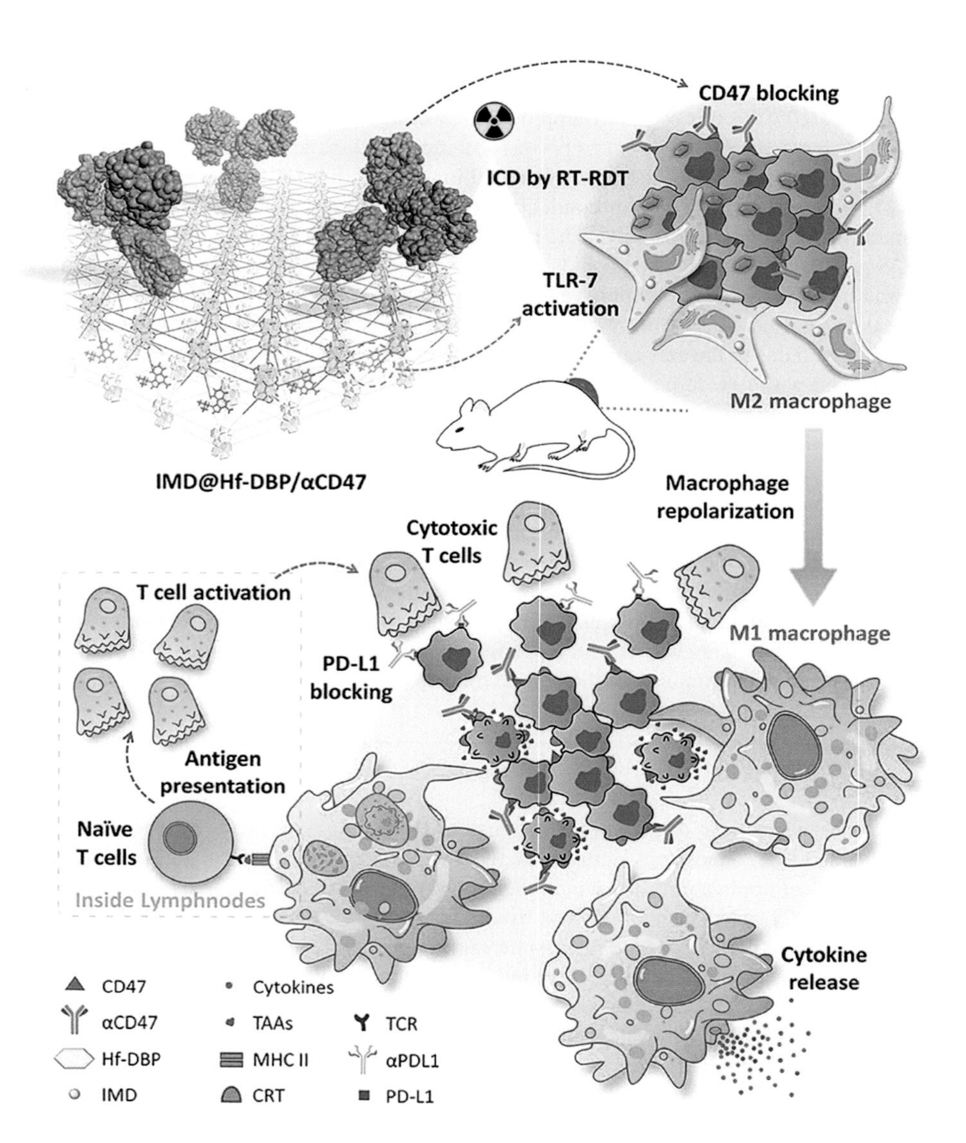

FIGURE 2.4 A schematic representation of the macrophage therapy initiated by the co-delivery of drug- and antibody-loaded Hf-DBP MOFs. CBI synergised tumour eradication. Reprinted with permission from ref (Ni et al. 2020a). Copyright (2020) American Chemical Society.

nanosheets to treat age-related macular degeneration (Fujie et al. 2014). Retinal pigment epithelial cells were grown and organised on micropatterned PLGA as a biodegradable polymer. The engineered cell monolayer was then delivered, locally, to subretinal space. After transplantation through a narrow tissue space, the cells were viable, and the layer was recovered into its original shape.

Transition metal dichalcogenides (TMDs) are another class of 2Dnanomaterials, which have widely been used in biomedicine. Biocompatibility, high photothermal conversion efficiency, and enzyme-mimic property are among several merits that give TMDs unprecedented capabilities in biomedicine. Liu et al. proposed an enzyme-responsive nanosystem to fight drug-resistant bacteria (Liu et al. 2019b). In their design, mesoporous ruthenium nanoparticles were loaded with ascorbic acid (AA) and coated with HA. In response to an enzyme called Hyal from the bacteria, the HA coating was degraded and AA was released. H_2O_2, as a pro-drug of AA, was converted to toxic hydroxyl radicals by the peroxidase mimic activity of MoS_2 nanosheets. The chemotherapeutic effect was synergised with the photothermal property of ruthenium nanoparticles. In vivo results of wound healing proved the efficiency of the chemo/phototherapeutic system within ten days of treatment. In an attempt to develop biomaterial scaffolds, Jaiswal et al. fabricated self-assembled nanocomposite hydrogels driven by the atomic vacancies in the structure of MoS_2 nanoassemblies (Jaiswal et al. 2017). The chemically cross-linked hydrogels consisted of defect-rich MoS_2 and polymeric binders. The atomic vacancies in the MoS_2 structure acted as an active site for the chemisorption of thiolated PEG. The gelation proceeded without UV exposure or the use of chemical initiators. Therefore, it provided a safe and non-toxic strategy to encapsulate cells in the hydrogels. More than 85% of cell viability was observed after encapsulation. The authors proposed that the vacancy-driven gelation process is a facile route for the preparation of bioactive hydrogels for use in regenerative medicine and therapeutic delivery. A summary of the applications of 2D nanomaterials in regenerative medicine is given in Table 2.1.

TABLE 2.1

A Summary of the Applications of 2D Nanomaterials in Regenerative Medicine

Nanomaterial	Delivered Cargo	Target Tissue	Cell Type	Function	Reference
GO/HAP/Au	–	Bone	C3H10T1/2 (differentiation and toxicity evaluation)	Osteogenesis; Antibacterial property	(Prakash et al. 2020)
ADM-GO-PEG/Que	Quercetin	Skin	C57BL/6 (differentiation)	Enhancing diabetic wound healing	(Chu et al. 2018)
GO-PLGA	IGF-1	Nervous	Neural stem cells	Promoting stem cell proliferation and differentiation	(Qi et al. 2019)
Graphene-PCL	Inositol phosphate and Niclosamide	Cardiac	HUVEC (toxicity evaluation)	Personalised drug-eluting stents	(Misra et al. 2017)
fGO_{VEGF}/GelMA	VEGF-165 pro-angiogenic gene	Cardiac	H9c2 (transfection); HUVEC (cell proliferation assay)	Transfection of myocardial tissues for the treatment of acute myocardial infarction	(Paul et al. 2014)
GO	TGF-β3	Cartilage	Human adipose-derived stem cells	Enhancing the chondrogenic differentiation of stem cells	(Yoon et al. 2014)
Gold-doped grapheme	Metformin	–	–	Wireless glucose sensing and drug delivery to diabetic mice	(Lee et al. 2016)
Peptide/GO hydrogel	Nucleus pulposus cells	Intervertebral disc	Bovine nucleus pulposus cell	Injectable hydrogel for Cell delivery	(Ligorio et al. 2019)
Quaternised chitosan /GO/ Ag hydrogel	Voriconazole	Ocular	Human corneal epithelial cells (toxicity evaluation)	Therapeutic contact lens against fungal keratitis	(Huang et al. 2016)
Pd/PPy/rGO	–	Bone	Saos-2	Promoting cell proliferation; Antibacterial property	(Murugesan et al. 2020)

(Continued)

TABLE 2.1 (Continued)

Nanomaterial	Delivered Cargo	Target Tissue	Cell Type	Function	Reference
GO/PNIPAM/GelMA	VEGF	Vein	NIH 3T3 cells (toxicity evaluation); HUVECs (for tube formation experiments)	NIR-responsive drug delivery for vessel formation	(Zhao et al. 2019)
GO on Ti implants	BMP-2 and substance P	Bone	Human bone marrow-derived MSCs (cell migration assay)	Promoting bone formation	(La et al. 2014)
Laponite-alginate	IGF-1 mimetic protein	Achilles tendon	NIH3T3 cells (bioactivity assessment)	pH-stimulated drug delivery	(Li et al. 2018a)
Laponite-alginate	VEGF	Vein	HUVEC (tubulogenesis assays)	Angiogenesis stimulation	(Page et al. 2019)
Laponite-gelatin	Secretome	Cardiac	HUVEC (biocompatibility and proangiogenic tests)	Treatment of myocardial infarction	(Waters et al. 2018)
Laponite-alginate	Dental pulp stem cells and VEGF	Dental pulp	Human dental pulp stem cells	Vascularised dental pulp regeneration	(Zhang et al. 2020c)
Polyurethane-MMT	Dexamethasone acetate	Ocular	ARPE-19	A potential support for retinal cell transplantation	(Da Silva et al. 2011)
$Ti_3C_2T_z$-PLA	–	Bone	MC3T3-E1	in vitro adhesion, proliferation, and osteogenic differentiation of	(Chen et al. 2018b)
Boron nitride-PVA	–	Cartilage	HeLa	Potentially can act as cartilage substitutes	(Jing et al. 2017)
Chitosan–gelatin/zinc oxide	Naproxen	Skin	HFF2	Antibacterial property	(Rakhshaei et al. 2019)

ADM: acellular dermal matrix; **BMP-2**: bone morphogenetic protein-2; **HAP**: hydroxyapatite; **IGF-1**: insulin-like growth factor 1; **GelMA**: gelatin methacryloyl; **PNIPAM**: poly (N-isopropylacrylamide); **PLA**: poly(lactic acid); **PLGA**: poly(lactic-co-glycolic acid); **PCL**: poly-l-caprolactone; **Qce**: quercetin; **TGF**: transforming growth factor; **VEGF**: vascular endothelial growth factor.

Future Prospects

The massive potential of 2D nanomaterials in different aspects of regenerative medicine is discussed in this chapter. Both sides of 2D nanomaterials are available for surface modification. Due to such topology, plenty of spaces are accessible for assembling various cargos, including genes, proteins, drugs, and other small nanomaterials. With the use of 2D nanomaterials in bioactive scaffolds, drug loading and release performance improved significantly. Besides, the inherent physicochemical properties of 2D nanomaterials have opened new horizons towards exploring novel functionalities and higher therapeutic outcomes. Furthermore, 2D nanomaterials can be used in 3D printing technology by appropriate mixing with shear-thinning biomaterials.

Despite their unique properties, the full potential of some subtypes of 2D nanomaterials, including TMDs and MXene in regenerative medicine, has not been understood. To better achieve the purpose, progress in the synthesis (Mohammadpour et al. 2018; Mohammadpour et al. 2020) and surface functionalisation of 2D nanomaterials beyond graphene should be made. Robust and effective therapeutic delivery strategies demand understanding of the interaction between 2D nanomaterials and the immune system. Therefore, the interactions between cells and 2D nanomaterials, specifically the newly emerged ones, are yet to be explored. Besides, their biosafety and toxicology profiles need to be well documented.

REFERENCES

Altinbasak I., R. Jijie, A. Barras, et al. 2018. *ACS Appl Mater Interfaces 10*(48):41098–41106. doi:10.1021/acsami.8b14784.

Ambrogi, Valeria, Donatella Pietrella, Morena Nocchetti, Serena Casagrande, Veronica Moretti, Stefania De Marco, and Maurizio Ricci. 2017. *Journal of Colloid and Interface Science 491*: 265–72. doi:10.1016/j.jcis.2016.12.058.

Askari, Esfandyar, Seyed Morteza Naghib, Amir Seyfoori, Ali Maleki, and Mehdi Rahmanian. 2019. *Ultrasonics Sonochemistry 58* (November): 104615. doi:10.1016/J.ULTSONCH.2019.104615.

Cao, Xianyi, Arnab Halder, Yingying Tang, Chengyi Hou, Hongzhi Wang, Jens Øllgaard Duus, and Qijin Chi. 2018. *Materials Chemistry Frontiers 2* (11): 1944–86. doi:10.1039/C8QM00356D.

Chen, Hang, Tianjiao Liu, Zhiqiang Su, Li Shang, and Gang Wei. 2018a. *Nanoscale Horizons 3* (2): 74–89. doi:10.1039/C7NH00158D.

Chen, Ke, Youhu Chen, Qihuang Deng, Seol Ha Jeong, Tae Sik Jang, Shiyu Du, Hyoun Ee Kim, Qing Huang, and Cheol Min Han. 2018b. *Materials Letters 229*: 114–17. doi:10.1016/j.matlet.2018.06.063.

Chen, Yi Xuan, Rong Zhu, Qin Fei Ke, You Shui Gao, Chang Qing Zhang, and Ya Ping Guo. 2017a. *Nanoscale 9* (20): 6765–76. doi:10.1039/c7nr00601b.

Chen, Yongjiu, Yakun Wu, Bingbing Sun, Sijin Liu, and Huiyu Liu. 2017b. *Small 13* (10): 1603446. doi:10.1002/smll.201603446.

Chimene, David, Daniel L. Alge, and Akhilesh K. Gaharwar. 2015. *Advanced Materials 27* (45): 7261–84. doi:10.1002/adma.201502422.

Choi, Jane Ru, Kar Wey Yong, Jean Yu Choi, Azadeh Nilghaz, Yang Lin, Jie Xu, and Xiaonan Lu. 2018. *Theranostics 8* (4): 1005–26. doi:10.7150/thno.22573.

Chu, Jing, Panpan Shi, Wenxia Yan, Jinping Fu, Zhi Yang, Chengmin He, Xiaoyuan Deng, and Hanping Liu. 2018. *Nanoscale 10* (20): 9547–60. doi:10.1039/c8nr02538j.

Cidonio, G., M. Cooke, M. Glinka, J.I. Dawson, L. Grover, and R.O.C. Oreffo. 2019a. *Materials Today Bio 4* (June): 100028. doi:10.1016/j.mtbio.2019.100028.

Cidonio, Gianluca, Cesar R. Alcala-Orozco, Khoon S Lim, Michael Glinka, Isha Mutreja, Yang Hee Kim, Jonathan I. Dawson, Tim B.F. Woodfield, and Richard O.C. Oreffo. 2019b. *Biofabrication 11* (3). doi:10.1088/1758-5090/ab19fd.

Cidonio, Gianluca, Michael Glinka, Yang Hee Kim, Janos M. Kanczler, Stuart A. Lanham, Tilman Ahlfeld, Anja Lode, Jonathan I. Dawson, Michael Gelinsky, and Richard O.C. Oreffo. 2020. *Biofabrication 12* (3). doi:10.1088/1758-5090/ab8753.

Ciriza, J, L. Saenz Del Burgo, M. Virumbrales-Muñoz, I. Ochoa, L.J. Fernandez, G. Orive, M.R. Hernandez, and J.L. Pedraz. 2015. *International Journal of Pharmaceutics 493* (1–2): 260–70. doi:10.1016/j.ijpharm.2015.07.062.

Silva, Gisele Rodrigues Da, Armando Da Silva-Cunha, Francine Behar-Cohen, Eliane Ayres, and Rodrigo L. Oréfice. 2011. *Materials Science and Engineering C 31* (2): 414–22. doi:10.1016/j.msec.2010.10.019.

Fayyazbakhsh, Fateme, Mehran Solati-Hashjin, Abbas Keshtkar, Mohammad Ali Shokrgozar, Mohammad Mehdi Dehghan, and Bagher Larijani. 2017. *Materials Science and Engineering: C 76*: 701–14.

Fernandes, A.C., M.L. Pinto, F. Antunes, J. Pires2016. *RSC Advances 6* (47): 41195–203. doi:10.1039/c6ra05794b.

Fernandes, Ana C., Moisés L. Pinto, Fernando Antunes, and João Pires. 2013. *Journal of Materials Chemistry B 1* (26): 3287–94. doi:10.1039/c3tb20535e.

Figueiredo, Mariana P., Vanessa R.R. Cunha, Fabrice Leroux, Christine Taviot-Gueho, Marta N. Nakamae, Ye R. Kang, Rodrigo B. Souza, Ana Maria C.R.P.F. Martins, Ivan Hong Jun Koh, and Vera R.L. Constantino. 2018. *ACS Omega 3* (12): 18263–74. doi:10.1021/acsomega.8b02532.

Fu, Jinping, Yue Zhang, Jing Chu, Xiao Wang, Wenxia Yan, Qiong Zhang, and Hanping Liu. 2019. *ACS Biomaterials Science and Engineering 5* (8): 4054–66. doi:10.1021/acsbiomaterials.9b00485.

Fujie, Toshinori, Yoshihiro Mori, Shuntaro Ito, Matsuhiko Nishizawa, Hojae Bae, Nobuhiro Nagai, Hideyuki Onami, Toshiaki Abe, Ali Khademhosseini, and Hirokazu Kaji. 2014. *Advanced Materials 26* (11): 1699–1705. doi:10.1002/adma.201304183.

Gaharwar, Akhilesh K., Lauren M. Cross, Charles W. Peak, Karli Gold, James K. Carrow, Anna Brokesh, and Kanwar Abhay Singh. 2019. *Advanced Materials 31* (23): 1–28. doi:10.1002/adma.201900332.

Gao, Yanshan, Tian Wei Teoh, Qiang Wang, and Gareth R. Williams. 2017. *Journal of Materials Chemistry B 5* (46): 9165–74. doi:10.1039/c7tb01825h.

Goenka, Sumit, Vinayak Sant, and Shilpa Sant. 2014. *Journal of Controlled Release 173* (1): 75–88. doi:10.1016/j.jconrel.2013.10.017.

Gooneh-Farahani, S., S.M. Naghib, and M.R. Naimi-Jamal. 2020. *Fibers and Polymers 21* (9). doi:10.1007/s12221-020-1095-y.

Gooneh-Farahani, Sahar, M. Reza Naimi-Jamal, and Seyed Morteza Naghib. 2019. *Expert Opinion on Drug Delivery 16* (1): 79–99. doi:10.1080/17425247.2019.1556257.

Guo, Zhong, Jing Xi He, Surendra H. Mahadevegowda, Shu Hui Kho, Mary B. Chan-Park, and Xue Wei Liu. 2020. *Advanced Healthcare Materials 9* (10): 1–11. doi:10.1002/adhm.202000265.

Gurunathan, Sangiliyandi, Min Hee Kang, Muniyandi Jeyaraj, and Jin Hoi Kim. 2019. *International Journal of Molecular Sciences 20* (2). doi:10.3390/ijms20020247.

Hashemi, Zahra Sadat, Mehdi Forouzandeh Moghadam, and Masoud Soleimani. 2015.*In Vitro Cellular and Developmental Biology – Animal 51* (5): 495–506. doi:10.1007/s11626-014-9854-y.

Hou, Junjun, Hui Wang, Zhilei Ge, Tingting Zuo, Qian Chen, Xiaoguo Liu, Shan Mou, Chunhai Fan, Yi Xie, and Lihua Wang. 2020. *Nano Letters 20* (2): 1447–54. doi:10.1021/acs.nanolett.9b05218.

Huang, Jian Fei, Jing Zhong, Guo Pu Chen, Zuan Tao Lin, Yuqing Deng, Yong Lin Liu, Piao Yang Cao, et al.2016. *ACS Nano 10* (7): 6464–73. doi:10.1021/acsnano.6b00601.

Jaiswal, Manish K., James K. Carrow, James L. Gentry, Jagriti Gupta, Nara Altangerel, Marlan Scully, and Akhilesh K. Gaharwar. 2017. *Advanced Materials 29* (36): 1–9. doi:10.1002/adma.201702037.

Jalani, Ghulam, Dhanalakshmi Jeyachandran, Richard Bertram Church, and Marta Cerruti. 2017. *Nanoscale 9* (29): 10161–66. doi:10.1039/c7nr00378a.

Jiao, Delong, Lingyan Cao, Yang Liu, Jiannan Wu, Ao Zheng, and Xinquan Jiang. 2019. doi:10.1021/acsbiomaterials.8b01264.

Jing, Lin, Hongling Li, Roland Yingjie Tay, Bo Sun, Siu Hon Tsang, Olivier Cometto, Jinjun Lin, Edwin Hang Tong Teo, and Alfred Iing Yoong Tok. 2017. *ACS Nano 11* (4): 3742–51. doi:10.1021/acsnano.6b08408.

Kalkhoran, Amir Hossein Zeinali, Seyed Morteza Naghib, Omid Vahidi, and Mehdi Rahmanian. 2018. *Biomedical Physics and Engineering Express*. doi:10.1088/2057-1976/aad745.

Kang, Heemin, Minkyu Kim, Qian Feng, Sien Lin, Kongchang Wei, Rui Li, Chan Ju Choi, et al.2017. *Biomaterials 149*: 12–28. doi:10.1016/j.biomaterials.2017.09.035.

Kevadiya, Bhavesh D., Shalini Rajkumar, Hari C. Bajaj, Shiva Shankaran Chettiar, Kalpeshgiri Gosai, Harshad Brahmbhatt, Adarsh S. Bhatt, Yogesh K. Barvaliya, Gaurav S. Dave, and Ramesh K. Kothari. 2014. *Colloids and Surfaces B: Biointerfaces 122*: 175–83. doi:10.1016/j.colsurfb.2014.06.051.

Kiaee, Gita, Pooria Mostafalu, Mohamadmahdi Samandari, and Sameer Sonkusale. 2018. *A Advanced Healthcare Materials 7* (18): 1–8. doi:10.1002/adhm.201800396.

Kim, Myung Hun, Woojune Hur, Goeun Choi, Hye Sook Min, Tae Hyun Choi, Young Bin Choy, and Jin Ho Choy. 2016. *Advanced Healthcare Materials 5* (21): 2765–75. doi:10.1002/adhm.201600761.

Kim, Seok Joo, Hye Rim Cho, Kyoung Won Cho, Shutao Qiao, Jung Soo Rhim, Min Soh, Taeho Kim, et al.2015. *ACS Nano 9* (3): 2677–88. doi: 10.1021/nn5064634.

Ku, Sook Hee, and Chan Beum Park. 2013. *Biomaterials 34* (8): 2017–23. doi:10.1016/j.biomaterials.2012.11.052.

La, Wan Geun, Min Jin, Saibom Park, Hee Hun Yoon, Gun Jae Jeong, Suk Ho Bhang, Hoyoung Park, Kookheon Char, and Byung Soo Kim. 2014. *International Journal of Nanomedicine 9* (SUPPL.1): 107–16. doi:10.2147/IJN.S50742.

Laurenti, Marco, Andrea Lamberti, Giada Graziana Genchi, Ignazio Roppolo, Giancarlo Canavese, Chiara Vitale-Brovarone, Gianni Ciofani, and Valentina Cauda. 2019. *ACS Applied Materials and Interfaces 11* (1): 449–56. doi:10.1021/acsami.8b20728.

Lee, Hyunjae, Tae Kyu Choi, Young Bum Lee, Hye Rim Cho, Roozbeh Ghaffari, Liu Wang, Hyung Jin Choi, et al. 2016. *Nature Nanotechnology 11* (6): 566–72. doi:10.1038/nnano.2016.38.

Lee, Sang Jin, Haram Nah, Dong Nyoung Heo, Kyoung Hwa Kim, Ji Min Seok, Min Heo, Ho Jin Moon, et al. 2020. *Carbon 168*: 264–77. doi:10.1016/j.carbon.2020.05.022.

Lee, Seunghyeon, Taesik Eom, Min Kyoung Kim, Su Geun Yang, and Bong Sup Shim. 2019. *Electrochimica Acta 313*: 79–90. doi:10.1016/j.electacta.2019.04.099.

Li, Jianyu, Eckhard Weber, Sabine Guth-Gundel, Michael Schuleit, Andreas Kuttler, Christine Halleux, Nathalie Accart, et al. 2018a. *Advanced Healthcare Materials 7* (9): 1–10. doi:10.1002/adhm.201701393.

Li, Keheng, Zhenfang Zhang, Dapeng Li, Wensi Zhang, Xiaoqing Yu, Wei Liu, Coucong Gong, Gang Wei, and Zhiqiang Su. 2018b. *Advanced Functional Materials 28* (29): 1–10. doi:10.1002/adfm.201801056.

Ligorio, Cosimo, Mi Zhou, Jacek K. Wychowaniec, Xinyi Zhu, Cian Bartlam, Aline F. Miller, Aravind Vijayaraghavan, Judith A. Hoyland, and Alberto Saiani. 2019. *Acta Biomaterialia 92*: 92–103. doi:10.1016/j.actbio.2019.05.004.

Liu, Bin, Junqin Li, Xing Lei, Sheng Miao, Shuaishuai Zhang, Pengzhen Cheng, Yue Song, et al. 2020. *RSC Advances 10* (43): 25652–61. doi:10.1039/d0ra03040f.

Liu, Han, Yuchen Du, Yexin Deng, and Peide D. Ye. 2015. *Chemical Society Reviews 44* (9): 2732–43. doi:10.1039/c4cs00257a.

Liu, Heng Wen, Shang Hsiu Hu, Yu Wei Chen, and San Yuan Chen. 2012. *Journal of Materials Chemistry 22* (33): 17311–20. doi:10.1039/c2jm32772d.

Liu, Ting, Weihua Dan, Nianhua Dan, Xinhua Liu, Xuexu Liu, and Xu Peng. 2017a. *Materials Science and Engineering C 77*: 202–11. doi:10.1016/j.msec.2017.03.256.

Liu, Xiangwei, Naiwen Tan, Yuchao Zhou, Hongbo Wei, Shuai Ren, Fan Yu, Hui Chen, Chengming Jia, Guodong Yang, and Yingliang Song. 2017b. *International Journal of Nanomedicine 12*: 7089–7101. doi:10.2147/IJN.S124584.

Liu, Xifeng, A. Lee Miller, Sungjo Park, Matthew N. George, Brian E. Waletzki, Haocheng Xu, Andre Terzic, and Lichun Lu. 2019a. *ACS Applied Materials & Interfaces 11* (26): 23558–72. doi:10.1021/acsami.9b04121.

Liu, Yanan, Ange Lin, Jiawei Liu, XuChen, Xufeng Zhu, Youcong Gong, Guanglong Yuan, Lanmei Chen, and Jie Liu. 2019b. *ACS Applied Materials and Interfaces 11* (30): 26590–606. doi:10.1021/acsami.9b07866.

Loftus, Christian, Mezida Saeed, Daniel M. Davis, and Iain E. Dunlop. 2018. *Nano Letters 18* (5): 3282–89. doi:10.1021/acs.nanolett.8b01089.

Luong, John H.T., and Sandeep Kumar Vashist. 2017. *Biosensors and Bioelectronics 89* (March): 293–304. doi:10.1016/j.bios.2015.11.053.

Magaz, Adrián, Mark D. Ashton, Rania M. Hathout, Xu Li, John G. Hardy, and Jonny J. Blaker. 2020. *Pharmaceutics 12* (8): 1–12. doi:10.3390/pharmaceutics12080742.

Mahanta, Arun Kumar, Dinesh K. Patel, and Pralay Maiti. 2019. *ACS Biomaterials Science and Engineering 5* (10): 5139–49. doi:10.1021/acsbiomaterials.9b00829.

Mamaghani, K.R., S.M. Naghib, A. Zahedi, A.H.Z. Kalkhoran, and M. Rahmanian. 2018. *Micro and Nano Letters 13* (2). doi:10.1049/mnl.2017.0461.

Mellado, Constanza, Toribio Figueroa, Ricardo Báez, Rosario Castillo, Manuel Melendrez, Berta Schulz, and Katherina Fernández. 2018. *ACS Applied Materials and Interfaces 10* (9): 7717–29. doi:10.1021/acsami.7b16084.

Misra, Santosh K., Fatemeh Ostadhossein, Ramya Babu, Joseph Kus, Divya Tankasala, Andre Sutrisno, Kathleen A. Walsh, Corinne R. Bromfield, and Dipanjan Pan. 2017. *Advanced Healthcare Materials 6* (11): 1–14. doi:10.1002/adhm.201700008.

Mohammadpour, Zahra, Seyyed Hossein Abdollahi, Akbar Omidvar, Afshan Mohajeri, and Afsaneh Safavi. 2020. *Journal of Molecular Liquids 309* (July): 113087. doi:10.1016/j.molliq.2020.113087.

Mohammadpour, Zahra, Seyyed Hossein Abdollahi, and Afsaneh Safavi. 2018. *ACS Applied Energy Materials 1* (11): 5896–5906. doi:10.1021/acsaem.8b00838.

Mohammadpour, Zahra, and Keivan Majidzadeh-A. 2020. *ACS Biomaterials Science and Engineering 6* (4): 1852–73. doi:10.1021/acsbiomaterials.9b01894.

Morariu, Simona, Maria Bercea, Luiza Madalina Gradinaru, Irina Rosca, and Mihaela Avadanei. 2020. *Materials Science and Engineering C 109* (May 2019): 110395. doi:10.1016/j.msec.2019.110395.

Munhoz, Davi R., Marcela P. Bernardo, João O.D. Malafatti, Francys K.V. Moreira, and Luiz H.C. Mattoso. 2019. *International Journal of Biological Macromolecules 141*: 504–10. doi:10.1016/j.ijbiomac.2019.09.019.

Murugesan, Balaji, Nithya Pandiyan, Mayakrishnan Arumugam, Jegatheeswaran Sonamuthu, Selvam Samayanan, Cai Yurong, Yao Juming, and Sundrarajan Mahalingam. 2020. *Applied Surface Science 510* (August 2019). doi:10.1016/j.apsusc.2020.145403.

Naghib, S.M., F. Behzad, M. Rahmanian, Y. Zare, and K.Y. Rhee. 2020a. *Nanotechnology Reviews 9* (1). doi:10.1515/ntrev-2020-0061.

Naghib, Seyed Morteza. 2019. *Micro and Nano Letters 14* (4): 462–65. doi:10.1049/mnl.2018.5320.

Naghib, Seyed Morteza, Yasser Zare, and Kyong Yop Rhee. 2020b. *Nanotechnology Reviews 9* (1): 53–60.

Nguyen, Hanh Thuy, Jeong Hoon Byeon, Cao Dai Phung, Le Minh Pham, Sae Kwang Ku, Chul Soon Yong, and Jong Oh Kim. 2019. *ACS Applied Materials and Interfaces 11* (28): 24959–70. doi:10.1021/acsami.9b04632.

Ni, Kaiyuan, Theint Aung, S. Li, Nina Fatuzzo, Xingjie Liang, and Wenbin Lin. 2019. *Chem 5* (7): 1892–1913. doi:10.1016/j.chempr.2019.05.013.

Ni, Kaiyuan, Taokun Luo, August Culbert, Michael Kaufmann, Xiaomin Jiang, and Wenbin Lin. 2020a. *Journal of the American Chemical Society 142* (29): 12579–84. doi:10.1021/jacs.0c05039.

Ni, Kaiyuan, Taokun Luo, Guangxu Lan, August Culbert, Yang Song, Tong Wu, Xiaomin Jiang, and Wenbin Lin. 2020b. *Angewandte Chemie - International Edition 59* (3): 1108–12. doi:10.1002/anie.201911429.

Nie, Wei, Cheng Peng, Xiaojun Zhou, Liang Chen, Weizhong Wang, Yanzhong Zhang, Peter X. Ma, and Chuanglong He. 2017. *Carbon 116*: 325–37.

Ou, Wenquan, Jeong Hoon Byeon, Raj Kumar Thapa, Sae Kwang Ku, Chul Soon Yong, and Jong Oh Kim. 2020. *ACS Nano 12* (10): 10061–74. doi:10.1021/acsnano.8b04658.

Page, Daniel J., Claire E. Clarkin, Raj Mani, Najeed A. Khan, Jonathan I. Dawson, and Nicholas D. Evans. 2019. *Acta Biomaterialia 100*: 378–87. doi:10.1016/j.actbio.2019.09.023.

Park, Junggeon, Jang Hee Choi, Semin Kim, Inseok Jang, Sungho Jeong, and Jae Young Lee. 2019. *Acta Biomaterialia 97*: 141–53. doi:10.1016/j.actbio.2019.07.044.

Paul, Arghya, Anwarul Hasan, Hamood Al Kindi, Akhilesh K. Gaharwar, Vijayaraghava T.S. Rao, Mehdi Nikkhah, Su Ryon Shin, et al.2014. *ACS Nano 8* (8): 8050–62. doi:10.1021/nn5020787.

Prabha, Rahul D., Bindu P. Nair, Nicholas Ditzel, Jorgen Kjems, Prabha D. Nair, and Moustapha Kassem. 2019. *Materials Science and Engineering C 94* (September 2018): 509–15. doi:10.1016/j.msec.2018.09.054.

Prakash, J, D. Prema, K.S. Venkataprasanna, and K. Balagangadharan. 2020. *International Journal of Biological Macromolecules 154*: 62–71.

Qi, Zhiping, Wenlai Guo, Shuang Zheng, Chuan Fu, Yue Ma, Su Pan, Qinyi Liu, and Xiaoyu Yang. 2019. *RSC Advances 9* (15): 8315–25. doi:10.1039/c8ra10103e.

Rakhshaei, Rasul, Hassan Namazi, Hamed Hamishehkar, Hossein Samadi Kafil, and Roya Salehi. 2019. *Journal of Applied Polymer Science 136* (22): 1–9. doi:10.1002/app.47590.

Ren, Xiansheng, Qinyi Liu, Shuang Zheng, Jiaqi Zhu, Zhiping Qi, Chuan Fu, Xiaoyu Yang, and Yan Zhao. 2018. *RSC Advances 8* (56): 31911–23. doi:10.1039/c8ra05250f.

Saifullah, Bullo, Arulselvan Palanisamy, Mohamed Ezzat El Zowalaty, Sharida Fakurazi, Thomas J. Webster, Benjamin M. Geilich, and Mohd Zobir Hussein. 2014. *International Journal of Nanomedicine 9* (1): 4749–62. doi:10.2147/IJN.S63608.

Salahandish, R., A. Ghaffarinejad, S.M. Naghib, K. Majidzadeh-A, and A. Sanati-Nezhad. 2018a. *IEEE Sensors Journal 18* (6). doi:10.1109/JSEN.2018.2789433.

Salahandish, R., A. Ghaffarinejad, S.M. Naghib, K. Majidzadeh-A, H. Zargartalebi, and A. Sanati-Nezhad. 2018b. *Biosensors and Bioelectronics* 117. doi:10.1016/j.bios.2018.05.043.

Salahandish, Razieh, Ali Ghaffarinejad, Eskandar Omidinia, Hossein Zargartalebi, Keivan Majidzadeh-A, Seyed Morteza Naghib, and Amir Sanati-Nezhad. 2018c. *Biosensors and Bioelectronics 120* (November): 129–36. doi:10.1016/j.bios.2018.08.025.

Salguero, Yadira, Laura Valenti, Ricardo Rojas, and Mónica C. García. 2020. *Materials Science and Engineering C 111* (February): 110859. doi:10.1016/j.msec.2020.110859.

Samadi, Morasae, Navid Sarikhani, Mohammad Zirak, Hua Zhang, Hao Li Zhang, and Alireza Z. Moshfegh. 2018. *Nanoscale Horizons 3* (2): 90–204. doi:10.1039/c7nh00137a.

Sedghi, Roya, Nastaran Sayyari, Alireza Shaabani, Hassan Niknejad, and Tahereh Tayebi. 2018. *Polymer 142*: 244–55. doi:10.1016/j.polymer.2018.03.045.

Shim, Gayong, Mi-Gyeong Kim, Joo Yeon Park, and Yu-Kyoung Oh2016. *Advanced Drug Delivery Reviews 105* (October): 205–27. doi:10.1016/j.addr.2016.04.004.

Shin, Su Ryon, Yi-Chen Li, Hae Lin Jang, Parastoo Khoshakhlagh, Mohsen Akbari, Amir Nasajpour, Yu Shrike Zhang, Ali Tamayol, and Ali Khademhosseini. 2016. *Advanced Drug Delivery Reviews 105* (October): 255–74. doi:10.1016/j.addr.2016.03.007.

Sinha, Arjyabaran, Bong Geun Cha, Youngjin Choi, Thanh Loc Nguyen, Pil J. Yoo, Ji Hoon Jeong, and Jaeyun Kim. 2017. *Chemistry of Materials 29* (16): 6883–92. doi:10.1021/acs.chemmater.7b02197.

Sinha, Arjyabaran, Youngjin Choi, Minh Hoang Nguyen, Thanh Loc Nguyen, Seung Woo Choi, and Jaeyun Kim. 2019. *Advanced Healthcare Materials 8* (5): 1–10. doi:10.1002/adhm.201800571.

Sun, Baohong, Zhenhua Xi, Fan Wu, Saijie Song, Xinrong Huang, Xiaohong Chu, Zhixuan Wang, et al.2019. *Langmuir 35* (47): 15275–86. doi:10.1021/acs.langmuir.9b02821.

Tanum, Junjira, Hyejoong Jeong, Jiwoong Heo, Moonhyun Choi, Kyungtae Park, and Jinkee Hong. 2019. *Applied Surface Science 486* (December 2018): 452–59. doi:10.1016/j.apsusc.2019.04.260.

Tao, Yu, Enguo Ju, Jinsong Ren, and Xiaogang Qu. 2014. *Biomaterials 35* (37): 9963–71. doi:10.1016/j.biomaterials.2014.08.036.

Tenci, M., S. Rossi, C. Aguzzi, E. Carazo, G. Sandri, M. C. Bonferoni, P. Grisoli, C. Viseras, C. M. Caramella, and F. Ferrari. 2017. *International Journal of Pharmaceutics 531* (2): 676–88. doi:10.1016/j.ijpharm.2017.06.024.

Tong, Liping, Qing Liao, Yuetao Zhao, Hao Huang, Ang Gao, Wei Zhang, Xiaoyong Gao, et al.2019. *Biomaterials 193* (November 2018): 1–11. doi:10.1016/j.biomaterials.2018.12.008.

Wang, Shenqiang, Hua Zheng, Li Zhou, Fang Cheng, Zhao Liu, Hepeng Zhang, and Qiuyu Zhang. 2020a. *Biomaterials 260* (August): 120314. doi:10.1016/j.biomaterials.2020.120314.

Wang, Wei, Yang Liu, Chao Yang, Weitao Jia, Xin Qi, Changsheng Liu, and Xiaolin Li. 2020b. doi:10.1021/acsbiomaterials.0c00558.

Wang, Xuzhu, Jundong Shao, Mustafa Abd El Raouf, Hanhan Xie, Hao Huang, Huaiyu Wang, Paul K. Chu, et al. 2018. *Biomaterials 179*: 164–74. doi:10.1016/j.biomaterials.2018.06.039.

Wang, Yuzhen, Zhiqiang Chen, Gaoxing Luo, Weifeng He, Kaige Xu, Rui Xu, Qiang Lei, Jianglin Tan, Jun Wu, and Malcolm Xing. 2016. *ACS Applied Materials and Interfaces 8* (11): 7411–21. doi:10.1021/acsami.5b11332.

Wang, Zhuqing, Lucio Colombi Ciacchi, and Gang Wei. 2017. *Applied Sciences 7* (11): 1175. doi:10.3390/app7111175.

Waters, Renae, Perwez Alam, Settimio Pacelli, Aparna R. Chakravarti, Rafeeq P.H. Ahmed, and Arghya Paul. 2018. *Acta Biomaterialia 69*: 95–106. doi:10.1016/j.actbio.2017.12.025.

Xavier, Janet R., Teena Thakur, Prachi Desai, Manish K. Jaiswal, Nick Sears, Elizabeth Cosgriff-Hernandez, Roland Kaunas, and Akhilesh K. Gaharwar. 2015. *ACS Nano 9* (3): 3109–18. doi:10.1021/nn507488s.

Yamagishi, Kento, Akihiro Nojiri, Eiji Iwase, and Michinao Hashimoto. 2019. *ACS Applied Materials and Interfaces 11* (44): 41770–79. doi:10.1021/acsami.9b17567.

Yang, Bowen, Yu Chen, and Jianlin Shi. 2018a. *Chem 4* (6): 1284–1313. doi:10.1016/j.chempr.2018.02.012.

Yang, Letao, Sy Tsong Dean Chueng, Ying Li, Misaal Patel, Christopher Rathnam, Gangotri Dey, Lu Wang, Li Cai, and Ki Bum Lee. 2018b. *Nature Communications 9* (1). doi:10.1038/s41467-018-05599-2.

Yang, Linnan, Jing Sun, Qiang Liu, Rongrong Zhu, Qiannan Yang, Jiahui Hua, Longpo Zheng, Kun Li, Shilong Wang, and Ang Li. 2019. *Advanced Science 6* (8). doi:10.1002/advs.201802012.

Yasaei, Mana, Mehrdad Khakbiz, Ebrahim Ghasemi, and Ali Zamanian. 2019. *Applied Surface Science 467–468* (June 2018): 782–91. doi:10.1016/j.apsusc.2018.10.202.

Yoon, Hee Hun, Suk Ho Bhang, Taeho Kim, Taekyung Yu, Taeghwan Hyeon, and Byung Soo Kim. 2014. *Advanced Functional Materials 24* (41): 6455–64. doi:10.1002/adfm.201400793.

Zeng, Yuyang, Muran Zhou, Lifeng Chen, Huimin Fang, Shaokai Liu, Chuchao Zhou, Jiaming Sun, and Zhenxing Wang. 2020. *Bioactive Materials 5* (4): 859–70. doi:10.1016/j.bioactmat.2020.06.010.

Zhang, Huijie, Ting Yan, Sha Xu, Shini Feng, Dandi Huang, Morihisa Fujita, and Xiao Dong Gao. 2017a. *Materials Science and Engineering C 73*: 144–51. doi:10.1016/j.msec.2016.12.072.

Zhang, Jingyang, Haolin Chen, Meng Zhao, Guiting Liu, and Jun Wu.2020a. *Nano Research 13* (8): 2019–34. doi:10.1007/s12274-020-2835-4.

Zhang, Li, Qing Zhou, Wen Song, Kaimin Wu, Yumei Zhang, and Yimin Zhao. 2017b. *ACS Applied Materials and Interfaces 9* (40): 34722–35. doi:10.1021/acsami.7b12079.

Zhang, Ling Xiao, Dong Qun Liu, Shao Wei Wang, Xiao Lin Yu, Mei Ji, Xi Xiu Xie, Shu Ying Liu, and Rui Tian Liu. 2017c. *Journal of Materials Chemistry B 5* (31): 6266–76. doi:10.1039/c7tb00819h.

Zhang, Ling Xiao, Xia Mei Sun, Ying Bo Jia, Xiao Ge Liu, Mingdong Dong, Zhi Ping Xu, and Rui Tian Liu. 2020b. *Nano Today 35*: 100923. doi:10.1016/j.nantod.2020.100923.

Zhang, Ling Xiao, Xi Xiu Xie, Dong Qun Liu, Zhi Ping Xu, and Rui Tian Liu. 2018. *Biomaterials 174*: 54–66. doi:10.1016/j.biomaterials.2018.05.015.

Zhang, Ruitao, Li Xie, Hao Wu, Ting Yang, Qingyuan Zhang, Yuan Tian, Yuangang Liu, et al.2020c. *Acta Biomaterialia 113* (14): 305–16. doi:10.1016/j.actbio.2020.07.012.

Zhang, Wendi, Yunpeng Zhao, Wenhan Wang, Jiangfan Peng, Yuanming Li, Yangtao Shangguan, Gege Ouyang, Mingyang Xu, Shuping Wang, and Jingjing Wei. 2020d. *2000092*: 1–11. doi:10.1002/adhm.202000092.

Zhang, Wenjie, Guangzheng Yang, Xiansong Wang, Liting Jiang, Fei Jiang, Guanglong Li, Zhiyuan Zhang, and Xinquan Jiang. 2017d. *Advanced Materials 29* (43): 1–9. doi:10.1002/adma.201703795.

Zhang, Yuanhao, Mingjiao Chen, Zhaobo Dai, Hongliang Cao, Jin Li, and Weian Zhang. 2020e. *Biomaterials Science 8* (2): 682–93. doi:10.1039/c9bm01455a.

Zhao, Xin, Yuxiao Liu, Changmin Shao, Min Nie, Qian Huang, Jieshou Li, Lingyun Sun, and Yuanjin Zhao. 2019. *Advanced Science 6* (20). doi:10.1002/advs.201901280.

Zhu, Chengzhou, Du Dan, and Yuehe Lin. 2015. *2D Materials 2* (3): 032004. doi:10.1088/2053-1583/2/3/032004.

3

Potential of Nanoparticles as Next Generation Therapeutics in Tissue Regeneration

Madhyastha Radha, Madhyastha Harishkumar, Nakajima Yuichi, and Maruyama Masugi
University of Miyazaki, Miyazaki, Japan

CONTENTS

Introduction

In the new millennium year 2000, the President of the USA, in his union speech, announced a $ 475 million budget for the "new" technology Research & development (R&D) initiative. Two years later, the US National Institute of Health (NIH) announced a five-year plan of nanoscience and nanotechnology initiative in medicine (Bogunia-Kubik and Sugisaka 2002). Today, nanomedicine is tremendously developing in all areas of sciences, and its subsequent incorporation into medical, aquaculture, energy, and agriculture arena is seen beneficial for the betterment of human health. Nanomedicine is a fast growing and latest revolutionary tool to overcome degenerative diseases through transformative and diagnostic approach. Physics and chemistry including biotechnology underpin the development of nanomedicine. Several nanohybrids are being synthesized for application in biomedical imaging, biosensing, *in situ* diagnosis, and therapy. Advances in nanomedicine require a basic understanding of chemo-biological interactions and influence on the physiological responses in the body, like uptake, endocytosis, cellular internalisation, and body clearance (Kuiken 2011). Many products are being tried today, with some already showing fruition and successfully developed into translational products. Integrated approach to adapt nanotechnology to translational medicine for combating regenerative diseases is depicted in Figure 3.1. Effective control of diseases with the help of engineering technology is the prime focus of future medicine.

Immuno-isolation using micro-matched nanochambers of crystalline silicon wafers of 20nm size is the oldest medical device (Desai et al. 1998). The small pores facilitate easy passage of oxygen, glucose, and insulin, but impede large particles like immunoglobins and large cells. Thereby, it can be used to control the body's insulin regulation. Gated nanosieves are the first artificial voltage-gated device developed by Nishizawa for the regulation of ion transports in intercellular spaces (Nishizawa et al. 1995). Fullerenes such as C_{60} have shown great promise as anti-microbial agents against *Escherichia coli, Streptococcus aureus, Mycobacterium tuberculosis*, and human immunodeficiency virus (Krokosz 2007). C_{60} derivatives exhibit cytotoxic and enzyme-inhibiting mechanisms with antioxidant properties and are thereby desired for use in photodynamic therapy. Gold-conjugated nanoshells can detect virus particles in single cells and are being used by clinicians to decide the strategy of medication (Jimenez Jimenez et al. 2019).

DOI: 10.1201/9781003153504-3

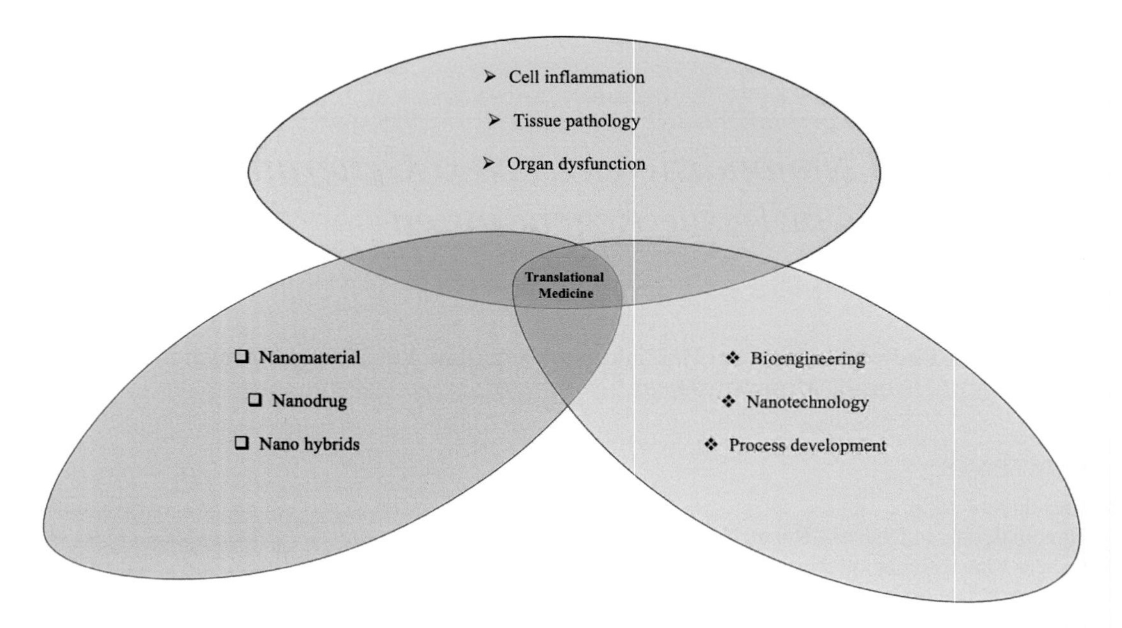

FIGURE 3.1 Integrated approach to link Nanotechnology to translational medicine for combating regenerative diseases.

Another type of nanomaterials is tecto-dendrimers. These smart design tree-shaped synthetic materials with regular branches are extensively studied in the area of cancer detection and treatment (Palmerston Mendes et al. 2017). Engineered microrobots can be designed with re-arrangement of desired genes (~125,000–200,000) for mico-production of vitamins, hormones, and enzymes in the patient's body as a strategy to combat the cell decay phenomenon. These miniature tools have enormous potential in biomedical applications such as drug delivery, *in vivo* bio-sensing, invasive surgery, and cell engineering (Halder and Sun 2019). Advancement of nanotechnology application emerged after the discovery of micelles as drug delivery agents, especially for drugs with poor bioavailability and distribution. Polymeric micelles are small-sized materials that spontaneously penetrate into tumours and act long term to destroy cancer cells (Yousefpour Marzbali and Yari Khosroushahi 2017). Some nanomaterials employ the principle of Trojan horse strategy, which helps in targeting cancer cells without affecting the adjuvant healthy cells. Apart from the above-mentioned materials, others such as liposomes, carbon nanotubes, paramagnetic nanoparticles, nanoburrs, respirocytes, micro-biovores, nanopores, nano-quantum dots, smart coating, and nanoband-aid contribute to curative aspects and can serve as a boon for the management of cancer, cardiovascular problems, diabetes-related health complications, osteoporosis, and overall systemic health. Despite many optimistic opportunities that nanotechnology can create history, it also possesses several risks since many nanomaterials are cytotoxic and can affect stem cell differentiation and reprogram the genetic signature of the cells, posing great challenge to the scientific community (Harishkumar et al. 2019, Sahoo et al. 2020).

Nanomaterials in Diabetes-Related Wound Complications

Diabetes imparts major complications in patients, leading to kidney failure, microvascular dysfunction, retinopathy, neuropathy, skin aetiology, depressions, dementia, and immune malfunction (Forbes and Cooper 2013). The more we study about the pathophysiology of diabetes mellitus, the more is the information that emerges about this health complication, including diabetic ketoacidosis, hyper glycaemic hyperosmomolar non-ketotic coma, severe hypoglycaemia, and acute stage of diabetes-related complications due to decreased insulin regulation (Brunton 2016). Among the skin pathogenesis, non-healing diabetic wound and its chronic complications can result in amputation. Normal wound healing involving

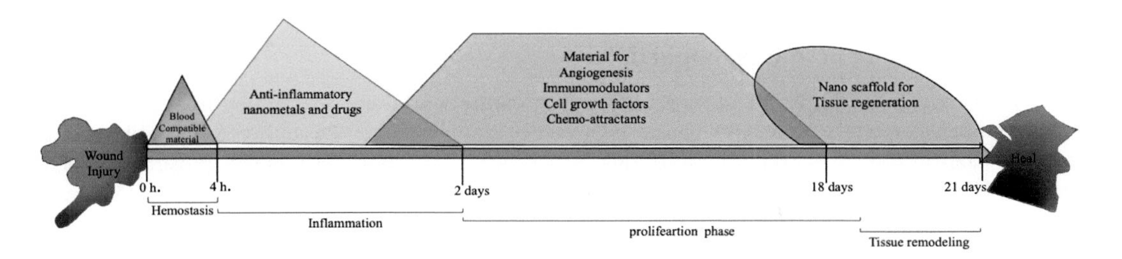

FIGURE 3.2 Trajectory of wound healing consists of haemostasis, inflammation, cell migration and tissue remodeling phases. Timely application of different nanomaterials helps in faster and healthy healing.

the overlapping and well-orchestrated molecular events like blood haemostasis, inflammation, cell proliferation, angiogenesis, collagen deposition, and re-epithelisation generally heals within 3–4 weeks, ending in new tissue formation (Figure 3.2). Each phase consists of overlapping events that are crucial for co-ordinated trans-phase communication (Robson et al. 2000, Steed 2003). The prime and major requirement of wound healing is prevention of infection and sepsis formation with suitable metal-based nanodrugs to control infections. However, in diabetic patients, the healing physiology is greatly disturbed due to the impaired and delayed actions of many regulators at each stage, thereby igniting the molecular alarm in the cells like neutrophils, monocytes macrophages, fibroblasts, keratinocytes, T cells, B cells, and endothelial cells (Madhyastha et al. 2012, Madhyastha et al. 2020). Wounds are of two types based on their origin: external and internal. External wounds like cuts, injuries, burns, and bruises are frequently unnoticed by diabetic patients because of peripheral nerve injury. Internal wounds like ulcers and calluses can cause destruction of skin and nearby tissues with higher chances of bacterial infection, ultimately leading to cellulitis, abscess, osteomyelitis, gangrene, and finally septicaemia (Patel et al. 2019). Despite various options available to manage wound healing by conventional methods like topical gel application, systematic growth factor-mediated treatment, cytokine stimulation, artificial skin replacement therapy, negative pressure wound closure, and local loading of anti-bacterial application, complete healing remains at less than 50% in diabetic patients. Critical and crucial need for more realistic state-of-the art efficient therapy is called for, especially for old age group, to overcome the limitations of the conventional technologies (Makrantonaki et al. 2017).

Non-invasive and smart healing nanotechnologies can provide advanced information about diabetic wound healing. Biopolymers synthesized from natural origins like animal, plant, and microbial sources have several advantages like anti-microbial, immune-modulatory, cell-chemotactic, and angiogenic properties to naturally create a healing micro-environment at the wound site. Biopolymers most widely used for wound biology management include collagen, cellulose, chitosan, hyaluronan, fucoidans, silk fibroin, and carrageen. Nanocollagens like de-cellularised matrix or depolymerised scaffold conjugates also showed benefits in the ECM crosslinking with proper *de-novo* wound bed formation (Bajpai et al. 2015, Nethi et al. 2019). Metals like silver and gold played vital roles in wound healing therapies since ancient time because of their potent anti-bacterial properties. In chronic wound microbial environment, the positively charged silver ions bind to negatively charged proteins of microbes and disturb the microbial respiratory chain and cell membrane function. Chronic wound microbial load includes yeast, moulds, MRSA, VRE, *E.coli, Enterobacter cloacae, Klebsiella pneumoniae, Acinetobacter baumannii, Salmonella typhimurium*, and *Pseudomonas* (Stallard 2018). In spite of promising anti-bacterial efficacy of AgNPs, silver has serious disadvantages like skin colour changes, initiation of eschar formation, skin irritation, and frequent dressing replacement due to reaction of chloride ions with bacterial exudes. However, surface-modified AgNPs can overcome the disadvantages of treatment with silver. Surface corona formation phenomenon and nanoparticle's reactions with serum proteins are other major challenges of drug efficacy during diabetic wound healing. In a mouse study, the thermo-responsive biocompatible gold particles conjugated with Pluronic^{127} and hydroxypropyl methylcellulose nanomaterials were found to be effective to treat burn wounds (Arafa et al. 2018).

Cancer Management and Nanoparticles

Cancer remains the major cause of morbidity and mortality worldwide. According to WHO's Global Cancer Observatory (GLOBOCAN), cancer is placed in the first rank in Disability-Adjusted Life Years (DALYs) statistics, with 244.6 million suffering from cancer of the lung, liver, breast, prostrate, colorectum, blood, cervix, uterus, thyroid, bladder, lymph nodes, pancreas, kidney, colon, and rectum (Mattiuzzi and Lippi 2019, Pesec and Sherertz 2015). Figure 3.3 summarises the cancer microenvironment niche with coordinated communications of various lymphocytes and myeloid cells. Lymphocytes consist of T cells, B cells, and NK cells. Myeloid population consists of macrophages, dendritic cells, and myeloid-derived suppressor cells. Conventional cancer cure strategies like pharmaceutical drug, radiotherapy, and chemotherapy are not very satisfactory in terms of reduced mortality rate, thus, prompting researchers to work overtime in their efforts to develop nanotechnology-based medicine, termed as cancer nanomedicine. The heterogenicity of tumour microenvironment is considered as a major challenge for cancer nanomedicine. The first class of nanomedicine-based drugs like liposomal doxorubicin, Doxil, and Myocet considerably improved the survival rate of cancer patients. Besides, chemotherapeutics conjugated NPs for example, anti-sense oligonucleotides, siRNA, miRNA conjugates, and DNA inhibitor oligonucleotides, developed by chemical and bio-engineering techniques show potential benefits (von Roemeling et al. 2017). However, the anti-tumour efficacy of nanomedicine, calibrated on the enhanced permeability and retention (EPR) effect, yielded poor results, due to clearance from blood stream through opsonisation and accelerated uptake by mononuclear phagocytotic system (MPS) (Sanada et al. 2009). The theranostic principle of visualisation, quantification, and treatment of tumour mass is being developed by many researchers as a smart strategy (Sahoo et al. 2020). Nanocarriers with integrated therapeutic agents with dual advantage mode are essential features of theranostic materials.

Several carriers like lipids, polymeric, and inorganic chemicals functioning as main linker molecules with chemo-drugs, photo-therapy agents, radio-pharmaceuticals, photo dynamic, and immunotherapeutic materials are being used in various imaging techniques like FRET, HIFU, SDT, IMXT, IMPT, and SELEX. In cancer clinical operation system, surgical resection is the usual and inevitable procedure. Theranostic-guided technology can possibly guide the surgeon with intra-operative directions for easy localisation and monitoring of the drug effect. Drugs linked with fluorescent dyes having properties of high molar extinction coefficient, higher quantum yield, good bio-compatibility, lower toxicity, and

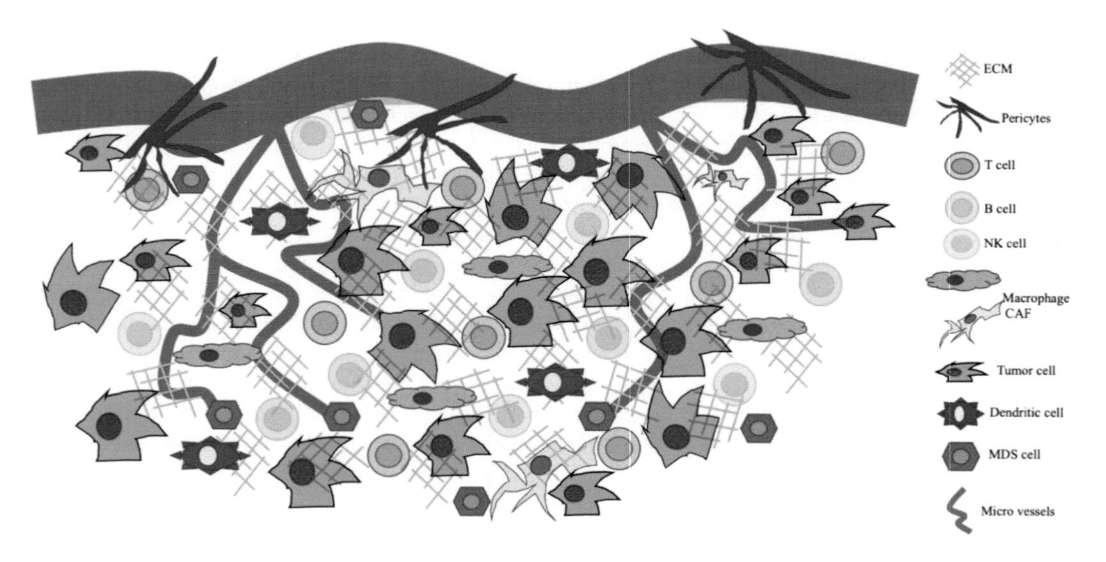

FIGURE 3.3 Angiogenesis-aided cellular cross talk in cancer microenvironment. Lymphocytes consists of T cells, B cells, and NK cells. Myeloid population consists of macrophages, dendritic cells, and myeloid-derived suppressor cells.

tunable molecular structure like cyanine, rhodamine, and squaraine show favourable advantages in cancer biology (DeLong et al. 2016). Various gold nanomaterials for example, nanorods, nanocages, nanopopcorns, nanostars, and nanoflowers with high sensitivity and remarkable enhancement of conjugated molecules on the rough surfaces are also attractive as campaign materials for theranostic therapy in SERS-guided cancer biology. Morphological changes of the internal organs can be monitored by high special resolution, real-time responsive, and non-invasive MRI techniques in conjugation with Fe, Gd, and Mn probes (Navyatha and Nara 2019). Acoustical, PET, and CT imaging therapy are also useful biomedical techniques in cancer treatment because of deep tissue penetration, cost-effective nature, and powerful post-processing (Pillarsetty et al. 2019). Nanotheranostics in combination with other biotechnologies will become a key and prosperous strategy in personalised precision medicine and finally translate the basic concept to mega medical application.

Nanomaterials in Orthopaedics

Although the term orthopaedics was coined in the 1700s, ortho-surgical techniques have been practised since ancient times. With the advent of an ageing society and their accompanying chronic diseases, realignment of critical-size bone defects remains a tough challenge for orthopaedic surgeons. Doctors face major challenges to treat bone defects, especially large segmental bone defects that are caused by trauma, tumours, infections, and congenital malformations. Figure 3.4 explains the applications of biocompatible materials in various bone defects like spinal regeneration, osteosarcoma, foot, ankle, hand, and joint fractures including the chondrogenesis regenerations like tendon and ligaments fracture management. Disc discectomy and fusion are often related with spinal injury, post-discectomy spondylosis and disc herniation syndrome. The domains of bioengineering are widely discussed and practised in the area of orthopaedics with research on cell-based therapies and in vertebral disc (IVD) replacement therapy, but have failed to give complete restoration and function of spinal disc (Stergar et al. 2019). Carbon, silicon,

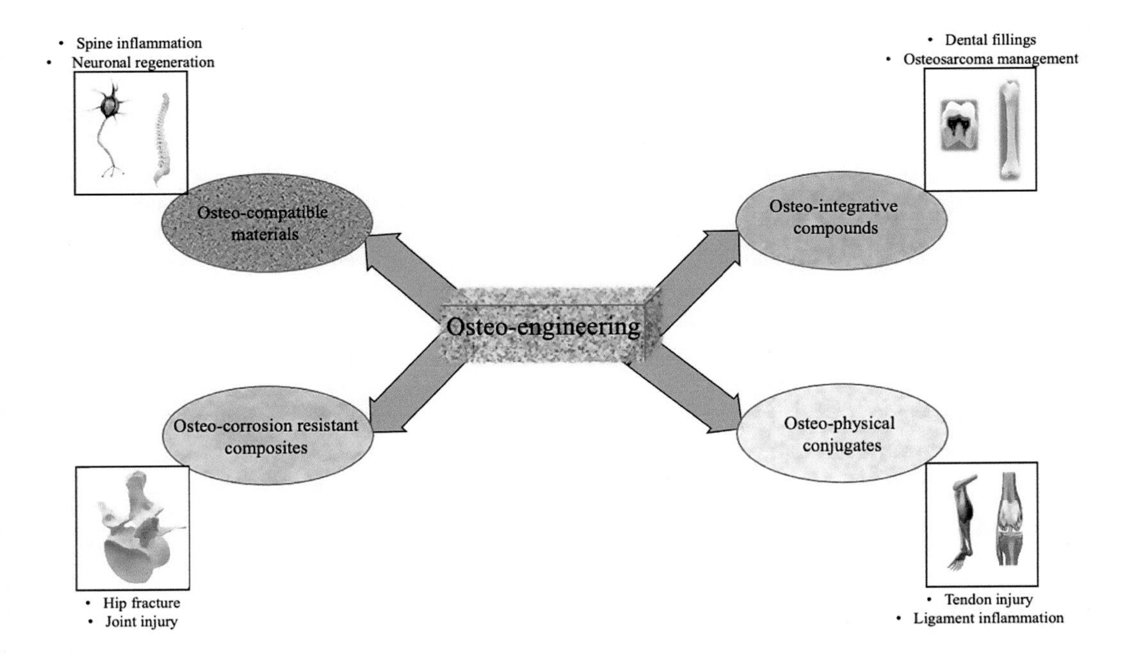

FIGURE 3.4 Schematic diagram of nanomaterials in controlling osteo-dysregulation. Osteo-compatible, anti-corrosive, osteo-integrative materials play vital roles in controlling neuronal and spine injury, dental and bone defects, and recovery of injured tendons and ligaments. Fractures in bone is also healed by nanoconjugated drugs.

and collagen combinations are demonstrated to promote axonal growth and mimic the nerve regeneration through nanoscaffold techniques as nerve conduit agents. Cervical nanocages prepared with silicon nitride NPs have demonstrated multiple biomechanical advantages over the traditional titanium-PEEK (poly-ether-ether-ketone) material (Vogel et al. 2018) with greater osteogenic and angiogenic improvements. The advances in nanotechnology engineering may offer an efficient method to overcome the lacuna of spinal injury.

Arthoplasty or joint replacement surgery in implant prosthesis plays a vital role in hip and musculoskeletal joint operation. Surface-modified nanotitanium materials proved to be a more favourable agent between native bone and implant surface and have strategic advantage than the ultra-high molecular weight polyethylene (UHMWPE) implants. Addition of nanocarbon composites to the titanium nanometal demonstrated the translational success to give better support as artificial bone to acetabular lining or tribal component (Szymański et al. 2020). Currently various types of bio-cements such as polymethyl methacrylate (PMMA) are being considered in the replacement techniques. Studies have revealed that natural agents like anthracyclic glycoside aloin have advantages as osteogenesis agents (Pengjam et al. 2016). In chondrogenesis, repair of cartilage defects in sportsmen has received extensive investigation. Mesenchymal stem cell (MSC)-incorporated biocompatible nanomesh prepared by polycaprolactone and gelatin showed early success (Lobo et al. 2018). Although application of nanotechnology in cartilage regeneration is not yet fully elaborated, the utilisation of nanomaterial as scaffolds for regenerative tissue engineering has shown to favourably affect cell adhesion, proliferation, and phenotypic selection of chondrocytes. Controlled and sustained release of mitomycin-C, a chemotherapy agent with the ability to decrease post-operative adhesions using hydrosol nanoparticles as drug carriers have shown significant results in the field of tendon injury. This allowed the reduction of the tendon adhesion phenomenon and maintained the tendon mechanical strength (Oragui et al. 2012). Extensive research is however still required to overcome the secondary infection in the implants and native tissue interface. Application of nanotechnology is still in the infancy stage but the concept has revolutionised into a promising avenue to go ahead with main focus on the safety and non-toxic aspects.

Conclusion

One of the most promising results of nanomaterial in medicine is translational use with diverse applications. While there are many advantages of nanomedicines, there are concerns over their potential side-effects and bio-risks when they are used in patients, thereby limiting wider applications. Further research on the physicochemical nature of nanomaterials like size, shape, surface chemistry, hydrophobicity and hydrophilicity, chirality, aggregation, morphology, bio-interface mechanism, bio-degradability, tissue excretion, and catalytic ability should be studied in detail. The gap between bench and bed or conception to implementation should be narrowed down; this is only possible through the shape-by-design approach of nanomaterials. Process scale-up of the proven nanomedicine is yet another challenge for technologists and needs further attention.

Acknowledgement

Authors immensely thank the authorities of University of Miyazaki for the facilities provided to conduct research on nanomedicine as a multidisciplinary activity.

ABBREVIATIONS

MRSA Methicillin-resistant *Staphylococcus aureus*
VRE Vancomycin-resistant *enterococci*
AgNP Silver nanoparticle
WHO World Health Organization

MRI	Magnetic Resonance Imaging
PET	Positron Emission Tomography
SPECT	Single Photon Emission Computed Tomography
CT	Computed Tomography
CAT	Computed Axial Tomography
miRNA	Micro Ribonucleic Acid
siRNA	Small Interfering Ribonucleic Acid
LHRH-PE40	Luteinizing Hormone Releasing Hormone-*Pseudomonas aeruginosa* Exotoxin 40
NP	Nanoparticles
MNP	Magnetic Nanoparticles
FITC	Fluorescein Isothiocyanate
FRET	Fluorescence Resonance Energy Transfer
HIFU	High Intensity Focused Ultrasound
SDT	Sonodynamic Therapy
IMXT	Intensity Modulated X-ray Therapy
IMPT	Intensity Modulated Proton Therapy
SELEX	Systematic Evolution of Ligands by Exponential Enrichment
ECM	Extra Cellular Matrix
SERS	Surface Enhanced Raman Spectroscopy
CAF	Cancer Associated Fibroblast
MDS	Myeloid Derived Suppressor Cell

REFERENCES

Arafa MG, El-Kased RF, Elmazar MM. 2018. *Sci Rep. 8*: 13674.
Bajpai SK, Pathak V, Soni B. 2015. *Int J Biol Macromol. 79*: 76–85.
Bogunia-Kubik K, Sugisaka M. 2002. *Biosystems 65*: 123–38.
Brunton S. 2016. *J Fam Pract. 65*(4 Suppl):supp_az_0416.
DeLong JC, Hoffman RM, Bouvet M. 2016. *Expert Rev Anticancer Ther. 16*: 71–81.
Desai TA, Chu WH, Tu JK, Beattie GM, Hayek A, Ferrari M. 1998. *Biotechnol Bioeng. 57*: 118–20.
Forbes JM, Cooper ME. 2013. *Physiol Rev. 93*: 137–88.
Halder A, Sun Y. 2019. *Biosens Bioelectron. 139*: 111334.
Harishkumar M, Radha M, Yuichi N, Kumar DH, Paduvarahalli N, Masugi M. 2019. *Material Today Proceedings. 10*: 100–5.
Jimenez Jimenez AM, Moulick A, Bhowmick S, Strmiska V, Gagic M, Horakova Z, et al. 2019. *Talanta. 205*: 120111.
Krokosz A. 2007.*[Fullerenes in biology].PostepyBiochem. 53*: 91–6.
Kuiken T. 2011. *Wiley Interdiscip Rev Nanomed Nanobiotechnol. 3*: 111–8.
Lobo AO, Afewerki S, de Paula MMM, Ghannadian P, Marciano FR, Zhang YS, et al. 2018. *Int J Nanomedicine. 13*: 7891–903.
Madhyastha H, Halder S, Queen Intan N, Madhyastha R, Mohanapriya A, Sudhakaran R, et al. 2020. *RSC Advances. 10*: 37683–94.
Madhyastha R, Madhyastha H, Nakajima Y, Omura S, Maruyama M. 2012. *Int Wound J. 9*: 355–61.
Makrantonaki E, Wlaschek M, Scharffetter-Kochanek K. 2017. *J Dtsch Dermatol Ges. 15*: 255–75.
Mattiuzzi C, Lippi G. 2019. *J Epidemiol Glob Health. 9*: 217–22.
Navyatha B, Nara S. 2019. *Nanomedicine (Lond). 14*: 766–96.
Nethi SK, Das S, Patra CR, Mukherjee S. 2019. *Biomater Sci. 7*: 2652–74.
Nishizawa M, Menon VP, Martin CR. 1995. *Science. 268*: 700–2.
Oragui E, Sachinis N, Hope N, Khan WS, Adesida A. 2012. *J Stem Cells. 7*: 121–6.
Palmerston Mendes L, Pan J, Torchilin VP. 2017. *Molecules. 22*(9): 1401.
Patel S, Srivastava S, Singh MR, Singh D. 2019. *Biomed Pharmacother. 112*: 108615.
Pengjam Y, Madhyastha H, Madhyastha R, Yamaguchi Y, Nakajima Y, Maruyama M. 2016. *Biomol Ther (Seoul). 24*: 123–31.
Pesec M, Sherertz T. 2015. *Future Oncol. 11*: 2235–45.

Pillarsetty N, Jhaveri K, Taldone T, Caldas-Lopes E, Punzalan B, Joshi S, et al. 2019. *Cancer Cell. 36*: 559-73.e7.
Robson MC, Hill DP, Woodske ME, Steed DL. 2000. *Arch Surg. 135*: 773–7.
Sahoo Pravas Ranjan, Madhyastha Harishkumar, Madhyastha Radha, Maruyama Masugi, Nakajima Y. 2020. *Recent Progress in Nanotheranostic Medicine. in Nanopharmaceuticals: Principles and Applications*, Yata VK, Ranjan S, Dasgupta N, Lichtfouse E (ed.), Springer Nature UK, Vol *3*: pp. 317–334.
Sanada Y, Yoshida K, Itoh M, Okita R, Okada M. 2009. *Hepatobiliary Pancreat Dis Int. 8*: 97–102.
Stallard Y. 2018. *J Wound Ostomy Continence Nurs. 45*: 179–86.
Steed DL. 2003. *Surg Clin North Am. 83*: 547–55, vi–vii.
Stergar J, Gradisnik L, Velnar T, Maver U. 2019. *Bosn J Basic Med Sci. 19*: 130–7.
Szymański T, Mieloch AA, Richter M, Trzeciak T, Florek E, Rybka JD, et al. 2020. *Materials (Basel). 13*(18): 4039.
Vogel D, Schulze C, Dempwolf H, Kluess D, Bader R. 2018. *Proc Inst Mech Eng H. 232*: 1030–8.
Von Roemeling C, Jiang W, Chan CK, Weissman IL, Kim BYS. 2017. *Trends Biotechnol. 35*: 159–71.
Yousefpour Marzbali M, Yari Khosroushahi A. 2017. *Cancer Chemother Pharmacol. 79*: 637–49.

4

Nanotechnology in Stem Cell Regenerative Therapy and Its Applications

Jainey P. James and Pankaj Kumar
Nitte (Deemed to be University), NGSM Institute of Pharmaceutical Sciences (NGSMIPS), Mangaluru, Karnataka, India

CONTENTS

DOI: 10.1201/9781003153504-4

Introduction

Stem cells are undifferentiated, specialised cells capable of differentiating into many other diverse cells (Weissman 2000). They originate from differentiated organs and are influential in repairing organs and their injury in the post-natal and adult phases of life. Formerly, the belief was that stem cells could only differentiate into adult cells of the same types of organs and tissues. However, current studies have proof that stem cells can differentiate into other kinds of cells. The significant properties of stem cells are self-renewal, clonality, and potency. Broadly, they are classified based on their differentiation capability and origin. Numerous researchers have reported the medicinal applications of stem cells in various serious illnesses, where they can regenerate and restore tissues.

Regenerative medicine is a promising area that focuses on therapeutics that restores and improves body functions (Daar and Greenwood 2007). The enormous development in stem cell regenerative therapeutics has given new hope in the field of medicine. It would be expected that there will be a large pool of tissue, organoid, and organs from adult stem cells. It has been speculated that in the coming future, a pharmaceutical compound may be processed such that it may promote cell differentiation to specific tissue or organ. At the same time, due to strict law in certain countries, it prevents the funding and limits the studies only to the less risky projects like UCSCs, BMSCs, and TSPSCs from biopsies.

Nanoparticles have emerged in research extensively due to their particle size and compatibility, which can interface with living cells and possess healing properties (Wolfram and Ferrari2019). They have contributed to the medical field, allowing the expansion of targeting, marking, tracing, and labelling along with theranostic uses. National Institutes of Health, USA, (Squillaro et al. 2016) has reported its application in various organ transplantations and disease management. Therefore, nanoparticles have attracted the attention of many researchers due to their enormous applications. The present aim of nanoparticle-based therapy is to provide patient-specific treatment in releasing drugs with curtailed systemic toxicity. The stem cells in nanoparticles have many advantages, as they function as carriers for active biological moieties. Nanoparticle-based therapy provides an appropriate atmosphere and improves transplantation. In nanoparticle stem cell research, the movement of the transplanted stem cells can be readily traced and targeted, which is used as the drug delivery system.

By considering the above facts, in this chapter, we have discussed in detail stem cell classification and their applications in various disease conditions, mainly, elaborating the advancements in regenerative medicine. Moreover, the emerging and burgeoning role of nanotechnology in stem cell research and its extensive impact are discussed in this chapter.

Stem Cell Classification

This section provides the stem cell types, proliferation, and their derivation (Figure 4.1). The classification of stem cells depends on their differentiation ability and origin.

Classification based on Differentiation Ability

The classification is based on their separation potential and sub-divided as totipotent, pluripotent, multipotent, and oligopotent.

Totipotent Stem Cells

Totipotent stem cells are competent to form a whole organism such as a zygote. They possess an incredible ability to extend into germ cell layers of the embryonic and extra-embryonic tissues, thus shaping into embryo and placenta (Rossant 2001). Totipotent cells originate from the early development stage, and these undifferentiated cells can develop into any type of cells further.

Pluripotent Stem Cells

Pluripotent stem cells self-renew themselves and differentiate only into three germ layers cells; however, it does not form embryo or placental cells. Induced pluripotent stem cells (iPSCs) are generated from restructuring somatic cells, which have the same features as embryonic stem cells (ESCs), also called as ESCs (Takahashi and Yamanaka 2006).

Multipotent Stem Cells

They are present in most tissues, such as bone marrow, adipose tissue, and umbilical cord blood (Augello et al. 2010). The proliferative capacity of multipotent cells is less than that of the aforementioned cells. They can synthesise different cells from germinal or specific layers such as mesenchymal stem cells

Stem Cells Classification

Differentiation ability
- Totipotent stem cells
- Pluripotent stem cells
- Multipotent stem cells
- Oligopotent Cells

Origin
- Embryonic Stem Cells
- Human embryonic germ cells
- Amniotic epithelial cells
- Umbilical cord blood stem cells
- Adult Stem Cells
- Hematopoietic stem cells
- Induced Pluripotent Stem Cells
- Mesenchymal stem cells

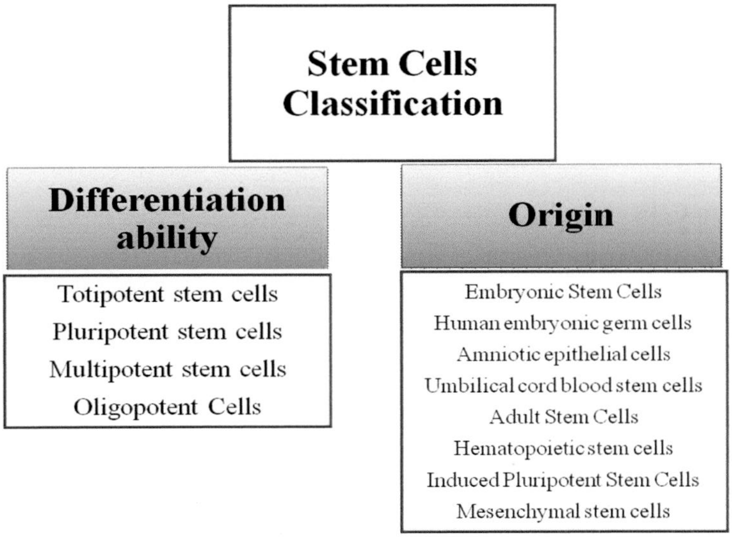

FIGURE 4.1 Outline of stem cell classification.

(MSCs) or haematopoietic stem cells (HSCs), respectively. The MSCs are the recognised multipotent stem cell among others.

Oligopotent Cells

Oligopotent stem cells self-renew and form two or more lineages within a specific tissue. To illustrate, the pig's ocular surface, including the cornea is accounted to have oligopotent stem cells, that can develop colonies of corneal and conjunctival cell individually (Majo et al. 2008).

Unipotent Stem Cells

Unipotent stem cells have not received much attention compared to that of other stem cells. As they lack the self-renewal property, they produce cells with only one lineage differentiation and generate only one cell type. Despite their limited capacity, they are potent and used for the treatment of various diseases (Rajabzadeh et al. 2019).

Classification based on Origin

Stem cells are categorised based on their origin as embryonic and non-embryonic stem cells (non-ESCs). ESCs divide into three germ layers, hence totipotent and the source is from an embryo, whereas non-ESCs stem cells or adult stem cells have limited separating ability, which is multipotent in nature and derived from mature/adult cells. Moreover, myriad specific stem cells are used clinically under various conditions that include mesenchymal, neural, pancreatic stem cells, iPSCs and so on, discussed in this section.

Embryonic Stem Cells

ESCs originate from the blastocyst stage and divide the tissue to become derivatives of germ layers, further leading to the formation of all types of cells. Transcription factors such as octamer-binding transcription factor-4 (OCT4)and SRY-related high-mobility group box protein-2 (SOX2) are responsible for the pluripotency and self-renewal nature. The blastocyst forms the inner and outer cell mass; the inner cell mass forms embryos and the external cell mass forms the placenta. Specific conditions are maintained in growing ESC lines to separate the cells from the inner cell layer of trophoblasts and transfer them to a culture dish (Bongso 2006). In 1998, Thomson isolated human ESCs and divided them into more than 200 categories of cells, which is promising for the treatment of various diseases, described in the next session of this chapter.

Embryonic Germ Cells

Primordial germline cells are the sources of embryonic germ cells (EGC). Only a few animal experiments on germ cells are available when compared to ESCs. Researchers assume that the efficiency of germ cells is less than that of ESCs because EGCs develop further (5–9 weeks). Shamblott et al. 1998 reported the isolation of germ cells from the gonadal ridge of human tissue.

Amniotic Epithelial Cells

These cells originate from the amniotic membrane in the human placenta; they also express the markers found on pluripotent ESCs and EGCs such as Oct-4, Nanog, and alkaline phosphatase. The three germ layers differentiate as ESCs and EGCs, consisting of pancreatic endocrine cells, hepatocytes, cardiomyocytes, and neural cells *in vitro* (Miki et al. 2005). When administered in immunodeficient animals, they were a safer substitute for human ESCs (hESCs) as they increase at a higher rate without pluripotency or teratogenicity loss. However, these must be verified if they have the pluripotency degree similar to hESCs.

Umbilical Cord Blood Stem Cells

Barker and Koh identified the abundance of HSCs in umbilical cord blood in the late 1980s (Barker and Wagner 2003). Umbilical cords are usually discarded after childbirth, enriched with HSCs and MSCs. The restoration of each blood and immune cell is through the HSCs. The public and private sector units can preserve the umbilical cord blood stem cells(USCs) in stem cell banks and they are available in need. This technique has umpteen merits when compared with other adult stem cells, for example, it is collected at every birth through a simple procedure; storage and processing is comfortable with low risk of infection.

Adult Stem Cells

Adult stem cells are derived from adult tissues as undifferentiated cells and found in various body tissues. Current studies report that these cells can change from diverse germ layers into cell types, i.e. mesoderm-derived bone marrow stem cells can transform into mesoderm and endoderm-derived cell lineages like the lung, liver, the GI tract, skin, and so on. *In vivo* studies have revealed that restoration of damaged organs occurs with the transplantation of adult stem cells (Chimutengwende and Khan 2012). Other studies have shown that cultured adult stem cells secrete various molecular mediators with properties that promote repair.

Haematopoietic Stem Cells

HSCs can be labelled and tracked with their morphologic nature and cell-surface markers in the bloodstream and target tissues (Lagasse et al. 2000). They exist in the bone marrow and can reform blood-forming lineages. This potential makes it the best-characterised stem cell niche. The missing parts of the haematopoietic system can be replenished by HSCs and can endure freezing for many years.

Induced Pluripotent Stem Cells

Adult somatic cells produce iPSCs and are hereditarily restructured to ESC-like. Mouse iPSCs were reported in 2006 with genetic transcription factors (Takahashi and Yamanaka 2006). With the advent of the technology, diverse adult cells staging through ESCs have developed iPSCs. Takahashi et al. in 2007 depicted the human iPSCs with the same four elements, which are morphologically similar to human ESCs.

Mesenchymal Stem Cells/Stromal Cells

MSCs are multipotent in nature and differentiate to mesodermal-originated tissues like bone, ligaments, muscles, and so forth. Presently, the primary source of MSCs is bone marrow (Zvaifler et al. 2000). They are also isolated from circulating blood and non-haematopoietic tissues like adipose tissue, dental pulp, and the lung.

Neural Stem Cells

Multipotent neural stem cells (NSCs) are self-restore that constantly produces glial progenitor cells during embryonic development stage (Baizabal et al. 2003). They were isolated from the embryos and adult brains and used as transplants in Parkinson's disease. Skeletal muscles can be generated by injecting NSC-cultured cell lines into the muscles (Galli et al. 2000).

Pancreatic Stem Cells

Multipotent pancreatic stem cells (PSCs) are isolated from the human foetal pancreas, derived from the foregut endoderm and reformed into progenitor lineages, essential for the growing pancreas (Jiang and Morahan 2014). Islet cell transplantation can reverse the damage of pancreatic islets cells in Type 1 diabetes (Ma et al. 2008).

Skin Stem Cells

They are multipotent in nature, found to be in matured skin, renew themselves, and capable of differentiating into diverse skin cell lineages. The skin stem cells' existence helps restore the epidermis and expand the hair follicles after injury. The epithelial stem cells generate the follicular epithelium and sebaceous glands after severe damage (Flores et al. 2005). They are active during skin renewal and repair skin damage all throughout life.

Cancer Stem Cells

In the late 1990s, John Dick identified cancer stem cells (CSCs) in acute myeloid diseases. The injured and aged tissues renew and regenerate with the assistance of stem cells. Many investigations have reported that the capability of a tumour to proliferate relies on a small cell mass characterised by stem-likeness, named as CSCs (Soltysova et al. 2005).

Stem Cell Therapy

The applications of stem cells in modern therapies are extensive as it provides novel treatment regimens in clinical practice. Due to their ability to restore damaged tissues and organs, their contributions in the medical field are enormous. This therapy has been used in the treatment of degenerative disorders which showed positive results in preclinical and clinical levels. Currently, clinical trials focus on significant diseases such as diabetes, leukaemia, liver cirrhosis, heart failure, etc.

Stem Cell Therapy in Neurological Disorders

Stem cells can act as resources of neurons in treating central nervous system disorders with different techniques by stimulating and restoring damaged neurons. Cell replacement therapy strategies were used to treat Parkinson's and Alzheimer's disease (AD) (Ramakrishna et al. 2011).

Parkinson's Disease

It's a condition of exhaustion of striatal dopamine due to the deterioration of dopaminergic neurons in the substantia nigra. In the treatment of such a disease, dopaminergic neurons must be generated in a large amount and supplied regularly. Dopamine neurons were developed by Kawasaki et al. 2000from mouse ES cells without embroid body formation. One of the clinical trials reports that cell replacement with human foetal dopaminergic neurons can show enduring progress in some patients (Lindvall et al. 2004).

Alzheimer's Disease

The characterisation of synaptic and neuronal loss across the brain with impairment in memory and cognition leading to dementia is Alzheimer's disease. Acetylcholinesterase inhibitors are used to treat AD, which enhances cholinergic function and partially alleviates the symptoms. The competency of the stem cells in delivering the factors will modify the disease stages. Nerve growth factor (NGF) has resulted in advanced cell function, and memory restoration in animal models have also benefited a few patients (Ramakrishna et al. 2011).

Cardiac Problems

Stem cell therapy in cardiac problems aims to restore the cardiovascular functions. It works by the principle that cardiac stem cells enhance the performance of the cardiac system by *de novo* cardiomyogenesis. Transplantation of cardiac stem cells has obtained blended data in human clinical trials (Liu et al. 2008). Recent research proves that adult and embryonic SCs can substitute injured heart muscle cells by instituting new blood vessels.

Stem Cells in Wildlife Conservation

This field has even been employed in preserving endangered species on the verge of extinction. By this technology, wildlife conservation can be deployed by preserving the endangered species' gene pool and gene plasm. This can be collected through dead or live animals from its DNA, sperms, eggs, embryo, gonads and skin, or other body parts (Comizzoli and Holt 2014). The conserved tissues can be further used to regenerate other cells, which can resurrect life.

Stem Cell Therapy in Regenerative Medicine

Various ongoing researches are focusing on stem cell regenerative therapies. The following section discusses in detail on stem cells that are reported with biological actions in clinical trials and animal experiments (Figure 4.2 and Table 4.1).

ESCs

In spinal injuries, ESCs transplantation in injury sites can regenerate spinal tissue and improve balance and sensation. In retinal pigment epithelium, destruction transplantation of ESCs-derived cones to the eye can restore vision and recover macular defects (Zhou et al. 2015). Factors such as diabetes, genetic factors, and lifestyle can induce cardiac problems. ESCs-derived cardiovascular progenitors, cardiomyocytes, and bone marrow-derived mononuclear cells (BMDMNCs) can regenerate cardiac tissues by suppressing heart arrhythmias and amalgamate to heart as pacemakers (Shiba et al. 2012). Liver injuries result from genetic factors, toxins, drugs, and various infections. ESC-derived hepatocytes can regenerate liver tissues by stimulating the elements essential for ESCs-hepatocyte conversion. Heart problems, genetic disorders, and lifestyle can cause diabetes. ESCs-derived pancreatic progenitor cells (PPCs)

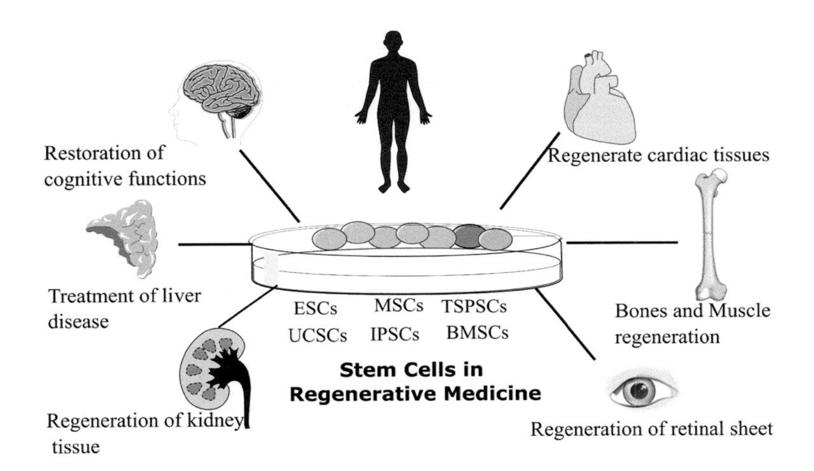

FIGURE 4.2 Stem cell therapy in regenerative medicine.

TABLE 4.1
Applications of Stem Cell Regenerative Therapy

Stem Cells	Stem Cell Applications
ESCs	Restoration of spinal tissues and cartilage tissues
	Regeneration of retinal sheet, cardiac tissues, and liver tissues
	Improvement in glucose level
TSPSCs	Diabetes, retinopathy, cancer therapy
	Restorative applications in neurodental problems
	Cognitive and ear functional acoustic restorations
	Intestinal mucosa and muscle regeneration
MSCs	Bladder and teeth tissues regeneration
	Repairing of orthopaedic and muscle damages
UCSCs	T1DM and T2DM treatment
	Krabbe's disease treatment
	Improves renal function
	Treatment of Hodgkin's disease
BMSCs	Blood cancer treatment
	AIDS treatment
	Alveolar bone tissue renewal
IPSCs	Kidney tissue restoration
	Retinal regeneration in AMD
	Placental, brain cortex defects improvement
	Liver and lung tissue regeneration

transplantation can improve glucose level (Bruin et al. 2015).The various injuries faced by athletes can be treated by restoring the cartilage tissues by transplanting chondrocyte organoids (Cheng et al. 2014).

TSPSCs

Accidents, age, and genetic-related factors result in neurodental issues; dental pulp stem cells' (DPSCs) transplants as neurons are applied in the treatment of neurodental disorder (Ye et al. 2016). Inner ear stem cell (IESC)-derived hair cell transplants can be used in acoustic problems caused by noise, infection, and age-related factors; IESCs restore acoustic roles by cochlear regeneration (Mizutari et al. 2013).

Food poison and genetic disorders can deteriorate the intestinal system, herein; intestinal progenitor cells (IPCs) can renew goblet mucosa and treat intestinal defects (Shaffiey et al. 2016). Limbal progenitor stem cells' (LPSCs) transplantation can treat corneal diseases by reviving corneal tissues (Ksander et al. 2014). In muscular deformities, PEG fibrinogen coaxed mesoangioblasts can restore muscle fibrils in the management of muscle abnormalities (Fuoco et al. 2015). Adipose-derived stem cells (AdSCs) are used in eye diseases and diabetic retinopathy by producing and restoring vasoprotective factors (Cronk et al. 2015).

MSCs

Human mesenchymal stem cells (hMSCs) have expressed excellent features as immunosuppressants at the sites of injury and inflammation. It has proved as an excellent method for treating numerous diseases. The clinical application is peculiar because of its potential to release several biologically active molecules to prevent and repair inflammation and injured cells, respectively. It has even proved its importance in the target-based release of anti-cancer genes at cancer-causing sites. The combination of MSCs with biologically bioactive hydrogen collagen-based polymer and electrospun fibre scaffolds improves the clinical management of wound healing (Weinstein et al. 2017). National Institutes of Health USA has reported about 500 different MSC-based clinical studies completed worldwide for numerous organ transplantations. These studies of stem cells in neurosciences assembled the attention of many researchers.

Certain studies have proved the neurogenesis and showed the presence of NSCs in the adult rat brain (Charrier et al. 2006). Therefore, it provided hope for the recovery of spinal cord tissue and injured brain using selective targets. The bone marrow-derived MSCs from baboon resemble bladder stem cells in morphology and are used in the regeneration of bladder tissue (Sharma et al. 2011). In dental problems, epithelial-MSCs coaxed with dental stem cells' polymer tissue revive oral tissues by giving rise to mature teeth units (Oshima and Tsuji 2015). Coaxed MSC transplantation and infusion act in bone and muscle deterioration by regenerating bones and muscles, respectively (Csaki et al. 2007). Cell retention time increases as MSCs coaxed with alginate gel results in restoring muscle tissues in a controlled fashion. Alginate encapsulated cells' transplantation to mice heart decreases the size of the scar. Thus, the vascularisation raises and heart functions regenerate. GAG-coated promotes dermal papillary cells' (DPCs) transplants, mimics ECM microenvironment, and promotes DPCs and hair follicle regeneration treatment applied for alopecia—the condition of loss of hair (Lin et al. 2016).

UCSCs

Coaxed fibrin with amniotic fluid stem cells (AFSCs) assist in tissue repair for treating congenital heart defects in newborns; PEG on the addition of vascular endothelial growth factor (VEGF) promotes organogenesis (Benavides et al. 2015). The gelatinous substance present within the umbilical cord is Wharton's jelly (WJ), which is loaded with mucopolysaccharides, fibroblast, and macrophages, and stem cells that can be differentiated to β-cells. Transplantation of these cells improves the function of β-cells, which can be applied in the treatment of diabetes (Hu et al. 2013). Intravenous injection of transplanted WJ-SCs can improve renal functions and stop degeneration of tissues (Mahla 2016). Neuroblastoma patients on receiving UCSCs who survived without any side-effects were reported in a case study. The coaxed organoids can be used to treat all the diseases involving lysosomal defects, like Krabbe's disease, hurler syndrome, adrenoleukodystrophy (ALD), metachromatic leukodystrophy (MLD), Tay-Sachs disease (TSD), and Sandhoff disease (Mahla 2016). As the HA gel factors promote regeneration, transplantation of UCB-MSCs with HA hydrogel regenerates cartilage tissues, therefore effective in regenerative therapy as candidates for healing cartilage and ligament injuries. UCSCs' transplantation is used in the treatment of Hodgkin's lymphoma and injecting WJ-SCs is found to be successful in the management of peritoneal fibrosis, as it prevents cell death and peritoneal wall thickness (Fan et al. 2016).

BMSCs

A mixture of lymphoid and myeloid BMSCs can improve aplastic anaemia and haematological malignancies. The physiological aspects are that haploidentical BMCs can rebuild the host immune system. Transplantation of H1V1-resistant $CD4^+$ cells, that express H1V1 anti-RNA, can be applied to treat AIDS (Herrera-Carrillo and Berkhout 2015). The megakaryocyte organoid transplantation can be used in burns and blood clotting diseases. In treating diabetic and neurodegenerative disorders, lipoic acid (LA) and BMSCs are combined to induce vascularisation, which directs microglia for colonisation. Bone marrow-derived stem and progenitors accelerate alveolar jaw bone degeneration, which further regenerates oral bone, skin, and gum defects what arise in congenital disabilities (Kaigler et al. 2013). Decullalarised diaphragm implantation can be utilised in replacement therapy. BMSCs hemidiaphragm has the same spirometry and myography (Mahla 2016).

iPSCs (induced Pluripotent Stem Cells)

Layered photoreceptor nuclei formed from NPCs restores visual activity; therefore, iPSCs-derived NPCs' transplantation can be utilised in the cure of age-related macular degeneration (ARMD) and other eye-related defects (Yang et al. 2016a). The molecular signature of telencephalic INs from GABAergic-INs (interneurons) (iGABA-INs) cells releases GABA, which inhibits to host granule neuronal activity (Pasca et al. 2015). When these cells are transplanted in the growing embryo, it helps cure seizures and related neurodegenerative disorders. Also, it heals genetic and acquired seizures by transplanting INs in developing embryos. A1AD (α-1 antitrypsin deficiency) is a condition that increases the risk for liver and lung diseases. The mutated gene SERPINA1 results in this deficiency. This gene guides the protein α-1 antitrypsin, which is responsible for guarding the lungs against the enzyme neutrophils elastase that is powerful in damaging the lung connective tissues. The known fact is that a single base-pair mutation of gene SERPINA1 causes A1AD, which can be rectified by transplanted A1AD mutation corrected iPSCs (Wilson et al. 2015).

Stem Cell Nanotechnology

When nanotechnology science is incorporated with stem cell, it brings tremendous advancement in its applications. In recent years, differentiation and regeneration for stem cell therapy have been processed using nanotechnology. This technology comprises biocompatible and decomposable nanoscaffolds/nanofibers, including collagen nanofibre, carbon nanotube, oxide of graphene, poly-ε caprolactone, phosphate and silicate of tricalcium, and self-organised peptide (Naskar et al. 2017). However, apart from these specific nanoparticles like iron oxide, quantum dots (QDs), and superparamagnetic iron oxide NPs, polymer or chitosan-based scaffolds has arisen as an excellent contender for marking, tracing, and labelling of transplanted cells (Lin et al. 2017). Along with nanotechnology, nanopatterning has gained importance in the present technology of stem cell. Nanopatterned surface helps to prescribe the bonding, distribution, self-regeneration, and focused differentiation of PSCs. Many researchers prove that therapeutic cells are a suitable vector for dynamic directed drug delivery. When the therapeutic cells are coated with drug-loaded NPs, it provides new hope for tracking, distribution, and delivery of the cells. With this, it has become essential to develop NPs with appropriate surface coatings to follow better monitoring and distribution of stem cell without distressing their proliferation.

Nanomaterials

Nanomaterials used for stem cell renewing therapy, drug transport, cell labelling, tracking and treatments are graphene and graphene-based nanomaterials (Table 4.2). Scaffold-based nanomaterials and their combinations for stem cell differentiation and proliferation always raise curiosity among the researchers. Nanotubes coated with titanium dioxide (TiO_2) possess excellent mechanical strength and become a choice as scaffold fabrication in replacement therapy. The binding pattern of biological molecules with scaffold materials decides the destiny of the stem cell. However, there are many types of scaffolds that can be utilised in the stem cell differentiation, but in this section, our discussion is limited to graphene and its oxide (GO) concerning its unique mechanical, electrical, optical, and large surface topographical

TABLE 4.2
Examples of Nanoparticles Employed in Stem Cell Differentiation

Stem Cell	Nanoparticle	Differentiation Capacity
Human mesenchymal stem cells	Graphene-coated Silicon/silicon dioxide	Improved and fastened osteogenic differentiation
	Graphene oxide/gelatin hydroxyapatite	Improved osteogenic differentiation
	Single-coated graphene	Growth and neural differentiation synchronisation
	Laminin-layered graphene film	Superior differentiation into neurons
	Polyvinyl alcohol, poly(ε-caprolactone) 3D scaffold	Advanced osteogenic differentiation
Human neural stem cells	Graphene oxide-patterned substrate	Augmentation in intregrin clustering, focal adhesion, and neuronal differentiation
Placental-derived mesenchymal stem cells	Gold Nanoparticle-coated collagen nanofibre	Advanced neuronal differentiation and proliferation
Human embryonic stem cells	Nanopore-moulded polystyrene	Differentiation into endodermal cells

properties. These biocompatible scaffolds used in the improved, controlled, and accelerated osteogenic differentiation of human bone marrow result from MSCs on graphene and its oxide (GO) layers, hMSCs on graphene-layered Si/SiO_2, graphene oxide nanoflakes frameworks unified with hydroxyapatite of gelatin. Human neuronal differentiation outcomes were enhanced using laminin-layered graphene and other neuronal differentiation technology.

The generally used nanomaterials for differentiation, labelling, tracking, and therapy of stem cells are discussed here and will provide ample information. It is structured as:

(i) Graphene, inorganic NPs, biodegradable polymer scaffolds, and nanofibres;
(ii) Nanomaterials for stem cell regenerative therapy, drug delivery, cell labelling, tracking, and therapy

Osteogenic Differentiation and Neuronal Differentiation using Graphene Scaffolds

Graphene scaffolds in osteogenic differentiation fasten the bone cell formation process of hMSCs even in the absence of growth factor BMP-2 (Zhao et al. 2013). Cells developed on graphene-layered with glass or silica or silicon dioxide are excellent since they don't lose integrity (Nayak et al. 2011). This shows improved reports than development carried out on polyethylene terephthalate (PET) or polydimethylsiloxane (PDMS)-layered graphene. The comparative study using immune fluorescent staining images proves that graphene acts as a motivating factor and accelerates osteogenic differentiation into bone cells.

Apart from the osteogenic differentiation, graphene scaffolds have even shown a role in neuronal differentiation. It promotes neurosphere development and accelerates the movement of NSCs; it also regulates the development and differentiation of hMSC (Yang et al. 2016b). Patterned graphene oxide-based substrate is useful in creating functional hNSCs and offers enrichment in integrin gathering, focal bond, and neuronal development. Immune staining images have proved that cells grown on graphene for differentiation were more stable than grown on glass, thus accelerating adhesion and speedup differentiation of hNSCs into neurons. Therefore, it can be used as neural prosthetics and regenerative medicine (Park et al. 2011).

Micro/nanopatterned graphene oxide shows the effect on cell spreading and morphology. Specific geometries prove to regulate stem cell growth and development into specific cell lineages. Human adipose mesenchymal cells have extremely extended shape and structure on nanosized graphene oxide patterns, thus producing augmented osteogenic differentiation. With the efficiency of 30% using NGO grid patterns, ectodermal neuronal cells can be achieved from the differentiation of mesodermal stem cells, and thus it imitates an extended and interconnected neuronal network.

Stem Cell Differentiation using Inorganic Nanoparticles

Stem cell differentiation can be achieved using inorganic nanoparticles' fusion with core-shell shape and size. These consist of graphene quantum dots and silicate nanoplatelets, used in osteogenic development and fused nanostructures. Graphene, upconversion, and core-shell nanoparticles along with retinoic acid-laden are used in neuronal development (Shah et al. 2015).

Osteogenic Differentiation using Inorganic NPs

With the application of NPs in diagnostics, drug release, and treatment, the interaction of NPs with the cell membrane is crucial. Therefore, its dimension, charge, size, and surface chemistry state their physiological barrier constraints. Different dimension Au-NPs were evaluated for osteogenic differentiation of hMSCs. These Au-NRs (gold nanorods) were coated either with peptides, or poly-sodium 4-styrene sulphonate (PSS) and poly-allylamine hydrochloride (PAH) (Zhao et al. 2015).

Using phenotypic marker alkaline phosphatase (ALP), it has been reported that Au spheres and rods enhanced osteogenic differentiation; however, this study is unclear in its signalling pathway. A similar analysis using 20 nm Au-NPs was performed; it states that MSCs demarcated the p38 mitogen-activated protein kinase pathway (MAPK) and the up-regulated Runx2 gene for osteogenic development; whereas PPAR gene down-regulated control adipogenesis. Similarly, up-regulation of Runx2 and osteoblast marker genes along with the down-regulated PPAR gene assists the effect of MSCs into osteoblast cells (Yi et al. 2010).

GQDs-NPs has developed as an exciting class of fluorescent NPs. The culture in the differentiation medium incubated with GQDs-NPs accelerates the osteogenic differentiation of MSCs. The other NPs metallofullerenol and functionalised selenium (Se) using the BMP pathway fasten the osteogenic development of MSCs. In contrast, silicon nanowires (SiNW) have also shown its effectiveness for osteogenesis and chondrogenesis of MSCs through the activation of Ras/Raf/MEK/ERK signalling cascades to control sticking, development, proliferation, and differentiation of MSCs.

NSCs Differentiation using NPs

Lee and the team designed a magnetic core-shell nanoparticle to deliver genetic materials within cells for particular differentiation, consisting of $ZnFe_2O_4$ magnetic nanoparticle core coated with the slim gold shell. These magnetic core-shell NPs are vital in delivering genetic content into NSCs to control its differentiation. However, these magnetic core-shell NPs don't affect the biological functions of NSCs. MCNPs with Cy3-labelled siRNA complex were transported into the cytoplasm of NSCs to set the compound in the existence of a magnetic field (MF) (Shah et al. 2013). The magnetic core-shell develops functional neuronal differentiation in specific green fluorescent protein (GFP) or Sox9 typical polyamides.

Stem Cell Differentiation using Biodegradable Polymer Scaffolds and Nanofibres

Another class of biodegradable material utilised in medical applications are polymer of lactic acid, L-lactic-co-glycolic acid, vinyl alcohol, and ε-caprolactone along with natural materials like polymers collagen, gelatin, and chitosan (Newman and McBurney 2004). These polymers are generally non-toxic, show beneficial effects, grow into three dissimilar lineages (osteogenic, chondrogenic, and adipogenic), and undergo a neural revival. This differentiation is determined by cytochemical staining. Apart from these biomimetic peptide-based scaffolds, hydrogels are utilised for the growth of neural cell-like, umbilical vein-HEC, and murine adult. The injectable hydrogels and PVA-based hydrogels form steady three-dimensional cultures, which differentiate to extended neuritis and osteoblast growth to human amniotic fluid-originated stem cells. Apart from these, hydrogels show pluripotency and growth potentials of human-originated embryonic cells, induced pluripotent cells, and human mesenchymal cells into distinct cell lineages.

Nanomaterials for Cell Labelling, Tracking, and Therapy

This part will outline the uses of diverse nanoparticles like gold, quantum dots, magnetic NPs, AuNRs-labelled stem cells, and upconversion nanoparticles in stem cell labelling, tracking, and the application.

Gold NP Conjugates

Au is considered as the best among all the metallic NPs due to its firmness and non-toxic nature to the cells. Many studies concentrate on its properties, especially on a surface charge, size, and coating and unlike neural and microglial cells uptake for differentiation, labelling and tracking of MSC. Au complex poly-l-lysine is used for labelling and tracking human MSCs, having a recognition threshold of $\approx 2 \times 104$ cells. Au-NP preserved magnetic core properties and coated with iron oxide is used in magnetic hyperthermia. It is even used in mRNA detection in hMSCs when covered with polydopamine. Labelling of hMSCs with gold nanocages has also been reported, and it has produced minimal cytotoxicity marked with stem cells. Many modalities like photoacoustic imaging, ultrasound, and micro-CT use MSCs (labelled with AuNPs) for labelling and tracking. Silica-coated Au-NRs like PEGylated gold nanostars (GNS-PEG) are used in the diagnosis of gastric cancer and also play a role in understanding its bio-distribution in tumours and the reticuloendothelial system of spleen and liver (Liang et al. 2015).

Quantum Dots

Semiconducting quantum dots (QDs) such as cadmium selenide and zinc sulphide have played a vital role in the stem cells for labelling and *in vivo* applications due to its superior emission properties and photostability. Uniform-sized quantum dots-loaded hMSCs are utilised for capturing the histologic sections *in vivo* (Zhang et al. 2012). Some of the features of QDs-packed hMSCs are: it proliferates more than 6 weeks *in vitro* without relocating and affecting adjacent cells and differentiation of hMSCs. When inserted into the rat heart, it organises into three dimensions. Peptide-coated cadmium selenide and zinc sulphide QDs can be used for the marked labelling of hMSCs. Selvan and co-workers have proved that peptide-coated cadmium selenide and zinc sulphide QDs can be utilised for the organised marking of human MSCs (Narayanan et al. 2013). Peptide-functionalised QDs studies have been carried in mice having liver failure problem, and it helps to identify the accumulation of transplanted cells in many organs. Multifunctional QDs have also shown its role in delivering siRNA and a small molecule to accelerate osteogenic differentiation and extend duration tracking of hMSCs. Apart from these, nanoparticle-like graphene QDs and doped nanocrystals that are non-cadmium in nature, e.g. Mn-doped or Cu-doped ZnS are used in different cell labelling and tracking. Ag2S QDs novel near-infrared (NIR) fluorescent imaging probes had shown great potential in bio-imaging and sensing. By this, translocation and dynamic dissemination of transplanted human mesenchymal stem cells in lung and liver of mice had been monitored.

Magnetic NPs

Iron oxide NPs have revealed tremendous ability in labelling, tracking stem cell due to its magnetic and fluorescent properties. Since Iron oxide NPs are superparamagnetic, it enhances the potential to trace allocation and differentiation of progenitor and stem cells with the help of high-resolution magnetic resonance, and when encoded with Tat peptides, it follows the recovery of progenitor cells (Lewin et al. 2000). One of the reports states that superparamagnetic NPs (SPION) are used within concentration for cell transplantation to repair the carotid artery. Specific stem cells-labelled iron oxide coated with NIR fluorescent dye and dextran are useful in treating Alzheimer's disease.

Upconversion NPs

Luminescent upconversion NPs (UCNPs) have gained importance as excellent optical nanoprobes. It is multifunctional as it shows its importance *in vivo* capturing cells, and in tiny animals it monitors the growth and traces the movement of implanted cells *in vivo*. Luminescent UCNPs conjugated with peptides are reported as differentiation and long-standing tracking of hMSCs (Chen et al. 2014), and when coated with silica, it acts as a moderator for NIR-initiated discharge of KGN, stimulating the chondrogenic differentiation efficiently.

Photothermal Therapy of Cancer using Gold Nanorods-Labelled Stem Cells

In the combination of Au-NRs with MSCs, advanced tumour-targeting efficacy in photothermal cancer treatment is due to heat-producing nature near infra-red region. It kills cancer cells and shows a trivial effect on a normal cell. The pH-sensitive synthesised Au-NRs are used in targeted tumour treatment; these particles accumulated at acidic endosomes by MSCs and increase the tumour-targeting efficiency. Similarly, Au-NR rods coated with mesoporous silica and fused with G-protein transmembrane receptor improve the loading efficiency to target the tumour cell (Kang et al. 2015).

Regenerative Treatment and Drug Delivery using Nanomaterials

This section explains the utilisation of nanomaterials or nanocomposites in the regenerative treatment of organs such as bone, the heart, the liver along with targeted drug delivery.

Nanomaterial-Assisted Stem Cells for Bone Regeneration

Photon-absorbing carbon nitride (C_3N_4) nanosheets, biocompatible scaffolds treated on human bone marrow-derived MSCs, enhance differentiation for bone formation and help in bone regeneration and fracture curing. These activities get accelerated in the presence of red light. Further, this recuperates mechanical properties, controls drug release and biodegradability along with the bone repair after surgery, and results in bone density with less fibrous tissue development. The bone regeneration can be increased from combined human bone morphogenetic protein-2 along with hybrid hydroxyapatite-chitosan nanoparticles due to its osteoconductivity. Hydroxyapatite NPs3D scaffolds can be prepared by combining with gelatin that starts osteogenesis via autophagy activation, which leads to strong bone formation. The delivery of augmentation factors for bone repair is done using hMSCs, chitosan NPs, and fibrous scaffolds. For the bone formation of NRs and drug discharge of dexamethasone to support the osteogenic growth, nanocomposites produced from nanodiamonds and gelatin methacrylamide (GelMA) hydrogels can be utilised. The biomechanical strength of bone fractures increases with siRNA/NP hydrogel, and the healing capacity enhances through magnetofection (Pacelli et al. 2017).

Nano/Biomaterial-Treated Stem Cells in Regeneration of Heart and Liver

Stem cell therapy has shown potential results in treating ischaemic cardiomyopathy and liver failure. A study reports that a clinical trial for a myocardial infarction (MI) patient has been carried with hMSCs' derived cardiac progenitor cells (Higuchi et al. 2017). The outcome of stem cells in the myocardium can be governed by biomaterials and supervised by NP-labelled stem cells due to the enhancement of cell retention. Liposomes-based NPs are used for skeletal myoblast cardiac repair through transporting therapeutic gene in stem cells to promote angiomyogenesis. Amidoamine NPs deliver siRNA and silencing prolyl hydroxylase domain protein 2 in MSC transplantation for infracted myocardium. This helps in angiogenesis elevation in the diseased myocardium and escalates heart function. Functionalised NPs like polyethyleneimine-modified mesoporous silica act as an excellent vehicle to drive advancements in the differentiation of mouse-ESCs into hepatocytes for liver regeneration. Another study states the role of magnetic NPs in augmentation magnetic resonance images in transplanted adipose-originated MSCs in the damaged liver during the repair (Yi et al. 2010). Chitosan NPs (CNPs) coated with hepatocyte growth factor (HGF) magnify growth of MSCs into hepatocytes and improve hepatic cirrhosis in mice. Another NP silver sulphide QDs was utilised in tracing hMSCs transplanted in mice due to their near-infrared emitting nature.

NPs-Labelled Stem Cells for Drug Delivery

Drug delivery is a challenging parameter in cancer treatment due to its inherent therapeutic resistance; therefore, to overcome this, strategies had been designed with AuNPs with facile bioconjugation of chemotherapeutic drugs. This method has improved the drug delivery and efficacy of anti-cancer drug such as doxorubicin. Specific polymeric NPs like N-isopropyl acrylamide nanogels attached with fluorescent tags internally track NSCs, advance the solubility discharge performance, and when loaded with retinoic acid, they have even raised the number of neurons. A study proves that when anti-cancer drug paclitaxel was loaded with poly-D, L-lactic-co-glycolic acid(PLGA) NPs and incorporated into MSCs, it increased the drug load without altering its viability and was successful in the destruction of A549 lung cancer and MA148 ovarian cancer cell lines (Meir et al. 2014).

Conclusion and Future Perspectives

The use of stem cells for the treatment of various clinical conditions has been mentioned; however, substantial clinical studies are required to confirm its exact role. MSCs have responded positively in many animal and clinical studies. The use of umbilical cord and amniotic fluid cells has received a lot of consideration as it can be used as an alternative effectively. Presently, several animal and human trials are ongoing to analyse the chances of applying stem cell therapy for regeneration and their promising results assist in understanding the regeneration potential of the body itself. However, the molecular mechanism of stem cell differentiation and its biological function should be researched thoroughly.

Dissimilar nonmaterials can play a role in the differentiation and activity of different stem cells. Biocompatible graphene has shown its importance in regenerative medicine by revival of the central nervous system using neural stem cell (NSC) engineering and directing stem cell differentiation. This also accelerated osteoblasts and neurons differentiation. AuNRs have shown their effect as photothermal as therapeutical agents for stem cells. Silica-coated gold nanoparticles help in tracking and monitoring labelled MSCs for a more extensive duration and also enhance the target efficiency in photothermal cancer treatment. Colloidal silver sulphide QDs and iron oxide nanoparticles have shown their importance in labelling and *in vivo* cell tracking of stem cells. Iron oxide nanoparticles have improved the efficacy of myocardial infarction by recuperating the therapeutic effect of MSCs.

In the near future, most of the terminal illness can be treated using stem cell therapy and can bring enormous benefits in the patient, and there is high optimism for the use of BMSCs, TSPSCs, and iPSCs. Overall, for development of this field, there must be advancement at each level including clinical trials that requires a lot of funding; therefore, public and private organisations should come forward to support these studies. The regulatory bodies should conduct the critical evaluation in each step for the comprehending success and efficacy in time frame along with any misuse of studies.

Although there is advancement in nanomaterials, their application in differentiation, tracking, and treatment of cell and in imaging procedures in combination with one or two in a multimodal setting (e.g. MRI, PET, and ultrasound) has been achieved, still, clinical applications have to be established. There is a lot of understanding required for inorganic NPs to eliminate the body because of inhomogeneous nanohybrids' nature within the body.

ABBREVIATIONS

UCSCs Umbilical Cord Stem Cells
BMSCs Bone Marrow Stem Cells
iPSCs Induced Pluripotent Stem Cells
MSCs Mesenchymal Stem Cells
ESCs Embryonic Stem Cells
TSPSCs Tissue Specific Progenitor Stem Cells
HSCs Haematopoietic Stem Cells
OCT4 Octamer-Binding Transcription Factor-4
EGC Embryonic Germ Cells
hESCs Human Embryonic Stem Cells
NSCs Neural Stem Cells
PSCs Pancreatic Stem Cells
CSCs Cancer Stem Cells
AD Alzheimer's Disease
NGF Nerve Growth Factor
BMDMNCs Bone Marrow Derived Mononuclear Cells
PPCs Pancreatic Progenitor Cells
DPSCs Dental Pulp Stem Cells
IESCs Inner Ear Stem Cells
IPCs Intestinal Progenitor Cells
LPSCs Limbal Progenitor Stem Cells
AdSCs Adipose Derived Stem Cells
hMSCs Human Mesenchymal Stem Cells
DPCs Dermal Papillary Cells
AFSCs Amniotic Fluid Stem Cells
PEG Polyethylene glycol

VEGF	Vascular Endothelial Growth Factor
WJ	Wharton's Jelly
WJ-SCs	Wharton's Jelly Stem Cells
ALD	Adrenoleukodystrophy
MLD	Metachromatic Leukodystrophy
TSD	Tay-Sachs Disease
BMCs	Bone Marrow Cells
ARMD	Age-Related Macular Degeneration
iGABA-INs	GABAergic-INs interneurons
AIAD	α-1 Antitrypsin Deficiency
QDs	Quantum Dots
PET	Polyethylene Terephthalate
PDMS	Polydimethylsiloxane
NGO	Nanosized Graphene Oxide
Au-NRs	Gold nanorods
PSS	Poly-Sodium 4-Styrene Sulphonate
PAH	Poly-Allylamine Hydrochloride
ALP	Alkaline Phosphatase
MAPK	Mitogen-Activated Protein Kinase Pathway
GQDs	Graphene Quantum Dots
SiNW	Silicon Nanowires
BMP	Bone Morphogenetic Protein
MCNPs	Magnetic Core-Shell NPs
MF	Magnetic Field
GFP	Green Fluorescent Protein
PVA	Polyvinyl Alcohol
NPs	Nanoparticles
AuNRs/AuNPs	Gold Nanorods/Gold Nanoparticles
GNS-PEG	PEGylated Gold Nanostars
SPION	Superparamagnetic NPs
UCNPs	Upconversion NPs
Au-NRs	Au Nanorods
NIR	Near-Infrared Window
KGN	Kartogenin
ASCs	Adipose Tissue-Derived Stem Cells
GelMA	Gelatin Methacrylamide
PU-PEI	Polyethyleneimine
MCNPs	Magnetic Core-Shell NPs

REFERENCES

Augello A., Kurth T.B. and De Bari C., 2010. *Eur Cell Mater 20*(121): e33.doi:10.22203/ecm.v020a11.

Baizabal J.M., Furlan-Magaril M., Santa-Olalla J. and Covarrubias L., 2003. *Archives of Medical Research 34*(6): 572–588. doi:10.1016/j.arcmed.2003.09.002.

Barker J.N. and Wagner J.E., 2003. *Critical Reviews in Oncology/Hematology 48*(1): 35–43. doi:10.1016/s1040-8428(03)00092.

Benavides O.M., Brooks A.R., Cho S.K., Petsche C.J., et al. 2015. *Journal of Biomedical Materials Research Part A 103*(8): 2645–2653. doi:10.1002/jbm.a.35402.

Bongso A., 2006. In *Human Embryonic Stem Cell Protocols*, 1–34. Humana Press. doi:10.1385/1-59745-046-4:13.

Bruin J.E., Saber N., Braun N., Fox J.K., et al. 2015. *Stem Cell Reports 4*(4): 605–620. doi:10.1016/j.stemcr.2015.02.011.

Charrier C., Coronas V., Fombonne J., Roger M., et al. 2006. *Neuroscience 138*(1): 5–16. doi:10.1016/j.neuroscience.2005.10.046.

Chen G., Qiu H., Prasad P.N. and Chen X., 2014. *Chemical Reviews 114*(10): 5161–5214. doi:10.1021/cr400425h.

Cheng A., Kapacee Z., Peng J., Lu S., et al. 2014. *Stem Cells Translational Medicine 3*(11): 1287–1294. doi:10.5966/sctm.2014-0101.

Chimutengwende-Gordon M. and Khan S.W., 2012. *Current Stem Cell Research & Therapy 7*(2): 122–126. doi:10.2174/157488812799219036.

Comizzoli P. and Holt W.V., 2014. *Reproductive Sciences in Animal Conservation*, 331–356. doi:10.1007/978-1-4939-0820-2_14.

Cronk S.M., Kelly-Goss M.R., Ray H.C., Mendel T.A., et al. 2015. *Stem Cells Translational Medicine 4*(5): 459–467. doi:10.5966/sctm.2014-0108.

Csaki C., Matis U., Mobasheri A., Ye H., et al. 2007. *Histochemistry and Cell Biology 128*(6): 507–520. doi:10.1007/s00418-007-0337-z.

Daar A.S. and Greenwood H.L., 2007. *Journal of Tissue Engineering and Regenerative Medicine 1*(3): 179–184. doi:10.1002/term.20.

Fan Y.P., Hsia C.C., Tseng K.W., Liao C.K., et al. 2016. *Stem Cells Translational Medicine 5*(2): 235–247.

Flores I., Cayuela M.L. and Blasco M.A., 2005. *Science 309*(5738): 1253–1256. doi:10.1126/science.1115025.

Fuoco C., Rizzi R., Biondo A., Longa E., et al. 2015. *EMBO Molecular Medicine 7*(4): 411–422. doi:10.15252/emmm.201404062.

Galli R., Borello U., Gritti A., Minasi M.G., et al. 2000. *Nature Neuroscience3*(10): 986–991. doi:10.1038/79924.

Herrera-Carrillo E. and Berkhout B., 2015. *Viruses 7*(7): 3910–3936. doi:10.3390/v7072804.

Higuchi A., Kumar S.S., Ling Q.D., Alarfaj A.A., et al. 2017. *Progress in Polymer Science 65*: 83–126. doi:10.1016/j.progpolymsci.2016.09.002.

Hu J., Yu X., Wang Z., Wang F., et al. 2013. *Endocrine Journal 60*(3): 347-357. doi:10.1507/endocrj.ej12-0343.

Jiang F.X. and Morahan G., 2014. *Stem Cells and Development 23*(23): 2803–2812. doi:10.1089/scd.2014.0214.

Kaigler D., Pagni G., Park C.H., Braun T.M., et al. 2013. *Cell Transplantation 22*(5): 767–777. doi:10.3727/096368912x652968.

Kang S., Bhang S.H., Hwang S., Yoon J.K., et al. 2015. *ACS Nano 9*(10): 9678–9690. doi:10.1021/acsnano.5b02207.

Kawasaki H., Mizuseki K., Nishikawa S., Kaneko S., et al. 2000. *Neuron 28*(1): 31–40. doi:10.1016/s0896-6273(00)00083-0.

Ksander B.R., Kolovou P.E., Wilson B.J., Saab K.R., et al. 2014. *Nature 511*(7509): 353–357. doi:10.1038/nature13426.

Lagasse E., Connors H., Al-Dhalimy M., Reitsma M., et al. 2000. *Nature Medicine 6*(11): 1229–1234. doi:10.1038/81326.

Lewin M., Carlesso N., Tung C.H., Tang X.W., et al. 2000. *Nature Biotechnology 18*(4): 410–414. doi:10.1038/74464.

Liang S., Li C., Zhang C., Chen Y., et al. 2015. *Cells Theranostics 5*(9): 970–984. doi:10.7150%2Fthno.11632.

Lin B.J., Wang J., Miao Y., Liu Y.Q., et al. 2016. *Journal of Materials Chemistry B 4*(3): 489–504. doi:10.1039/c5tb02265g.

Lin B.L., Zhang J.Z., Lu L.J., Mao J.J., et al. 2017. *Nanomaterials 7*(5): 107. doi:10.3390/nano7050107.

Lindvall O., Kokaia Z. and Martinez-Serrano A., 2004. *Nature Medicine 10*(7): S42–S50. doi:10.1038/nm1064.

Liu Y.H., Karra R. and Wu S.M., 2008. *Drug Discovery Today: Therapeutic Strategies 5*(4): 201–207. doi:10.1016/j.ddstr.2008.12.003.

Ma M., Sha J., Zhou Z., Zhou Q., et al. 2008. *Journal of Nanjing Medical University 22*(3): 135–142. doi:10.1016/s1007-4376(08)60052-0.

Mahla R.S., 2016. *International Journal of Cell Biology 2016*: 1–24. doi:10.1155/2016/6940283.

Mayo F., Rochat A., Nicolas M., Jaoudé G.A., et al. 2008. *Nature 456*: 250–254. doi:10.1038/nature07406.

Meir R., Motiei M. and Popovtzer R., 2014. *Nanomedicine 9*(13): 2059–2069. doi:10.2217/nnm.14.129.

Miki T., Lehmann T., Cai H., Stolz D.B., et al. 2005. *Stem Cells 23*(10): 1549–1559. doi:10.1634/stemcells.2004-0357.

Mizutari K., Fujioka M., Hosoya M., Bramhall N., et al. 2013. *Neuron 77*(1): 58–69. doi:10.1016/j.-neuron.2012.10.032.

Narayanan K., Yen S.K., Dou Q., Padmanabhan P., et al. 2013. *Scientific Reports 3*(1): 1–6. doi:10.1038/srep02184.

Naskar D., Ghosh A.K. and Mandal M., 2017. *Biomaterials 136*: 67–85. doi:10.1016/j.biomaterials.2017.05.014.

Nayak T.R., Andersen H., Makam V.S., Khaw C., et al. 2011. *ACS Nano 5*(6): 4670–4678. doi:10.1021/nn200500h.

Newman K.D. and McBurney M.W., 2004. *Biomaterials 25*(26): 5763–5771. doi:10.1016/j.biomaterials.2004.01.027.

Oshima M. and Tsuji T., 2015. *Engineering Mineralized and Load Bearing Tissues*, 255–269. doi:10.1007/978-3-319-22345-2_14.

Pacelli S., Maloney R., Chakravarti A.R., Whitlow J., et al. 2017. *Scientific Reports 7*(1): 1–15. doi:10.1038/s41598-017-06028-y.

Park S.Y., Park J., Sim S.H., Sung M.G., et al. 2011. *Advanced Materials 23*(36): H263–H267. doi:10.1002/adma.201101503.

Paşca A.M., Sloan S.A., Clarke L.E., Tian Y., et al. 2015. *Nature Methods 12*(7): 671–678. doi:10.1038/nmeth.3415.

Rajabzadeh N., Fathi E. and Farahzadi R., 2019. *Stem Cell Investigation 6*: 19. doi:10.21037/sci.2019.06.04.

Ramakrishna V., Janardhan P.B. and Sudarsanareddy L., 2011. *Annual Research & Review in Biology 1*(4): 79–110.

Rossant J., 2001. *Stem Cells 19*(6): 477–482. doi:10.1634/stemcells.19-6-477.

Shaffiey S.A., Jia H., Keane T., Costello C., et al. 2016. *Regenerative Medicine 11*(1): 45–61. doi:10.2217/rme.15.70.

Shah B., Yin P.T. and Ghoshal S., 2013. *Angewandte Chemie International Edition 52*(24): 6190–6195. doi:10.1002/anie.201302245.

Shah S., Liu J.J., Pasquale N., Lai J., et al. 2015. *Nanoscale 7*(40): 16571–16577. doi:10.1039/C5NR03411F.

Shamblott M.J., Axelman J., Wang S., et al. 1998. *Proceedings of the National Academy of Sciences 95*: 13726–13731. doi:10.1073/pnas.95.23.13726.

Sharma A.K., Bury M.I., Marks A.J., Fuller N.J., et al. 2011. *Stem Cells 29*(2): 241–250. doi:10.1002/stem.568.

Shiba Y., Fernandes S., Zhu W.Z., Filice D., et al. 2012. *Nature 489*(7415): 322–325. doi:10.1038/nature11317.

Soltysova A., Altanerova V.and Altaner C., 2005. Cancer stem cells. *Neoplasma 52*(6): 435–440.

Squillaro T., Peluso G. and Galderisi U., 2016. *Cell Transplantation 25*(5): 829–848. doi:10.3727/096368915x689622.

Takahashi K., Tanabe K., Ohnuki M., Narita M., et al. 2007. *Cell 131*(5): 861–872. doi:10.1016/j.cell.2007.11.019.

Takahashi K. and Yamanaka S., 2006. *Cell 126*(4): 663–676. doi:10.1016/j.cell.2006.07.024.

Weinstein-Oppenheimer C.R., Brown D.I., Coloma R., Morales P., et al. 2017. *Materials Science & Engineering C-Materials for Biological Applications 79*: 821–830. doi:10.1016/j.msec.2017.05.116.

Weissman I.L., 2000. *Cell 100*(1): 157–168. doi:10.1016/S0092-8674(00)81692-X.

Wilson A.A., Ying L., Liesa M., Segeritz C.P., et al. 2015. *Stem Cell Reports 4*(5): 873–885. doi:10.1016/j.stemcr.2015.02.021.

Wolfram J. and Ferrari M., 2019. *Nano Today 25*: 85–98. doi:10.1016/j.nantod.2019.02.005.

Yang J., Cai B., Glencer P., Li Z., et al. 2016a. *Stem Cells International 2016*: 1–6. doi:10.1155/2016/2850873.

Yang K., Lee J. and Lee J.S., 2016b. *ACS Appl. Mater. Interfaces 8*: 17763–17774. doi:10.1021/acsami.6b01804.

Ye L., Robertson M.A., Mastracci T.L. and Anderson R.M., 2016. *Developmental Biology 409*(2): 354–369. doi:10.1016/j.ydbio.2015.12.003.

Yi C., Liu D., Fong C.C., Zhang J. and Yang M., 2010. *ACS Nano 4*(11): 6439–6448. doi:10.1021/nn101373r.

Zhang M., Bai L., Shang W., Xie W., et al. 2012. *Journal of Materials Chemistry 22*(15): 7461–7467. doi:10.1039/C2JM16835A.

Zhao C., Tan A., Pastorin G. and Ho H.K., 2013. *Biotechnology Advances 31*(5): 654–668. doi:10.1016/j.biotechadv.2012.08.001.

Zhao X., Huang Q. and Jin Y., 2015. *Materials Science and Engineering C 54*: 142–149. doi:10.1016/j.msec.2015.05.013.

Zhou S., Flamier A., Abdouh M., Tétreault N., et al. 2015. *Development 142*(19): 3294–3306. doi:10.1242/dev.125385.

Zvaifler N.J., Marinova-Mutafchieva L., Adams G., Edwards C.J., et al. 2000. *Arthritis Research & Therapy 2*(6): 477. doi:10.1186/ar130.

5

The Emerging Role of Exosome Nanoparticles in Regenerative Medicine

Zahra Sadat Hashemi
Motamed Cancer Institute, ACECR, Tehran, Iran

Mahlegha Ghavami
Cancer Research Institute, Biomedical Research Center, Bratislava, Slovakia

Saeed Khalili
Shahid Rajaee Teacher Training University, Tehran, Iran

Seyed Morteza Naghib
Iran University of Science and Technology, Tehran, Iran

CONTENTS

Introduction

Exosomes are nanosized EVs (50–150 nm) originating from multivesicular bodies (MVBs). Various cells could release them into the extracellular environment through membrane fusion. These lipid bilayer nanovesicles are loaded with different cargos such as miRNA, DNA, RNA, lipids, and proteins. Exosomes are involved in different biological pathways such as intercellular communications, signal transferring, antigen presentation, and tumour progression. Their uptake occurs through endocytosis, direct fusion, or receptor–ligand interaction. Exosomes could be isolated and characterised by various methods such as

DOI: 10.1201/9781003153504-5

Nanoparticle Tracking Analysis, Dynamic Light Scattering, Electron Microscopy, and Tunable Resistive Pulse Sensing (according to their size, density, surface charge, distinctive biomarkers, and membrane antigens).

Contemporarily, the therapeutic potential of exosomes as delivery systems has garnered a lot of attention in regenerative medicine. Exosomes present unique characteristics such as small size, higher biocompatibility, biodegradability, immune tolerability, better safety profile, higher penetration (especially across the blood–brain barrier), and stability while increasing specific targeting. Exosomes as next generation therapeutics exert their regenerative features by trafficking towards injured tissues, promotion of angiogenesis, and reduction of inflammation. Moreover, the ability of different cells to release them into the extracellular environment makes them potential and accessible diagnostic tools for early detection of diseases. Therapeutic exosomes that are derived from stem cells (SCs) have been demonstrated to be more effective than cell-based and tissue engineering therapies. This property of SC-derived exosomes should be rooted in the self-renewal and multi-potency of SCs. Exosomes circumvent the limitation related to direct cell therapy such as tumour formation, immune rejection, and cell clumping.

In this chapter, the applications of the exosomes as natural drug carriers will be discussed in regenerative medicine (free or in combination with other biomaterials by housing and delivering their tissue repairing cargoes). Moreover, their therapeutic contributions in degenerative and inflammatory diseases will be delineated. Engineering strategies to increase the regenerative ability and efficiency of exosome-based therapeutics are also summarised. Additionally, the most recent *in vivo* and clinical studies of exosome-mediated regeneration therapy in the bone, kidneys, heart, wounds, brain, cancer, and other organs/tissues, as well as the associated limitations, are highlighted. The exosome-based clinical trials regarding the recent COVID-19 outbreak are also described.

What are Extracellular Vesicles?

Different cells communicate with each other in their microenvironment via different ways by soluble factors and also EVs as a paracrine signalling. EVs are exocytic lipid bilayers that are heterogeneous populations of vesicles naturally released from cells and are detectable in all body fluids (e.g. serum, blood, saliva, lymph, breast milk, bile, faeces, seminal plasma, urine, and amniotic, bronchoalveolar, uterine, synovial, uterine, nasal, cerebrospinal fluids) (Yáñez-Mó et al. 2015). The highest number of EVs is known to exist in serum. It's estimated that approximately 3 million EVs is present in each microlitre of the serum (Masyuk et al. 2013). EVs range in diameter from 30 nm to 10 micrometres. This property resulted in different EV subtypes based on their function, size, internal, and external content including microvesicles, exosomes, and apoptotic bodies (Figure 5.1). Exosomes are mostly secreted by mesenchymal stem cells (Yeo et al. 2013), macrophages, dendritic cells, B and T cells, endothelial and epithelial cells, and a variety of cancer cells (Batrakova and Kim 2015). In the mid-20th century, the existence of EVs was proved following a combined application of ultracentrifugation and electron microscopy. They were primarily assumed to be members of the cellular waste disposal system (Jella et al. 2018). But, in fact, in the past decade, the importance of EVs became apparent when they were introduced as potent vehicles to transport biological materials such as proteins (such as cytokines, receptors, or their ligands), nucleic acids (mRNA, miRNA, and DNA), lipids, and metabolites from the parent cells. As non-coding RNA molecules, microRNAs (miRNAs/miRs) are involved inRNA silencing and post-transcriptional regulation of gene expression (Hashemi et al. 2018a; Hashemi et al. 2019; Hashemi et al. 2018b; Rezaei et al. 2020; Choghaei et al. 2016).

The Biogenesis Pathway of Exosomes and Their Distinction with other Cell-Derived Vesicles

Biogenesis of exosomes is related to the endosome/lysosome pathway. In an endosome, budding from the perimeter membrane into the endosome lumen forms the membrane-bound intraluminal vesicles (ILVs). This late endosome is a multi-vesicular compartment also called multi-vesicular bodies (MVBs). Finally

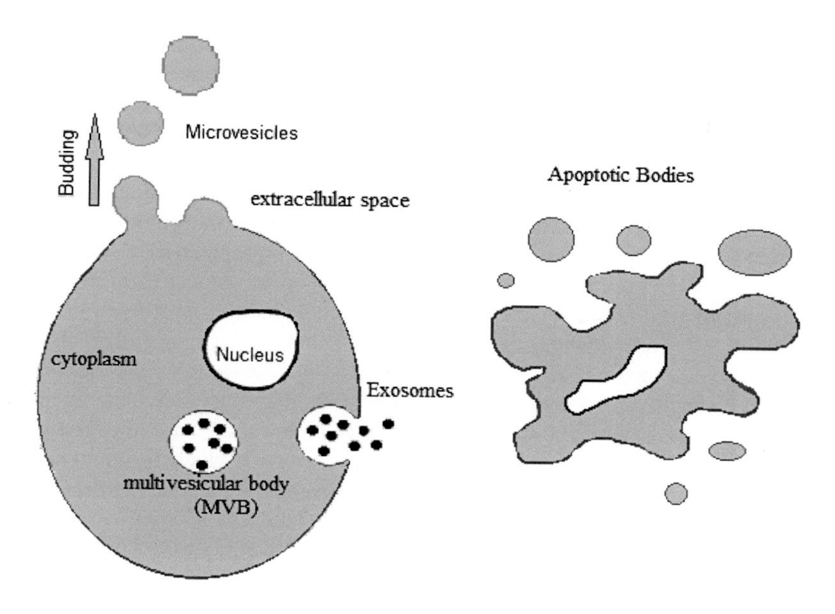

FIGURE 5.1 Three main classes for biogenesis of extracellular vesicles. The main biogenesis schemes are illustrated.

the MVB will be fused with the plasma membrane or the lysosomes. In the case of fusion with the plasma membrane, these ILVs will be released as exosomes (Johnsen et al. 2014; Raposo and Stoorvogel 2013; Yáñez-Mó et al. 2015). Consequently, the exosomes have endosome-derived membrane. This fusion is executed via a series of Rab GTPase proteins including RAB11, RAB27A, RAB27B, and RAB35 (Ostrowski et al. 2010). The micro-vesicles could directly bud from the plasma membrane and their size is larger than EVs (100–500 nm), so they have plasma-derived membrane. The third kind of EVs are apoptotic bodies that are 500–1000 nm in size and are released during the late phase of apoptosis (György et al. 2011).

It should be noted that the biogenesis pathway of exosomes and their distinction fromother cell-derived vesicles were identified due to the existence of lysosomal surface protein (LAMP), tetraspanins (CD81, CD9, and CD63), heat shock proteins (Hsc70), and also some fusion proteins such as Annexin, CD9, and flotillin in the exosomal membrane (Caby et al. 2005; Andaloussi et al. 2013; Conde-Vancells et al. 2008; Mohammadpour and Majidzadeh-A 2020). The transport (ESCRT) process requires the endosomal sorting complex, which is a collection of proteins necessary for formation and sorting of cargo into exosomes.

It has been revealed that the exosomes contain functional miRNAs and mRNAs (Valadi et al. 2007). Other studies show more information about the functional role of exosomes due to their difference in cargo content. For example, the transfer of exosomes from glioblastoma as parent cells showed malignancy enhancement in other cell populations and creation of a metastatic niche. Therefore, these exosomes can be employed as diagnostic biomarkers for promotion of tumour growth or diseases (Skog et al. 2008; Hashemi et al. 2017). These intercellular communications are both local and distant, which could promote the growth and metastasis of tumour progression and neoplastic lesions (Zhang and Grizzle 2014). Moreover, exosomes are known for their part in immune-related functions such as presentation of B lymphocyte antigens (Raposo et al. 1996). The signalling from the parent to the recipient cells is mediated by exosomes through direct fusion, endocytosis, or receptor–ligand interaction (via cell surface or endosomal receptors).

Methods of Isolation and Characterisation

Due to very small size of the EVs, the easiest method to observe and detect them was to use microscopes with high resolution power. It should be pointed out that apoptotic bodies can be identified by the detection of histones and DNA. Electron microscopy (EM) measures the scattered electron beam to acquire the morphology and size of the particles. There solving power of this electron beam is higher than other microscopes owing to its shorter wavelength (up to 100,000 times) compared to visible light photons (Erni et al. 2009). Some EM methods are widely used in biological studies such as scanning electron microscopy (SEM) (for imaging surfaces at the atomic level) and transmission electron microscopy (TEM) in which an image is formed by transmission of an electron beam through a specimen. X-ray microscopy circumvents the diffractive limit of classic light microscopes by reducing its wavelength to a few nanometres. This method could be harnessed to detect the size and morphology of EVs.

A pre-processing step is often necessary for biological sample preparation, to be visible under EM (Mehdizadeh et al. 2014). These sample preparations consisted of some treatments like dehydration (organic solvents are replaced by water), chemical fixation (chemical crosslinking of lipids with osmium tetroxide and proteins with formaldehyde or glutaraldehyde), and cryofixation (called cryogenic electron microscopy (cryo-EM)). Sometimes electron beam may damage the sample to overcome this problem. The cryo-EM could be applied for EVs analysis. It is the method used to study heterogeneous shapes of exosomes close to their native state. The specimen rapidly (in milliseconds) becomes cooled and frozen to cryogenic temperatures (usually at the temperature of liquid nitrogen), which are enough to form icecrystals. The water molecules are not regularly arranged in a hexagonal lattice and a vitreous ice is formed, which is an amorphous solid form of water (Tivol et al. 2008; Debenedetti 2003).

One of the popular super resolution fluorescence microscopy techniques is known as the stimulated emission depletion (STED) microscopy. It is developed to avoid the diffraction limit of light microscopy and create images by selectively deactivating fluorophores using stimulated emission. The distribution of fluorescently labelled antigens present at the surface of EVs and the size of EVs could be measured by STED (Tønnesen et al. 2011). Some studies have reported STED resolution limit as 50 nm (even 10 nm) (Hein et al. 2008). The STED microscopy images managed to show the size of synaptic vesicles. The synaptic vesicle contains neurotransmitters and is approximately 40 nm in diameter (Willig et al. 2006).

A 1000 times higher resolution than the optical diffraction limit could be achieved by Atomic Force Microscopy (AFM), which is a type of Scanning Probe Microscopy (SPM). This method measures the interaction forces between the surface and the probing tip. Its resolution is shown to be in the order of fractions of nm. Three dimensional (3D) topography and diameter of EVs could be acquired via SPM. The superiority of this method is its capability to measure samples in their native conditions with minimised pre-processing for sample preparation (Yuana et al. 2010).

Other than EM, which meanly is used for direct observation of EVs morphology, different methods are used for their characterisation. As explained, exosomes can contain markers on their surface as their parent cells. The immuno-affinity-based approaches such as Enzyme-Linked Immunosorbent Assays (ELISA), Flow Cytometry, and antibody-coated magnetic beads rely on surface markers of EVs for their immuno-isolation and characterisation. Amongst these, the ELISA method has most commonly been applied for detection of EVs. The plates of ELISA kit could be coated by pan-exosome antibodies (anti-CD63 and anti-caveolin-1 melanoma-derived exosomes) to capture the exosomes existing in the samples (Logozzi et al. 2009; Khodashenas et al. 2019). As a high throughput approach of label-free protein analysis, the electrochemical sensing platforms and the plasmonic were combined with immuno-affinity methods (Contreras-Naranjo et al. 2017). However, EVs can be collected by their protein contents and applying conventional detection method, such as total protein analysis and western blots (Bradford assays, bicinchoninic acid (BCA)) (Khodashenas et al. 2019). Aside from the protein content, the nucleic acid contents (DNA, RNA, or miRNA) can be searched by PCR (polymerase chain reaction) or DNA sequencing methods (Eldh et al. 2012; Li et al. 2014; Zarovni et al. 2015).

The EVs (especially the exosomes) are most commonly isolated by the ultracentrifugation method. This conventional technique is based on the size of the particle (commonly by >1,00,000g as speed and ~5 hours as time). The exosomes could also be precipitated using special reagents (e.g. polymeric additives). The basis for commercial exosome isolation kits is the ~30-min precipitation by a standard centrifuge (~10,000g). Another commercial exosome isolation method is the membrane filters (e.g. polyvinylidene difluoride (PVDF) or polycarbonate with pore size of 50–450 nm). The chromatography technique could also be employed for exosome isolation. The commercial size-exclusion chromatography columns are provided to induce the isolation of the exosomes within eluted fractions, which contain larger to smaller particles.

The scattered and fluorescent lights can be measured by flow cytometry method. The approximate size of EVs can be detected by the forward scatter (FSC) light (0.5–5°) and the granularity of internal structures can be measured by the side scatter (SSC) light (15–150°) (Szatanek et al. 2017). With the advancement of the flow cytometry technique, pulsed laser activated cell sorter (PLACS) was developed. A single layer of PDMS (polydimethylsiloxane) microfluidic channel is used to create a high-speed liquid flow for the detection of fluorescent samples. Eventually, it leads to high purity (with >90%) and high throughput sorting (with >30,000 particles/sec) of EVs (Chen et al. 2014b; Chen et al. 2014a; Heinzelman et al. 2016).

Microfluidic platforms are the new generation of exosome isolation approaches. The inner surface of these platforms could be functionalised by modifying or employing capture beads through immuno-affinity capture methods. Chen et al. were the first group to report the application of microfluidic platforms for exosome isolation. They have used a channel with herringbone groves to process the samples (conditioned medium from cells in culture or serum blood samples) of 400-mL volumes and biotinylated anti-CD63 as surface modification (Chen et al. 2010). Other surface markers such as CD41 (Ashcroft et al. 2012), EpCAM with rapid inertial solution exchange (RInSE) (Dudani et al. 2015), CD24 and EpCAM with nano-plasmonic exosome (nPLEX) assay (Im et al. 2014), Her2 and CD9 with the method of tunable alternating current electrohydrodynamic (ac-EHD) (Vaidyanathan et al. 2014), CD81 as Nano-IMEX (Zhang et al. 2016b), CD63 and CD9 with gold surface channels (Sina et al. 2016), and CD81 as µMED (Ko et al. 2016) have also been analysed. For feasibility of the isolation, combination of these anti-surface markers with magnetic beads as immuno-magnetic microbeads were used in several studies such as iMER fluidic chip (EGFR) (Shao et al. 2015), ExoSearch (CD9) (Zhao et al. 2016), and CD63 (Fang et al. 2017). However, label-free passive microfluidic approaches are approved for EVs isolation. In membrane-based filtration as the other microfluidic platform, a double filtration method (employing a 200-nm pore size membrane followed by a 30-nm pore size membrane) allows proteins to pass through and enrich the small EVs (Liang et al. 2017). Wang et al. have performed EVs trapping (<120 nm) on an array of ciliated (nanowires) micropillars (Wang et al. 2013). Lee et al. have showed the isolation of vesicles smaller than 200 nm by continuous contact-free acoustic nanofilter (Lee et al. 2015). Deterministic Lateral Displacement (DLD) applying nanopillar arrays for exosome sorting (20 nm) was first reported by Wunsch et al. (Wunsch et al. 2016). Meanwhile, using poly-(oxyethylene) (PEO) as a biocompatible polymer to control the viscoelastic forces resulted in 80% recovery and 90% purity of exosomes in field-free microfluidic sorting (Liu et al. 2017).

In dynamic light scattering (DLS) or photon correlation spectroscopy, a monochromatic and coherent laser beam passes through a suspension of EVs. The basis of this technique is that the laser beam hits a particle and the light is dispersed and scattered in all directions. Thereafter, the intensity of the scattered light is recorded as a function of time in Brownian motion. The measurement of particles ranging from 1 nm to 6 µm is the biggest advantage of the DLS (Hoo et al. 2008). Analysing the fluorescence intensity fluctuations is an alternative technique to DLS. In this method, known as Fluorescence Correlation Spectroscopy (FCS), the fluorescently labelled EVs (under 50 nm) are tractable (Starchev et al. 1999). Distinctive biomarkers or membrane antigens could also be used for EVs isolation. In this novel approach, the CD63-specific exosomal marker (in EVs derived from cultured HEK293T) was labelled by anti-CD63-FITC antibody. An immuno-fluorescently labelled EVs population was highly concentrated from unbound fluorescent antibodies and soluble proteins. It showed the relative expression level of a specific marker and heterogeneity of EVs (Wyss et al. 2014). The basis of the ExoScreen technique is the interaction of an exosomal marker with its antibody for the detection of tumour-derived exosomes in colorectal cancer patients' serum. The anti-CD63 biotinylated antibody could be captured by streptavidin-coated beads (Yoshioka et al. 2014). The most important limitation of DLS could be the intensity-weighted skewing (Anderson et al. 2013). This could be eliminated by another approach that uses Tunable Resistive Pulse Sensing (TRPS) for characterisation of EVs. The EVs crossing through a size tunable nanopore results in a temporal change in an ionic current flow, which in turn generates a pulse commensurate to the physical properties of EVs (such as concentration, density, size, ζ potential, or surface charge) (Vogel et al. 2011).

Raman spectroscopy is a spectroscopy light scattering technique and identifies the vibrational states of sample particles. The laser light interacts with particles and results in a shift in the energy of the laser photons. Finally, it gives molecular fingerprints and bond structures of the sample. According to this approach, the chemical composition of EV will be achieved (for example human urinary exosomes were reported (Tatischeff et al. 2012)).

The Brownian motion will also be used in the nanoparticle tracking analysis (NTA) method study of the EVs. The modal value, the average size, and the size distribution (approximately 10–1000 nm) of EVs (as nanoparticles in liquid suspension) will be calculated in this technique. Actually, the light scattered by these nanoparticles is captured using a CCD (charge-coupled device), EMCCD (electron-multiplying CCD), or CMOS (complementary metal–oxide–semiconductor) camera over multiple frames (Filipe et al. 2010; Sokolova et al. 2011). The different light scattering patterns as a function of time indicates the movement of each particle, which is traced and recorded using a camera. Applying the 2D Stokes–Einstein equation, the particle size distribution could be determined (Filipe et al. 2010).

EV Usages in Medicine

EV application in medicine is classified into two groups: (1) as biomarkers and (2) as drug carriers. As described previously, EV subgroups carried different cargos such as nucleic acids, proteins, and lipids. Logically, the EVs-derived normal cells are distinguishable from EVs derived tumour/abnormal cells. This distinction is the basics of EV application as a biomarker in cancer and diseases related to regenerative medicine. Different exosomes with functions in cell signalling pathways has therapeutic potential.

The proteomic analysis of urinary samples from the glomerular podocytes, the proximal tubule, kidney cortical collecting duct cells, the distal convoluted tubule, and the epithelium of the urinary bladder have identified the existence of the EVs. These exosomal transmitters as potential signalling mediators frequently contained aquaporin-2 and CD24 (glycosylated protein). Moreover, the presence of phosphorylated fetuin-A in the calciprotein vesicles of serum is associated with predialysis chronic kidney disease (CKD) (Fang et al. 2013; Dear et al. 2013). Tissue transplantation is undoubtedly one of the most important categories of regenerative medicine. It is observed that differentiation in mRNAs and proteins of the urinary EVs could predict the consequences of renal transplantation. The expression of exosomal mRNAs has already been compared for IL-18 (interleukin-18), cystatin C, NGAL (neutrophil gelatinase-associated lipocalin), KIM-1 (kidney injury molecule-1), and 18S RNA. Following the transplantation, it has been shown that the exosomal NGAL mRNA decreases, the exosomal cystatin C and KIM-1 mRNAs remain unchanged, and the exosomal 18S RNA increases (Peake et al. 2014).

The significance of eight microRNAs of the tumour-derived exosomes (miR-21, miR-141, miR-200a, miR-200c, miR-200b, miR-203, miR-205 and miR-214) as diagnostic biomarkers of ovarian cancer has already been demonstrated (Taylor and Gercel-Taylor 2008). Beheshti et al. have performed proteomic and lipidomic analyses on the prostate-derived exosomes from a panel of prostate cell lines. These analyses have provided novel targets as biomarkers and therapeutic agents for prostate cancer treatment (Hosseini-Beheshti et al. 2012; Nilsson et al. 2009).

It should be noted that a control group is required for all studies. In fact, if the expression of each variable (such as protein, mRNA, miRNA, and so on) was distinctive in the desired fluid (such as blood) of control group, the diagnosis can be detected by comparing the control and patient groups. In this regard, next-generation deep sequencing (NGS) was used to analyse the exosomal miRNA expression profile of peripheral blood fluid. The results of NGS analysis could potentially be used to introduce comparative biomarkers for such diseases as Alzheimer's and neurodegenerative diseases (deregulated exosomal miR-9, miR-20a, and miR-132) (Cheng et al. 2014). In this regard, exosomal EGFRvIII mRNA was specifically obtained in serum EVs of 28% of glioblastoma patients (Skog et al. 2008).

Previously, oncogene amplifications (c-myc), mitochondrial DNA, single-stranded DNA (ssDNA), and retrotransposon RNA transcripts have been discovered in EVs. However, Thakur et al. have indicated the existence of the double-stranded DNA (dsDNA) cargos in the tumour-derived exosomes. These dsDNA cargos can be used as novel biomarkers in cancer detection (Thakur et al. 2014).

Currently, liposomes, polymeric nanoparticles, and exosomes are assembled into various sizes as drug delivery carriers. The exosomes are deemed to be superior over other delivery carriers due to their biocompatibility, the intrinsic ability to target tissues, long circulating half-life, and no toxicity. Meanwhile, the blood–brain barrier (BBB) could be easily crossed by exosomes. This property could be exploited to treat brain cancers or neurological disorders, whereas nearly 98% of central nervous system drugs can't cross (Ha et al. 2016).

Methods for accommodation of small molecules, protein, or nucleic acids into the exosomes are: (1) physical methods such as electroporation (in the range of 150–700 V), sonication, extrusion procedures, freeze–thaw cycles, incubation at RT (with or without the use of saponin permeabilisation), or other temperature. The small-sized molecules can cross the lipid bilayer of exosome by simple incubation. (2) Chemical methods such as transfection by Lipofectamine 2000. This method is frequently used for siRNA (small interference RNA) packaging into the exosomes. (3) Biological methods called transfection of exosome-producing cells. In this method, the parent cells are genetically modified to overexpress a certain gene. The overexpressed gene would ultimately be collected into the parent cell-derived exosomes. For example, various sources of Mesenchymal Stem Cells (MSCs) were transfected with miR-146b and the exosomes containing the miR-146a cargo were collected from culture media (Johnsen et al. 2014).

Small molecules such as curcumin, doxorubicin, and paclitaxel are well known to be loaded in exosomes. The curcumin is a polyphenol molecule with pronounced anti-inflammatory effects. The production of inflammatory cytokines (IL-6 and TNF-α) of macrophages have been reduced byEL-4 (murine tumour cell line) derived exosomes, which carry the curcumin (commonly after only 5 min of incubation of curcumin with the exosomes at 22 °C (Johnsen et al. 2014)) (Sun et al. 2010).

The antitumour doxorubicin drug was loaded into dendritic cell-derived exosomes. These exosomes make efficient delivery to MDA-MB-231 cells (metastatic breast cancer cells) as their target (Tian et al. 2014).

Sometimes the loaded exosomes were labelled by some traceable materials such as fluorescent dye and radiolabelling (called exolabelling methods). These substances depend on whether they are water-soluble (hydrophilic) or fat-soluble (hydrophobic), and they would be trapped in the interior space or in phospholipid bilayer membrane, respectively. In a study, an antioxidant protein catalase was incorporated into exosomes and finally incubated with lipophilic fluorescent dye. This sorting and packaging of catalase cargos in exosomes have confirmed their uptake and indicated their regenerative effects on Parkinson's disease (PD) (Haney et al. 2015).

Exosomes are natural carriers of nucleic acids, and therefore via changing their cargo as targeted vehicles, the genetic modifications could be induced to the target cells. They are mostly used for transporting siRNA and miRNA (called exomiRs). Alvarez-Erviti et al. were the first to report the successful delivery of siRNA into the brain of mice (Alvarez-Erviti et al. 2011). Other studies confirmed the delivery of siRNAs into the other cells using human-derived exosomes (Limoni et al. 2019). Thereafter, antitumour microRNAs against the epidermal growth factor receptor (EGFR) were loaded into the exosomes. This cargo was targeted to breast cancer cells (Ohno et al. 2013). Simply, the miR-150 was loaded into exosomes after 1 hour of incubation at 37 °C (Johnsen et al. 2014).

Regenerative Medicine and Stem Cells

Stem cells are undifferentiated cells that could differentiate into specialised cell types (potency) and proliferate indefinitely with numerous cell growth cycles (self-renewal) (Hashemi et al. 2015; Hashemi et al. 2013; Molaabasi et al. 2020; Ghorbanzade et al. 2020). The use of stem cells is a highly established approach in regenerative medicine (Askari and Naghib 2020; Ghorbanzade and Naghib 2019). For instance, Embryonic Stem Cells (ESCs) can differentiate into more than 200 types of cells which could be used to restore a patient's tissue from severe injuries or chronic diseases (Mahla 2016). The application of regenerative medicine could encompass the cell therapy (using the patient's own cells or non-native donor cells), treatment with growth factors, applications of recombinant proteins, small molecules, and finally tissue engineering and gene therapy. The cell therapy method could be defined as the introduction of new cells into the tissue for disease treatment. These new cells often focus on the stem cells or mature, functional cells with or without genetic modification (gene therapy) for both kinds of cells (Wei et al. 2013).

Mesenchymal stem cells (MSCs) have already been extensively applied for cell therapy in the clinic (Ghorbanzade et al. 2020). They are safe and effective for migration to damaged tissue sites with inflammation. Their therapeutic effects are aligned with production of trophic factors and homing efficiency, which are correlated by matrix metalloproteinases (MMPs), chemokines, and adhesion molecules such as related trafficking molecules. Due to their multi-potency and high potential for differentiation in tissue engineering application, they are contemplated as amenable candidates for MSC-mediated clinical trials (Li et al. 2019). Furthermore, MSCs are capable of releasing several anti-inflammatory factors including IL-10and TGF-β to suppress the immune responses of dendritic cells (DCs), B cells, T cells, macrophages, and natural killer (NK) cells. Therefore, they could play a significant role in the mechanisms of immunomodulation (Putra et al. 2018).

The stem cells could be applied both directly (using cells) or indirectly (using"derived" products from stem cells) in regenerative medicine. The derived products include the secreting growth factors, proteins, cytokines, and also EVs. It should be noted that the direct use of stem cells had a number of limitations. For example, adult stem cells are pre-directed to differentiate a narrow range of cell types, for instance, blood stem cells make only blood cells. In long-term culture proliferation, some epigenetic changes and chromosomal abnormalities may occur. Since the stem cells as complete cell structures contain nuclei, manipulations in each laboratory may lead to acquisition of potential mutations (Choumerianou et al. 2008). Moreover, the political and ethical considerations are always correlated with stem cell manipulations. However, exosomes lack nuclei and they can easily penetrate tissues such as BBB due to their small sizes.

Exosomes as cell-free therapy are exactly similar to their parent cells while lacking the nucleus. Given these circumstances, they simply injected, targeted, labelled, and penetrated the injured tissues. Accordingly, based on the benefits that are mentioned about the stem cells, these cells should be the most important source of exosome production. The main donors of exosomes in cell-free therapy are adipose-MSCs (AMSCs), bone marrow-MSCs (BMSCs), and prenatal MSCs.

Exosomes in Tissue Repairing and Regeneration

Several biomaterials and nanostructures are employed in tissue engineering and regenerative medicine (Kalantari et al. 2019; Rahmanian et al. 2019; Kalantari and Naghib 2019; Kalantari et al. 2018a; Kalantari et al. 2018b; Kalantari et al. 2018c). Considering the utmost importance of MSCs and their secretomes in tissue regeneration and remodelling, exosomes released from MSCs are potential substitutes in trophic cell-free therapies (Sarvar et al. 2016). Exosomes are identified as the principal form of paracrine signalling. They are implicated to play different roles such as pro-inflammation, anti-inflammation, neovascularisation, and coagulation (Jing et al. 2018). Hence, the most striking evidence for possible utilisation of exosomes in the promotion of tissue repair and regeneration is highlighted (Figure 5.2). MSC-derived exosomes implement their functions through regenerative potential, immune-modulatory and anti-inflammatory features, the same as their parents. i.e. MSCs (Harding et al. 2013).

Neural Regeneration

Exosomes play a crucial role in promotion of the regeneration, protection of the nervous system, and the meditation of communication amongst nervous system cells (Kumar et al. 2018). The neurons of central nervous system (CNS) lose their intrinsic ability to regenerate as they mature, whereas peripheral nervous system (PNS) neurons have relatively good regenerative capacity (Lai and Breakefield 2012).

The regenerative effects of exosomes have been extensively studied in the repair and detection of neurological diseases. Ischaemic strokes occur when the brain supplying blood vessels become blocked. Clinically, the early detection of ischaemic strokes is a critical factor for successful therapy (Delcayre et al. 2005). Studies explored the diagnostic values of miR-9 and miR-124 in exosomal serum as indicating biomarkers for the detection of acute ischaemic stroke (AIS) and damage determination (Ji et al. 2016). A preclinical study reported that MSC-derived exosomes loaded with miR-124 could improve the neurovascular recovery after stroke and inhibit the post-ischaemic immunosuppression in mice (Yang et al. 2017). A clinical trial (NCT03384433) conducted by Zali et al., genetically manipulated MSC-exosomes enriched by miR-124, were administered in patients suffering from acute ischaemic stroke to

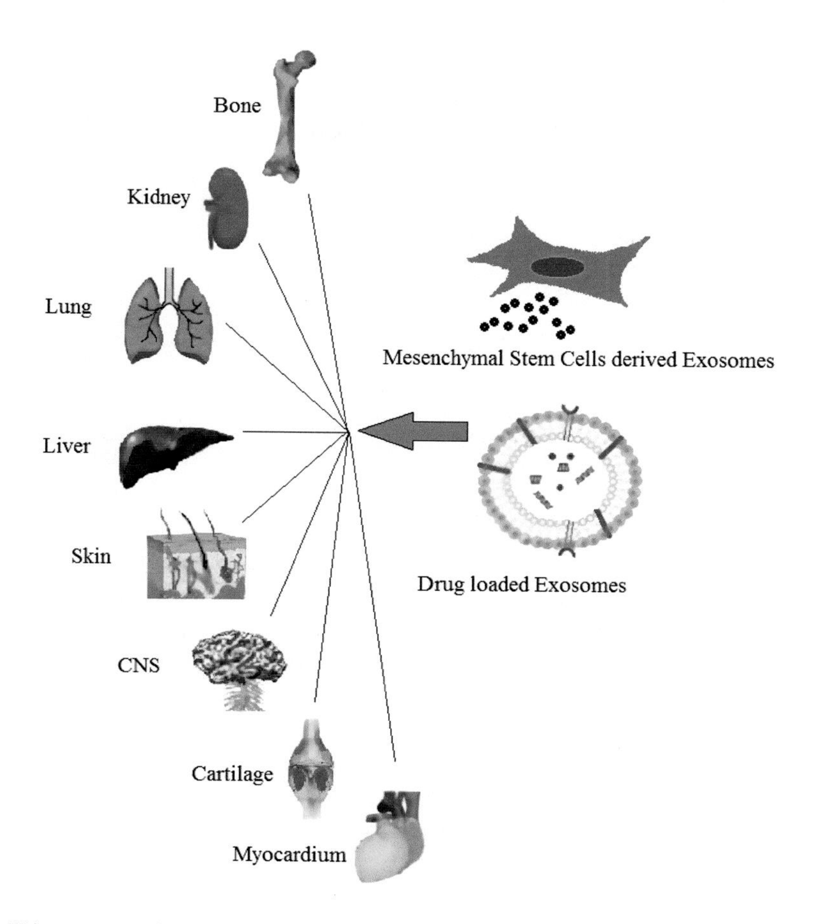

FIGURE 5.2 Using exosomes in regenerative medicine.

improve the disability. The preclinical assays had shown that the exosome treatment has improved the neurite remodelling, neurogenesis, and angiogenesis (Table 5.1).

Exosomes are involved in tissue regeneration, possibly via the delivery of cargo miRNAs. MSCs exert their therapeutic effect on the brain parenchymal cells, either by direct interaction or via the paracrine secretome, thereby improving functional recovery. Xin et al. have suggested the communication of MSCs with the brain parenchymal cells by miRNAs. They have reported that after treating the brain parenchymal cells of rats (subjected to middle cerebral artery occlusion (MCAo)) with MSCs and their exosomes, the expression of miRNA-133b elevated in neural cells. Further verification affirmed the presence of exosome as the delivery means of miRNA-133b to improve the post-stroke neurological functions, neurite remodelling, and neurogenesis (Xin et al. 2012). MiR-125b contributes significantly in the differentiation of neuronalstem cells. Takeda et al. proposed that neuronal progenitor cells-derived exosomes enriched with miRNAa could participate in neuronal differentiation. MSCs managed to develop neuron-like morphology following the treatment with neuronal cell lines. Microarray analysis indicated the upregulation of miR-125b as a potential mechanism in neural differentiation of MSCs (Takeda and Xu 2015). Exosomes derived from MSCs could promote the neurite growth of cortical neurons in a miRNA-dependent manner. Tailored MSC-exosomes containing elevated levels of the miR-17-92 cluster have improved the axonal growth considerably compared to native MSC-exosomes by activation of selectivity for their target genes (Zhang et al. 2017).

Studies also supported the effectiveness of BMSC-derived exosomes for traumatic and degenerative ocular diseases. Traumatic (optic neuropathy) and degenerative (glaucoma) eye diseases, which cause irreversible blindness, are the results of the loss and dysfunction of retinal ganglion cells (RGC) and their

TABLE 5.1

Studies Involving MSC EVs and Ongoing Clinical Trials That Apply EVs to Treat COVID-19 Patients

Tissue Injury/Disease	Stem Cell Source	Trial Phase	Trial ID
Lung injury	Allogenic adipose MSC	Phase I	NCT04313647
Lung disease (pneumonia)	Allogenic adipose MSC	Phase I	NCT04276987
Chronic lung Disease	Bone marrow MSC	Phase I	NCT03857841
Cartilage injury (Osteoarthritis)	Adipose MSC	Phase I	NCT04223622
Skin injury (Dystrophic Epidermolysis Bullosa)	MSC	Phase II	NCT04173650
Kidney disease (CDK)	Allogenic umbilical cord MSC	Phase II/III (completed)	Nassar et al. 2016
Dystrophic Epidermolysis Bullosa	Allogenic MSC	Phase I/IIA	NCT04173650
Refractory macular holes (MHs)	MSC and MSC-derived exosomes (MSC-Exos)	Phase II	NCT03437759
Pancreatic Adenocarcinoma	MSC	Phase I	NCT03608631
Diabetes Mellitus Type 1	MSC and its microvesicles	Phase III	NCT02138331
COVID-19 pneumonia	Allogenic adipose MSC (MSCs-Exo)	Phase I	NCT04276987
COVID-19	MSCs from allogenic bone marrow	Phase II	NCT04493242 ARDS
COVID-19 pneumonia	Allogenic COVID-19 Specific T cell (cstc)	Phase I	NCT04389385
COVID-19	Human amniotic fluid (HAF)	Phase I/II	NCT04384445
COVID-19	Allogenic-cardio-sphere-derived cells	Phase I	NCT04338347
COVID-19 pneumonia	MSC	Phase I/II	NCT04491240
COVID-19 pneumonia	Allogenic adipose MSC	Phase I	NCT04276987
Brain injury (Acute Ischaemic Stroke)	Allogeneic MSC	Phase II	NCT03384433

axons. The neuroprotective and axogenic effects of MSCs have been well-established on RGC models (Mead et al. 2015). Further research conducted by this group has suggested that BMSC-derived exosomes may be beneficial for the survival of RGCs via miRNA-dependent mechanisms in a rodent optic nerve crush (ONC) model (Mead and Tomarev 2017).

Spinal cord injuries (SCIs) could lead to failure of axonal regeneration. Schwann cells (SCs) significantly sustain the axonal regeneration in the peripheral nervous system. After peripheral nerve damage, the bilateral crosstalk between SCs and regenerating axons is a key mechanism to promote the axonal regrowth and restore the function (Yamauchi et al. 2008). In one such study, Lopez-Verrilli et al. have defined a case study suggesting the potential role of exosomes secreted by dedifferentiated SCs on SCIs. They have demonstrated that the selective internalisation and uptake of SCs exosomes by axons could mediate axonal regeneration. SCs supported the axonal regeneration and maintenance after nerve damage via exosomes. Furthermore, *in vivo* studies have demonstrated that the uptake of SC-exosomes by axons could markedly increase the regenerative capacity of injured sciatic nerves (Lopez-Verrilli et al. 2013; Lopez-Verrilli 2012). The other failures that are associated with traumatic SCI include loss of motor function and disruption of microvascular stability. Moreover, the enhanced disruption and permeability of the blood–spinal cord barrier (BSCB) were observed in SCI, which is believed to be mediated by the migration of pericytes. These cells are one of the critical constitutions of neurovascular unit, preserving the integrity of BSCB (Sweeney et al. 2016). Notably, treatment with EVs released from BMSCs could increase the survival rate and regeneration of neural cells, hinder the pericyte migration, and improve the motor function recovery in spinal cord injury model (Lu et al. 2019). More recently, Bucan et al. have assessed the therapeutic effects of adipose-derived MSCs on neurite elongation and sciatic nerve regeneration in traumatic peripheral nerve injury. ADMSCs exosomes have facilitated the neurite outgrowth *in vitro* and encouraged the post-sciatic nerve damage regeneration *in vivo* (Bucan et al. 2019). In a similar study, the effect of SCs modulation has been validated. It has been found that the internalisation of the ADSC-EVs could lead to the proliferation of SCs on sciatic nerve repair (Haertinger et al. 2020).

Prior studies have showed that the dysregulation of miRNAs is another factor that should be taken into account in SCI. Owing to the vital role of miR-133b in the differentiation of neurons and the outgrowth of neuritis (after the administration of miR-133b-modified MSCs), miR-133b containing exosomes considerably improved the functional recovery of hindlimb in SCI rat model. Besides, miR-133b treatment has decreased the lesion volume and enhanced the axonal regeneration following the SCI. Furthermore, it is suggested to activate some signalling pathways involved in axonal regeneration (Li et al. 2020).

Accumulating evidence has proven the efficacy of applying exosome-based nano-therapeutics in the treatment of autoimmune and central nervous system disorders. Multiple sclerosis (MS) is an inflammatory demyelination disorder of the CNS. MSCs-derived exosomes have been exploited for the treatment of MS in the experimental model of autoimmune encephalomyelitis (EAE). The intravenous administration of the exosomes resulted in decreased demyelination and reduced neuro-inflammation (Riazifar et al. 2019). The differentiation of the endogenous oligodendrocyte precursor into mature myelinating oligodendrocytes has been encouraged by MSC-EVs in a mouse model. MSC-EVs stimulated the myelin regeneration in the spinal cord of treated mice and led to boosted motor function (Clark et al. 2019). Similarly, the neuroprotective ability of MSC-derived exosomes was evaluated in a mouse model of Alzheimer's disease. This cell-free treatment stimulated neurogenesis, promoted cognitive impairment, and improved neural plasticity, which are corroborated with therapeutic effects of MSCs (Ferreira et al. 2018; Reza-Zaldivar et al. 2020). Increasing evidence and observations have supported the efficacy of MSCs as a therapy for brain damages, including traumatic brain injury in experimental models and clinical settings (Li and Chopp 2009; Zhang et al. 2008). Zhang and co-workers have suggested that exosomes released from MSCs have promoted the functional recovery and neurovascular plasticity and reduced inflammation in rat model of traumatic brain injury (Zhang et al. 2015d).

Myocardial Regeneration

Exosomes can stimulate the myocardium regeneration by shuttling of miRNAs and other functional signals to restore the function of injured heart tissue as a replacement to cell therapy. The potential regenerative options of exosomes have already been implicated as an innovative therapy in numerous myocardial ischaemia-reperfusion injury (IRI) studies (the worst serious form of cardiovascular diseases (Lazar et al. 2018)). Cardiosphere-derived cells (CDCs) secrete plenty of cytokines and growth factors that stimulate regeneration and angiogenesis via its secretome (Stastna et al. 2010). In a recent study, the exosomes isolated from CDCs have been injected into ischaemia injury mouse model. Upon treatment, the cardiomyocyte cells were proliferated, and apoptosis was inhibited. Therefore, the released exosomes recapitulate the regeneration and protective effect of CDC transplantation (Ibrahim et al. 2014). Here, the caridoprotective effects of exosomes appear to primarily be attributable to the enriched miR-146a, which plays an imperative role in the pathology of myocardial infarction (MT). The impairment of blood flow occurs in ischaemic diseases during angiogenesis. miR-132 regulates the endothelial cell behaviours including the angiogenic responses. Thus, it can be a promising therapeutic agent to cure ischaemic diseases by promotion of angiogenesis. Teng et al. have investigated the delivery of miR-132 by MSC-derived exosomes. miR-132-loaded exosomes can effectively improve the angiogenic potency of endothelial cells by downregulation of its target gene. Moreover, the transplantation of miR-132 exosomes in the ischaemic hearts of mice has profoundly increased the neovascularisation in the peri-infarct area and restored cardiac functions (Ma et al. 2018).

Considering the cardio-protective role of MSCs in various studies, the preconditioning and incubation of cardiac stem cells (CSCs) with MSC secretomes could improve the regenerative and angiogenic potency of CSCs and promote the myocardial repair. MSCs have stimulated the amplification, migration, and angiotube formation of CSCs in a rat model of MI (a major cardiovascular disease (CVD)). Performed miRNA array revealed that the preconditioning by MSCs could cause changes in the micro-RNA profile of CSC (Zhang et al. 2016c). The miRNA expression profile array confirmed a set of differentially expressed miRNAs in CSCs after MSC-Exo exposure. Furthermore, the miRNAs shuttled by MSC-exosomes may regulate the proliferation and differentiation of CSCs. Since exosomes enclose the contents of the parental cell, choosing the right donor cells is very important. ESCs with the pluripotency advantages and their unique microRNA and protein contents can hold a good promise in myocardial cellular therapy. They could increase the endogenous cardiomyocyte/CPC proliferative and survival/differentiation responses after myocardial injury. Khan et al. have pointed out that exosomes derived from mouse embryonic stem cells (mES-Ex) augmented restoration and cardiac function in accordance with reduced fibrosis in the MI model. It has been demonstrated that the significant enrichment for a set of miRNA, specifically miR-294 in ESCs-derived exosomes, and their subsequent uptake by CPCs was the underlying mechanism for tissue repair (Khan et al. 2015).

Pericardial fluid (PF), the plasma ultrafiltrate inside the pericardium sac around the heart, contains exosomes that are enriched with miRNAs secreted by heart. Co-culturing of PF with endothelial cells (ECs) induces its survival and proliferation. PF exosomes restore the blood flow recovery and angiogenesisinto a mouse model of ischaemic injury. Of note, internalisation of pro-angiogenic miRNA let-7b-5p into ECs by PF exosomes might coordinate the vascular repair (Beltrami et al. 2017).

For clinical translation of cell-free therapy, the determination of optimal culture conditions is very important. The hypoxic precondition of MSCs can augment their therapeutic potentials. It also could enhance the performance of MSCs transplantation for MI treatment by miRNA regulation (Hu et al. 2016). The myocardial repair benefits of exosomes derived from hypoxia-treated MSCs (Exo^H) was compared with the normoxia-treated MSC (Exo^N) culture condition. Exo^H has considerably increased the angiogenesis, cardiomyocyte viability, and trigger of cardiac progenitor cells (CPC) in the infarcted heart. The microRNA expression analysis demonstrated the upregulation of miR-210 levels in exosome and their parental cells. This property can largely be attributed to the superior pro-angiogenic and cardio-protective capability of exosomes cultured in hypoxia condition. Therefore, hypoxia preconditioning offers an efficient method to maximise the restorative effects of MSC-derived exosomes for CVD (Zhu et al. 2018). In a similar study, Agarwal and co-workers investigated the impact of donor age (neonates, infant, and child) and hypoxia level in the regenerative capability of exosomes derived from human paediatric CPC in a rat model of myocardial IRI. The findings have demonstrated that the exosomes derived from hypoxic and normal conditions boosted cardiac function through decreased fibrosis and promoted angiogenesis. However, exosomes obtained from older ages only implemented restoration ability when CPCs were cultured in hypoxic conditions (Agarwal et al. 2017).

The short-lived retention rate and quick clearance of exosomes are their major limitations in clinical translation studies. It has been reported that exosomes became undetectable 3 hours after myocardial injection (Gallet et al. 2017). To increase the retention rate, the human umbilical cord MSCs (hUCM-SCs)-derived exosomes were loaded in peptide hydrogels and administered in a MI of rat model. It has been observed that functional recovery was improved. The PGN hydrogel encapsulation has effectively provided prolonged and stable retention for exosomes. The administration of exosome/PGN hydrogel complex confers myocardial functional recovery through extended delivery (Zhou et al. 2019). This treatment reduces inflammation, fibrosis, and apoptosis. The functional biomaterials such as hydrogels mimic the extracellular matrix for sustainable delivery of EVs directly into an infarcted rat heart (Liu et al. 2018). Huang and co-workers examined the sequential intramyocardial delivery of MSCs and their secreted exosomes to treat acute MI (Huang et al. 2019). Combinatorial treatment has improved the recruitment and retention of transplanted MSCs; thereby, reduced scar size and restored heart function was observed following the MI compared to single treatment of exosomes or MSCs. The administered exosomes prepared the ground for MSC recruitment and survival to replenish lost cells.

Hepatic Regeneration

The liver is one of the most essential organs having the highest intrinsic regenerative capacity. It has several indispensable functions in metabolic homeostasis and detoxification. However, the prolonged exposure of hepatic tissue to various damaging conditions could lead to cell death and hepatic dysfunction (Sookoian and Pirola 2015). Liver regeneration is a preparatory process to restore the function of liver mass loss following damage, or diseases (Michalopoulos 2013). Regenerative medicine has opened new horizons for the acute and chronic liver disease therapy and restoring homeostasis after liver injury, dysfunction, or disease. Hepatocyte-derived exosomes can be used as a diagnostic tool to indicate an injury. They may also serve as a novel mechanism to enhance repair and regeneration (Cho et al. 2017; Momen-Heravi et al. 2015; Thulin et al. 2017; Povero et al. 2014). The modulatory effect of exogenous-ly-applied exosomes has already been delineated. These exosomes could trigger liver repair and regeneration via intercellular communication. The exosome level of alcoholic hepatitis patients was compared to healthy counterparts. The obtained results revealed that a higher amount of exosomes exists in alcoholic hepatitis with distinct miRNA profiles (Momen-Heravi et al. 2015). Hepatocyte-derived exosomes can potentially mediate the regenerative and reparative effect of hepatocytes *in vitro* and liver regeneration *in vivo* (Nojima et al. 2016). The regenerative potency could be rooted in the exosome-mediated delivery of neutral ceramidase and sphingosine kinase 2 (SK2) towards hepatocytes. This transfer could induce the higher production of sphingosine-1-phosphate (S1P) inside the hepatocytes. The sphingomyelin-ceramide pathway is crucial for exosome production (Trajkovic et al. 2008). This can address the increased circulating exosomal levels after liver injury. The exosomes derived from other liver cells did not have the same effect. These exosomes lack the key SK2 cargo and are incapable of evoking the proliferation of hepatocytes. These observations affirm the particular features of hepatocyte exosomes. Consistently, Ying dong et al. showed the hepato-protective impacts of exosomes derived from human-induced pluripotent stem cell-derived mesenchymal stromal cells (hiPSC-MSCs-Exo) on hepatic IRI (Du et al. 2017).

Drug and toxicants injury is a significant cause of liver diseases (Fisher et al. 2015). Oxidative stress-related molecules are associated with liver damage and may arise after CCl_4, H_2O_2, and acetaminophen treatment (de la Rosa et al. 2006). Yan and co-workers have demonstrated the necessity of glutathione peroxidase1 (GPX1) from the hUCMSC for the restoration of hepatic oxidant injury. The antioxidant and anti-apoptotic effects of administered exosomes saved mice from hepatic injury caused by carbon tetrachloride (CCl4). The underlying mechanism could be attributed to the detoxifying effect of hUCMSC-exosome-derived glutathione peroxidase1 (GPX1) CCl_4, H_2O_2, and finally prohibited oxidative stress and apoptosis (Yan et al. 2017).

Hepatic IRI occurs during liver resection, transplantation, and trauma (Huguet et al. 1994; Lemasters and Thurman 1997; Kim 2003). Hepatic IRI can lead to serious impairment of the liver function and even trigger a cascade of multiple organ malfunctions by producing irreversible failures (Cannistrà et al. 2016). Nong et al. have used the benefits of exosomes (produced from hiPSC-MSC) against hepatic IRI in rats. The employed exosome-based treatment has suppressed the hepatocyte injury and inflammatory markers, lowered the apoptotic markers in liver tissues from the experimental group, and relieved hepatic IRI (Nong et al. 2016).

Yao and co-workers have discovered another mechanism for alleviation of hepatic IRI through EVs derived from hUCMSCs. The protection against IRI has been conferred through the reduction of neutrophil infiltration and alleviation of oxidative stress. Oxidative stress in the hepatic IRI model was decreased by the exosomes enriched with manganese superoxide dismutase (MnSOD), which is an antioxidant enzyme within mitochondria (Li et al. 2019).

Liver is the most frequent target for MSC-EVs after systemic administration (Zhao et al. 2018). An effective EVs therapy requires a certain amount of EV accumulation in the target tissue over the defined time period. MSC-EVs were used to improve hepatic regeneration in chronic liver failure in a sustained manner to prolong retention and delivery. EV-encapsulated PEG hydrogels could be achieved through mixing of exosomes with polyethylene glycol (PEG) macromeres by a quick click reaction. Upon bio-distribution, the EV-encapsulated PEG hydrogels were delivered sustainably over 1 month. The sustained systemic diffusion and further accumulation of EV-encapsulated PEG inside the liver cells was prolonged in a rat model of chronic liver fibrosis. One month later, the regenerative, anti-apoptosis, and anti-fibrosis effects of GEL-EV were found in the collected liver by histopathological examinations. The sustained systemic release of EVs has enhanced their beneficial effects up to 50% compared to EV-free delivery (Mardpour et al. 2019). The effectiveness of EVs-containing clickable PEG solutions was demonstrated in prevention of fibrogenesis compared to Free-EV group.

Renal Regeneration

Increasing evidence suggests the therapeutic benefits of using MSC-EVs for kidney regeneration in the treatment of acute kidney injuries (AKI) and chronic kidney disease (CDK).

Acute Kidney Injury

The major complication of cisplatin chemotherapy is related to AKI. Cisplatin-induced oxidative stress leads to renal cell damage and high mortality rates (Antunes et al. 2000). Tamasoni et al. have reported that human BMSCs-derived exosomes increased the proliferation of cisplatin-damaged proximal tubular epithelial cells by lateraltransfer of the insulin-like growth factor-1 (IGF-1) mRNA and local release of growth factors (Tomasoni et al. 2013). Another study exploited the hUCMSC-derived exosomes for the treatment of cisplatin-induced nephrotoxicity. They have suggested that the renal oxidative stress ameliorated by decreasing the formation of several detrimental products, suppressing the apoptosis caused by oxidative stress in AKI model, and promoting the renal epithelial cell proliferation (Zhou et al. 2013).

Renal ischaemia/reperfusion (IR) is thought to be as the leading cause of AKI. It primarily affects the function and structure of tubular epithelial cells and renal endothelial function (Basile 2007; Bonventre and Yang 2011). IR results in major vascular damage, which leads to capillary loss. Therefore, administration of endothelial lineage cells would be a possible therapeutic approach for this condition (Patschan et al. 2011). In this regard, Vinas et al. studied an ischaemic AKI model, employing the regenerative potency of human umbilical cord blood-derived endothelial colony-forming cells (ECFCs) and ECFC-derived exosomes. Their findings demonstrated that ECFC administration could alleviate renal injuries in ischaemic AKI. Moreover, it has been shown that the ECFC-derived exosomes can calm the ischaemic AKI by delivery of miR-486-5p, which targets the phosphatase and tensin homolog PTEN (Viñas et al. 2016).

One of the most common kidney complications is the diabetic nephropathy (DN) in patients with long-term diabetes, which has a high mortality rate (Yu and Bonventre 2018). Currently, stem cells and their secretomes are considered as the ideal therapeutic approach for DN. Hyperglycaemia is a critical factor in DN development. It mediates tissue damage by the generation of oxidative stress and reactive oxygen species elements, inflammatory cytokines, and lipid mediators. This results in clinical manifestation of DN, which is related to glomerulosclerosis and tubule-interstitial fibrosis. Human urine stem cell-derived exosomes (USCs-Exo) are shown to inhibit the apoptosis of podocytes (the visceral glomerular epithelial cells) and promote vascular regeneration and survival of diabetic cells (Forbes et al. 2008).

USCs-Exo implemented the therapeutic efficiency of kidney damage repair. Intravenous injection of USCs-Exo from the tail vein could significantly suppress apoptosis and boost cell survival and vascular regeneration. Analysing the contents of USCs-Exo revealed that its potential components might be cytokines, vascular endothelial growth factor (VEGF), TGF-β1, angiogenin, and bone morphogenetic protein 7 (BMP-7) (Jiang et al. 2016).

In miRNA therapy, the engineered MSCs-derived exosomes, which were overexpressing the miRNA-let7c, were targeted to injured kidneys. As a consequence of this treatment, the damage was alleviated by up-regulation of miR-let7c gene expression (Wang et al. 2016a). Exosomes derived from human BMSCs and human liver stem-like cells (HLSCs) stopped the development of fibrosis in a mouse model of DN via activation of regenerative processes. The stem cell treatment considerably prevents renal fibrosis. miRNA array analysis showed the modulation of targeting pro-fibrotic genes (Grange et al. 2019). In a recent publication, Zhao and co-workers have reviewed various studies, which provide evidence for the renoprotective effect of miRNAs in stem cell-derived EVs during AKI (Zhao et al. 2019). Multiple miRNAs are the key post-transcriptional regulators and play prominent roles in kidney protection with different mechanisms. They will be delivered within damaged renal cells through EVs secreted by stem cells; therefore, they would ameliorate AKI by anti-fibrotic, anti-apoptotic, and immune modulatory effects. Moreover, the vesicles secreted by injured renal cells could promote renal recovery due to their role in exerting phenotypic and functional changes while communicating.

The therapeutic efficiency of MSC-EVs is hampered by their quick elimination after systemic administration. A hydrogel slow-released system was developed to prolong the EVs retention and maintain their stability. A designed matrix metalloproteinase-2 (MMP2) sensitive self-assembling peptide (KMP2) hydrogel loaded EVs was exploited to deliver locally in mice with renal IRI. The MSC-EVs-loaded KMP2 hydrogel exerted cell apoptosis, inflammation, and angiogenesis effects due to its biologically compatible and nanosize characteristics. These properties promote the microvascular endothelial cell regeneration after IRI (Zhou et al. 2019). Matrix metalloproteinases (MMPs) are zinc end peptidases that act on the degradation of extracellular matrix (ECM) proteins involved in tissue remodelling (Kunugi et al. 2011). MMP-cleavable peptides have been applied in the production of MMP-degradable biomaterial for drug release and tissue repair (Wang et al. 2018b).

Chronic Kidney Disease (CKD)

Human liver stem cell-derived extracellular vesicles (HLSC-EVs) were investigated for their therapeutic efficiency on tubular injury and interstitial fibrosis in aristolochic acid (AA)-induced nephropathy (a kind of CKD). The HLSC-EV-mediated treatment has significantly attenuated the tubular necrosis and interstitial fibrosis. Furthermore, HLSC-EVs down regulate the expression of the pro-fibrotic genes (Kholia et al. 2018).

MSC-EVs derived from umbilical cord ameliorated the CDK development in grade III-IV CKD patients and promoted the kidney function (Nassar et al. 2016). A phase II/III clinical trial has reported the results of using EVs derived from cord tissue MSCs to attenuate the development of CKD (Nassar et al. 2016). Twenty patients with diagnosed CKD have received two consecutive doses of (two doses1 week apart) MSC-derived EVs (100 μg/kg/dose). The initial dose was injected by intravenous route and the second one was infused into the renal artery.

Cutaneous Regeneration

Over a decade ago, cell therapy has gained interest as a novel treatment strategy for wound healing and regeneration. An optimal scarless wound healing needs the orchestration of various complicated and physiological processes with the involvement of different cells, ECM, and growth factors (Eming et al. 2014). Exosomes derived from stem cells have got various implications in inflammation and wound healing. One of the main processes in wound healing and tissue regeneration is angiogenesis. The exosome cross-communication in the tumour microenvironment promotes angiogenesis and cancer metastasis (Salem et al. 2016; Wang et al. 2016b).

hUCMSC-derived exosomes could potentially suppress the inflammatory process caused by burn. Additionally, it has been elucidated that miR-181C is a key component in regulation of inflammation by targeting proteins involved in inflammatory signalling pathways. Overexpression of miR-181C more significantly mitigates the inflammation in severely burnt rats. This observation could be rooted in the downregulation of the TLR4 signalling pathway (Li et al. 2016). Zhang et al. have elucidated the underlying mechanism in the regenerative potential of hUCMSC-exosomes. Their findings suggested the role of hUCMSC-exosomes in the activation of Wnt/β-catenin signalling pathway that leads to healing at the beginning of an in-depth second-degree burn repairing process. However, hUCMSC-exosomes could inhibit the Wnt/β-catenin signalling by stimulation of YAP phosphorylation which could overcome the extreme skin cell growth and collagen deposition following the remodelling stage (Zhang et al. 2015a).

Human amniotic epithelial cells (hAECs) have recently gained a lot of interest in wound healing studies. This is due to their dual characteristics as both embryonic cells and MSCs and their ability for differentiation. Zhao et al. have verified the regenerative capacity of exosomes on the repair of full-thickness skin deficiencies in rats by triggering the delivery and proliferation of fibroblasts. hAECs-exosomes-downregulated the expression of ECM, such as collagen-I and III, boosted the wound repair, and prevented the scar formation (Zhao et al. 2017).

One of the critical complications of diabetic mellitus is the impaired and delayed skin wound healing. This issue could result in chronic ischaemic skin damage and even limb amputation (Kharroubi and Darwish 2015; Lioupis 2005). Therefore, the treatment modalities that augment the local angiogenesis may be of great importance in the diabetic wound healing process. The distinct characteristics of endothelial progenitor cells make them an attractive source in wound closure and healing in chronic diabetic wounds. They can differentiate into endothelial cells, which play a dominant role in angiogenesis and neovascularisation. However, they stimulate the regenerative potency of target cells by releasing different mixtures of trophic factors (Critser and Yoder 2010). Zhang et al. investigated the healing and angiogenesis effects of human umbilical cord blood-derived endothelial progenitor cells exosomes (EPC-Exos) in diabetic rat models (Zhang et al. 2016a). EPSc facilitate diabetic cutaneous wound repair by improving neovascularisation and stimulation of capillaries formation, which usually gets compromised in diabetes patients. Further analysis demonstrated that these therapeutic effects had been exerted by altering the expression of a class of genes related to extracellular signal-regulated kinases 1 and 2 (Erk1/2) signalling pathway upon exosomes treatment. Since the preclinical studies demonstrated the efficacy of MSC-derived exosomes in the diabetes model, the first clinical trial with the NCT0213833 register number examined the impact of allogeneic cord tissue MSC-derived EVs in type 1 diabetes mellitus patients. In this study, 20 patients with type 1 diabetes have been administered intravenously with consecutive doses of MSC-derived exosomes and MSC-derived microvesicles. At the end of the study, functional parameters of the liver have been analysed.

Another study emphasised the healing potential of platelet-rich plasma (PRP-Exos) exosomes in recruiting and proliferation of endothelial cells and fibroblasts to enhance angiogenesis and re-epithelialisation in the healing process of chronic cutaneous wounds (Guo et al. 2017). PRP encapsulates various extensive growth factors released by platelets, so it can be considered as an enriched and valuable alternative in the treatment of chronic wounds.

A plethora of studies have demonstrated that MSC-derived exosomes involve in the regeneration of cutaneous wound repairing process. MSC exosomes promote the wound healing through activation of signalling pathways involved in wound closure (Akt, ERK, and STAT3) and stimulation of several growth factors such as hepatocyte growth factor (HGF, insulin-like growth factor-1 (IGF1), nerve growth factor (NGF), and stromal-derived growth factor-1 (SDF1) (Shabbir et al. 2015). The adipose tissue MSC-derived EVs revealed their wound healing facilitation via activation of the AKT pathway (Ferreira et al. 2017).

The efficacy of hiPSC-MSCs in the wound healing process was assessed by local injection around the wound site in a rat model. HiPSC-MSC administration has reduced the scar width and accelerated re-epithelialisation and neovascularisation. Taken together, hiPSC-MSCs significantly improved the wound healing through enhanced collagen synthesis and angiogenesis (Zhang et al. 2015b).

Aside from the growth factors, cytokines, and other compounds, it is reported that human ADMSCs exosomes containing a long noncoding (lnc) RNA, MALAT1 (metastasis-associated lung adenocarcinoma transcript 1) represented a profound effect on the migration of human dermal fibroblast and ischaemic wound repair process. They attributed the increased cell migration to the lncRNA (Cooper et al. 2018).

Other Organs

The secretome of MSCs has been widely used as a novel cell-free candidate in the treatment of bone defects after injury in skeletal regeneration and bone tissue engineering. Exosomes are involved in the healing and bone reconstruction through promotion of angiogenesis and activation of osteogenesis (Yu et al. 2009). The combined effect of the exosomes, which were derived from human ADMSCs and immobilised on poly lactic-co-glycolic acid (PLGA) scaffolds, has elaborately restored the critical-sized mouse calvarial defects. The slow and consistent release of human ADMSC-derived exosomes have significantly accelerated the bone regeneration by promotion of osteo-inductive capability, migration of MSC homing, and tropism towards the newly formed bone tissue (Li et al. 2018). The same group previously has unveiled the rapid and efficient osteogenic differentiation of human ADMSCs (Zhang et al. 2015c).

For dental defects, human MSC-derived exosomes (loaded with collagen sponge) have improved periodontal regeneration and promoted periodontal ligament cell functions. Exosomes have been shown to exert more efficient regenerative response in an immunocompetent rat model of periodontal defect in comparison to control rats. The exosome-mediated regeneration could be due to the increased cell proliferation and migration via CD73-mediated adenosine receptor activation of pro-survival AKT and ERK signalling (Chew et al. 2019).

Preclinical experiments have revealed that MSCS can restore cartilage function by differentiation into chondrocytes (Huang et al. 2016; Driessen et al. 2017; Bhadra et al. 2019). Considering the various limitations related to MSC transplantation such as aberrant cell phenotype, weak differentiation ability after multiple passages, and complicated issues related to clinical translation, the paracrine secretion of trophic factors could be used as an alternative strategy in osteochondral regeneration and cartilage repair. In a novel insight, the UCMSC-derived exosomes, obtained from 3D culture in a hollow-fibre bioreactor, have exerted chondroprotective effects and activated steochondral regeneration activity. They have confirmed the higher yield of 3D-culture exosomes compared to 2D-culture exosomes (7.5-fold higher). This is promising outcome for clinical-grade usage of 3D-culture exosomes. The *in vitro* results demonstrated the higher impact of 3D exosomes in the proliferation of chondrocytes, migration, and matrix synthesis. The underlying mechanism can be due to the transforming growth factor beta 1 and Smad2/3 signalling induction (Anju et al. 2019).

Extracellular Vesicles and COVID-19

COVID-19 virus infection, a kind of acute respiratory distress syndrome (ARDS), has recently raised a globally critical public health emergency. EVs, including exosomes, are the major transporters of viral components from infected cells to healthy ones (Nolte et al. 2016). Exosomes and viruses have some common characteristics such as size, structure, and entrance mechanism. SARS-CoV-2 broke out first in the Wuhan city of China in December 2019. This virus is characterised by severe respiratory impairment. The elevated rate of cytokines and chemokines targeting lung tissue is the primary reason of morbidity and mortality associated with COVID-19 infection. Herein, we describe the regenerative potential of MSC-derived EVs, including exosomes. However, more studies are warranted in preclinical and clinical evaluations to unravel the mechanism of action, safety, stability, and effectiveness of established methods.

Once the cells are infected, the first stage in viral infection is the explicit identification of the angiotensin I that converts the enzyme 2 receptor (ACE2) by the viral spike protein (Leng et al. 2020; Payandeh et al. 2020). The immune system overreaction through production of different cytokines (cytokine storm) and inflammatory factors, such as natural killer cells and effector T cells, leads to body damage (Mangalmurti and Hunter 2020; Meftahi et al. 2020). The cytokine storm is due to increased plasma levels of granulocyte colony-stimulating factor (GCSF) and tumour necrosis factor alpha (TNF-α) (Leng et al. 2020). Consequently, the elevated level of cytokines leads to lung damage, which might end in death. Thus, suppressing super-inflammatory immunological response and eliminating the cytokine storm as well as the lung regenerative therapy could be deemed as an efficient therapeutic strategy for COVID-19 patients (Leng et al. 2020). The combination therapy of antiviral drugs with the regenerative potency of stem cells and their secretomes may attenuate the severity of the COVID-19. Owing to their regenerative potential in mitigation of the inflammation and repair of the COVID 19 symptoms, MSC-derived exosomes can be compassionately employed for COVID-19 therapy. In preclinical studies, exosome-based therapies have been used against acute ARDS (Abraham and Krasnodembskaya 2020). The cytokine storm and inflammation signals had been attenuated after exosomal administration. Furthermore, the elevated level of anti-inflammatory signalling molecules reduced the severity of lung damage due to increased permeability of alveolar epithelium. This change facilitates the exchange of oxygen. It should be noted that exosome-mediated delivery of mitochondria to alveolar cells would boost the survivability leading to cellular regeneration.

Several studies confirmed the role of miRNA of exosomes as a critical effector for enhanced lung restoration of patients suffering from viral infections such as influenza, hypoxia-induced pulmonary hypertension, and ventricular induced lung injury. Wang et al. reported that specific miRNAs like miR-290, miR-21, let-7, and miR-200 play a significant role in lung regeneration (Wang et al. 2018c). Alipoor et al. have discussed that stem cell-derived exosomes could facilitate the reduction of hypertension and inflammation by deactivation of the signalling pathways associated with hypoxia, especially in respiratory disorder (Alipoor and Mortaz 2020). The intra-tracheal delivery of MSC-derived exosomes in a pig model of influenza has significantly decreased viral shredding (12-hour post-infection) (Khatri et al. 2018). Exosomes share similar physicochemical properties with viruses and novel findings unraveled the contribution of exosomes and viruses to the infection process (Dongen et al. 2016). Viruses take advantage of exosomes for cellular release, and exosomes exploit the entry mechanism of viruses to deliver their cargo. Thus, in the near future, the exosome-virus interplay can be used for antiviral vaccine development. During the viral infections, exosomes package the virus-derived nucleic acids, proteins, and lipids to spread viral components (Gould et al. 2003).

A clinical trial (NCT04276987) was established to examine the efficacy and safety of allogeneic adipose MSC-derived exosomes for the treatment of severe pneumonia of patients diagnosed with

COVID-19. Patients received the aerosol inhalation of MSC-derived EVs once a day for five consecutive days (Clinical Trials.gov. A pilot clinical study on inhalation of mesenchymal stem cells exosomes treating severe novel coronavirus pneumonia. Registration number: NCT04276987" 2020).

In another clinical trial (ChiCTR2000030261), MSC-derived exosomes were delivered to the lungs of people suffering from COVID-19 pneumonia by the atomisation method ("Chinese Clinical Trial Register. A study for the key technology of MSCs exosomes atomisation in the treatment of novel coronavirus pneumonia (COVID-19). Registration number: ChiCTR2000030261."2020). In another clinical trial (ChiCTR2000030484), the participants received a combination therapy of HUMSCs and EVs to treat the severe lung injury due to COVID-19 ("Chinese Clinical Trial Register. HUMSCs and exosomes treating patients with lung injury following novel coronavirus pneumonia (COVID-19). Registration number: ChiCTR2000030484."2020).

The safety and efficacy of BMSC-derived EVs (ExoFlo™) were evaluated for the treatment of COVID-19 patients (Sengupta et al. 2020). This study reported the successful improvement of clinical symptoms 3–4 days after treatment.

A recent therapeutic option showed improvements in COVID-19 patients by employing convalescent plasma therapy. This approach uses the enriched EVs and other products such as whole blood, plasma or serum, pooled human immunoglobulin IgG, and polyclonal and monoclonal antibodies (Shen et al. 2020). Plenty of improvements including elevated oxygen saturation, increased IgG and IgM antibodies titration, and generally, the decreased viral load, have been observed exploiting this approach (Rajendran et al. 2020; Sullivan and Roback 2020). To prevent the side-effects of systemic treatment options in COVID-19 pneumonia, the intrinsic tropism of platelets to inflammation sites was used. In this regard, the engineered platelet-derived EVs (PEV) enriched with TPCA-1 (anti-inflammation therapeutics) were used as a targeted delivery vehicle to treat pneumonia. The PEV delivery method exhibited a significant accumulation at the inflammation site as well as suppression of the cytokine storm. In addition, it has been demonstrated that this targeted therapy could extensively be used to target the other inflammatory regions. Considering the high stability and low immunogenicity of exosomes, Kuate et al. have exploited them as a vaccine for the treatment of SARS coronavirus in a murine model. The exosome donor cells were transfected to express an engineered chimeric S protein. Afterwards, the elevation of neutralising antibodies has been tested. The results showed that exosomes could activate antibody production by two injections. But, the highest neutralising activity was achieved by the combination therapy of engineered exosomes with an adenovirus vector expressing the chimeric S protein (György et al. 2015). Still, a lot of studies are required to warrant the safety of using exosome-based therapeutics in COVID-19 therapy.

Engineering Strategies for Tissue Repair and Regeneration

Strategies for engineering of exosomes can increase the regenerative and therapeutic ability of naïve exosomes. There are different strategies to elevate the distinct features of this natural delivery system.

Preloading of exosomes with therapeutic agents: Due to unique characteristics of exosomes such as low immunogenicity, tolerability, and small size, they can be preloaded with different agents and delivered effectively to target tissues (Rayner et al. 2017; Sun et al. 2010). Sun and co-workers have established an anti-inflammatory exosome-based delivery system by incorporation of curcumin into EL-4 (a murine lymphoma cell line)-derived exosomes (Zhuang et al. 2011). After treatment of the above-mentioned exosomes with RAW 264.7 cells (a murine macrophage cell line), the lower production of pro-inflammatory cytokines (IL-6 and TNF-α) was observed. Furthermore, they have confirmed the higher stability and solubility of exosome-mediated curcumin delivery. The exosome-mediated siRNA delivery can be an efficient tool to suppress or inactivate the inflammation process that occurs after CNS injury. Numerous studies have evaluated the delivery of siRNA inside the target tissues. Alvarez et al. have used engineered targeted exosome towards brain cells. The targeted exosomes were loaded with siRNA against GAPDH or BACE1. Another study included the embryonic cortical neuronal cells exosomes loaded with apoptosis speck-like protein containing a caspase recruitment domain (ASC) in the spinal cord-injured model (de Rivero Vaccari et al. 2016).

Bioengineering of exosomes to introduce targeting ligands: For specific targeting of exosomes, two strategies are developed: (1) parental cell engineering: As already discussed, exosomes are packaged by a lipid bilayer with distinct surface markers like CD9, CD63, and CD81. One of the most efficient strategies for introducing ligands on the surface of exosome is the expression of ligand with a fusion protein abundantly expressed in the exosomes. In this method, the parent cells were genetically modified with DNA encoding a ligand (or a receptor), which is specified for a receptor (or a ligand). Therefore, the secretion of parent cell-derived exosomes expressing the targeting peptide would be occurred. For instance, Gomari and colleagues transfected HEK293T cell lines with pLEX-LAMP2b-DARPin vector. The condition media containing viral particles was collected. The MSCs as a preference source of exosomes were transduced by these lenti viral particles. The LAMP2b-DARPin MSC-derived exosomes were harvested, which are effectively targeting HER2-positive breast tumour mice models (Gomari et al. 2019). (2) Post-isolation engineering: It is also known as post-insertion modification. In this procedure, the naive purified EVs were incubated with modified surface proteins that result in surface modification of EVs (Hood 2016; Kooijmans et al. 2016). Two methods of directed-EVs generation are compared by Wang et al. (Wang et al. 2018a). In this study, the HEK293 cells were transfected by pEVC1C2HER plasmid that codes for EVHB (anti-HER2 scFv antibody). The transfected HEK293-derived exosomes were obtained (called transfection EVs). In another experiment, the intact HEK293 cell-derived exosomes were harvested and incubated with pure EVHB (called reconstitution EVs). The ELISA results have showed that in comparison to transfection EVs, tenfold higher amount of the reconstitution EVs were bound to the HER2 receptor. Loading-modified exosomes with therapeutics that stimulate the regeneration capacity towards injured cells and tissues would be a promising strategy in regenerative therapy.

Limitations and Prospects

As summarised here, exosomes derived from multiple origins of stem cells exhibit a potential approach for the treatment of various disorders due to strong immunosuppressive and regenerative characteristics. Therefore, exosomes could open new horizons in the field of regenerative medicine. However, there are still limitations in clinical translational studies. Amongst the various issues to be addressed, technically, the presence of safe, robust, and reproducible isolation and purification method feels needed. Such a method is required to prepare exosomes on a large scale for clinical use. There are some other limitations regarding the yield, isolation method and purity of exosomes. The gold standard method for exosome isolation is differential ultracentrifugation; however, there are still some concerns about contaminating compounds, exosome aggregation, and proteins, which decrease the purity of exosomes. To address these limitations, commercial kits are introduced to isolate higher amounts of exosome faster and more reproducible. However, the heterogeneous population of exosomes, contamination with microvesicles, and apoptotic bodies compromise the outcome. Further studies are also warranted to set up the proper administration route and dosage to optimise their clinical utility. Besides, there are controversial discussions about tumour-derived exosomes in tumour suppression or activation. Therefore, the contents of exosomes must be compulsively evaluated and screened for potential negative feedbacks in clinical translation studies for tissue regeneration. Although there remains many unanswered questions in the area of exosome research, the distinct characteristics of exosomes apparently serve as new effective, safe, and powerful therapeutic approach. The reported preclinical and clinical trials in multiple target tissues and organs provide evidences for further translational research. Moreover, the regenerative potential of exosomes could be further accelerated by virtue of exosome engineering.

REFERENCES

Abraham, Aswin, and Anna Krasnodembskaya. 2020. *Stem Cells Translational Medicine 9* (1): 28–38.

Agarwal, Udit, Alex George, Srishti Bhutani, Shohini Ghosh-Choudhary, Joshua T Maxwell, Milton E Brown, Yash Mehta, Manu O Platt, Yaxuan Liang, and Susmita Sahoo. 2017. *Circulation Research 120* (4): 701–712.

Alipoor, Shamila D, and Esmaeil Mortaz. 2020. In *Exosomes*, 383–414. Academic Press.

Alvarez-Erviti, Lydia, Yiqi Seow, Hai Fang Yin, Corinne Betts, Samira Lakhal, and Matthew JA Wood. 2011. *Nature Biotechnology 29* (4): 341–345.

Andaloussi, Samir EL, Imre Mäger, Xandra O Breakefield, and Matthew JA Wood. 2013. *Nature Reviews Drug Discovery 12* (5): 347–357.

Anderson, Will, Darby Kozak, Victoria A Coleman, Åsa K Jämting, and Matt Trau. 2013. *Journal of Colloid and Interface Science 405*: 322–330.

Anju, VP, PR Jithesh, and Sunil K Narayanankutty. 2019. *Sensors and Actuators A: Physical 285*: 35–44.

Antunes, Lusânia M Greggi, Joana D'arc C Darin, and Maria De Lourdes P Bianchi. 2000. *Pharmacological Research 41* (4): 405–411.

Ashcroft, BA, J De Sonneville, Y Yuana, S Osanto, R Bertina, ME Kuil, and TH Oosterkamp. 2012. *Biomedical Microdevices 14* (4): 641–649.

Askari E, Naghib SM. 2020. *Methods Mol Biol. 2125*: 197–204.

Basile, DP. 2007. *Kidney International 72* (2): 151–156.

Batrakova, Elena V, and Myung Soo Kim. 2015. *Journal of Controlled Release 219*: 396–405.

Beltrami, Cristina, Marie Besnier, Saran Shantikumar, Andrew IU Shearn, Cha Rajakaruna, Abas Laftah, Fausto Sessa, Gaia Spinetti, Enrico Petretto, and Gianni D Angelini. 2017. *Molecular Therapy 25* (3): 679–693.

Bhadra, Jolly, Anton Popelka, Asma Abdulkareem, Marian Lehocky, Petr Humpolicek, and Noora Al-Thani. 2019. *Sensors and Actuators A: Physical 288*: 47–54.

Bonventre, Joseph V, and Li Yang. 2011. *The Journal of Clinical Investigation 121* (11): 4210–4221.

Bucan, Vesna, Desiree Vaslaitis, Claas-Tido Peck, Sarah Strauß, Peter M Vogt, and Christine Radtke. 2019. *Molecular Neurobiology 56*(3): 1812–1824.

Caby, Marie-Pierre, Danielle Lankar, Claude Vincendeau-Scherrer, Graça Raposo, and Christian Bonnerot. 2005. *International Immunology 17* (7): 879–887.

Cannistrà, Marco, Michele Ruggiero, Alessandra Zullo, Giuseppe Gallelli, Simone Serafini, Mazzitelli Maria, Agostino Naso, Raffaele Grande, Raffaele Serra, and Bruno Nardo. 2016. *International Journal of Surgery 33*: S57–S70.

Chen, Chihchen, Johan Skog, Chia-Hsien Hsu, Ryan T Lessard, Leonora Balaj, Thomas Wurdinger, Bob S Carter, Xandra O Breakefield, Mehmet Toner, and Daniel Irimia. 2010. *Lab on a Chip 10* (4): 505–511.

Chen, Yue, Aram J Chung, Ting-Hsiang Wu, Michael A Teitell, Dino Di Carlo, and Pei-Yu Chiou. 2014a. *Small 10* (9): 1746–1751.

Chen, Yue, Ting-Hsiang Wu, Aram Chung, Yu-Chung Kung, Michael A Teitell, Dino Di Carlo, and Pei-Yu Chiou. 2014b. *Optical Trapping and Optical Micromanipulation XI*.

Cheng, Lesley, Robyn A Sharples, Benjamin J Scicluna, and Andrew F Hill. 2014. *Journal of Extracellular Vesicles 3* (1): 23743.

Chew, Jacob Ren Jie, Shang Jiunn Chuah, Kristeen Ye Wen Teo, Shipin Zhang, Ruenn Chai Lai, Jia Hui Fu, Lum Peng Lim, Sai Kiang Lim, and Wei Seong Toh. 2019. *Acta Biomaterialia 89*: 252–264.

"Chinese Clinical Trial Register. A study for the key technology of mesenchymal stem cells exosomes atomization in the treatment of novel coronavirus pneumonia (COVID-19). Registration number: ChiCTR2000030261." 2020.

"Chinese Clinical Trial Register. HUMSCs and exosomes treating patients with lung injury following novel coronavirus pneumonia (COVID-19). Registration number: ChiCTR2000030484." 2020.

Cho, Young-Eun, Sang-Hyun Kim, Byung-Heon Lee, and Moon-Chang Baek. 2017. *Biomolecules & Therapeutics 25* (4): 367.

Choghaei, Encieh, Gholamreza Khamisipour, Mojtaba Falahati, Behrooz Naeimi, Majid Mossahebi-Mohammadi, Rahim Tahmasebi, Mojtaba Hasanpour, Shakib Shamsian, and Zahra Sadat Hashemi. 2016. *Oncology Research Featuring Preclinical and Clinical Cancer Therapeutics 23* (1–2): 69–78.

Choumerianou, Despoina M, Helen Dimitriou, and Maria Kalmanti. 2008. *Tissue Engineering Part B: Reviews 14* (1): 53–60.

Clark, Kaitlin, Sheng Zhang, Sylvain Barthe, Priyadarsini Kumar, Christopher Pivetti, Nicole Kreutzberg, Camille Reed, Yan Wang, Zachary Paxton, and Diana Farmer. 2019. *Cells 8* (12): 1497.

"Clinical Trials.gov. A pilot clinical study on inhalation of mesenchymal stem cells exosomes treating severe novel coronavirus pneumonia. Registration number: NCT04276987." 2020.

Conde-Vancells, Javier, Eva Rodriguez-Suarez, Nieves Embade, David Gil, Rune Matthiesen, Mikel Valle, Felix Elortza, Shelly C Lu, Jose M Mato, and Juan M Falcon-Perez. 2008. *Journal of Proteome Research 7* (12): 5157–5166.

Contreras-Naranjo, Jose C, Hung-Jen Wu, and Victor M Ugaz. 2017. *Lab on a Chip 17* (21): 3558–3577.

Cooper, Denise R, Chunyan Wang, Rehka Patel, Andrea Trujillo, Niketa A Patel, Jamie Prather, Lisa J Gould, and Mack H Wu. 2018. *Advances in Wound Care 7* (9): 299–308.

Critser, Paul J, and Mervin C Yoder. 2010. *Current Opinion in Organ Transplantation 15* (1): 68.

Dear, James W, Jonathan M Street, and Matthew A Bailey. 2013. *Proteomics 13* (10–11): 1572–1580.

Debenedetti, Pablo G. 2003. *Journal of Physics: Condensed Matter 15* (45): R1669.

Delcayre, Alain, Angeles Estelles, Jeffrey Sperinde, Thibaut Roulon, Pedro Paz, Barbara Aguilar, Janeth Villanueva, SuSu Khine, and Jean-Bernard Le Pecq. 2005. *Blood Cells, Molecules, and Diseases 35* (2): 158–168.

van Dongen, Helena M, Niala Masoumi, Kenneth W Witwer, and D Michiel Pegtel. 2016. *Microbiology and Molecular Biology Reviews 80* (2): 369–386.

Driessen, Britta JH, Colin Logie, and Lucienne A Vonk. 2017. *Cell Biology and Toxicology 33* (4): 329–349.

Du, Yingdong, Dawei Li, Conghui Han, Haoyu Wu, Longmei Xu, Ming Zhang, Jianjun Zhang, and Xiaosong Chen. 2017. *Cellular Physiology and Biochemistry 43* (2): 611–625.

Dudani, Jaideep S, Daniel R Gossett, Henry TK Tse, Robert J Lamm, Rajan P Kulkarni, and Dino Di Carlo. 2015. *Biomicrofluidics 9* (1): 014112.

Eldh, Maria, Jan Lötvall, Carina Malmhäll, and Karin Ekström. 2012. *Molecular Immunology 50* (4): 278–286.

Eming, Sabine A, Paul Martin, and Marjana Tomic-Canic. 2014. *Science Translational Medicine 6* (265): 265sr6-265sr6.

Erni, Rolf, Marta D Rossell, Christian Kisielowski, and Ulrich Dahmen. 2009. *Physical Review Letters 102* (9): 096101.

Fang, Doreen YP, Hamish W King, Jordan YZ Li, and Jonathan M Gleadle. 2013. *Nephrology 18* (1): 1–10.

Fang, Shimeng, Hongzhu Tian, Xiancheng Li, Dong Jin, Xiaojie Li, Jing Kong, Chun Yang, Xuesong Yang, Yao Lu, and Yong Luo. 2017. *PLoS One 12* (4): e0175050.

Ferreira, Ana Catarina, Nuno Sousa, João M Bessa, João Carlos Sousa, and Fernanda Marques. 2018. *Neuroscience & Biobehavioral Reviews 95*: 73–84.

Ferreira, Andrea da Fonseca, Pricila da Silva Cunha, Virgínia Mendes Carregal, Priscila de Cássia da Silva, Marcelo Coutinho de Miranda, Marianna Kunrath-Lima, Mariane Izabella Abreu de Melo, Camila Cristina Fraga Faraco, Joana Lobato Barbosa, and Frédéric Frezard. 2017. *Stem Cells International* 9841035. doi: 10.1155/2017/9841035.

Filipe, Vasco, Andrea Hawe, and Wim Jiskoot. 2010. *Pharmaceutical Research 27* (5): 796–810.

Fisher, Kurt, Raj Vuppalanchi, and Romil Saxena. 2015. *Archives of Pathology and Laboratory Medicine 139* (7): 876–887.

Forbes, Josephine M, Melinda T Coughlan, and Mark E Cooper. 2008. *Diabetes 57* (6): 1446–1454.

Gallet, Romain, James Dawkins, Jackelyn Valle, Eli Simsolo, Geoffrey De Couto, Ryan Middleton, Eleni Tseliou, Daniel Luthringer, Michelle Kreke, and Rachel R Smith. 2017. *European Heart Journal 38* (3): 201–211.

Ghorbanzade S, Naghib SM, Sadr A, et al. 2020. Multifunctional magnetic nanoparticles-labeled mesenchymal stem cells for hyperthermia and bioimaging applications. *Methods in Molecular Biology 2125*: 57–72. doi:10.1007/7651_2019_271.

Ghorbanzade S, Naghib SM. 2019. *Methods Mol Biol 2125*: 181–192.

Gomari, Hosna, Mehdi Forouzandeh Moghadam, Masoud Soleimani, Mahlegha Ghavami, and Shabanali Khodashenas. 2019. *International Journal of Nanomedicine 14*: 5679.

Gould, Stephen J, Amy M Booth, and James EK Hildreth. 2003. *Proceedings of the National Academy of Sciences 100* (19): 10592–10597.

Grange, Cristina, Stefania Tritta, Marta Tapparo, Massimo Cedrino, Ciro Tetta, Giovanni Camussi, and Maria Felice Brizzi. 2019. *Scientific Reports 9* (1): 1–13.

Guo, Shang-Chun, Shi-Cong Tao, Wen-Jing Yin, Xin Qi, Ting Yuan, and Chang-Qing Zhang. 2017. *Theranostics 7* (1): 81.

György, Bence, Michelle E Hung, Xandra O Breakefield, and Joshua N Leonard. 2015. *Annual Review of Pharmacology and Toxicology 55*: 439–464.

György, Bence, Tamás G Szabó, Mária Pásztói, Zsuzsanna Pál, Petra Misják, Borbála Aradi, Valéria László, Eva Pállinger, Erna Pap, and Agnes Kittel. 2011. *Cellular and Molecular Life Sciences 68* (16): 2667–2688.

Ha, Dinh, Ningning Yang, and Venkatareddy Nadithe. 2016. *Acta Pharmaceutica Sinica B 6* (4): 287–296.
Haertinger, Maximilian, Tamara Weiss, Anda Mann, Annette Tabi, Victoria Brandel, and Christine Radtke. 2020. *Cells 9* (1): 163.
Haney, Matthew J, Natalia L Klyachko, Yuling Zhao, Richa Gupta, Evgeniya G Plotnikova, Zhijian He, Tejash Patel, Aleksandr Piroyan, Marina Sokolsky, and Alexander V Kabanov. 2015. *Journal of Controlled Release 207*: 18–30.
Harding, Clifford V, John E Heuser, and Philip D Stahl. 2013. *Journal of Cell Biology 200* (4): 367–371.
Hashemi, Zahra Sadat, Saeed Khalili, Mehdi Forouzandeh Moghadam, and Esmaeil Sadroddiny. 2017. *Expert Review of Respiratory Medicine 11* (2): 147–157.
Hashemi, Zahra Sadat, Mahdi Forouzandeh Moghadam, and Masoud Soleimani. 2013. *Iranian Journal of Basic Medical Sciences 16* (10): 1075.
Hashemi, Zahra Sadat, Mehdi Forouzandeh Moghadam, and Masoud Soleimani. 2015. *In Vitro Cellular & Developmental Biology-Animal 51* (5): 495–506.
Hashemi, Zahra Sadat, Mehdi Forouzandeh Moghadam, Samila Farokhimanesh, Masoumeh Rajabibazl, and Esmaeil Sadroddiny. 2018a. *Iranian Journal of Basic Medical Sciences 21* (4): 427.
Hashemi, Zahra Sadat, Mehdi Forouzandeh Moghadam, and Esmaeil Sadroddiny. 2018b. *International Journal of Cancer Management 11* (8): e63540. doi: 10.5812/ijcm.63540.
Hashemi, Zahra Sadat, Mehdi Forouzandeh Moghadam, Saeed Khalili, Mahlegha Ghavami, Fatemeh Salimi, and Esmaeil Sadroddiny. 2019. *Breast Cancer 26* (2): 215–228.
Hein, Birka, Katrin I Willig, and Stefan W Hell. 2008. *Proceedings of the National Academy of Sciences 105* (38): 14271–14276.
Heinzelman, Pete, Tina Bilousova, Jesus Campagna, and Varghese John. 2016. *International Journal of Alzheimer's Disease* 8053139. doi: 10.1155/2016/8053139.
Hoo, Christopher M, Natasha Starostin, Paul West, and Martha L Mecartney. 2008. *Journal of Nanoparticle Research 10* (1): 89–96.
Hood, Joshua L. 2016. *Nanomedicine 11* (13): 1745–1756.
Hosseini-Beheshti, Elham, Steven Pham, Hans Adomat, Na Li, and Emma S Tomlinson Guns. 2012. *Molecular & Cellular Proteomics 11* (10): 863–885.
Hu, Xinyang, Yinchuan Xu, Zhiwei Zhong, Yan Wu, Jing Zhao, Yingchao Wang, Haifeng Cheng, Minjian Kong, Fengjiang Zhang, and Qi Chen. 2016. *Circulation Research 118* (6): 970–983.
Huang, Brian J, Jerry C Hu, and Kyriacos A Athanasiou. 2016. *Biomaterials 98*: 1–22.
Huang, Peisen, Li Wang, Qing Li, Jun Xu, Junyan Xu, Yuyan Xiong, Guihao Chen, Haiyan Qian, Chen Jin, and Yuan Yu. 2019. *Stem Cell Research & Therapy 10* (1): 1–12.
Huguet, Claude, Adolfo Gavelli, and Stefano Bona. 1994. *Journal of the American College of Surgeons 178* (5): 454–458.
Ibrahim, Ahmed Gamal-Eldin, Ke Cheng, and Eduardo Marbán. 2014. *Stem Cell Reports 2* (5): 606–619.
Im, Hyungsoon, Huilin Shao, Yong Il Park, Vanessa M Peterson, Cesar M Castro, Ralph Weissleder, and Hakho Lee. 2014. *Nature Biotechnology 32* (5): 490–495.
Jella, Kishore Kumar, Tahseen H Nasti, Zhentian Li, Sudarshan R Malla, Zachary S Buchwald, and Mohammad K Khan. 2018. *Vaccines 6* (4): 69.
Ji, Qiuhong, Yuhua Ji, Jingwen Peng, Xin Zhou, Xinya Chen, Heng Zhao, Tian Xu, Ling Chen, and Yun Xu. 2016. *PloS One 11* (9): e0163645.
Jiang, Zhen-zhen, Yu-mei Liu, Xin Niu, Jian-yong Yin, Bin Hu, Shang-chun Guo, Ying Fan, Yang Wang, and Nian-song Wang. 2016. *Stem Cell Research & Therapy 7* (1): 24.
Jing, Hui, Xiaomin He, and Jinghao Zheng. 2018. *Translational Research 196*: 1–16.
Johnsen, Kasper Bendix, Johann Mar Gudbergsson, Martin Najbjerg Skov, Linda Pilgaard, Torben Moos, and Meg Duroux. 2014. *Biochimica et Biophysica Acta (BBA)-Reviews on Cancer 1846* (1): 75–87.
Kalantari, Erfan, and Seyed Morteza Naghib. 2019. *Materials Science and Engineering: C 98*: 1087–1096.
Kalantari, Erfan, Seyed Morteza Naghib, Narges Jafarbeik Iravani, Atousa Aliahmadi, M. Reza Naimi-Jamal, and Masoud Mozafari. 2018a. *Ceramics International Ceramics 44* (12): 14704–14711.
Kalantari, Erfan, Seyed Morteza Naghib, Narges Jafarbeik Iravani, Rezvan Esmaeili, M Reza Naimi-Jamal, and Masoud Mozafari. 2019. *Materials Science and Engineering: C 105*: 109912.
Kalantari, Erfan, Seyed Morteza Naghib, M Reza Naimi-Jamal, Atousa Aliahmadi, Narges Jafarbeik Iravani, and Masoud Mozafari. 2018b. *Ceramics International 44* (11): 12731–12738.
Kalantari, Erfan, Seyed Morteza Naghib, M Reza Naimi-Jamal, Rezvan Esmaeili, Keivan Majidzadeh-A, and Masoud Mozafari. 2018c. *Materials Today: Proceedings 5* (7): 15744–15753.

Khan, Mohsin, Emily Nickoloff, Tatiana Abramova, Jennifer Johnson, Suresh Kumar Verma, Prasanna Krishnamurthy, Alexander Roy Mackie, Erin Vaughan, Venkata Naga Srikanth Garikipati, and Cynthia Benedict. 2015. *Circulation Research 117* (1): 52–64.
Kharroubi, Akram T, and Hisham M Darwish. 2015. *World Journal of Diabetes 6* (6): 850.
Khatri, Mahesh, Levi Arthur Richardson, and Tea Meulia. 2018. *Stem Cell Research & Therapy 9* (1): 1–13.
Khodashenas, Shabanali, Saeed Khalili, and Mehdi Forouzandeh Moghadam. 2019. *Biotechnology Letters 41* (4–5): 523–531.
Kholia, Sharad, Maria Beatriz Herrera Sanchez, Massimo Cedrino, Elli Papadimitriou, Marta Tapparo, Maria Chiara Deregibus, Maria Felice Brizzi, Ciro Tetta, and Giovanni Camussi. 2018. *Frontiers in Immunology 9*: 1639.
Kim, Yang-Il. 2003. *Journal of Hepato-Biliary-Pancreatic Surgery 10* (3): 195–199.
Ko, Jina, Matthew A Hemphill, David Gabrieli, Leon Wu, Venkata Yelleswarapu, Gladys Lawrence, Wesley Pennycooke, Anup Singh, Dave F Meaney, and David Issadore. 2016. *Scientific Reports 6*: 31215.
Kooijmans, SAA, LAL Fliervoet, R Van Der Meel, MHAM Fens, HFG Heijnen, PMP Van Bergen En Henegouwen, P Vader, and RM Schiffelers. 2016. *Journal of Controlled Release 224*: 77–85.
Kumar, Dhwani, Rachna Manek, Vijaya Raghavan, and Kevin K Wang. 2018. *Molecular Neurobiology 55* (3): 2112–2124.
Kunugi, Shinobu, Akira Shimizu, Naomi Kuwahara, Xuanyi Du, Mikiko Takahashi, Yasuhiro Terasaki, Emiko Fujita, Akiko Mii, Shinya Nagasaka, and Toshio Akimoto. 2011. *Laboratory Investigation 91* (2): 170–180.
Lai, Charles Pin-Kuang, and Xandra Owen Breakefield. 2012. *Frontiers in Physiology 3*: 228.
Lazar, Erzsebet, Theodora Benedek, Szilamer Korodi, Nora Rat, Jocelyn Lo, and Imre Benedek. 2018. *World Journal of Stem Cells 10* (8): 106.
Lee, Kyungheon, Huilin Shao, Ralph Weissleder, and Hakho Lee. 2015. *ACS Nano 9* (3): 2321–2327.
Lemasters, John J, and Ronald G Thurman. 1997. *Annual Review of Pharmacology and Toxicology 37* (1): 327–338.
Leng, Zikuan, Rongjia Zhu, Wei Hou, Yingmei Feng, Yanlei Yang, Qin Han, Guangliang Shan, Fanyan Meng, Dongshu Du, and Shihua Wang. 2020. *Aging and Disease 11* (2): 216.
Li, Chenggang, Xiao Li, Bichun Zhao, and Chunfang Wang. 2020. *Archives of Physiology and Biochemistry*: 1–7.
Li, Mu, Emily Zeringer, Timothy Barta, Jeoffrey Schageman, Angie Cheng, and Alexander V Vlassov. 2014. *Philosophical Transactions of the Royal Society B: Biological Sciences 369* (1652): 20130502.
Li, Wei, Wei Ding, Yao Nie, Qian He, Jinxia Jiang, and Zidong Wei. 2019. *ACS Applied Materials & Interfaces 11* (25): 22290–22296.
Li, Wenyue, Yunsong Liu, Ping Zhang, Yiman Tang, Miao Zhou, Weiran Jiang, Xiao Zhang, Gang Wu, and Yongsheng Zhou. 2018. *ACS Applied Materials & Interfaces 10* (6): 5240–5254.
Li, Xiao, Lingying Liu, Jing Yang, Yonghui Yu, Jiake Chai, Lingyan Wang, Li Ma, and Huinan Yin. 2016. *E Bio Medicine 8*: 72–82.
Li, Yi, and Michael Chopp. 2009. *Neuroscience Letters 456* (3): 120–123.
Liang, Li-Guo, Meng-Qi Kong, Sherry Zhou, Ye-Feng Sheng, Ping Wang, Tao Yu, Fatih Inci, Winston Patrick Kuo, Lan-Juan Li, and Utkan Demirci. 2017. *Scientific Reports 7*: 46224.
Limoni, Shabanali Khodashenas, Mehdi Forouzandeh Moghadam, Seyed Mohammad Moazzeni, Hosna Gomari, and Fatemeh Salimi. 2019. *Applied Biochemistry and Biotechnology 187* (1): 352–364.
Lioupis, C. 2005. *Journal of Wound Care 14* (2): 84–86.
Liu, Bohao, Benjamin W Lee, Koki Nakanishi, Aranzazu Villasante, Rebecca Williamson, Jordan Metz, Jinho Kim, Mariko Kanai, Lynn Bi, and Kristy Brown. 2018. *Nature Biomedical Engineering 2* (5): 293–303.
Liu, Chao, Jiayi Guo, Fei Tian, Na Yang, Fusheng Yan, Yanping Ding, Jing Yan Wei, Guoqing Hu, Guangjun Nie, and Jiashu Sun. 2017. *ACS Nano 11* (7): 6968–6976.
Logozzi, Mariantonia, Angelo De Milito, Luana Lugini, Martina Borghi, Luana Calabro, Massimo Spada, Maurizio Perdicchio, Maria Lucia Marino, Cristina Federici, and Elisabetta Iessi. 2009. *PloS One 4* (4): e5219.
Lopez-Verrilli, María Alejandra. 2012. *Frontiers in Physiology 3*: 205.
Lopez-Verrilli, María Alejandra, Frederic Picou, and Felipe A Court. 2013. *Glia 61* (11): 1795–1806.
Lu, Yanhui, Yan Zhou, Ruiyi Zhang, Lulu Wen, Kaimin Wu, Yanfei Li, Yaobing Yao, Ranran Duan, and Yanjie Jia. 2019. *Frontiers in Neuroscience 13*: 209.

Ma, Teng, Yueqiu Chen, Yihuan Chen, Qingyou Meng, Jiacheng Sun, Lianbo Shao, Yunsheng Yu, Haoyue Huang, Yanqiu Hu, and Ziying Yang. 2018. *Stem Cells International* 3290372. doi: 10.1155/2018/3290372.

Mahla, Ranjeet Singh. 2016. *International Journal of Cell Biology* 6940283. doi: 10.1155/2016/6940283.

Mangalmurti, N, CA Hunter. 2020. *53*(1): 19–25. doi:10.1016/j.immuni.2020.06.017.

Mardpour, Soura, Mohammad Hossein Ghanian, Hamid Sadeghi-Abandansari, Saeid Mardpour, Abdoreza Nazari, Faezeh Shekari, and Hossein Baharvand. 2019. *ACS Applied Materials & Interfaces 11* (41): 37421–37433.

Masyuk, Anatoliy I, Tatyana V Masyuk, and Nicholas F LaRusso. 2013. *Journal of Hepatology 59* (3): 621–625.

Mead, Ben, Martin Berry, Ann Logan, Robert AH Scott, Wendy Leadbeater, and Ben A Scheven. 2015. *Stem Cell Research 14* (3): 243–257.

Mead, Ben, and Stanislav Tomarev. 2017. *Stem Cells Translational Medicine 6* (4): 1273–1285.

Meftahi, Gholam Hossein, Zohreh Jangravi, Hedayat Sahraei, and Zahra Bahari. 2020. *Inflammation Research*: 1–15.

Mehdizadeh, Kashi Abolfazl, Kobra Tahermanesh, Shahla Chaichian, Mohammad Taghi Joghataei, Fatemeh Moradi, Seyed Mohammad Tavangar, Najafabadi Ashraf Sadat Mousavi, Nasrin Lotfibakhshaiesh, Beyranvand Shahram Pour, and Anvari Yazdi Abbas Fazel. 2014. *Galen Medical Journal 3* (2): 63–80.

Michalopoulos, George K. 2013. *Comprehensive Physiology 3* (1): 485–513.

Mohammadpour, Zahra, and Keivan Majidzadeh-A. 2020. *ACS Biomaterials Science & Engineering 6* (4): 1852–1873.

Molaabasi, F, B Hajipour-Verdom, M Alipour, SM Naghib. *Methods Mol Biol*. 2020; *2125*: 27–37. doi:10.1007/7651_2019_273.

Momen-Heravi, Fatemeh, Banishree Saha, Karen Kodys, Donna Catalano, Abhishek Satishchandran, and Gyongyi Szabo. 2015. *Journal of Translational Medicine 13* (1): 261.

Nassar, Wael, Mervat El-Ansary, Dina Sabry, Mostafa A Mostafa, Tarek Fayad, Esam Kotb, Mahmoud Temraz, Abdel-Naser Saad, Wael Essa, and Heba Adel. 2016. *Biomaterials Research 20* (1): 21.

Nilsson, Jonas, Johan Skog, Annika Nordstrand, Vladimir Baranov, Lucia Mincheva-Nilsson, XO Breakefield, and Anders Widmark. 2009. *British Journal of Cancer 100* (10): 1603–1607.

Nojima, Hiroyuki, Christopher M Freeman, Rebecca M Schuster, Lukasz Japtok, Burkhard Kleuser, Michael J Edwards, Erich Gulbins, and Alex B Lentsch. 2016. *Journal of Hepatology 64* (1): 60–68.

Nolte, Esther, Tom Cremer, Robert C Gallo, and Leonid B Margolis. 2016. *Proceedings of the National Academy of Sciences 113* (33): 9155–9161.

Nong, Kate, Weiwei Wang, Xin Niu, Bin Hu, Chenchao Ma, Yueqing Bai, Bo Wu, Yang Wang, and Kaixing Ai. 2016. *Cytotherapy 18* (12): 1548–1559.

Ohno, Shin-ichiro, Masakatsu Takanashi, Katsuko Sudo, Shinobu Ueda, Akio Ishikawa, Nagahisa Matsuyama, Koji Fujita, Takayuki Mizutani, Tadaaki Ohgi, and Takahiro Ochiya. 2013. *Molecular Therapy 21* (1): 185–191.

Ostrowski, Matias, Nuno B Carmo, Sophie Krumeich, Isabelle Fanget, Graça Raposo, Ariel Savina, Catarina F Moita, Kristine Schauer, Alistair N Hume, and Rui P Freitas. 2010. *Nature Cell Biology 12* (1): 19–30.

Patschan, D, S Patschan, and GA Müller. 2011. *International Journal of Nephrology 2011*.

Payandeh, Zahra, Mohammad Reza Rahbar, Abolfazl Jahangiri, Zahra Sadat Hashemi, Alireza Zakeri, Moslem Jafarisani, Mohammad Javad Rasaee, and Saeed Khalili. 2020. *Journal of Theoretical Biology 505*: 110425.

Peake, Philip W, Timothy J Pianta, Lena Succar, Mangalee Fernando, Debbie J Pugh, Kathleen McNamara, and Zoltan H Endre. 2014. *PLoS One 9* (6): e98644.

Povero, Davide, Akiko Eguchi, Hongying Li, Casey D Johnson, Bettina G Papouchado, Alexander Wree, Karen Messer, and Ariel E Feldstein. 2014. *PloS One 9* (12): e113651.

Putra, Agung, Fatkhan Baitul Ridwan, Allisha Irwaniyanti Putridewi, Azizah Retno Kustiyah, Ken Wirastuti, Nur Anna Chalimah Sadyah, Ika Rosdiana, and Delfitri Munir. 2018. *Open Access Macedonian Journal of Medical Sciences 6* (10): 1779.

Rahmanian, Mehdi, Mohammad Mehdi Dehghan, Leila Eini, Seyed Morteza Naghib, Hossein Gholami, Saeed Farzad Mohajeri, Kaveh Rahimi Mamaghani, and Keivan Majidzadeh-A. 2019. *Journal of the Taiwan Institute of Chemical Engineers 101*: 214–220.

Rajendran, K, N Krishnasamy, J Rangarajan, J Rathinam, M Natarajan, and A Ramachandran. 2020. *Journal of Medical Virology 92*(9): 1475–1483. doi:10.1002/jmv.25961.

Raposo, Graca, Hans W Nijman, Willem Stoorvogel, R Liejendekker, Clifford V Harding, CJ Melief, and Hans J Geuze. 1996. *Journal of Experimental Medicine 183* (3): 1161–1172.

Raposo, Graça, and Willem Stoorvogel. 2013. *Journal of Cell Biology 200* (4): 373–383.

Rayner, Simon, Sören Bruhn, Helen Vallhov, Anna Andersson, R Blake Billmyre, and Annika Scheynius. 2017. *Scientific Reports 7* (1): 1–9.

Rezaei T, Amini M, Hashemi ZS, Mansoori B, Rezaei S, Karami H, Mosafer J, Mokhtarzadeh A, Baradaran B. 2020. *Free Radic Biol Med. 152*: 432–454.

Reza-Zaldivar, Edwin E, Mercedes A Hernández-Sapiéns, Yanet K Gutiérrez-Mercado, Sergio Sandoval-Ávila, Ulises Gomez-Pinedo, Ana L Márquez-Aguirre, Estefanía Vázquez-Méndez, Eduardo Padilla-Camberos, and Alejandro A Canales-Aguirre. 2020. *Neural Regeneration Research 14* (9): 1626.

Riazifar, Milad, M Rezaa Mohammadi, Egest J Pone, Ashish Yeri, Cecilia Lasser, Aude I Segaliny, Laura L McIntyre, Ganesh Vilas Shelke, Elizabeth Hutchins, and Ashley Hamamoto. 2019. *ACS Nano 13* (6): 6670–6688.

de Rivero Vaccari, Juan Pablo, Frank Brand III, Stephanie Adamczak, Stephanie W Lee, Jon Perez-Barcena, Michael Y Wang, M Ross Bullock, W Dalton Dietrich, and Robert W Keane. 2016. *Journal of Neurochemistry 136*: 39–48.

de la Rosa, Laura Conde, Marieke H Schoemaker, Titia E Vrenken, Manon Buist-Homan, Rick Havinga, Peter LM Jansen, and Han Moshage. 2006. *Journal of Hepatology 44* (5): 918–929.

Salem KZ, Moschetta M, Sacco A, Imberti L, Rossi G, Ghobrial IM, Manier S, Roccaro AM. 2016. *Methods Mol Biol 1464*: 25–34.

Sarvar, Davod Pashoutan, Karim Shamsasenjan, and Parvin Akbarzadehlaleh. 2016. *Advanced Pharmaceutical Bulletin 6* (3): 293.

SenguptaV, Sengupta S, Lazo A, Woods P, Nolan A, Bremer N. 2020. *Stem Cells Dev 29*(12): 747–754. doi:10.1089/scd.2020.0080.

Shabbir, Arsalan, Audrey Cox, Luis Rodriguez-Menocal, Marcela Salgado, and Evangelos Van Badiavas. 2015. *Stem Cells and Development 24* (14): 1635–1647.

Shao, Huilin, Jaehoon Chung, Kyungheon Lee, Leonora Balaj, Changwook Min, Bob S Carter, Fred H Hochberg, Xandra O Breakefield, Hakho Lee, and Ralph Weissleder. 2015. *Nature Communications 6* (1): 1–9.

Shen, Chenguang, Zhaoqin Wang, Fang Zhao, Yang Yang, Jinxiu Li, Jing Yuan, Fuxiang Wang, Delin Li, Minghui Yang, and Li Xing. 2020. *Jama 323* (16): 1582–1589.

Sina, Abu Ali Ibn, Ramanathan Vaidyanathan, Shuvashis Dey, Laura G Carrascosa, Muhammad JA Shiddiky, and Matt Trau. 2016. *Scientific Reports 6*: 30460.

Skog, Johan, Tom Würdinger, Sjoerd Van Rijn, Dimphna H Meijer, Laura Gainche, William T Curry, Bob S Carter, Anna M Krichevsky, and Xandra O Breakefield. 2008. *Nature Cell Biology 10* (12): 1470–1476.

Sokolova, Viktoriya, Anna-Kristin Ludwig, Sandra Hornung, Olga Rotan, Peter A Horn, Matthias Epple, and Bernd Giebel. 2011. *Colloids and Surfaces B: Biointerfaces 87* (1): 146–150.

Sookoian, Silvia, and Carlos J Pirola. 2015. *World Journal of Gastroenterology: WJG 21* (3): 711.

Starchev, Konstantin, Jacques Buffle, and Elías Pérez. 1999. "Applications of fluorescence correlation spectroscopy: polydispersity measurements." *Journal of Colloid and Interface Science 213* (2): 479–487.

Stastna, Miroslava, Isotta Chimenti, Eduardo Marbán, and Jennifer E Van Eyk. 2010. *Proteomics 10* (2): 245–253.

Sullivan, H Cliff, and John D Roback. 2020. *Transfusion Medicine Reviews 34*(3): 145–150.

Sun, Dongmei, Xiaoying Zhuang, Xiaoyu Xiang, Yuelong Liu, Shuangyin Zhang, Cunren Liu, Stephen Barnes, William Grizzle, Donald Miller, and Huang-Ge Zhang. 2010. *Molecular Therapy 18* (9): 1606–1614.

Sweeney, Melanie D, Shiva Ayyadurai, and Berislav V Zlokovic. 2016. *Nature Neuroscience 19* (6): 771–783.

Szatanek, Rafal, Monika Baj-Krzyworzeka, Jakub Zimoch, Malgorzata Lekka, Maciej Siedlar, and Jarek Baran. 2017. *International Journal of Molecular Sciences 18* (6): 1153.

Takeda, Yuji S, and Qiaobing Xu. 2015. *PloS one 10* (8): e0135111.

Tatischeff, Irène, Eric Larquet, Juan M Falcón-Pérez, Pierre-Yves Turpin, and Sergei G Kruglik. 2012. *Journal of Extracellular Vesicles 1* (1): 19179.

Taylor, Douglas D, and Cicek Gercel-Taylor. 2008. *Gynecologic Oncology 110* (1): 13–21.

Thakur, Basant Kumar, Haiying Zhang, Annette Becker, Irina Matei, Yujie Huang, Bruno Costa-Silva, Yan Zheng, Ayuko Hoshino, Helene Brazier, and Jenny Xiang. 2014. *Cell Research 24* (6): 766–769.

Thulin, Petra, Robert J Hornby, Mariona Auli, Gunnar Nordahl, Daniel J Antoine, Philip Starkey Lewis, Christopher E Goldring, B Kevin Park, Neus Prats, and Björn Glinghammar. 2017. *Biomarkers 22* (5): 461–469.

Tian, Yanhua, Suping Li, Jian Song, Tianjiao Ji, Motao Zhu, Gregory J Anderson, Jingyan Wei, and Guangjun Nie. 2014. *Biomaterials 35* (7): 2383–2390.

Tivol, William F, Ariane Briegel, and Grant J Jensen. 2008. *Microscopy and Microanalysis: The Official Journal of Microscopy Society of America, Microbeam Analysis Society, Microscopical Society of Canada 14* (5): 375.

Tomasoni, Susanna, Lorena Longaretti, Cinzia Rota, Marina Morigi, Sara Conti, Elisa Gotti, Chiara Capelli, Martino Introna, Giuseppe Remuzzi, and Ariela Benigni. 2013. *Stem Cells and Development 22* (5): 772–780.

Tønnesen, Jan, Fabien Nadrigny, Katrin I Willig, Roland Wedlich-Söldner, and U Valentin Nägerl. 2011. *Biophysical Journal 101* (10): 2545–2552.

Trajkovic, Katarina, Chieh Hsu, Salvatore Chiantia, Lawrence Rajendran, Dirk Wenzel, Felix Wieland, Petra Schwille, Britta Brügger, and Mikael Simons. 2008. *Science 319* (5867): 1244–1247.

Vaidyanathan, Ramanathan, Maedeh Naghibosadat, Sakandar Rauf, Darren Korbie, Laura G Carrascosa, Muhammad JA Shiddiky, and Matt Trau. 2014. *Analytical Chemistry 86* (22): 11125–11132.

Valadi, Hadi, Karin Ekström, Apostolos Bossios, Margareta Sjöstrand, James J Lee, and Jan O Lötvall. 2007. *Nature Cell Biology 9* (6): 654–659.

Viñas, Jose L, Dylan Burger, Joseph Zimpelmann, Randa Haneef, William Knoll, Pearl Campbell, Alex Gutsol, Anthony Carter, David S Allan, and Kevin D Burns. 2016. *Kidney International 90* (6): 1238–1250.

Vogel, Robert, Geoff Willmott, Darby Kozak, G Seth Roberts, Will Anderson, Linda Groenewegen, Ben Glossop, Anne Barnett, Ali Turner, and Matt Trau. 2011. *Analytical Chemistry 83* (9): 3499–3506.

Wang, Bo, Kevin Yao, Brooke M Huuskes, Hsin-Hui Shen, Junli Zhuang, Catherine Godson, Eoin P Brennan, Jennifer L Wilkinson-Berka, Andrea F Wise, and Sharon D Ricardo. 2016a. *Molecular Therapy 24* (7): 1290–1301.

Wang, Jing-Hung, Alexis V Forterre, Jinjing Zhao, Daniel O Frimannsson, Alain Delcayre, Travis J Antes, Bradley Efron, Stefanie S Jeffrey, Mark D Pegram, and AC Matin. 2018a. *Molecular Cancer Therapeutics 17* (5): 1133–1142.

Wang, Leo L, Jennifer J Chung, Elizabeth C Li, Selen Uman, Pavan Atluri, and Jason A Burdick. 2018b. *Journal of Controlled Release 285*: 152–161.

Wang M, Yuan Q, Xie L. 2018c. *Stem Cells Int.*; *2018*: 3057624. doi:10.1155/2018/3057624.

Wang, Zhen, Jun-Qiang Chen, Jin-lu Liu, and Lei Tian. 2016b. *Journal of Translational Medicine 14* (1): 297.

Wang, Zongxing, Hung-jen Wu, Daniel Fine, Jeffrey Schmulen, Ye Hu, Biana Godin, John XJ Zhang, and Xuewu Liu. 2013. *Lab on a Chip 13* (15): 2879–2882.

Wei, Xin, Xue Yang, Zhi-peng Han, Fang-fang Qu, Li Shao, and Yu-fang Shi. 2013. *Acta Pharmacologica Sinica 34* (6): 747–754.

Willig, Katrin I, Silvio O Rizzoli, Volker Westphal, Reinhard Jahn, and Stefan W Hell. 2006. *Nature 440* (7086): 935–939.

Wunsch, Benjamin H, Joshua T Smith, Stacey M Gifford, Chao Wang, Markus Brink, Robert L Bruce, Robert H Austin, Gustavo Stolovitzky, and Yann Astier. 2016. *Nature Nanotechnology 11* (11): 936–940.

Wyss, Romain, Luigino Grasso, Camille Wolf, Wolfgang Grosse, Davide Demurtas, and Horst Vogel. 2014. *Analytical Chemistry 86* (15): 7229–7233.

Xin, Hongqi, Yi Li, Ben Buller, Mark Katakowski, Yi Zhang, Xinli Wang, Xia Shang, Zheng Gang Zhang, and Michael Chopp. 2012. *Stem Cells 30* (7): 1556–1564.

Yamauchi, Junji, Yuki Miyamoto, Jonah R Chan, and Akito Tanoue. 2008. *Journal of Cell Biology 181* (2): 351–365.

Yan, Yongmin, Wenqian Jiang, Youwen Tan, Shengqiang Zou, Hongguang Zhang, Fei Mao, Aihua Gong, Hui Qian, and WenrongXu. 2017. *Molecular Therapy 25* (2): 465–479.

Yáñez-Mó, María, Pia R-M Siljander, Zoraida Andreu, Apolonija Bedina Zavec, Francesc E Borràs, Edit I Buzas, Krisztina Buzas, Enriqueta Casal, Francesco Cappello, and Joana Carvalho. 2015. *Journal of Extracellular Vesicles 4* (1): 27066.

Yang, Jialei, Xiufen Zhang, Xiangjie Chen, Lei Wang, and Guodong Yang. 2017. *Molecular Therapy-Nucleic Acids 7*: 278–287.

Yeo, Ronne Wee Yeh, Ruenn Chai Lai, Bin Zhang, Soon Sim Tan, Yijun Yin, Bao Ju Teh, and Sai Kiang Lim. 2013. *Advanced Drug Delivery Reviews 65* (3): 336–341.

Yoshioka, Yusuke, Nobuyoshi Kosaka, Yuki Konishi, Hideki Ohta, Hiroyuki Okamoto, Hikaru Sonoda, Ryoji Nonaka, Hirofumi Yamamoto, Hideshi Ishii, and Masaki Mori. 2014. *Nature Communications 5* (1): 1–8.

Yu, Haiying, Pamela J Vande Vord, Li Mao, Howard W Matthew, Paul H Wooley, and Shang-You Yang. 2009. *Biomaterials 30* (4): 508–517.

Yu, Samuel Mon-Wei, and Joseph V Bonventre. 2018. *Advances in Chronic Kidney Disease 25* (2): 166–180.

Yuana, Yuana, Tjerk H Oosterkamp, Svetlana Bahatyrova, Brian Ashcroft, P Garcia Rodriguez, Rogier M Bertina, and Susanne Osanto. 2010. *Journal of Thrombosis and Haemostasis 8* (2): 315–323.

Zarovni, Natasa, Antonietta Corrado, Paolo Guazzi, Davide Zocco, Elisa Lari, Giorgia Radano, Jekatarina Muhhina, Costanza Fondelli, Julia Gavrilova, and Antonio Chiesi. 2015. *Methods 87*: 46–58.

Zhang, Bin, Xiaodan Wu, Xu Zhang, Yaoxiang Sun, Yongmin Yan, Hui Shi, Yanhua Zhu, Lijun Wu, Zhaoji Pan, and Wei Zhu. 2015a. *Stem Cells Translational Medicine 4* (5): 513–522.

Zhang, Huang-Ge, and William E Grizzle. 2014. *The American Journal of Pathology 184* (1): 28–41.

Zhang, Jieyuan, Chunyuan Chen, Bin Hu, Xin Niu, Xiaolin Liu, Guowei Zhang, Changqing Zhang, Qing Li, and Yang Wang. 2016a. *International Journal of Biological Sciences 12* (12): 1472.

Zhang, Jieyuan, Junjie Guan, Xin Niu, Guowen Hu, Shangchun Guo, Qing Li, Zongping Xie, Changqing Zhang, and Yang Wang. 2015b. *Journal of Translational Medicine 13* (1): 49.

Zhang, Peng, Mei He, and Yong Zeng. 2016b. *Lab on a Chip 16* (16): 3033–3042.

Zhang, Xiao, Jing Guo, Gang Wu, and Yongsheng Zhou. 2015c. *Cell Proliferation 48* (6): 650–660.

Zhang, Yanlu, Michael Chopp, Yuling Meng, Mark Katakowski, Hongqi Xin, Asim Mahmood, and Ye Xiong. 2015d. *Journal of Neurosurgery 122* (4): 856–867.

Zhang, Yi, Michael Chopp, Xian Shuang Liu, Mark Katakowski, Xinli Wang, Xinchu Tian, David Wu, and Zheng Gang Zhang. 2017. *Molecular Neurobiology 54* (4): 2659–2673.

Zhang, ZX, ZX Zhang, LX Guan, K Zhang, Q Zhang, and LJ Dai. 2008. *Cytotherapy 10* (2): 134–139.

Zhang, Zhiwei, Junjie Yang, Weiya Yan, Yangxin Li, Zhenya Shen, and Takayuki Asahara. 2016c. *Journal of the American Heart Association 5* (1): e002856.

Zhao, Bin, Yijie Zhang, Shichao Han, Wei Zhang, Qin Zhou, Hao Guan, Jiaqi Liu, Jihong Shi, Linlin Su, and Dahai Hu. 2017. *Journal of Molecular Histology 48* (2): 121–132.

Zhao, Lingfei, Chenxia Hu, Ping Zhang, Hua Jiang, and Jianghua Chen. 2019. *Stem Cell Research & Therapy 10* (1): 1–9.

Zhao, Lu, Shanquan Chen, Xiaowei Shi, Hongcui Cao, and Lanjuan Li. 2018. *Stem Cell Research & Therapy 9* (1): 72.

Zhao, Zheng, Yang Yang, Yong Zeng, and Mei He. 2016. *Lab on a Chip 16* (3): 489–496.

Zhou, Qing, Guanghui Li, Kaiyang Chen, Hong Yang, Mengran Yang, Yuye Zhang, Yakun Wan, Yanfei Shen, and Yuanjian Zhang. 2019. *Analytical Chemistry 92* (1): 983–990.

Zhou, Ying, Huitao Xu, Wenrong Xu, Bingying Wang, Huiyi Wu, Yang Tao, Bin Zhang, Mei Wang, Fei Mao, and Yongmin Yan. 2013. *Stem Cell Research & Therapy 4* (2): 1–13.

Zhu, Jinyun, Kai Lu, Ning Zhang, Yun Zhao, Qunchao Ma, Jian Shen, Yinuo Lin, Pingping Xiang, Yaoliang Tang, and Xinyang Hu. 2018. *Artificial Cells, Nanomedicine, and Biotechnology 46* (8): 1659–1670.

Zhuang, Xiaoying, Xiaoyu Xiang, William Grizzle, Dongmei Sun, Shuangqin Zhang, Robert C Axtell, Songwen Ju, Jiangyao Mu, Lifeng Zhang, and Lawrence Steinman. 2011. *Molecular Therapy 19* (10): 1769–1779.

6

Bioceramic Nanoparticles for Tissue Engineering

Nitu Bhaskar
Indian Institute of Science, Bangalore, India

CONTENTS

Introduction

The intensifying ageing of world population, together with an accelerating incidence of diseases, is a leading driving force stimulating a rise in research studies to develop new materials (Fernandes et al. 2018). Tissue engineering (TE), a multidisciplinary field evolved decades ago, is a major revolution in the field of medicine, which combine the principles of engineering, material science, and life science focusing on the replacement, regeneration, and maintenance of various tissues of the human body. Hence, it is very important to restore the organ structure and functions that have been lost due to the destruction and loss of extracellular matrix (ECM) with the void of functional cells. This approach of tissue engineering has brought high expectations in the past several decades the perspective of regenerative diseased and damaged tissues using exciting synthetic materials with excellent cell compatibility. The concept of tissue engineering necessitates three main approaches: (i) the use of ideal three-dimensional scaffolds (3D scaffolds) acting as a porous substrate for cell proliferation and tissue development, (ii) an artificial extracellular matrix, (iii) the use of living cell or cell substitutes, and regulating biomolecules to replace limited functions of the tissue as shown in Figure 6.1 (Zhou and Lee 2011) (Duan et al. 2011). Three-dimensional scaffolds are used to mimic the essential extracellular matrix structure of a tissue in which cells reside, proliferate, and transform into different cell types (Hasan et al. 2018). The changes in cellular behaviours, which activate the mechanism of tissue engineering resulting in the formation of viable tissues, are due to intracellular cascade of biochemical and biomechanical signals originating from the close interaction between cell and ECM having the features (topography and structure) of nanometre size (Engel et al. 2008).

During the past few years, different varieties of materials (bioresorbable, bioactive, biodegradable, and permanent) have been designed and engineered to construct these TE applicable scaffolds. But among all of them, bioceramics have been considered extensively, as bone cement since 1892, due to their properties like corrosion resistance, a hard brittle surface, osteoconductivity, and excellent *in vivo* biological responses with minimal foreign body response, which are not shown by other materials such as metals and polymers. In general, bioceramics are divided into three categories depending on their integration with surrounding tissue after implantation (Table 6.1).

DOI: 10.1201/9781003153504-6

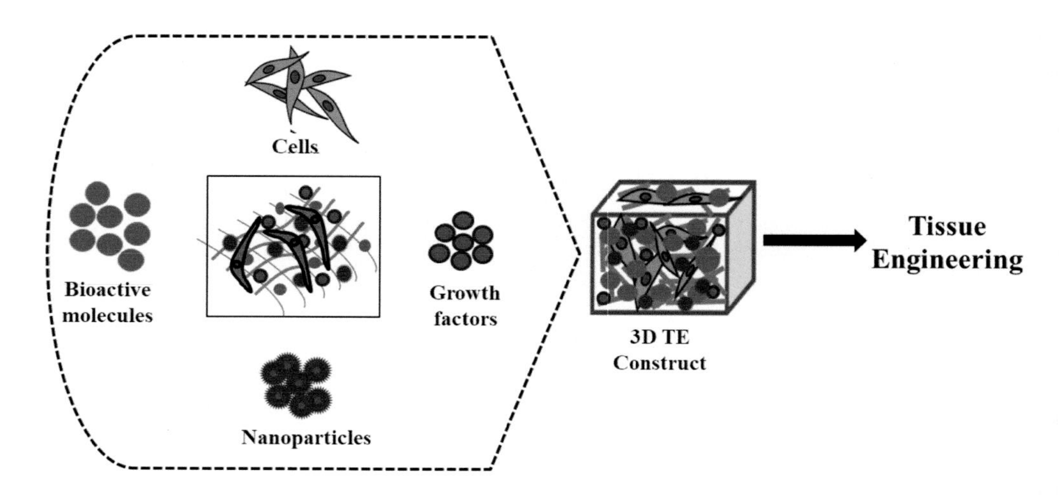

FIGURE 6.1 Schematic of Essential Components Used for 3D Construct/Scaffold-Based Tissue Engineering.

TABLE 6.1
Different Kinds of Bioceramic Used for Biomedical Applications

Type of Bioceramic	Properties	Tissue–Implant Interaction	Uses	Example
Bioinert (First generation)	Good mechanical properties; fracture toughness, wear and corrosion resistance	No interaction, mild inflammatory response with minor encapsulation	Used for structural-hold up for implants, such as femoral head and bone related devices	Alumina (Al_2O_3), Zirconia (ZrO_2), SiC, Si_3N_4, Sintered Hydroxyapatite (HA)
Bioactive (Second generation)	Brittle	Bond directly with living tissues through osteogenesis	Filling of small-bone defects and periodontal irregularities	Bioglasses and glass ceramics
Bioresorbable (Third generation)		Gradually absorbed *in vivo* and is replaced by bone over time	Bone tissue regeneration with minimum load bearing application	Calcium phosphates (CaPs), and calcium carbonates, calcium phosphate cements or calcium silicates

As compared to bulk material, the use of nanomaterials, which have constituent dimensions (smaller than 100 nm), has evoked a considerable amount of recognition because of their potential to mimic physical and chemical properties and immediate niche of tissues. For these reasons, over the years, nanotechnology has achieved tremendous progress, which includes synthesizing nanomaterials or nanoparticles (NPs) with improved mechanical and biological properties as compared to already existing traditional tissue engineering materials. The interest in nanoparticle research for tissue engineering is very widely spread. From a material point-of-view, the outstanding properties like high surface area with adjustable surface characteristics and high penetration ability make NPs among the finest potential and preferred candidates in tissue engineering.

NPs can be fabricated with different varieties of materials essentially metals, ceramics, polymers, organic materials, and composite. Bioceramic nanoparticles derived from inorganic, mechanically rigid, and non-metallic ceramics are the special set of fully, partially, or non-crystalline ceramics materials used to fill and restore damaged parts of hard tissues such as the musculoskeletal system and periodontal defects (Baino et al. 2015). Since these ceramics are brittle and have poor fracture toughness and elasticity with extremely high stiffness, their application in tissue engineering is limited only to hard tissue regeneration.

The aim of this chapter is to reconsider the available literature concerning the use of bioceramic nanoparticles used for various tissue engineering approaches. Other equally important nanoparticles, such as those derived from metals and polymers, fall out of the scope of this chapter and are not discussed.

Characteristic of Bioceramic Nanoparticles

The use of ceramic nanoparticles for tissue engineering is growing rapidly. It is better to fabricate scaffolds with ceramic nanoparticles as reinforcement to improve their mechanical and structural properties (Iron et al. 2019). The nanoparticles possess special physicochemical characteristics for example small dimensions, high reactivity, and wide surface area to mass ratio, which are distinct from their bulk composition (Moreno-Vega et al. 2012). The use of nanoparticles for medical applications has both advantageous and disadvantageous aspects. Fabrication of safe and non-toxic particles is very challenging, which significantly affects the way they come in contact with biological systems and influence their cellular functionality. The magnitude of the interaction of these particles with cellular components (proteins and lipids) depends on two factors: (a) the morphology and size and (b) the proficiency with which they will communicate or gather within the particular location of sedimentation (Moreno-Vega et al. 2012). Apart from this, the total time duration taken by the nanoparticles to saturate the biological site will also play a significant role. All these properties, which impart the special applications of nanoparticles in tissue engineering, are discussed.

As mentioned in the previous section, the nanoscale particles of ceramic are comprised of inorganic compounds and can be developed in various shapes, sizes, and porosities similar to those of the natural extracellular matrix. As mentioned previously, few regulatory authorities such as Scientific Committee on Emerging and Newly Identified Health Risks (SCENIHR), International Organisation of Standardisation (ISO), and American Society for Testing Materials (ASTM) have denoted nanoparticle as a distinct structure that can have any three dimensional (3D) form with a size varying from 1–100 nm. The size plays an important role as it provides these particles the characteristics of both complete materials and their molecular forms. The nanoparticles can be considered as "gapfiller" between the macroscopic and microscopic structures (Hasan et al. 2018). Undoubtedly, the size of nanoparticles, which provides them the most interesting inherent properties (such as an elevated surface to volume ratio), remarkably affects the way they interact with living systems such as cells and their tissues (Moreno-Vega et al. 2012). For biological applications, the size of the nanoparticles should range from 2 nm to 1 μm (Moreno-Vega et al. 2012); the size that is similar to peptides and small proteins suitably allows them to get into the body in a significantly speedy fashion. The entrapment of nanoparticles into a porous scaffold with different bioactive molecules such as growth factors, cytokines, inhibitors, genes, drugs etc., could enhance the success of the strategy involved for the TE and regenerative medicine greatly. The use of the desirable size of particles that enable the intracellular delivery of bioactive molecules is very critical for support and improvement of tissue growth. The interaction of nanoparticles with intracellular organelles ends in the activation of the various intracellular signalling pathways, which ultimately guide various cellular behavior. Various studies have reported a comparative analysis between surface, size and morphology of nanoparticles and their interactions with host reactions. In addition, little is known about how these properties (shape and size) of nanoparticles affect the body's defence system. *In vitro* experiments have manifested that nanoparticles with a size of 200 nm can effortlessly penetrate human mucus blockade, modifying their formation. It is thus important to investigate the complex consequences of the different physicochemical and biological characteristics of nanoparticles on the modulation of the immune reaction. In a free state, the nanoparticles are highly moving, which is why they have a significantly slow rate of sedimentation. In addition, they may show what is known as quantum effects, which is marked by different varieties of conformations varying from hard to soft materials depending on their area of use. Because of the quantum effect, nanoparticles have ultimate control over the surface energy which in turn will allow initial protein adsorption essential for cellular interactions and their ultimate cellular fates.

Nanoparticles embedded in 3D scaffolds may elucidate various kinds of cellular reactions, which may be beneficial for cells or may lead to toxicity considering the their harmful aspects, which totally depends on how these nanoscale entity are translocated, bio distributed, and eliminated in the host body. All these metabolic pathways are affected by many factors like chemical composition, size, morphology, surfaces, and time taken by nanoparticles to remain within the body, which will have different impacts on pathophysiological states in humans. Some groups of nanoparticles will be safe, while others will have different magnitude of toxicity. For example, the bulk ceramic composition of an implant can support the growth of mitochondrial-viable cells *in vitro* with good cell-to-cell and cell-to-substrate attachments, but their nanosized fundamental particles may lead to contemporary unpredicted clinical reactions in the patients. Ceramic nanoparticles possess many advantages, but they do have certain side effects, which causes toxic effects in the body. A lot of research has been carried out in assessing the toxicity of bioceramic nanoparticles, but still proper and adequate knowledge is missing, which needs to be filled (Kumar et al. 2015, Bhaskar et al. 2018). However, with so many benefits offered by ceramic nanoparticles, research is following in a positive direction to regulate manufacturing strategies so that they could be used for future prospects of tissue engineering.

In the following subsections, we discuss different kinds of bioceramic nanoparticles (inert, bioactive, or resorbable) categorized according to their tissue response.

Bioinert Nanoceramics

Bioinert ceramics such as alumina ceramic (Al_2O_3) and zirconia ceramic (ZrO_2) possess properties like better chemical stability and inertness *in vivo*, low friction, wettability, a higher compressive strength, cracking strength and bending strength, corrosion resistance, and better biocompatibility when they are implanted in animal tissues such as bone (Yamamuro 2004). They are specifically used for total knee and hip replacement surgery and for dental repair treatment. Contact osteogenesis occurs when bone comes in contact with the implants (especially all endosseous dental implants) made up of these bioinert ceramics. These engineering ceramic materials have been used since 1970 for dental implants and other osteosynthetic devices such as bone and joint prostheses (Pina et al. 2017, Kurtz et al. 2014). It has been known that alumina or zirconia has high mechanical properties combined with high wear-resistant character. These properties are related to surface smoothness and surface energy. Because of high abrasion resistance properties, the use of the alumina and zirconia is limited to the bearing surface of joint prostheses, for fabrication of porous scaffold and for biomimetic coatings. For maxillofacial applications, alumina ceramics are used for developing jaw bone, ear bone substitutes, and various dental implants (Greenspan 2016) (Pina et al. 2017).

The initial applications of both of these ceramics in total joint replacement was found to be associated with certain limitations (Thamaraiselvi and Rajeswari 2003, Roualdes et al. 2010). High fracture rates were found in the case of alumina, whereas zirconia was found to be unstable and get transformed catastrophically into the monoclinic phase, depending on its manufacturing conditions and hydrothermal effects *in vivo* (Kurtz et al. 2014). To address these clinical problems, the development of mixed oxides ceramic materials known as "composite" materials represents a major new advancement of clinically available orthopaedic biomaterials. Ceramic composites act synergistically to give properties, better than those provided by either component alone and also, enjoy the superiority due to similarity with bone minerals (Thamaraiselvi and Rajeswari 2003). Two types of composites known as zirconia-toughened alumina (ZTA), in which alumina matrix is embedded with zirconia particles, and alumina-toughened zirconia (ATZ) where zirconia matrix is dispersed with particles of alumina, exhibiting superior strength and toughness, can be fabricated from mixtures of alumina and zirconia (Affatato et al. 2006). The concerns related to hydrothermal firmness still persist in the case of ATZ (Affatato et al. 2006). The properties such as high strength, toughness, hardness, wear resistance, and low susceptibility of stress-assisted degradation of alumina make ZTA increasingly important as a structural material for orthopaedic applications (Kurtz et al. 2014) (Affatato et al. 2006). All these properties of the ZTA reduce the risk of dislocation and impingement and enhance the stability of implants. However, the magnitude of all these mechanical as well as biological properties is completely dependent on the different processing routes proposed for

ZTA composites. Affatato et al. (Affatato et al. 2006) reported the development of high-density ZTA nanocomposites with superior properties such as very homogenous microstructure, high crack resistance, increased fracture toughness, and hydrothermal stability. Further, they observed no significant differences between the wear behaviours and osteoblast growth onto nanocomposite samples of ZTA in comparison with commercial alumina and experimental alumina specimens. In another study by Roualdes et al. (Roualdes et al. 2010), the *in vitro* and *in vivo* biocompatibility of a ceramic composite composed of alumina-zirconia, processed by a standard powder-mixing technique, was investigated. The results showed that *in vitro*, the ZTA composite resulted in no deleterious effects on cell growth and its phenotypical features, and extra-cellular matrix production by fibroblasts and osteoblast cultured upon sintered ceramic discs and in the presence of submicron-sized alumina or zirconia particles at various dose concentrations. A very normal and non-specific response of the synovial membrane leading to the formation of granulomatous tissue, but no major pain or inflammation was observed at local or systemic site in Sprague Dawley rats after intra-articular injection of ZTA particulates.

The polymorphic crystalline structure-based zirconia and alumina nanoceramics have been substantially utilised in bone tissue engineering, not only due to their outstanding mechanical properties like high strength, fracture toughness, elastic modulus, and wear resistance, but also because of their ability to promote osteogenic pathways leading to osteoconduction and osseointegration (Afzal 2014) (Kurtz et al. 2014) (Pina et al. 2017). In a recent study, it was reported that incorporation of alumina particles together with selenite-doped carbonated hydroxyapatite (Se-CHAP) in the matrix of ε-polycaprolactone microfibres resulted in enhancement of mechanical and microstructural properties, like toughness, porous nature, and strain, and with no notable decline in cell growth confirming good biocompatibility (Ahmed et al. 2019).

Bioactive Glass Nanoceramics

The nanoparticles of bioactive glasses (BGs) are the subclass of glass-forming oxide-based bioceramics for example SiO_2 (e.g. 45S5, silicate), B_2O_3 (borate), or P_2O_5 (phosphate), which exhibit tuneable mechanical properties, high biocompatibility, and inorganic composition similar to those of natural bone. They strongly attach to both soft and hard tissues while stimulating new tissue development and the enhancement of bone proliferation (osteoproduction), which is the special feature of this class of ceramic (Kong et al. 2018). Professor Larry Hench's team had established the first BG 45S5, also referred by its commercial name Bioglass, with the SiO_2-CaO-P_2O_5-Na_2O core structure, in 1969 (Hench 2006). Due to its strong bone interaction, it has been treated as a gold grade for bioactive ceramic substances used for tissue engineering and regenerative medicines. The release of sodium and calcium ions on the dissolution of glass, before deposition of a layer of hydroxyapatite on it, is associated with the mechanism of bone bonding (Hench 2006). These ions once get released into immediate vicinity start degrading, and the release of ionic by-products induces favourable intracellular and extracellular responses stimulating proliferation and differentiation of osteoblast (Kong et al. 2018). Nanoparticles of BG (nBG) with improved porous structure and specific surface area can impart speedy ion delivery compared to bulk bioactive glasses, resulting in the enhancement of adsorption of proteins and cellular components osteoconduction, osteoinduction, and bioresorption (Fathi-Achachelouei et al. 2019). Due to the high specific surface area and faster release of ions in body fluid, these nanosized BG also exhibit stronger anti-microbial and anti-bacterial effects (Zysk et al. 2004). Different kinds of methods like sol-gel, laser spinning, micro-emulsion, and gas-phase synthesis have been employed to fabricate nanoparticles of bioactive glasses with various elements like potassium, sodium, silicone, magnesium, oxygen, calcium, and phosphorous. A lot of studies have been reported which show that nanocomposites fabricated from synthetic biopolymers with bioactive ceramics have emerged as an appropriate method for the growth of osseous tissue because of the stimulatory features of mineralisation (Nafary et al. 2017). Recently, Iron et al. have investigated PHB (polyhydroxyalkanoates; PHA family member) nanocomposite scaffolds containing 58S nBGs fabricated by the electrospinning process. 58S nBGs with particles size lesser than 100 nm were developed by the sol-gel method. They have shown that all scaffolds with interconnected porous architecture had increased the porosity volume. The increased level of tensile strength was gained by adding a higher amount of nBGs to the PHB scaffold (Iron et al. 2019). Eventually, the report of *in vitro* bioactivity demonstrated the generation of apatite similar to bone, on the nanocomposite scaffold indicating the high potential of nBGs for bone tissue engineering applications (Iron et al. 2019). Similarly, in an another study, which has been done to analyse the mechanical and biological properties of chitosan and gelatin based composite scaffolds embedded with nanoparticles of bioactive glasses like β-TCP, HAp, and 58s (Dasgupta et al. 2019), the authors have clearly reported that composite scaffolds with 58s glass nanoparticle reinforcement showed high porosity varying with higher mean pore size of 120 μm, the maximum average compressive strength, *in vitro* enhanced expression of osteogenic genes (osteocalcin and RUNX2) faster MSCs differentiation into osteoblast, and significantly improved bone-forming efficiency *in vivo*, when compared with GC scaffold and other composites scaffolds (Dasgupta et al. 2019).

A lot of studies have been reported on Ca-, Mg-, and Si-containing a new class of bioactive ceramic particles, which are capable of stimulating cell proliferation and activating osteoblast forming gene expression. Particularly, Mg-containing silicate bioactive ceramics such as magnesium hydroxide, diopside, and bredigite have shown exceptional abilities to promote osteoblast activity in bone tissue reconstruction supporting thermal and electrical requirements for bone implants and porous scaffolds (Xu et al. 2009, Janning et al. 2010, Wu et al. 2010, Kalantari et al. 2017, Sahmani et al. 2018). A novel calcium silicate bioceramic nanocomposite monticellite ($CaMgSiO_4$ with 33.33% of Mg) has been found to have increased *in vitro* cytocompatibility and osteogenic activity, and anti-biofilm and anti-bacterial activity because of the existence of elements Mg, Si, and Ca in the composition (Kalantari et al. 2017, Kalantari et al. 2018). Similarly, another silicate-based bioactive ceramic akermanite ($Ca_2MgSi_2O_7$) have been reported to have osteoinduction capability (Dehsheikh and Karamian 2016) when used in either form (powder or bulk). Akermanite is a newly reported bioceramic; a melilite mineral of the sorosilicate class containing magnesium, calcium, silicon, and oxygen elements which distinguished it from other inorganic bioceramics (Dehsheikh and Karamian 2016). In a recent study, akermanite nanoparticles have been used for coating surface-modified electrospun nanofibrous poly-L-lactic acid (PLLA)-based scaffolds, and their effect on the growth and osteogenic divergence of human adipose tissue-derived mesenchymal stem cells (hAMSCs) has been studied (Nafary et al. 2017). In their study, they have reported that surface modification of PLLA nanofibrous scaffolds coated with akermanite nanoparticles divulged elevated capabilities in the support growth and differentiation of hAMSc in osteogenic lineage, showing an increased level of bone-forming markers like ALP activity, Ca deposition, and osteogenic gene expression (Nafary et al. 2017).

Some studies have reported that spherical shape prolongs the solidity of the particles in a bioactive material with diffusion ability under dilute alcohol and aqueous circumstances (Labbaf et al. 2011, Hong et al. 2009, El-Kady et al. 2010). For example, developed new round-shaped bioactive glass nanoparticles with sizes below 50 nm, based on the SiO_2-CaO-P_2O_5 and SiO_2-CaO systems, were found to have the ability to prompt mineralisation when immersed in simulated body fluid (SBF), which will lead to enhanced bone bonding and biodegradability (Luz and Mano 2011). Similarly, BG nanoparticles with round shape and controlled particle dimensions of 87 ± 5 nm were acquired with the nominal confirmation of 60% SiO_2, 36% CaO, and 4% P_2O_5 (mass%) (de Oliveira et al. 2013). The prepared BG nanoparticles showed an increase in specific surface area and volume of pore and presented diffusion ability in aqueous conditions. The current study examined that the kinetics of HA on particles surface are influenced by particle size with increased HA nucleation on the surface of BG nanoparticles (de Oliveira et al. 2013). Further, similar-sized and spherically shaped nanoparticles with diameter more than <60 nm, having the ability to support primary human osteoblast (HOB) cell viability and differentiation with an elevated level of RUNX2 gene expression, were produced, which establishes the advantages of using the nanoparticles as nanofillers in the development of 3D scaffolds and/or for coating the implant for tissue engineering applications (Fan et al. 2014).

Bioresorbable Nanoceramics

Nanosized bioceramic particles of calcium phosphates (CaPs) like nano-hydroxyapatite (nHAp) and nano-tricalcium phosphates (nTCP) have obtained specific interest in the development of biomaterials used for clinical applications specifically as restorative dental and orthopaedic implants. Different compounds of calcium phosphates such as hydroxyapatite (HAp), calcium tetraphosphate ($Ca_4P_2O_9$), tricalcium phosphate (TCP) $Ca_3(PO_4)_2$, and calcium hydroxyapatite $Ca_{10}(PO_4)_6(OH)_2$ can be produced by varying the atomic ratio of Ca/P from 1.5 to 2 (Ramay and Zhang 2004). Since the structure of calcium phosphate resembles the inorganic component of natural hard tissue, *i.e.* bone and teeth, and also they have typical biological responses such as good bioaffinity and enhancement of osseo-integration as well as successful clinical history, these ceramics are in high demand in the clinical field (Ebrahimi et al. 2019). However, the biological performance such as protein adsorption, angiogenesis, and vascularisation will totally depend on CaP particle composition, structure, morphology, and crystallite sizes (Zhou et al. 2013). Also, calcium phosphates are soluble under aqueous solution below pH 4.2, the property which makes it exploited during resorption of bone mineral by osteoclasts (bone remodelling) leading to the conditions of lysosomal degradation. Therefore, calcium phosphate nanoceramics are often applied for making scaffolds for solid bone tissue engineering (Ebrahimi et al. 2019).

Various synthesis techniques, such as acid-base reaction, sol-gel techniques, precipitation and co-precipitation, hydrothermal methods, mechanic-chemical synthesis, plasma spray process, hydrolysis, microwave sintering, and laser-induced techniques, have been introduced for the production of HAp nanoparticles employing varieties of precursor raw materials (e.g. $CaHPO_4$, $Ca_2P_2O_7$, $CaCO_3$, CaO, $Ca(OH)_2$, etc.) with exact control over their nanostructure (Fathi-Achachelouei et al. 2019, Ebrahimi et al. 2019). It is reported that the synthesis of highly pure HAp like calcium phosphates is a very challenging process because of different kinds of reaction conditions and derivatives, which play important roles in controlling the final properties of the nHAp (Han et al. 2006). Since each of these HAp synthesis methods has some or other problems (time-consuming, tedious work, difficult quality control, high cost, poor reproducibility, and chemical contamination) associated with it, a combination of various processes can be used to overcome these challenges (Han et al. 2006, Poinern et al. 2014). The synthesis involving a combination of processes should be done in a very critical manner with all the experimental parameters, so that finally produced nHA mimic all the properties of natural HA present in bone. For example, nanosized HAp-based needle-shaped structure with size ranging from 20–50 nm in length and 4–15 nm in width and the spherical-shaped structure with size 10–30 nm in diameter was successfully developed using the microwave-hydrothermal method system from phosphoric acid and calcium hydroxide (Han et al. 2006). Further, non-homogeneous distribution pattern, high agglomeration rate, poor crystallinity, and low biodegradation rate are some of the disadvantages faced during the development of nHAp particles. These limitations often lead to potential heterogeneity and highly unpredictable biological responses. To overcome all these limitations, the concept of biphasic calcium phosphate ceramics has been introduced and growing up interestingly (Lobo and Livingston Arinzeh 2010, Tang et al. 2016). For example, biomimetic calcium-deficient nanoparticles were developed from carbonate-substituted biphasic bioceramic (nHAp/β-TCP) by the modified wet mechano-chemical method together with the solid state synthesis method (Ebrahimi et al. 2019). The synthesised nanoparticles from biphasic bioceramic (nHAp/β-TCP) showed better crystallinity and homogeneity, and decreased crystallite/particle size and reduced particle agglomeration size, which would improve the release of calcium ions and biodegradation resulting in bone regeneration and remodelling (Ebrahimi et al. 2019).

These ceramic nanoparticles can be utilised separately or in combination with other materials as blended scaffolds or gel with growth factors showing improved mechanical properties like fracture toughness and compression modulus, and biological activity depending on the preparation method and the nature of composite (Fathi-Achachelouei et al. 2019, Zhou and Lee 2011, Chen et al. 2018, Duan et al. 2008, Duan et al. 2010, Zhou et al. 2008, Zhou et al. 2015, Ramier et al. 2014, Kaur et al. 2017). A colloidal gel system composed of negatively charged inorganic HAp nanoparticles combined with positively charged (d,l-lactic-co-glycolic acid) (PLGA) nanoparticles were reported to have a special reaction to recoverable properties, external shear force, and minor toxicity towards human umbilical cord mesenchymal stem cells (hUCMSCs) (Wang et al. 2013). Porosity and pore size in 3D scaffold plays a key role in the interchange of nutrients and metabolites and in promoting cellular activity such as proliferation and differentiation. A nanocomposite porous scaffold developed from a hydroxyapatite (HAp) and β-tricalcium phosphate (β-TCP) matrix nanofibre showed enhancement in compressive strength and toughness on the addition of HAp nanoparticles (Ramay and Zhang 2004). Scaffold composed of polycaprolactone (PCL), β-tricalciumphosphate (β-TCP) nanoparticles, and salt porogens showed interconnected porosity, pore-size distributions, composition distributions on compressive properties and favourable cell adhesion and penetration with human foetal osteoblast (hFOB) cells along with their differentiation rates (Ozkan et al. 2010). HAp nanoparticles were prepared using the nanoemulsion method and used to fabricate highly porous polycaprolactone (PCL)-based scaffolds with an adequate distribution of pore (Hassan et al. 2012). Further, the 3D-printed PLA scaffold surface was prepared by incorporating HAp nanoparticles. It was found that Hap-modified PLA scaffolds influenced the enhancement of compressive stress property and cellular functionality. The interaction of HAp nanoparticles on PLA scaffold surface leads to the adsorption of proteins and facilitates cellular activity (Mondal et al. 2020). A poly(ε-caprolactone) (PCL) and silicon-substituted hydroxyapatite (Si-HAp) nanoparticle-based composite exhibited better cellular adhesion and growth with increased production of alkaline phosphatase (ALP) and calcium content in mouse calvarial preosteoblasts (MC3T3-E1) (Lei et al. 2020).

Because of the relatively poor mechanical properties of HAp in stress modes, its use has been limited for high-load applications. Therefore, ionic substitutions can be introduced in calcium phosphate-based ceramics which will closely mimic the chemical compositions of naturally occurring HAp. Hap has an open and flexible crystal structure, which allows outside ions to accommodate into it to obtain specific desired properties (Ahmed et al. 2019). In a physiological environment, bone apatite will efficiently exchange CO_3^{2-}, F^-, Cl^-, Mg^{2+}, Na^+, and K^+ ions with Ca^{2+} and PO_4^{3-}ions, which results in the change in crystal structure, solubility, and biological activity of bone apatite (Dittler et al. 2019, Khandan et al. 2018). After knowing about structural openness and flexibility, dopants such as Ba2$^+$, Mg^2+, Na^+, K^+, Sr^{2+}, Fe^{2+}/Fe^{3+} Mn^{2+}, Cu^{2+}, Si^{2+}, F^-, Cl^-, Zn^{2+}, and Ti^{4+} have been used to design single-doped and multi-doped HAp compositions, where these dopants do not possess marked biological roles (Ahmed et al. 2019, Youness et al. 2018, Wei et al. 2017, Dittler et al. 2019, Ullah et al. 2019). The addition of an optimised weight ratio of these dopants allows modifying the material properties for the desired clinical application. Magnesium-substituted calcium phosphate bioceramics and biphasic Mg-HA and Mg-whitlockite were produced through continuous hydrothermal flow synthesis technology, which could be extremely useful in bone regeneration and related clinical applications (Chaudhry et al. 2008). A new bioceramic nanostructured forsterite (Mg_2SiO_4) were prepared by sol-gel method and found to have much better hardness and fracture toughness and *in vitro* biocompatibility than hydroxyapatite (Kharaziha and Fathi 2010). Further, varieties of nanofillers such as TiO_2, carbon nanotubes, graphene, and transition metal dichalcogenides, have been employed in HAp-based nanocomposites to improve the hardness and elastic modulus (Riau et al. 2016, Lahiri et al. 2010, Que et al. 2008, Huang et al. 2011, Sahmani et al. 2018). Nanoparticles of Willemite (Zn_2SiO_4), a zinc-containing absorbable silica bioceramic, have also been reported to have potential applications in proliferation and osteogenesis differentiation of stem cells (Ramezanifard et al. 2016) (Halabian et al. 2019). In a study, willemite nanoparticles were coated on the surface of electrospun-fabricated poly(lactide-co-glycolide) (PLGA)-based scaffolds implanted in calvarial critical size defects using a rat model. After digital mammography, multislice spiral-computed tomography (MSCT) imaging, and histological evaluations, it was confirmed that ceramics supported efficient bone regeneration and highest bone reconstruction *in vi*vo (Jamshidi Adegani et al. 2014). The prepared biodegradable bioceramic nanoparticles of ceramics containing Mg and Ag (both) by quick alkali-mediated sol-gel method showed anti-microbial response against *methicillin-resistant Staphylococcus aureus* and acceptable cytocompatibility response towards MG 63 cell lines (Kaur et al. 2016). Recently, the development of HAp-based bioceramic systems was reported by using silica-coated reduced graphene oxide (S-rGO) hybrid nanosheets with simultaneously enhanced Young's modulus hardness and fracture toughness. Also, S-rGO incorporation was found to increase cell proliferation and higher cell viability and alkaline phosphatase activity of osteoblast-like MG-63 cells (Li et al. 2020). From all these research studies, one can easily conclude that nanoparticles of bioabsorable ceramic are very advanced and promising candidate for tissue engineering applications especially where osteoinduction and osteoconduction are concerned.

Possible Application of Bioceramic Nanoparticles

Bioceramic nanoparticles have been used in a diversity of clinical applications including in tissue engineering, drug delivery, gene delivery, gene transfections, biosensing, for treatment of bone cancer, implant coating, orthopaedic, maxillofacial, and dental surgery applications. Mostly, nanoscale particles of ceramics, such as bioglass, tricalcium phosphate, HAp, zirconia, alumina, titanium oxide and silica which are incorporated in biodegradable polymer-based scaffolds for repairing and regenerating the hard tissues, are mainly used for orthopaedics and maxillofacial clinical applications. Alongside is the possibility to use them for the enhancement of electrical and mechanical properties of implant/scaffolds with biological features to facilitate the important cues and signals promoting the proliferation of different kinds of tissues. The particles, which are used for building material for the scaffolds for tissue regeneration applications, are known as third-generation bioceramics. The use of the right type of bioceramic nanoparticles and combining them with bioactive signalling molecules such as growth factors, mitogens, and morphogens in the TE approach can significantly enhance the metabolic activity for the regeneration process in an animal model (Pina et al. 2017) (Hasan et al. 2018). Titanium oxide (TiO_2) nanoparticles and bone morphogenetic proteins (BMPs) growth factors have been tremendously utilised to enhance bone development and healing, preferably for long-bone defects, in spinal fusion, and for maxillofacial and oral surgery (Pina et al. 2017) (Hasan et al. 2018). In bone tissue engineering, it has been observed that most bioceramic particles have osteoinduction ability and promote osteogenic differentiation (Pina et al. 2017), which make them excellent candidates for scaffold fabrication. Bone formation is regulated by several factors, mainly chemical dissolution, resorption, and physical structure of scaffolds. The rate of solubilisation of nanoparticles in scaffold plays an important role in enhancing bioactivity. For example, biocermic akermanite ($Ca_2MgSi_2O_7$) nanoparticles have been reported to have important properties such as osteostimulatory features, mechanical stability, good bioactivity, and biodegradability. These nanoparticles have been coated on the PLLA nanofibrous scaffolds and studied for the osteogenic divergence of human adipose mesenchymal stem cell line (hAMSCs). The result showed their high ability of proliferation and an increased level of bone-forming markers like alkaline phosphatase and calcium deposition, and osteogenic related gene (osteonectin, osteocalcin) expression was observed on these akermanite coated scaffolds, confirming the osteoinductive ability of these nanoparticles (Nafary et al. 2017). In an ideal scenario, the tissue regeneration profile of scaffolds should match with the degradation rate of it. While not instigating any negative reaction and with temporary load tolerating ability, these scaffolds should have interconnected porosity with an optimum dispersal of pore size while offering a high surface area for the cell anchoring (Vallet-Regí 2019). However, there should always be a balance between the mechanical properties of the scaffold and the porosity needed for bone ingrowth after implanting it at a specific site (Burdick and Mauck 2010). Additionally, it has also been reported that the integration of bioceramic nanoparticles in 3D scaffolds used for bone TE affects the formation of osteoclast cells from their precursor, which in turn inhibits the mitochondrial(dys) function in osteoblast cells leading to the formation of more bone cells (Hasan et al. 2018). For bone tissue engineering, materials that have a close chemical composition with that of natural bone and the desired resistance to tensile and compressive forces under normal loading conditions should be preferred. A variety of bioceramic nanoparticles has been used either alone or in the form of composite, to attain the essential support framework (physical and biological) similar to that of natural bone, for the required bone regenerative responses. The process and success of bone defect regeneration by using these scaffolds will totally depend on how the surface of scaffold is adsorbing the essential proteins that are responsible for further cellular adhesion (Manzano and Vallet-Regí 2012). Also, the bone graft substitutes prepared from composite material to repair defects should not be highly crystalline. The bioceramics of tricalcium phosphate and hydroxyapatite, the major inorganic components of bone and teeth, are the most extensively investigated calcium phosphates due to their tissue-mimicking properties such as biocompatibility, osteoconductivity, osteoinductivity, and other physical and chemical properties (Moreno-Vega et al. 2012). Nanoparticles of these bioceramics act as the reinforcing phase of the polymer matrix. The excellent biocompatibility of HAp has made it a promising candidate for repairing hard tissues like bone. HAp as bulk material cannot be utilised for high load-bearing applications due to its low mechanical properties. Therefore, a lot of interest has reignited on the development of HAp nanoparticles so that these can be used for scaffold preparation, for coating the implants, and as fillers used for teeth and bone regeneration. The addition of HA nanoparticles was found to influence the composition of the apatite layer on biomaterials, which affects the viability and differentiation of preosteoblast cells showing higher bioactivity for bone tissue engineering (Burdick and Mauck 2010, Kong et al. 2006). All these properties make Ca-P-based nanoparticles irreplaceable and the most important material for bone regeneration and repair.

Bioceramic nanoparticles have also become integrated into dentistry with the introduction of scaffold fabrication and biomodulation methods to develop a functional bio-artificial tooth and dental tissue reconstruction (Chieruzzi et al. 2016). These engineered tissues significantly mimic the compositional, structural, and functional properties of the natural tooth, gem, and other surrounding structures. Enamel which is the utmost protective layer of tooth crown cannot develop again on its own because of loss of ameloblast cells during tooth formation. A solution made of nano-HAp powder was used for the regeneration of a damaged tooth enamel layer, to cure decayed teeth and as tooth whitening material (Zakaria et al. 2013, Sowmya et al. 2013). Recently, an *in situ* pilot study was conducted by two volunteers who took intraoral splints with staged samples from enamel to analyse the consequences of suspensions based on hydroxyapatite and their interactions with surfaces of distinct dental materials like ceramics, titanium, and polymethyl-methacrylate (PMMA) kept in oral conditions (Nobre et al. 2020). The result showed that nano-HAp solutions applied in oral conditions showed good coverage and adherence not only to enamel but also to titanium, ceramic, and PMMA surfaces (Nobre et al. 2020). Coming to other classes of bioceramic nanoparticles, bio-glass and zirconia have also been used in the field of dental repair and as dental fillers due to their bone-bonding ability (Kudo et al. 1990, Stanley et al. 1976) (Chevalier 2006). The addition of nanostructured yttria-stabilised tetragonal zirconia (Y-TZP) to dental filler and dental tissue layers markedly affects the mechanical stiffness of the material and can promote bone regeneration and growth (Xavier et al. 2015).

The development and use of bioceramic nanoparticles seem to be very interesting and promising strategies in tissue engineering as illustrated in various *in vitro* and *in vivo* studies. But the effective clinical applications have shown steady progress. This may be due to the lack of a comprehensive understanding of biological effects (bioaccumulation or toxicity) caused by NPs released in human patients. Therefore, there is a need for studies to confirm the potential effects of these nanoparticles once they enter the body and interact with biological tissues.

Conclusion and Future Trends

Knowing the fact that nanoparticles have the significant ability to improve well-established cell biocompatibility approaches, the use of bioceramic nanoparticles in tissue engineering have shown magnificent development. Since nanoparticles can be fabricated with different shapes, sizes, and with versatile functionality, various composite materials and scaffolds have been developed employing the platform of ceramic-based nanoparticles. These nanoparticles have been successfully utilised for developing 3D-based dense and porousfibres, hydrogels, and implants used for orthopaedic and dental tissue engineering applications, but they can be appropriately used for a wide variety of other necessary biomedical applications. Ongoing research involves the utilisation of these nanoparticles for enhancing the properties of tissue engineering scaffolds by altering the chemistry, composition, and structure of the composite materials. These biomimetic modifications in scaffold properties can help in delivering the sequential release of biological agents or growth factors in a manner that not only helps in improving the mechanical integrity upon implantation, but also in providing suitable porosity for tissue-specific reactions. Although till date, there has been a significant increase in scientific interest and subsequent studies focusing on bioceramic nanoparticles used in engineering new tissue, the concerns about the safety of these nanoparticles for clinical applications have not been addressed. Therefore, additional consistent and comprehensive research is required to ensure nanoparticle's innocuousness and convert the present studies into a viable strategy, which can be used for human health applications. The strategies should be devoted to the clear understanding of mechanical strength for long-term service under external stress, the reproducibility of well-characterised nanoparticles, bioceramic-tissue interactions, and their ultimate fate in the body. Thus, in the future, one can expect the development of new and superior scaffolds and implants, which will provide a strong toolbox and open up new avenues for future tissue engineering applications.

REFERENCES

Affatato, S., R. Torrecillas, P. Taddei, M. Rocchi, C. Fagnano, G. Ciapetti & A. Toni 2006. *Journal of Biomedical Materials Research Part B: Applied Biomaterials*, *78B*, 76–82.

Afzal, A. 2014. *Materials Express*, *4*(1):1–12.
Ahmed, M., R. Ramadan, S. El-Dek & V. Uskoković 2019. *Journal of Alloys and Compounds*, *801*, 70–81.
Baino, F., G. Novajra & C. Vitale-Brovarone 2015. *Frontiers in Bioengineering and Biotechnology*, *3*, 202.
Bhaskar, N., D. Sarkar & B. Basu 2018. *ACS Biomaterials Science & Engineering*, *4*, 3194–3210.
Burdick, J. A. & R. L. Mauck. 2010. *Biomaterials for tissue engineering applications: a review of the past and future trends*. Springer Science & Business Media, Germany.
Chaudhry, A. A., J. Goodall, M. Vickers, J. K. Cockcroft, I. Rehman, J. C. Knowles & J. A. Darr 2008. *Journal of Materials Chemistry*, *18*, 5900–5908.
Chen, Y., N. Kawazoe & G. Chen 2018. *Acta Biomaterialia*, *67*, 341–353.
Chevalier, J. 2006. *Biomaterials*, *27*, 535–543.
Chieruzzi, M., S. Pagano, S. Moretti, R. Pinna, E. Milia, L. Torre & S. Eramo 2016. *Nanomaterials (Basel)*, 6.
Dasgupta, S., K. Maji & S. K. Nandi 2019. *Materials Science and Engineering: C*, *94*, 713–728.
Dehsheikh, H. G. & E. Karamian 2016. *International Journal of Bio-Inorganic Hybrid Nanomaterials*, *5*, 223–227.
Dittler, M. L., I. Unalan, A. Grünewald, A. M. Beltrán, C. A. Grillo, R. Destch, M. C. Gonzalez & A. R. Boccaccini 2019. *Colloids and Surfaces B: Biointerfaces*, *182*, 110346.
Duan, B., W. L. Cheung & M. Wang 2011. *Biofabrication*, *3*, 015001.
Duan, B., M. Wang, W. Zhou & W. Cheung 2008. *Applied Surface Science*, *255*, 529–533.
Duan, B., M. Wang, W. Y. Zhou, W. L. Cheung, Z. Y. Li & W. W. Lu 2010. *Acta Biomaterialia*, *6*, 4495–4505.
Ebrahimi, M., M. Botelho, W. Lu & N. Monmaturapoj 2019. *Journal of Biomedical Materials Research Part A*, *107*, 1654–1666.
El-Kady, A. M., A. F. Ali & M. M. Farag 2010. *Materials Science and Engineering: C*, *30*, 120–131.
Engel, E., A. Michiardi, M. Navarro, D. Lacroix & J. A. Planell 2008. *Trends in Biotechnology*, *26*, 39–47.
Fan, J. P., P. Kalia, L. Di Silvio & J. Huang 2014. *Materials Science and Engineering: C*, *36*, 206–214.
Fathi-Achachelouei, M., H. Knopf-Marques, C. E. Ribeiro da Silva, J. Barthès, E. Bat, A. Tezcaner & N. E. Vrana 2019. *Frontiers in Bioengineering and Biotechnology*, *113*, 7.
Fernandes, H. R., A. Gaddam, A. Rebelo, D. Brazete, G. E. Stan & J. M. F. Ferreira 2018. *Materials (Basel, Switzerland)*, *11*, 2530.
Greenspan, D. C. 2016. *International Journal of Applied Glass Science*, *7*, 134–138.
Halabian, R., K. Moridi, M. Korani & M. Ghollasi 2019. *International Journal of Molecular and Cellular Medicine*, *8*, 24–38.
Han, J.-K., H.-Y. Song, F. Saito & B.-T. Lee 2006. *Materials Chemistry and Physics*, *99*, 235–239.
Hasan, A., M. Morshed, A. Memic, S. Hassan, T. J. Webster & H. E.-S. Marei 2018. *International Journal of Nanomedicine*, *13*, 5637–5655.
Hassan, M., M. Mokhtar, N. Sultana, Ceng Csci & T. Khan 2012. In: *2012 IEEE-EMBS Conference on Biomedical Engineering and Sciences, 2012*, pp. 239–242, doi:10.1109/IECBES.2012.6498024.
.Hench, L. L. 2006. The story of bioglass. *Journal of Materials Science: Materials in Medicine*, *17*, 967–978.
Hong, Z., R. L. Reis & J. F. Mano 2009. *Journal of Biomedical Materials Research Part A*, *88*, 304–313.
Huang, X., Z. Yin, S. Wu, X. Qi, Q. He, Q. Zhang, Q. Yan, F. Boey & H. Zhang 2011. *Small*, *7*, 1876–1902.
Iron, R., M. Mehdikhani, E. Naghashzargar, S. Karbasi & D. Semnani 2019. *Materials Technology*, *34*, 540–548.
Jamshidi Adegani, F., L. Langroudi, A. Ardeshirylajimi, P. Dinarvand, M. Dodel, A. Doostmohammadi, A. Rahimian, P. Zohrabi, E. Seyedjafari & M. Soleimani 2014. *Cell Biology International*, *38*, 1271–1279.
Janning, C., E. Willbold, C. Vogt, J. Nellesen, A. Meyer-Lindenberg, H. Windhagen, F. Thorey & F. Witte 2010. *Acta Biomaterialia*, *6*, 1861–1868.
Kalantari, E., S. M. Naghib, M. R. Naimi-Jamal, R. Esmaeili, K. Majidzadeh-A & M. Mozafari 2018. *Materials Today: Proceedings*, *5*, 15744–15753.
Kalantari, E., S. M. Naghib, M. Reza Naimi-Jamal & M. Mozafari 2017. *Journal of Sol-Gel Science and Technology*, *84*, 87–95.
Kaur, K., K. Singh, V. Anand, G. Bhatia, R. Kaur, M. Kaur, L. Nim & D. S. Arora 2017. *Materials Science and Engineering: C*, *71*, 780–790.
Kaur, K., K. Singh, V. Anand, G. Bhatia, S. Singh, H. Kaur & D. S. Arora 2016. *Ceramics International*, *42*, 12651–12662.
Khandan, A., S. Saber-Samandari & A. Montazeran 2018. Artificial intelligence investigation of three silicates bioceramics magnetite bio-nanocomposite: Hyperthermia and biomedical applications. *Nanomedicine Journal*, 5. doi:10.22038/nmj.2018.005.0006.

Kharaziha, M. & M. Fathi 2010. *Journal of the Mechanical Behavior of Biomedical Materials*, *3*, 530–537.
Kong, C. H., C. Steffi, Z. Shi & W. Wang 2018. *Journal of Biomedical Materials Research Part B: Applied Biomaterials*, *106*, 2878–2887.
Kong, L., Y. Gao, G. Lu, Y. Gong, N. Zhao & X. Zhang 2006. *European Polymer Journal*, *42*, 3171–3179.
Kudo, K., M. Miyasawa, Y. Fujioka, T. Kamegai, H. Nakano, Y. Seino, F. Ishikawa, T. Shioyama & K. Ishibashi 1990. *Oral Surgery, Oral Medicine, Oral Pathology*, *70*, 18–23.
Kumar, A., N. Bhaskar & B. Basu 2015. *Journal of the American Ceramic Society*, *98*, 3202–3211.
Kurtz, S. M., S. Kocagöz, C. Arnholt, R. Huet, M. Ueno & W. L. Walter 2014. *Journal of the Mechanical Behavior of Biomedical Materials*, *31*, 107–116.
Labbaf, S., O. Tsigkou, K. H. Müller, M. M. Stevens, A. E. Porter & J. R. Jones 2011. *Biomaterials*, *32*, 1010–1018.
Lahiri, D., V. Singh, A. K. Keshri, S. Seal & A. Agarwal 2010. *Carbon*, *48*, 3103–3120.
Lei, T., W. Zhang, H. Qian, P. N. Lim, E. San Thian, P. Lei, Y. Hu & Z. Wang 2020. *Colloids and Surfaces B: Biointerfaces*, *187*, 110714.
Li, Z., W. Zhu, S. Bi, R. Li, H. Hu, H. Lin, R. S. Tuan & K. A. Khor 2020. *Journal of Biomedical Materials Research Part A*, *108*, 1016–1027.
Lobo, S. E. & T. Livingston Arinzeh 2010. *Materials*, *3*, 815–826.
Luz, G. M. & J. F. Mano 2011. *Nanotechnology*, *22*, 494014.
Manzano, M. & M. Vallet-Regí 2012. *Progress in Solid State Chemistry*, *40*, 17–30.
Mondal, S., T. P. Nguyen, G. Hoang, P. Manivasagan, M. H. Kim, S. Y. Nam & J. Oh 2020. *Ceramics International*, *46*, 3443–3455.
Moreno-Vega, A.-I., T. Gomez-Quintero, R.-E. Nunez-Anita, L.-S. Acosta-Torres & V. Castaño 2012. *Journal of Nanotechnology*, *2012*, 936041, doi: 10.1155/2012/936041.
Nafary, A., E. Seyedjafari & A. Salimi 2017. *Journal of Biomaterials and Tissue Engineering*, *7*, 91–100.
Nobre, C. M. G., N. Pütz & M. Hannig 2020. *Scanning*, *2020*, 6065739.
de Oliveira, A. A., D. A. de Souza, L. L. Dias, S. M. de Carvalho, H. S. Mansur & M. de Magalhães Pereira 2013. *Biomedical Materials*, *8*, 025011.
Ozkan, S., D. M. Kalyon & X. Yu 2010. *Journal of Biomedical Materials Research Part A: An Official Journal of The Society for Biomaterials, The Japanese Society for Biomaterials, and The Australian Society for Biomaterials and the Korean Society for Biomaterials*, *92*, 1007–1018.
Pina, S. C. A., R. L. Reis & J. M. Oliveira 2017. Ceramic biomaterials for tissue engineering. In: Thomas S., Sreekala M. S., and Balakrishnan P. (eds.), *Fundamentals Biomaterials: Ceramics*, Elsevier.
Poinern, G. E. J., R. K. Brundavanam, X. Thi Le, P. K. Nicholls, M. A. Cake & D. Fawcett 2014. *Scientific Reports*, *4*, 6235.
Que, W., K. A. Khor, J. Xu & L. Yu 2008. *Journal of the European Ceramic Society*, *28*, 3083–3090.
Ramay, H. R. R. & M. Zhang 2004. *Biomaterials*, *25*, 5171–5180.
Ramezanifard, R., E. Seyedjafari, A. Ardeshirylajimi & M. Soleimani 2016. *Materials Science and Engineering: C*, *62*, 398–406.
Ramier, J., T. Bouderlique, O. Stoilova, N. Manolova, I. Rashkov, V. Langlois, E. Renard, P. Albanese & D. Grande 2014. *Materials Science and Engineering: C*, *38*, 161–169.
Riau, A. K., D. Mondal, M. Setiawan, A. Palaniappan, G. H. Yam, B. Liedberg, S. S. Venkatraman & J. S. Mehta 2016. *ACS Applied Materials & Interfaces*, *8*, 35565–35577.
Roualdes, O., M.-E. Duclos, D. Gutknecht, L. Frappart, J. Chevalier & D. J. Hartmann 2010. *Biomaterials*, *31*, 2043–2054.
Sahmani, S., A. Khandan, S. Saber-Samandari & M. Aghdam 2018. *Ceramics International*, *44*, 9540–9549.
Sowmya, S., J. D. Bumgardener, K. P. Chennazhi, S. V. Nair & R. Jayakumar 2013. *Progress in Polymer Science*, *38*, 1748–1772.
Stanley, H. R., L. Hench, R. Going, C. Bennett, S. J. Chellemi, C. King, N. Ingersoll, E. Ethridge & K. Kreutziger 1976. *Oral Surgery, Oral Medicine, Oral Pathology*, *42*, 339–356.
Tang, X., L. Mao, J. Liu, Z. Yang, W. Zhang, M. Shu, N. Hu, L. Jiang & B. Fang 2016. *Ceramics International*, *42*, 15311–15318.
Thamaraiselvi, T. V. & S. Rajeswari2003. *Trends in Biomaterials & Artificial Organs 18*: 9–17.
Ullah, I., A. Gloria, W. Zhang, M. W. Ullah, B. Wu, W. Li, M. Domingos & X. Zhang 2019. *ACS Biomaterials Science & Engineering*, *6*, 375–388.
Vallet-Regí, M. 2019. *Chimie pure et appliquee*, *91*, 687–706.

Wang, Q., Z. Gu, S. Jamal, M. S. Detamore & C. Berkland 2013. *Tissue Engineering. Part A*, *19*, 2586–2593.

Wei, L., D. Pang, L. He & C. Deng 2017. *Ceramics International*, *43*, 16141–16148.

Wu, C., Y. Ramaswamy & H. Zreiqat 2010. *Acta Biomaterialia*, *6*, 2237–2245.

Xavier, J., P. Desai, V. Varanasi, I. Al-Hashimi & A. Gaharwar 2015. *Nanotechnology in Endodontics*. 5–22. doi:10.1007/978-3-319-13575-5_2.

Xu, L., F. Pan, G. Yu, L. Yang, E. Zhang & K. Yang 2009. *Biomaterials*, *30*, 1512–1523.

Yamamuro, T. 2004. Bioceramics. In *Biomechanics and Biomaterials in Orthopedics*, ed. D. G. Poitout, 22–33. London: Springer London.

Youness, R. A., M. A. Taha, A. A. El-Kheshen & M. Ibrahim 2018. *Ceramics International*, *44*, 20677–20685.

Zakaria, S. M., S. H. Sharif Zein, M. R. Othman, F. Yang & J. A. Jansen (2013) Nanophase hydroxyapatite as a biomaterial in advanced hard tissue engineering: a review. *Tissue Engineering Part B: Reviews*, *19*, 431–441.

Zhou, C., C. Deng, X. Chen, X. Zhao, Y. Chen, Y. Fan & X. Zhang 2015. *Journal of the Mechanical Behavior of Biomedical Materials*, *48*, 1–11.

Zhou, C., Y. Hong & X. Zhang 2013. *Biomaterials Science*, *1*, 1012.

Zhou, H. & J. Lee 2011. *Acta Biomaterialia*, *7*, 2769–2781.

Zhou, W. Y., S. H. Lee, M. Wang, W. L. Cheung & W. Y. Ip 2008. *Journal of Materials Science: Materials in Medicine*, *19*, 2535–2540.

Zysk, S. P., H. Gebhard, W. Plitz, G. H. Buchhorn, C. M. Sprecher, V. Jansson, K. Messmer & A. Veihelmann. 2004. *Journal of Biomedical Materials Research Part B: Applied Biomaterials*, *71B*(1): 108–115.

7

Organoids as an Emerging Tool for Nano-Pharmaceuticals

Anupam Mittal and Gaurav Sharma
PGIMER, Chandigarh, India

CONTENTS

Introduction

Organoids are stem cell-derived, three-dimensional (3D) cultured structures that are generated *ex-vivo*. Mostly, organoids are composed of distinct types of cells that are derived from organ progenitors or pluripotent stem cells (Lancaster and Knoblich 2014). Over a period of time, advancement in technology to differentiate stem cells and drive them into specific lineages has revolutionised the field of three-dimensional models of development. Moreover, these developments helped in disease modelling, which led to the elucidation of etiologic pathways involved in human pathologies. Another important angle in organoid culture is the use of biomaterials to include the effect of all surrounding microenvironment and thus gives an opportunity to explore the purpose of various cues in the determination of cell fate. Specifically, mechanical perturbations imposed by entropy and dynamic turbulences in tissue/3D organoid systems play critical regulatory roles.

At a cellular level, after sensing the cues, these cues may be processed as external stimuli by modulating the microenvironment and inducing specific responses through activation of genetic programme, known as the mechanotransduction process (Chan et al. 2017, Davidson 2017). These biophysical signals are interpreted by a set of specialised proteins and their dysregulation, in turn, leads to various pathologies. Vinculin is an important mechanotransducter, serving as a linker for integrins to cytoskeleton, and pathogenic variations in this protein in humans are reported in various pathologies like cancer, cardiomyopathies, and neural defects (Olson et al. 2002, Vasile et al. 2006, Liang et al. 2014, Chinthalapudi et al. 2016).

Thus, it is very important to study factors in quasi-state with these mechanical microenvironments. This modulation of the microenvironment by mechanical perturbations is broadly divided into two categories: (1) active and (2) passive. Active mechanical modulation means changes in the microenvironment in a controlled manner. Passive modulation defines the setting up of a biomechanical milieu for cells in which it can interact with neighbouring cells. There are some premier studies that looked into the stiffness of the substrate and sufficient in driving cell fate (Engler et al. 2006). In both active and passive manipulation, there is a state of homogenous mechanical microenvironment, which is not the real picture faced

DOI: 10.1201/9781003153504-7

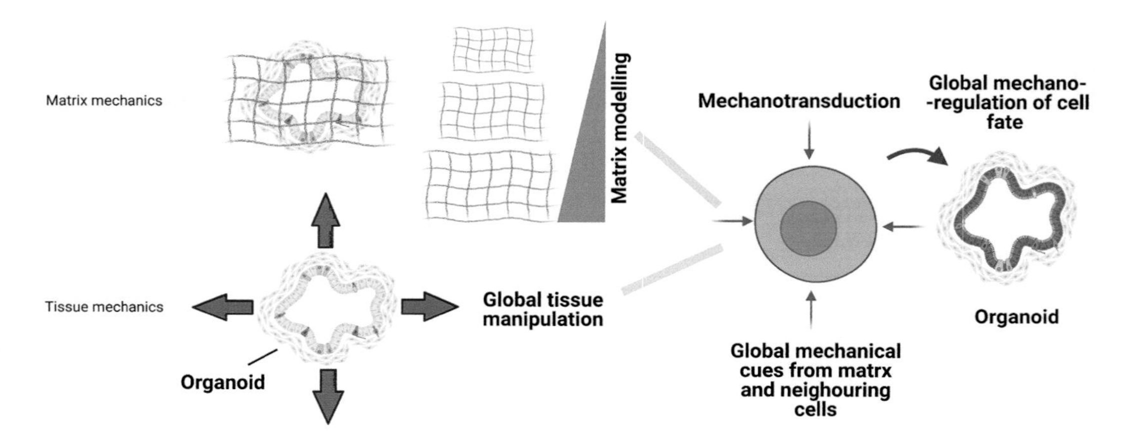

FIGURE 7.1 Matrix engineering done using synthetic microenvironment. These manipulations provide global cues to organoids. Image is redrawn with permission from Abdel Fattah and Ranga (2020).

in vivo scenario (Figure 7.1). To overcome such limitations, advancements in *in vitro* technologies for modelling tissue mechanics are looking into mechanical interplay between tissues.

Another challenge for the clinics is to expedite the drug discovery and identify novel therapeutic molecular targets because of limitations associated with two-dimensional *in vitro* cultures. To answer this problem, there have been significant advancements towards adopting more biomimetic platforms for drug screening platforms with higher fidelity for testing bioactivity and toxicity. To this end, in recent years, there has been an innovative concept of "Organoid as the Trojan horse system," which synergistically link advanced nanotechnology tools with novel host cell-mediated drug delivery. Briefly, this approach involves the loading of therapeutic nanoparticles into host cells (can be organoids) *ex vivo* utilising endocytosis.

Here, in this chapter, we brought a comprehensive picture of the role of mechanotransduction and biology in organoid development, and discussed that how nanoparticles (NPs) provide a wide array of opportunities to engineer the matrix for organoids and another important aspect of this chapter is how organoids act as biologically compatible systems for nanopharmacutical screening and nanoparticle-based drug delivery.

Efficient Organoid Derivation with Mechanical Perturbations

Continuous dynamic perturbations/movements contribute to the developmental aspect of cellular organisation and are brought about by the 3D space of the cell. The basic mechanisms for tissue morphogenesis include epithelial-to-mesenchymal transitions (EMT), as well as constrictions at the apical region, contributing to the mechanical forces generated intracellularly. It has been shown that cells of the same type can make a regularised construct, on the basis of specific catenins and cadherins (Steinberg 1963). Similarly, free energy manipulations can help in germ layer speciation (Davis et al. 1997). Moreover, interactions of extracellular matrix (ECM) with cells molecules mediated by integrins are crucial in enabling the movement of large-scale tissues (Bénazéraf et al. 2010). Various molecules of ECM also play a pivotal role in inducing cytoskeletal modifications (Bedzhov and Zernicka-Goetz 2014). Looking at all the levels, starting from a cell level to multicellular levels, these movements contribute in the inception of biophysical fields that signals to regulatory networks. Hippo pathway effector Yap plays an important role in the molecular signalling mechanisms of mechano sensing and mechano transduction (Benham-Pyle et al. 2015). Piezo1/2, gated ion channels produce their effect through Yap translocation (Hennes et al. 2019) or through the cytoskeleton structure (Stewart et al. 2015). The role of Yap translocation is already established in stem cell differentiation (Lian et al. 2010). Piezo1 and Piezo2 have been shown to sense stress directly (Lin et al. 2019). Piezo1 plays an important role in determining lineage in stem cells, activation led to neurogenesis and inhibition differentiated stem cells to astrocytes (Pathak et al. 2014).

Artificial Extracellular Matrices as a Tool for Microenvironmental Engineering

Organoids provide us with facile model systems to delineate the complexity of biological and mechanical responses. To this, mechanical engineering of the matrix showed its potential to manipulate the microenvironment. Collagens, which are extensively used as supportive matrices, are important for 3D organoid culture. However, these materials suffer from lot-to-lot variabilities, which is the major setback. Artificial extracellular matrices (aECM) are promising as these variabilities could be overcome by modelling specific constituents of natural occurring ECMs. For mechanical support, hydrogels give better remodelling capabilities. Manipulation of polymer density has been shown to direct cell fates in various types of stem cells. These aECM platforms are already in use to study the importance of the matrix in various other cell types such as in iPSCs reprogramming (Caiazzo et al. 2016) and mouse organoid model (Ranga et al. 2016). Polyethylene glycol, artificial extracellular matrices, are useful in organoid formation in intestine (Gjorevski et al. 2016). Matrices such as stiff matrices are useful in intestinal organoid formation. Similarly, hydrogels with biodegradation properties showed promising results for intestinal organoids (Cruz-Acuña et al. 2017). Moreover, in addition to mechanical modifications, pH (change in acidic to alkaline or *vice versa*) and temperature are manipulated to engineer matrix to have characteristically defined organoids. Tumoroids could be derived in thermos-reversible hydrogels and changing temperature to 4°C for liquefying the matrix (Li et al. 2016b). pH-derived matrix changes are also found beneficial for those based on the diseased and healthy tissue pH changes. The biggest disadvantage of these matrices is that it involves phase changes affecting the spatial organisation at cellular levels. Further optimisation of such aECMs will be useful in organoid cultures. Next is how NPs are useful as extrinsic cues for organoid morphogenesis and formation.

Role of Nanoparticles as External Stimuli For Matrix Modelling

Matrix engineering approaches have the potential to generate forces *in vitro*, but these forces lack the heterogeneity as seen in *in vivo* scenarios, whereas NPs non-proportionately impose mechanical changes. NPs can directly impose forces at a cellular level by internalisation into the cells by tethering to targeted cells in a receptor-dependent manner (Figure 7.2).

NPs are of various sizes, shapes, and materials, with the common characteristic of length <100 nm. It includes rod-shaped particles (Wong et al. 1997), spherical particles (Pankhurst et al. 2003), quantum

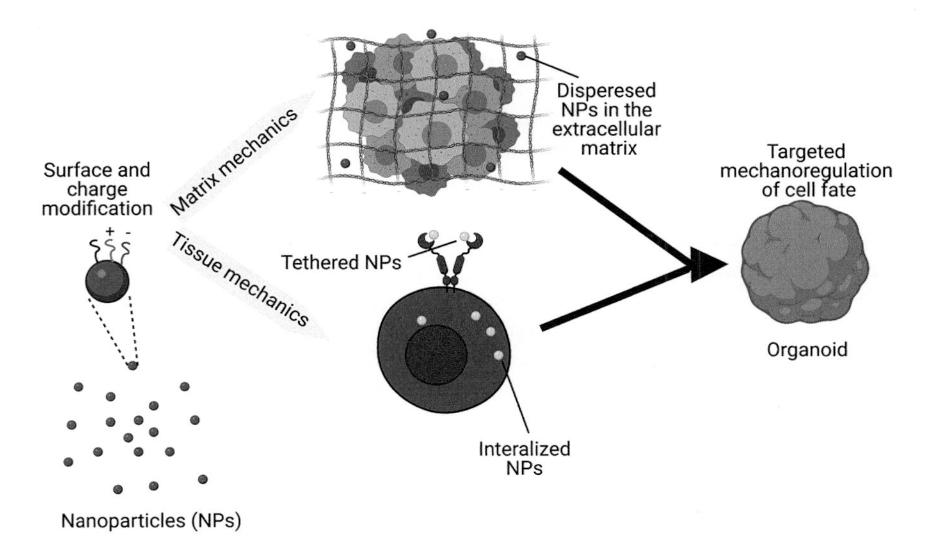

FIGURE 7.2 Nanoparticles (NPs) can be engineered for matrix modulation applications. Inactive NPs affect the local matrix stiffness while activated NPs generate local biophysical and mechanical forces. NPs are a good medium for mechanical stress transfer in surrounding tissue. Image is redrawn with permission from Abdel Fattah and Ranga (2020).

dots (Michalet et al. 2005), and 3D lattices (Ji et al. 2019). NPs have been shown to have the potential to deliver drugs to cancer cells (Breunig et al. 2008). NPs are extensively being used in disease modelling (Scialabba et al. 2019). Organoids are upcoming and good models for human pathologies/diseases (Davoudi et al. 2018). The intrinsic characteristics of NPs to easily cross the cellular permeability make them a new contributor in matrix engineering.

NPs in tissue engineering serve majorly as strengthening material for the matrix (Compton and Lewis 2014). For example, NPs with titanium oxide coating provides strength to collagen 3D scaffold and enhance the keratinocyte growth (Li et al. 2016a). Another example is a fibrous scaffold; silk fibroin scaffolds enriched with NPs enhanced the osteoblast culture capacity (Kim et al. 2014).

Active Manipulation of Matrix for Organoid Development

As we learnt in the previous section that NPs incorporated in matrices or organoids can change matrix properties. One of the important aspects is NPs through external media; NPs responsive through exogenous fields have some advantages that there will be less changes in the 3D space (Jestin et al. 2008), and mimic the *in vivo* conditions generating heterogeneous forces in the local 3D space (Abdel Fattah et al. 2016). NPs enhance the effect of consistent biophysical and mechanical perturbations on tissues through controlling the activation field. Magnetic fields can be engineered more precisely within the culture environment using big magnets. For example, hydrogel embedded with iron particles has shown great promise in reversibly manipulating the substrate stiffness (Abdeen et al. 2016). In another study, stromal cells from vasculature were activated by NP movement under the effect of magnetic field (Filippi et al. 2019). These activated cells exhibited enhanced cellular markers and activated pathways involved in metabolism and mechanotransduction. These studies are very important as they bring out the potential of matrices responsive to magnetic field that can regulate cell fate speciation and function. These provide a promising technology platform for matrix engineering in mechanobiology (Dupont et al. 2011).

Generating Forces from by NP Internalisation

While we learnt that mechanical modulations through external NPs into the cellular bodies can modulate the matrix and thus the organoid differentiation and maturation, NPs get internalised into the target cells to impart direct force. Synthetic NPs are pro-compatible; in order to make them less toxic/damaging to the cells, the surface needs modifications. The most obvious mechanism of NPs entering the cells is through endocytosis. But sometimes, over accumulation of these particles into the cells to toxic levels led to cell death (Schweiger et al. 2012). Nanoscale particles have the potential to interact with cytoskeletal rearrangement in the cell, dysregulating important cellular processes (Bouissou et al. 2004). For example, silver NPs disrupted ß-catenin pathway in differentiating neural stem cells (Cooper et al. 2019). To overcome the cytotoxic effects of NPs inside the cells, these can be made compatible through surface charge modifications, which allow smooth internalisation into the cell (Hsiao et al. 2019). NPs size is one of the main determining factors to decide for internalisation mechanism. For NPs>250 nm, internalisation occurs via phagocytosis, while below 100 nm, pinocytosis is the main process (Panariti et al. 2012). The shape of NP is another determining factor for uptake by the cell or organoid. For example, spherical NPs have higher internalisation capacity as compared to nanorods (Chithrani and Chan 2007).

We have discussed that how matrix could be engineered using the NPs as extrinsic or intrinsic stimuli to model the better culture conditions for organoids. Another important aspect is to look into organoids as a tool for nanoparticle-based drug discovery and drug delivery approaches.

Organoids as Tools for Drug Discovery

Developing novel drugs (drug discovery) and identifying novel therapeutic molecular targets is a long, costly, and challenging task because of limited success in the initial screening process at the *in vitro* level. As a result, there have been significant advancements towards adopting more biomimetic platforms for drug screening platforms with higher fidelity for testing bioactivity and toxicity. To this end, in recent years, there have been encouraging efforts towards switching from 2D assays to physiologically more acceptable 3D system of assays, including cell-based assays and multicellular spheroid models as well miniaturised organ on chip systems (Ranga et al. 2016). In particular, in last few years, there have been consistent attempts to develop complex multicellular constructs termed "Organoid" equivalents for organs towards providing high-value de-risking platforms for recapitulating properties of respective organs. These 3D assays can potentially fill the gap and connect the missing link between primary drug screening of compounds and forward lead optimisation into animal and human clinical trials (Figure 7.3).

There has been a growing number of organs having *in vitro* organoid models as shown in Figure 7.4. Organoid technology virtually recapitulates organ developmental processes and human diseases (infections, pathogenic variants), thereby providing a physiologically more acceptable tool to facilitate analysis of drug toxicity, pharmacokinetics, and efficacy evaluation. Further, organoids offer a translational strategy in regenerative medicine through overcoming the deficiency in the supply of healthy tissues and rescuing rejection causing an allogeneic immune response.

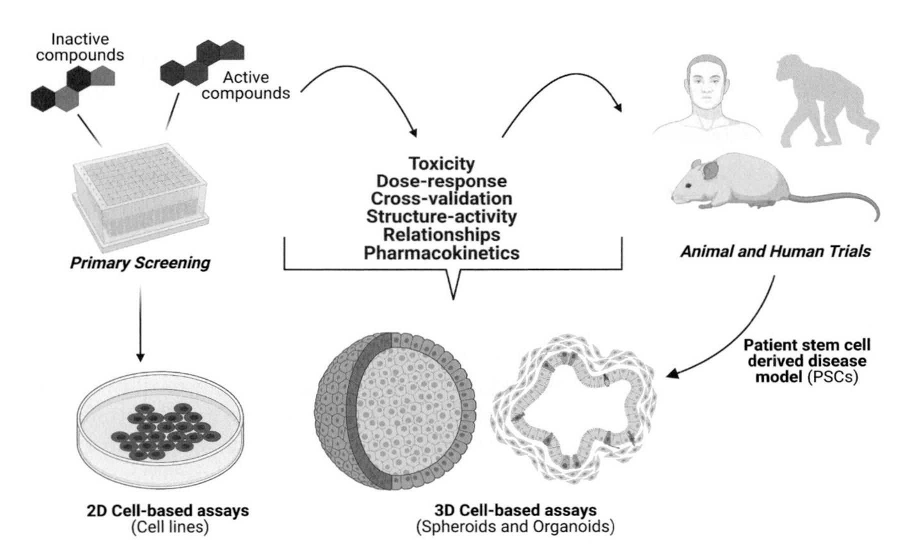

FIGURE 7.3 3D assays as an important link between primary screening to forward lead optimisation into animal and human clinical trials.

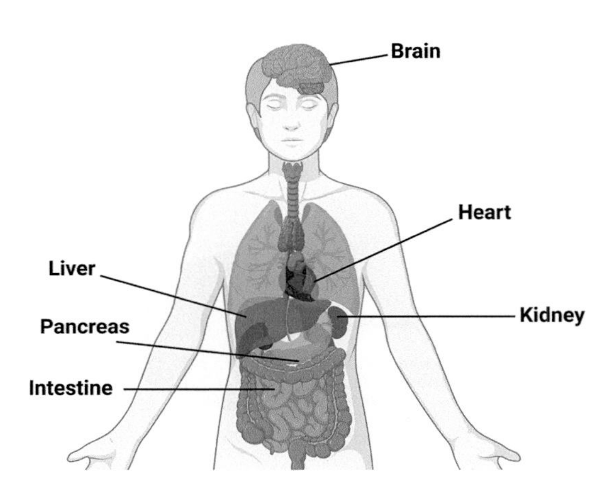

FIGURE 7.4 Current organoid models of various organs which can recapitulate physiological properties of their respective organs.

Organoids could serve as "live" clinical specimen for drug screening and testing (Liu et al. 2016). For example, literature support the role of organoids as drug screening platforms in various preclinical models for pharmacodynamic profiling of human tumours. For example, Chua et al. used prostate cancer organoids generated from Nkx3.1 expressing cells for drug response modelling (Chua et al. 2014). These tumour organoids reflected genetic and phenotypic characteristics of their progenitor. Similarly, Huang et al. used organoids from primary human pancreatic ductal adenocarcinoma to identify novel effective drugs restricting proliferation after treatment (Huang et al. 2015). These organoids reflected similar differentiation markers and histological architecture to progenitor tumour.

Trojan Horse System and Nanotechnology: A Novel Drug Delivery Approach

In recent years, an innovative concept of "the Trojan horse system" has emerged which synergistically link advanced nanotechnology tools with novel host cell-mediated drug delivery. These host cells, which encapsulate therapeutic nanoparticles, can serve as "Trojan horses" or delivery vectors and injected at the localised site for treatment.

Choi et al. (2007) successfully demonstrated that therapeutically inaccessible regions of tumours could become accessible by exploiting monocytes as Trojan horse for the nanoparticle-based drug delivery. Their results demonstrated that critical steps of trojan horse therapeutics were accomplished, *viz*. effective phagocytosis of Au nanoshell by macrophages/monocytes, competence of these loaded cells for tumour trafficking, and drug delivery *in situ*. An *in vitro* model of tumour hypoxic conditions, for example, human breast tumour spheroids, was analysed for cellular uptake efficacy as well as nanoshell-facilitated photomediated ablation at hypoxic tumour site (Figure 7.5). These results suggest that such an approach could be exploited for delivering a variety of nanoparticle-based drugs to otherwise inaccessible hypoxic tumour regions. However, the nontoxicity of loaded cargo to encapsulating cells or organoid equivalent till they are generated functionally and thereafter reach the site of delivery is an essential prerequisite.

Trojan horse, a way of drug delivery which offers several advantages, viz. higher drug concentration at the local site, increased retention time, enhanced distribution, longer dosing gaps, thus improved overall therapeutic efficacy. This way of drug delivery through cells of self-origin is highly specific and augments longer persistence, which could be attributed to rescuing immune response on one side while also facilitating homing to specific tissues as well as modulating microenvironment on the other side. Nevertheless, one of the prerequisites of this approach is ensuring inactivity and nontoxicity of drug cargos to the encapsulating cells till delivering their loads to the targeted site. Literature support several advantages of using stem cell-based drug delivery systems over the synthetic drug delivery methods (Müller et al. 2006, Scoville et al. 2008, Porada and Almeida-Porada 2010, Tang et al. 2010).

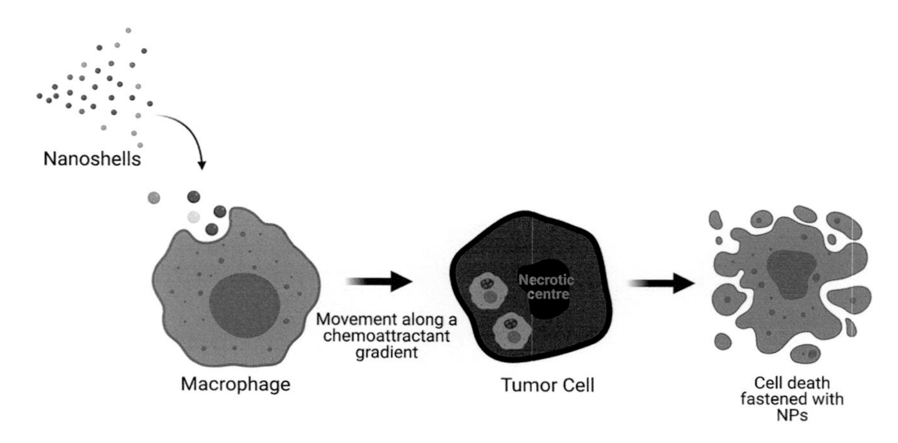

FIGURE 7.5 Trojan horse cell-based approach for therapeutic delivery of nanoparticles based on hypoxic conditions in the tumour hypoxic region.

The recent advancements in organoid culture have encouraged great expectation in proposing such drug delivery with synergy to advanced nanotechnology-based approaches. To this end, Peng et al. (2015) analysed the potential utility of self-assembled "organoid" (intestinal) to rectify enteric diseases, radiation injury, and inflammatory bowel disease. Incidentally, the physiological environment of the GI tract is harsh limiting the stability of targeting of nanocarriers to intestines. Therefore, they expanded the primary intestinal stem cells isolated from mice *ex vivo* and established self-renewing intestinal organoids through defined protocols using growth factors. The organoid reflected a cauliflower-like morphology with distinguishable crypts with multicellular mucosal architecture. After incubation of ISCs with DNA-functionalised gold nanoparticles, these nanoparticles were encapsulated in the lumen to develop a novel trojan horse, which suggested the possibility of gene regulation treatment strategies for inflammatory bowel disease. Besides this example of delivery of genes, this intestinal trojan horse delivery system could further be utilised for other cargos, viz. growth factors, peptides, imaging chemicals, etc. Similarly, Davoudi et al. (2018) utilised intestinal organoids as carriers of 5-amino salysilic acid (5-ASA)-loaded poly (lactic co glycolic acid) nanoparticles. These nanoparticles had no considerable impact on organoid development, assembly as well as viability, suggesting the feasibility of organoid trojan horse system for delivering drugs. Thus, the translational relevance of such organoid trojan horse system for intestine and other tissues can be exploited further in nanotechnology-based diagnosis and efficacious drug delivery for treatment of enteric and other diseases.

However, there are significant limitations of utilising the organoid system for drug discovery as well as drug delivery-based evolving approaches, which are as below:

1. In cultured organoids, for example, intestinal epithelial organoids, the limited number of cell types along with the absence of the nervous, vascular, and immune system could result in altered drug effects than in vivo scenario.
2. The 3D organisation could vary spatially and structurally compared to that observed in vivo.
3. The microenvironment, pH, growth factors, hormones, and other epigenetic factors can differentially affect as those cannot be exactly reproduced under in vitro scenario.

Through this chapter, we brought the importance of mechanobiology in tissue/organoid development and discussed that how nanoparticles (NPs) provide a wide array of opportunities to engineer the matrix for organoids. The other aspect of this chapter is to address how organoids act as biologically compatible systems for nanopharmaceutical screening and nanoparticle-based drug delivery. We have learnt that there is a lot of scope in addressing the problem of organoid maturation via matrix engineering using nanoparticles. There are few reports on organoid as trojan horses for drug delivery. Extensive research needs to be undertaken to exploit organoids for drug delivery.

REFERENCES

Abdeen, A.A., Lee, J., Bharadwaj, N.A., Ewoldt, R.H., and Kilian, K.A., 2016. *Advanced Healthcare Materials*, *5* (19), 2536–2544.

Abdel Fattah, A.R., Ghosh, S., and Puri, I.K., 2016. *ACS Applied Materials &Interfaces*, *8* (17), 11018–11023.

Abdel Fattah, A.R. and Ranga, A., 2020. *Frontiers in Bioengineering and Biotechnology*, *8* 240. doi:10.3389/fbioe.2020.00240.

Bedzhov, I. and Zernicka-Goetz, M., 2014. *Cell*, *156* (5), 1032–1044.

Bénazéraf, B., Francois, P., Baker, R.E., Denans, N., Little, C.D., and Pourquié, O., 2010. *Nature*, *466* (7303), 248–252.

Benham-Pyle, B.W., Pruitt, B.L., and Nelson, W.J., 2015. *Science*, *348* (6238), 1024–1027.

Bouissou, C., Potter, U., Altroff, H., Mardon, H., and Van Der Walle, C., 2004. *Journal of Controlled Release*, *95* (3), 557–566.

Breunig, M., Bauer, S., and Goepferich, A., 2008. *European Journal of Pharmaceutics and Biopharmaceutics*, *68* (1), 112–128.

Caiazzo, M., Okawa, Y., Ranga, A., Piersigilli, A., Tabata, Y., and Lutolf, M.P., 2016. *Nature Materials*, *15* (3), 344–352.

Chan, C.J., Heisenberg, C.-P., and Hiiragi, T., 2017. *Current Biology: CB*, *27* (18), R1024–R1035.

Chinthalapudi, K., Rangarajan, E.S., Brown, D.T., and Izard, T., 2016. *Proceedings of the National Academy of Sciences*, *113* (34), 9539–9544.
Chithrani, B.D. and Chan, W.C.W., 2007. *Nano Letters*, *7* (6), 1542–1550.
Choi, M.-R., Stanton-Maxey, K.J., Stanley, J.K., Levin, C.S., Bardhan, R., Akin, D., Badve, S., Sturgis, J., Robinson, J.P., Bashir, R., Halas, N.J., and Clare, S.E., 2007. *Nano Letters*, *7* (12), 3759–3765.
Chua, C.W., Shibata, M., Lei, M., Toivanen, R., Barlow, L.J., Bergren, S.K., Badani, K.K., McKiernan, J.M., Benson, M.C., Hibshoosh, H., and Shen, M.M., 2014. *Nature Cell Biology*, *16* (10), 951–961, 1–4.
Compton, B.G. and Lewis, J.A., 2014. *Advanced Materials*, *26* (34), 5930–5935.
Cooper, R.J., Menking-Colby, M.N., Humphrey, K.A., Victory, J.H., Kipps, D.W., and Spitzer, N., 2019. *Neurotoxicology*, *71*, 102–112.
Cruz-Acuña, R., Quirós, M., Farkas, A.E., Dedhia, P.H., Huang, S., Siuda, D., García-Hernández, V., Miller, A.J., Spence, J.R., Nusrat, A., and García, A.J., 2017. *Nature Cell Biology*, *19* (11), 1326–1335.
Davidson, L.A., 2017. *Philosophical Transactions of the Royal Society of London. Series B, Biological Sciences 372*(1720):20150516. doi:10.1098/rstb.2015.0516.
Davis, G.S., Phillips, H.M., and Steinberg, M.S., 1997. *Developmental Biology*, *192* (2), 630–644.
Davoudi, Z., Peroutka-Bigus, N., Bellaire, B., Wannemuehler, M., Barrett, T., Narasimhan, B., and Wang, Q., 2018. *Journal of Biomedical Materials Research. Part A*, *106* (4), 876–886.
Dupont, S., Morsut, L., Aragona, M., Enzo, E., Giulitti, S., Cordenonsi, M., Zanconato, F., Le Digabel, J., Forcato, M., Bicciato, S., Elvassore, N., and Piccolo, S., 2011. *Nature*, *474* (7350), 179–183.
Engler, A.J., Sen, S., Sweeney, H.L., and Discher, D.E., 2006. *Cell*, *126* (4), 677–689.
Filippi, M., Dasen, B., Guerrero, J., Garello, F., Isu, G., Born, G., Ehrbar, M., Martin, I., and Scherberich, A., 2019. *Biomaterials*, *223*, 119468.
Gjorevski, N., Sachs, N., Manfrin, A., Giger, S., Bragina, M.E., Ordóñez-Morán, P., Clevers, H., and Lutolf, M.P., 2016. *Nature*, *539* (7630), 560–564.
Hennes, A., Held, K., Boretto, M., De Clercq, K., Van den Eynde, C., Vanhie, A., Van Ranst, N., Benoit, M., Luyten, C., Peeraer, K., Tomassetti, C., Meuleman, C., Voets, T., Vankelecom, H., and Vriens, J., 2019. *Scientific Reports*, *9* (1), 1779.
Hsiao, I.-L., Fritsch-Decker, S., Leidner, A., Al-Rawi, M., Hug, V., Diabaté, S., Grage, S.L., Meffert, M., Stoeger, T., Gerthsen, D., Ulrich, A.S., Niemeyer, C.M., and Weiss, C., 2019. *Small*, *15* (10), 1805400.
Huang, L., Holtzinger, A., Jagan, I., BeGora, M., Lohse, I., Ngai, N., Nostro, C., Wang, R., Muthuswamy, L.B., Crawford, H.C., Arrowsmith, C., Kalloger, S.E., Renouf, D.J., Connor, A.A., Cleary, S., Schaeffer, D.F., Roehrl, M., Tsao, M.-S., Gallinger, S., Keller, G., and Muthuswamy, S.K., 2015. *Nature Medicine*, *21* (11), 1364–1371.
Jestin, J., Cousin, F., Dubois, I., Ménager, C., Schweins, R., Oberdisse, J., and Boué, F., 2008. *Advanced Materials*, *20* (13), 2533–2540.
Ji, M., Ma, N., and Tian, Y., 2019. *Small*, *15* (26), 1805401.
Kim, H., Che, L., Ha, Y., and Ryu, W., 2014. *Materials Science and Engineering: C*, *40*, 324–335.
Lancaster, M.A. and Knoblich, J.A., 2014. *Science (New York, N.Y.)*, *345* (6194), 1247125.
Li, N., Fan, X., Tang, K., Zheng, X., Liu, J., and Wang, B., 2016a. *Colloids and Surfaces. B, Biointerfaces*, *140*, 287–296.
Li, Q., Lin, H., Wang, O., Qiu, X., Kidambi, S., Deleyrolle, L.P., Reynolds, B.A., and Lei, Y., 2016b. *Scientific Reports*, *6* (1), 31915.
Lian, I., Kim, J., Okazawa, H., Zhao, J., Zhao, B., Yu, J., Chinnaiyan, A., Israel, M.A., Goldstein, L.S.B., Abujarour, R., Ding, S., and Guan, K.-L., 2010. *Genes & Development*, *24* (11), 1106–1118.
Liang, Q., Han, Q., Huang, W., Nan, G., Xu, B.-Q., Jiang, J.-L., and Chen, Z.-N., 2014. *PLoS One*, *9* (7), e102496.
Lin, Y.-C., Guo, Y.R., Miyagi, A., Levring, J., MacKinnon, R., and Scheuring, S., 2019. *Nature*, *573* (7773), 230–234.
Liu, F., Huang, J., Ning, B., Liu, Z., Chen, S., and Zhao, W., 2016. *Frontiers in Pharmacology*, *7*:334. doi:10.3389/fphar.2016.00334.
Michalet, X., Pinaud, F.F., Bentolila, L.A., Tsay, J.M., Doose, S., Li, J.J., Sundaresan, G., Wu, A.M., Gambhir, S.S., and Weiss, S., 2005. *Science (New York, N.Y.)*, *307* (5709), 538–544.
Müller, F.-J., Snyder, E.Y., and Loring, J.F., 2006. *Neuroscience*, *7* (1), 75–84.
Olson, T.M., Illenberger, S., Kishimoto, N.Y., Huttelmaier, S., Keating, M.T., and Jockusch, B.M., 2002. *Circulation*, *105* (4), 431–437.

Panariti, A., Miserocchi, G., and Rivolta, I., 2012. *Nanotechnology, Science and Applications*, *5*, 87–100.
Pankhurst, Q.A., Connolly, J., Jones, S.K., and Dobson, J., 2003. *Journal of Physics D: Applied Physics*, *36* (13), R167–R181.
Pathak, M.M., Nourse, J.L., Tran, T., Hwe, J., Arulmoli, J., Le, D.T.T., Bernardis, E., Flanagan, L.A., and Tombola, F., 2014. *Proceedings of the National Academy of Sciences of the United States of America*, *111* (45), 16148–16153.
Peng, H., Wang, C., Xu, X., Yu, C., and Wang, Q., 2015. *Nanoscale*, *7* (10), 4354–4360.
Porada, C.D. and Almeida-Porada, G., 2010. *Advanced Drug Delivery Reviews*, *62* (12), 1156–1166.
Ranga, A., Girgin, M., Meinhardt, A., Eberle, D., Caiazzo, M., Tanaka, E.M., and Lutolf, M.P., 2016. *Proceedings of the National Academy of Sciences*, *113* (44), E6831–E6839.
Schweiger, C., Hartmann, R., Zhang, F., Parak, W.J., Kissel, T.H., and Rivera_Gil, P., 2012. *Journal of Nanobiotechnology*, *10* (1), 28.
Scialabba, C., Sciortino, A., Messina, F., Buscarino, G., Cannas, M., Roscigno, G., Condorelli, G., Cavallaro, G., Giammona, G., and Mauro, N., 2019. *ACS Applied Materials & Interfaces*, *11* (22), 19854–19866.
Scoville, D.H., Sato, T., He, X.C., and Li, L., 2008. *Gastroenterology*, *134* (3), 849–864.
Steinberg, M.S., 1963. *Science*, *141* (3579), 401–408.
Stewart, R.M., Zubek, A.E., Rosowski, K.A., Schreiner, S.M., Horsley, V., and King, M.C., 2015. *Journal of Cell Biology*, *209* (3), 403–418.
Tang, C., Russell, P.J., Martiniello-Wilks, R., Rasko, J.E.J., and Khatri, A., 2010. *Stem Cells (Dayton, Ohio)*, *28* (9), 1686–1702.
Vasile, V.C., Ommen, S.R., Edwards, W.D., and Ackerman, M.J., 2006. *Biochemical and Biophysical Research Communications*, *345* (3), 998–1003.
Wong, E.W., Sheehan, P.E., and Lieber, C.M., 1997. *Science*, *277* (5334), 1971–1975.

8

Hyaluronan-Based Hydrogels as Functional Vectors for Standardised Therapeutics in Tissue Engineering and Regenerative Medicine

Alexandre Porcello
School of Pharmaceutical Sciences, University of Geneva, Switzerland
Institute of Pharmaceutical Sciences of Western Switzerland, University of Geneva, Switzerland

Alexis Laurent
LAM Biotechnologies SA, Switzerland
Lausanne University Hospital, Switzerland

Nathalie Hirt-Burri and Philippe Abdel-Sayed
Lausanne University Hospital, Switzerland

Anthony de Buys Roessingh
Lausanne University Hospital, Switzerland

Wassim Raffoul
Lausanne University Hospital, Switzerland

Olivier Jordan and Eric Allémann
School of Pharmaceutical Sciences, University of Geneva, Switzerland
Institute of Pharmaceutical Sciences of Western Switzerland, University of Geneva, Switzerland

Lee Ann Applegate
Lausanne University Hospital, Switzerland
University of Zurich, Switzerland
Oxford University, China

CONTENTS

DOI: 10.1201/9781003153504-8

Introduction: Hyaluronan for Tailorable Regenerative Medicine Products

Industrial Transposition and Clinical Translation of Standardised Complex Biologicals

Within design and development phases of novel medicinal products (e.g. combined advanced therapy medicinal products (cATMP), tissue engineering products (TEP), medical devices), primary therapeutic materials often require conjugation to ancillary inert, supportive, or potentiating delivery vehicles or scaffolds. For local delivery by transplantation of cellular materials for relatively soft tissue repair, or local diffusion of therapeutic components, hydrogels represent prime formulation choices (Hartmann-Fritsch et al. 2016; Fallacara et al. 2018). Finely tuned physico-chemical properties of hydrogels, intrinsic or environment-responsive, may play a preponderant role towards mediation, hindering, or stimulation of therapeutic potential borne by standardised cellular substrates or derivatives thereof. Optimised developmental workflows may enable pragmatic regulatory classification of cell-based or cell-derived hydrogel preparations, addressing both current regulatory hurdles and economic constraints (Bertram et al. 2012; Heathman et al. 2015). In particular, sustainable and highly scalable industrial transposition capacity is a prerequisite in modern developmental approaches and must be a critical factor when implementing therapeutic cell sources, hydrogel components, and manufacturing methodologies (Huerta-Ángeles et al. 2018; Laurent et al. 2020b). Extensive industrial experience and hindsight have been generated around the use of versatile hyaluronan-based hydrogels, which allow arrays of functionalisation possibilities and tailoring to specific topical (i.e. epidermal, dermal) or musculoskeletal medical indications (e.g. topical rehabilitation preparations, viscosupplementation, volumetric supplementation). Numerous human and veterinary applications (i.e. medicinal, aesthetic, cosmetic) have been considered for hyaluronic acid (HA), serving as a backbone for quasi-infinite structural modification and physico-chemical property modulation (Fallacara et al. 2018).

Harnessing Regenerative Potentials of Biologicals for Optimised Clinical Benefits

In the present chapter, an overview of critical therapeutic product design considerations will be addressed, as well as application examples of hyaluronan-based preparations for cell therapy delivery, combined with various autologous or allogeneic optimised therapeutic cell sources (e.g. platelet-rich plasma, stem cells, banked primary foetal progenitors). Particular focus is set on specific parameters directly impacting the potential of considered formulations in terms of clinical translation and biological effectiveness. The necessity for robust designs, scalable and sustainable industrial workflows, and standardised therapeutic efficacy are outlined. Overall, it may be shown that systematic approaches of biotechnological processing are key in sourcing and manufacturing both raw materials for hydrogel constitution (e.g. biofermented hyaluronan) and homogenous primary therapeutic cell sources (e.g. foetal progenitor cells consistently and simultaneously derived from a single organ donation). Such cell sources may be differentially and parallely established from multiple tissues (i.e. dermis, cartilage, tendon, muscle, bone, lung, intervertebral disc, etc.), undergo bioprocessing through *in vitro* propagation under inductive conditions for multi-tiered cell bank establishment, eventually being valorised after large-scale GMP manufacturing and release of clinical cell lots or cell-derived cell-free materials (Hohlfeld et al. 2005; Laurent et al. 2020b). These concepts of cell sourcing and delivery method optimisation have enabled the generation of over three decades of clinical experience in Switzerland, with the use of foetal progenitor cells and hyaluronan in particular, and shall continue to tangibly contribute to improvement of patient health in cutaneous and musculoskeletal regenerative medicine applications in general (Grognuz et al. 2016).

General Properties, Physiological, and Therapeutic Roles of Hyaluronan-Based Hydrogels

General Physical and Chemical Properties of Hydrogels

Hydrogels have been generically and commonly defined as three-dimensional hydrophilic polymeric networks suited for relatively important hydric captation (i.e. water, aqueous solution, or biological liquids), while maintaining a dispersion state, rather than a solute state (Peppas et al. 2000). Statutory definitions of the hydrogel form are devoid of minima regarding relative water contents. Relatively low values of water content typically revolve around 10%–20% of initial dry weight values (Hoffman 2012). Hydrogels are typically constituted by aqueously suspended, chemically or physically cross-linked polymers, inherently displaying macroscopic similarities (i.e. soft texture, flexible material, and rubbery properties) with extracellular matrix (ECM) components or native soft biological tissues (Peppas et al. 2012). Synthetic hydrogel products may be designed to display stimuli-responsive behaviours, through specific controlled reactions to environmental modifications (e.g. pH, temperature, ionic strength, pressure) (Peppas et al. 2000, 2012; Thomas et al. 2007; Hoffman 2012). Hydrogel differentiation and classification criteria include polymer identity and type (i.e. physico-chemical properties), sourcing (i.e. natural, synthetic, hybrid), combinations (i.e. homopolymer, heteropolymer, or co-polymer gels), functionalisation (e.g.

chemical groups, peptides, biologicals), and cross-linking, as well as the overall composition (Peppas et al. 2000, 2012; Thomas et al. 2007; Hoffman 2012; Ahmed 2015). Such diversity and potential for specific modulation and property tuning have promoted hydrogels to the forefront of several disciplines, among which biomedicine (Hoffman 2012; Mandal et al. 2020).

Hyaluronan-Based Hydrogel Definitions

Biopolymers (i.e. human, animal, plant, or bacteria-derived) are often characterised by optimal cyto- and biocompatibility. This class of materials comprises the natural polysaccharide HA or hyaluronan, a ubiquitous constituent of all vertebrates and a key component in a variety of medical, pharmaceutical, nutritional, and cosmetic applications (Liao et al. 2005; Ahmed 2015; Fallacara et al. 2018). An overview of the sources and potential applications of hyaluronan and derived products is presented in Figure 8.1. The natural, non-sulfated, polyanionic, and unbranched (i.e. linear chains of disaccharide

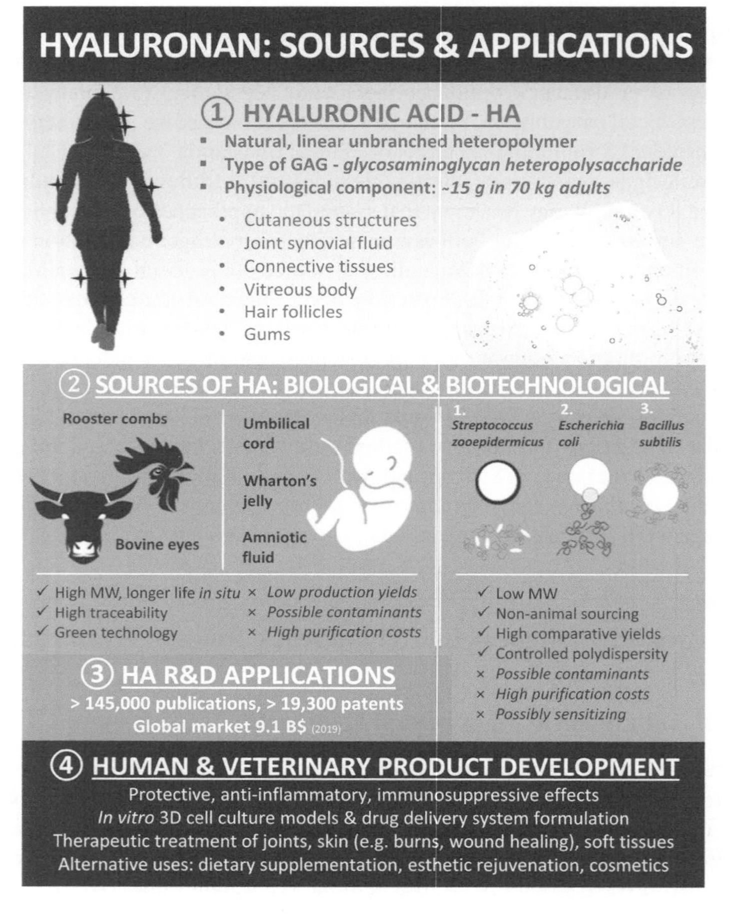

FIGURE 8.1 Overview of hyaluronan physiological anatomical locations, sources for therapeutic product development, current R&D statistics, and possible modern commercial applications. As a natural and ubiquitous physiological component, hyaluronan possesses diversified functions (e.g. structural, biomechanical, immunologic) (1). Historically, hyaluronan isolation and purification from food industry waste products were preferred for the obtention of large batches, while human gestational tissues and fluids have been studied and applied in some cases of burn wound treatment. Modern industrial processes have switched to bacterial fermentation for hyaluronan production, enabling obtention of homogenous and important yields of relatively lower molecular weight polymers. Respective advantages (i.e. ✓) and disadvantages (i.e. ×) are presented for biological and biotechnological sources of hyaluronan (2). The high interest around HA for R&D applications is attested by the considerable published literature, market statistics, and diversified modalities or indications for *in vitro* models and therapeutic products in development (3, 4). Market data was gathered from www.grandviewresearch.com (i.e. accessed November 2020).

units of N-acetyl-D-glucosamine and D-glucuronic acid linked by β-1,3 and β-1,4-glycosidic bonds) glycosaminoglycan (GAG) was coined as "hyaluronic acid" upon discovery by Meyer and Palmer in 1934, due to a mild acidic behaviour. Under physiological conditions, HA takes a polyelectrolyte form with associated ions, hence the alternative "sodium hyaluronate" or "hyaluronate" appelations, whereas "hyaluronan" refers to all possible forms of HA molecules (Dicker et al. 2014). Due to the presence of alcohol and carboxylic groups, HA is highly hydrophilic and polyanionic. Based on nuclear magnetic resonance spectroscopy (NMR), X-ray diffraction, and molecular modelling data, hydrogen bonds linking neighbouring disaccharides conform HA to an extended twofold helix secondary structure, with maximal overall molecular weights (MW) in the range of 10^8 Da (Liao et al. 2005; Fallacara et al. 2018). Overarching hyaluronan-based hydrogel network-related properties (i.e. viscosity and viscoelasticity) are directly related to MW, polymer concentration, and cross-link density.

Essential Physiological Roles of Hyaluronan and Related Bioengineering

Physiologically, hyaluronan is a highly hygroscopic ECM constituent found predominantly in the umbilical cord, connective tissues, synovial fluid, skin, and vitreous body of the eye, playing important roles in structural integrity and hydration maintenance. In cartilage tissues, hyaluronan represents up to two percent of total GAGs and plays essential lubrication and homeostasis functions (Kuiper and Sharma 2015; Fallacara et al. 2018). It is mainly synthesised by three transmembrane glycosyltransferase isoenzymes (i.e. hyaluronan synthases, HAS), and degradation is mediated by specific enzymes (i.e. hyaluronidases, HYAL) and reactive oxygen species (ROS) (Aruffo et al. 1990; Watterson and Esdaile 2000; Stern et al. 2007; Fallacara et al. 2018). Physiological turnovers are tissue-specific, wherein hyaluronan is renewed daily in the epidermis, once every three days in the dermis, and up to once every two days in cartilage (Fraser et al. 1993). In addition to water-retention and moisturising, specific and important hyaluronan functions may mediate and support cell migration, anti-inflammatory responses, angiogenesis, and tissue architectural organisation *via* interacting with collagen bundles. Such diversity in functions and effects may be partly due to the size variability of hyaluronan polymers. Natural location-specific decreases in HA content with ageing may play a role in diminishing tissue healing capacities. Additionally, in an inflammatory tissue context, HA degradation may be exacerbated, potentially mitigating the efficacy of various therapeutic products by restricting the residence time of the hydrogel in its original designed form. Nevertheless, due to important physiological functions of hyaluronan, high versatility, and emergence of modern physical and chemical processes, physiology-guided engineering has contributed to the democratisation of hyaluronan-based hydrogels in wide arrays of applications, specifically in the field of tissue engineering (Hoffman 2012; López-Ruiz et al. 2019).

Mechanisms Governing Physiological Hyaluronan Functions and Fate in the Skin

Hyaluronan synthesis in the skin by dermal fibroblasts and epidermal keratinocytes has been associated with CD44 receptors and shown to be under growth factor control (e.g. FGF, TGF-α/β, EGF, PDGF) (Aya and Stern 2014; Kavasi et al. 2017). Apart from structural purposes, additional cell environment control and protection roles have been attributed to hyaluronan, notably with the modulation of epidermal permeability and ECM composition, mediated by CD44 receptors (Fraser et al. 1997; Bourguignon et al. 2006). Physiological dermal hyaluronan is present in high molecular weight (HMW) form (i.e. around 10^6 Da), naturally producing cytoprotective, anti-angiogenic, and anti-inflammatory effects, as well as cytokine and growth factor bioavailability modulation, or induction of regulatory T cells (Dicker et al. 2014; Rayahin et al. 2015). Conversely, lysed hyaluronan in low molecular weight (LMW) forms (i.e. < 500 kDa) binds to toll-like receptors (e.g. TLR2 and TLR4), inducing the expression of pro-inflammatory genes, release of cytokines and chemokines by macrophages, and activation of dendritic cells and T cells (Gariboldi et al. 2008; Albano et al. 2016). Specifically, hyaluronan fragments smaller than 12 units induce an activated macrophage state, fragments between 8 and 32 units exhibit inflammatory and angiogenic properties, and the presence of fragments between 20 and 200 kDa reflects a state of tissular stress (Dicker et al. 2014; Rayahin et al. 2015). Physiological hyaluronan clearance is mediated by the liver and kidneys, wherein the normal blood half-life is of two to five minutes (Fraser et al. 1997). Along with effective size, polymer purity must also be taken into account to fully understand differential physiological effects of hyaluronan (Dong et al. 2016).

Pivotal Physiological Roles of Hyaluronan in Tissular Repair

In the context of skin lesions, hyaluronan accumulates at the repair site over three days to initiate tissue responses, maintains cell integrity, and finalises skin repair before being eliminated by HYAL enzymes after ten days (Mesa et al. 2002; Kavasi et al. 2017). Around 20%–30% of the hyaluronan turnover takes place locally in the skin, while the rest is eliminated by the lymphatic system. Interactions with specific receptors (i.e. CD44 and RHAMM) induce mitotic, migratory, and angiogenic responses (Dicker et al. 2014). Various specific hyaluronan polymer sizes establish and mediate the dialogue between damaged ECM and resident cells (e.g. fibroblasts, phagocytes), modulating both inflammatory and immune responses during tissue repair (Šafránková et al. 2010). Pivotal and specific roles of hyaluronan-mediated repair processes may be observed in developing foetuses, wherein high hyaluronan contents in the amniotic fluid prevent fibrosis and scarring (Nyman et al. 2013; Aya and Stern 2014). Specifically, longer persistence of hyaluronan may reduce collagen deposition and prevent scarring, leading to scar-free tissue repair in the foetus (Weindl et al. 2004). Such activities are drastically hindered in elderly organisms, functionally delaying or impairing the tissue repair process (Price et al. 2005).

Biotechnological Applications of Hyaluronan in Cell Culture Systems

Hyaluronan may be found in various modern cell culture systems or bio-printing workflows, wherein *ex vivo* cell culture is increasingly used for studies in the field of regenerative medicine (Murphy et al. 2013). Cell-supporting matrices are not limited to passive transport roles, but constitute essential microenvironments for cell signalling and proliferation (Bourguignon et al. 2006). Whereas monolayer cultures remain standard for *in vitro* cellular expansion and experimental systems, three-dimensional culture environments (i.e. static or dynamic) have been shown to optimally mimic native cell behaviours through physical scaffold provision (Turner et al. 2004). Natural and synthetic hydrogel sources have been characterised and proposed for cell culture, wherein the former types (e.g. collagen, hyaluronan, chitosan, alginate) are particularly biocompatible and functional for supporting cell viability, proliferation, and further development (Tibbitt and Anseth 2009). In view of tissue reconstruction, hyaluronan-based scaffolds are of great interest for mechanistic investigation of cell behaviour in the context of proof-of-concept establishment, as well as product biofabrication, as they mimic the natural composition and disposition of biological living tissues (Gurski et al. 2009; Jeffery et al. 2014).

Diversified Industrial Applications of Hyaluronan-Based Hydrogels

Numerous and diversified industrial applications have been proposed and investigated for native hyaluronan and related derivatives, non-exhaustively comprising drug delivery, cancer therapy, wound treatment, ophthalmology, arthrology, pneumology, urology, otolaryngology, odontology, or cosmetic and dietary applications (Fallacara et al. 2018). Specifically, drug carrier functions have been proposed for hyaluronan in prodrugs and liposomal, nanoparticle, microparticle, or hydrogel formulations, mediating controlled release, targeted, or enhanced pharmacodynamics of various therapeutic agents (e.g. anti-diabetics, steroids, anti-tumorals) (Kong et al. 2010; Chen et al. 2018; Fallacara et al. 2018). Vast clinical hindsight has been generated in orthopaedic applications, wherein arrays of marketed hyaluronan-based viscosupplementation products have been suggested as effective for managing symptoms and evolution of osteoarthritis (OA) and rheumatoid arthritis (Bowman et al. 2018).

Designing, Manufacturing, and Characterising Hyaluronan-Based Therapeutic Hydrogels

Hyaluronan-Based Therapeutic Hydrogels: Design Considerations

Intrinsic, modulable, and integrated characteristics or physico-chemical properties of hyaluronan (e.g. viscoelasticity, hygroscopicity, biocompatibility, tissue repair stimulation capacity) have favoured its adoption in a hydrogel form as an optimal bioengineering matrix, outperforming carbomers or sodium alginate, with recognised benefits in cell therapies and regenerative medicine (Liao et al. 2005; Prestwich 2011; Guo et al. 2015; Mandal et al. 2020). Summarised advantages of this unique polymer are presented in Table 8.1. Numerous methods have been proposed for chemical modifications of native hyaluronan in view of cell encapsulation, drug delivery, volumetric or viscous supplementation, and related biomedical applications (Prestwich 2011; Schanté et al. 2011a). It was suggested that innovative hyaluronan derivatives constituted tools enabling vast improvement of industrial therapeutic developments, assorted to a specific claim that chemically modified biopolymers allow for exploitation of biological degradation pathways, unlike synthetic materials (Prestwich 2011; Schanté et al. 2011a). A synthetic workflow for hyaluronan-based combination product (e.g. cATMP) design and development is presented in Figure 8.2. It was further shown that design considerations (e.g. choice of MW, hyaluronan type and concentration) directly impacted therapeutic functionality (i.e. anti-apoptotic, anti-inflammatory, and cytoprotective effects) and clinical efficacy in a dose-dependent manner (Neuman et al. 2015). Specifically, and outside the controlled context of clinical trials, the restricted number of marketed ready-to-use cATMPs using hydrogel vectors may be partly explained by the low efficiency of cell viability maintenance, prompting the design of therapeutic kits comprising off-the-freezer cell sources to be extemporaneously suspended in an adequate hydrogel for improved clinical effectiveness (Schmidt et al. 2008; Prestwich 2011).

Background and Modern Processes for Hyaluronan Sourcing

Hyaluronan was discovered as a main viscous component in bovine eyes in 1934 (i.e. Dr. Karl Meyer, Ophthalmology laboratory, Columbia University) and was soon proposed for diverse therapeutic uses. Food chain availabilities and vast industrial potential led to the development and patenting of optimised extraction procedures for purification of hyaluronan from rooster combs (e.g. USA4141973A patent).

TABLE 8.1

Summary of Hyaluronan Properties and Related Benefits for Tissue Engineering. Adapted from Selected References

Property	Advantage	Comments	References
Natural ECM component	Complete biodegradability	High relative amounts of hyaluronan are found in the dermal layer of the skin	Lataillade et al. 2010
Non-immunogenic	Xenogeneic application in humans Complete biodegradability	Same polymeric structure of backbones across all species	Aya and Stern 2014
Non-allergenic	Generally safe for topical application or injection	May be free of endotoxins or be purified	Guo et al. 2015
High hygroscopicity	Maintenance of tissue hydration Facilitate the exchange of nutrients	Significant water retention Prevention of tissue dehydration and subsequent cell death May absorb 1,000 times its initial weight in water and attain 10,000 times its initial volume	Aya and Stern 2014 Guo et al. 2015
Viscoelasticity	Ideal biological lubrication	Viscosity variates as a function of oscillatory movements Rapid movements reduce viscosity by increasing elasticity, restoring deformation, and minimising cell distortion	Fraser et al. 1997
Non-adhesive	Ease of topical application and removal Easy monitoring	Painless wound dressing exchanges and patients have a pleasant feeling on the skin	Madaghiele et al. 2014 Shevchenko et al. 2010 Guo et al. 2015 Mahedia et al. 2016
Antioxidant	Protects the epidermis	Hyaluronan traps free radicals generated by UV rays	Weindl et al. 2004

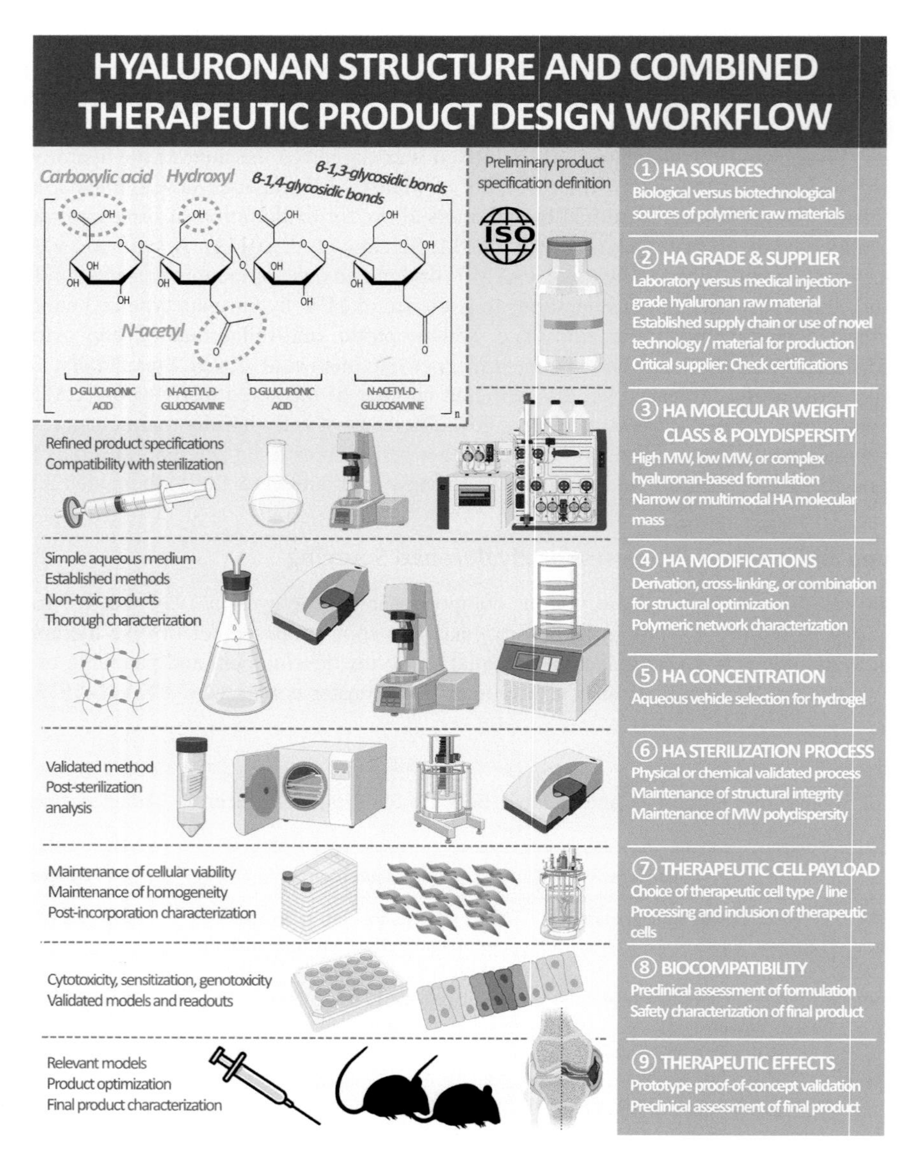

FIGURE 8.2 Chemical structure overview of hyaluronan (i.e. including modifiable functional groups) and comprehensive workflow for optimal hyaluronan-based combination product design and development. Supply chain considerations should ensure continuity of raw material availability. Based on development advancement (i.e. R&D or preclinical stage), various types and grades of hyaluronan may be used. Final product specifications partly depend on molecular weight class and polydispersity thereof (i.e. low or high molecular weight, combinations), which must be chosen appropriately for the considered sterilisation process, intended applications, and claimed effect. In particular, raw material specifications (e.g. analysis, acceptance criteria, storage) and product specifications (e.g. rheology, cohesivity, adhesivity, injectability, particle sizing) must be defined early in the development process. Hyaluronan structural modifications (i.e. conjugation or cross-linking) should be chosen to improve product parameters and effects, preferably using green chemistry and non-toxic reagents, while allowing characterisation (e.g. IR, NMR, SEC for process evaluation) and processing (e.g. lyophilisation). Hydrogel sterilisation should take into account the sensitivity of the polymeric network and of the therapeutic payload (e.g. cells or biologicals), and should be chosen among validated methods (e.g. ethylene oxide gas, steam, filtration, electron beam, gamma radiation) allowing post-sterilisation characterisation (e.g. rheology, SEC, NMR). Inclusion of therapeutic cells in the formulation must be performed while maintaining viability and homogeneity, with post-incorporation characterisation (e.g. injectability, cell viability at extrusion, stability testing). Following appropriate regulatory requirements, biocompatibility of products needs to be assessed after proper sample preparation and by using validated models. Finally, therapeutic proof-of-concept needs to be established on relevant models (e.g. *in vitro*, *ex vivo*, *in vivo*) for the intended product application.

The large red testosterone-responsive skin flap developed by roosters had been identified as the most prominent and ubiquitous tissue source (i.e. poultry waste product) for HMW hyaluronan (i.e. around 0.9 g of HA *per* kg of tissue, 4–6 MDa). Starting in the 1970s, hyaluronan was exploited for viscosupplementation of arthritic knees in the competitive horse industry to help the overworked equines reduce inflammation, and for veterinary ophthalmic uses. In human medicine, the original work of Dr. Balazs on the non-inflammatory fraction of hyaluronan extracted from biological tissues led to the development of various products. Non-exhaustive examples comprise Healon® for intraocular cataract surgery and lens replacement, Hylashield® for ocular pain and irritations, Synvisc® for viscosupplementation, and dermal visco-augmentation solutions for depressed scars, fine lines, and wrinkles. Such marketed treatments have since then been used in hundreds of millions of patients world-wide (Selyanin et al. 2015; Fallacara et al. 2018).

Major interest from the pharmaceutical industry has continuously driven the development of hyaluronan sourcing from rooster combs and related products. The Swedish company Pharmacia benefited from the patents of Dr. Balazs to develop Healon®, before transferring the technology to Pfizer. Nevertheless, hyaluronan production continued in Sweden (i.e. Swedish white leghorn roosters), with such intensive output-driven breeding that chickens eventually could not maintain an upright head posture, due to excessive comb weight. Alternative poultry species have been bred in Northeastern USA farms and valorised by Genzyme for viscosupplementation and aesthetic medicine product development. Based on the extensive use of animal-sourced HA, it was demonstrated that processing of biological tissues was readily and scalably applicable in industrial settings (Kang et al. 2010; Kulkarni et al. 2018). Specifically, quantities of rooster combs reaching a metric ton could be processed by trituration and soaking, before the water-soluble fraction was filtered and treated with alcohol for obtention of hyaluronan in powder form. Alternative industrial hyaluronan sourcing methodologies comprise bacterial fermentation (i.e. around 6–10 g of HA *per* kg of culture, 1.5–2.5 MDa), avoiding the dependency to animal tissue supply (Agerup et al. 2005). Production yields are thereby greatly improved by hyaluronan derivation from *Streptococcus equi*, resulting in excellent immune tolerance and low risk of contamination by animal pathogens, and continue to be applied by many key suppliers (e.g. Lifecore Biomedical LLC, USA; Contipro a.s., Czech Republic; Givaudan International SA, Switzerland). Some concerns were raised around the potential pro-inflammatory effects of LMW hyaluronan and fragments thereof, wherein comparative studies have shown that endotoxin-free preparations are important to avoid inflammatory stimuli by the product (Dong et al. 2016).

Establishing Hyaluronan Raw Material Specifications and Characterisation Workflows

In view of industrial manufacturing for various types of products intended for human applications (e.g. medical devices, combination drug products), clear specifications and acceptance criteria must be defined for raw materials to be used in good manufacturing practice (GMP) production (Huerta-Ángeles et al. 2018). Raw material analysis using validated preparation and analytical methods should comprise macroscopic description, physico-chemical properties measurement, structural identification, stability testing, and impurities and biological contaminants analysis (Baeva et al. 2017). Specifications constitute an integral part of the product master file and define the acceptability criteria for inclusion in the production process. Based on predefined material specifications, adequately validated multifactorial characterisation (i.e. following Pharmacopoeial methods, ASTM standards, ISO norms) must be implemented to release hyaluronan batches. In addition to ensuring quality and adequation of raw materials, international standards and norms parallely enable holistic analysis of manufacturing processes from a risks and hazards point of view (e.g. ISO 14971 for medical devices). Applied to the design of a hyaluronan-based medical device for instance, such analysis should take into account the finished product and individual components, respectively, with a thorough assessment of sourcing and purity issues to exclude safety concerns (Huerta-Ángeles et al. 2018; Šafránková et al. 2018).

Hyaluronan Grade, Molecular Weight, and Polydispersity

Hyaluronan is a highly versatile biomolecule, as polymer MW has been shown to influence biological effects, therapeutic activity and efficacy, as well as product biodistribution, residence time, and clearance parameters. Differential effects of HMW versus LMW hyaluronan are presented in Table 8.2.

TABLE 8.2

Comparison of the Effects of HMW versus LMW Hyaluronan. Adapted from Gao et al. 2008, Pardue et al. 2008, Dong et al. 2016, and Fouda et al. 2016

HMW HA (>500 kDa)	LMW (<500 kDa)
Expressed in healthy tissues	Expressed in tissues under stress Produced under inflammatory processes where LMW and oligomers result from HMW HA degradation
Cytoprotector, influences the bioavailability of cytokines and growth factors	Increases the release of β-defensins by keratinocytes
Resolves inflammation by induction of regulatory T cells	Stimulates TLR receptors Induces the production of local TGF-β Activates dendritic cells and phagocytes
Puts the cells into quiescence (i.e. *via* CD44 receptor)	Stimulates angiogenesis (i.e. *via* CD44 and RHAMM receptors) Activates cytoskeletal remodelling (i.e. *via* RHAMM receptor) by stimulating cell migration

Native hyaluronan in healthy human synovial fluid exists in the form of HMW (i.e. over 1 MDa). In contrast, relatively LMW hyaluronan (i.e. < 500 kDa) has been shown to induce inflammatory responses in various cell types (Jiang et al. 2007). HMW hyaluronan is characterised by relatively high viscosity values, long biological retention times, and appropriate therapeutic effects (López-Ruiz et al. 2019). Due to the aforementioned considerations and modern development practices, products may contain hyaluronan of various MWs for tuning of physico-chemical properties and/or effect, while ultra HMW hyaluronan may be used to preemptively compensate for the degradation occurring during product heat-sterilisation. Initial and final polymer MW and polydispersity must in all cases be studied and validated. While laboratory-grade hyaluronan is suitable for design and product development steps, medical-grade raw material is required for final formulation and product registration and commercialisation phases. In any case, hyaluronan sourcing should comprise precise MW distribution parameters, definition of acceptable ranges for the specific considered application, and refining (e.g. using size-exclusion chromatography, flow field-flow fractionation) if necessary.

Vast Potential for Hyaluronan Derivation by Conjugation or Cross-Linking

Hyaluronan as a raw material may be obtained from different manufacturers in the form of a white powder, producing a highly viscous liquid (i.e. comparable to egg white) upon aqueous suspension, termed free HA, unmodified HA, or uncrosslinked HA, which is an excellent lubricant (Tezel and Fredrickson 2008). Soluble hyaluronan has been used in various clinical applications, such as viscosupplementation for OA, eye surgery, or wound healing, despite suboptimal intrinsic mechanical properties and rapid *in vivo* degradation. In view of chemical and biological properties alteration, hyaluronan may be modified on three functional sites presented in Figure 8.2, namely, the carboxylic acid group, the hydroxyl group, and the N-acetyl group (Fallacara et al. 2018). Structural manipulations may consist in grafting compounds to the hyaluronan backbone (i.e. conjugation, single chemical bond) or coupling of multiple polymeric chains (i.e. cross-linking, two or more chemical bonds) (Schanté et al. 2011a). Thereafter, hyaluronan derivatives may be classified as "monolithic," wherein the terminal form of the polymer does not allow new bonds to be created with cells or molecules, or as "living," wherein derivatives are able to form covalent bonds in the presence of cells, tissues, or molecules, thus allowing their incorporation prior to *in situ* gel formation by cross-linking (Prestwich 2011). By definition, living hyaluronan derivatives are therefore majorly favoured for combination product clinical applications or for preclinical uses, such as three-dimensional cell culture or formulation of therapeutic cells in view of *in vivo* delivery.

Overview of Hyaluronan Conjugation Processes

Exploiting the aqueous solubility of hyaluronan, various conjugation reactions have been proposed in water (Schanté et al. 2011a; López-Ruiz et al. 2019). Hyaluronan carboxylic acid modification in aqueous solution with active amino groups *via* carbodiimides (e.g. 1-ethyl-3-[3-(dimethylamino)-propyl]-carbodiimide, or EDC) has been widely used since almost 50 years and was well described (Danishefsky and Siskovic 1971). Activation of hyaluronan carboxyl groups with EDC in the pH range of 4–7 forms an active intermediate (i.e. O-acylisourea) which reacts promptly with primary amines (i.e. nucleophiles) to form stable amide bonds. To prevent rapid hydrolysis of the active intermediate, N-hydroxysuccinimide (NHS), 1-hydroxybenzotriazole (HOBt), or highly reactive N-hydroxysulfosuccinimide (Sulfo-NHS) may be added to the reaction, favouring the formation of an NHS ester intermediate. This intermediate is more stable and reacts slowly with primary amines to eventually form stable amide bonds. Depending on the specific manufacturing needs, reaction yields in water may be modulated by adjustment of pH, reagent quantities, and availability of amino groups (Schanté et al. 2011b; Maudens et al. 2018). To attain high relative reaction yields (i.e. >80%), the use of solvents such as dimethyl sulfoxyde (DMSO) or dimethylformamide in anhydrous conditions is necessary (Schneider et al. 2007; Schanté et al. 2011a). Hyaluronan carboxyl groups are furthermore known to be specifically recognised by HYAL-2 enzymes. This characteristic allows for degradation protection by means of chemical carboxyl conjugation (Aruffo et al. 1990; Schanté et al. 2011a).

Overview of Hyaluronan Cross-Linking Processes

To avoid *in vivo* solubilisation and subsequent expedited clearance, hyaluronan properties may be improved through cross-linking processes (Segura et al. 2005). Varying degrees of cross-linking enable the formation of polymeric hydrogel networks, conferring a uniform behaviour to the material, which might be considered as a chemical and physical barrier against enzyme and free radical degradation (Tezel and Fredrickson 2008). Specifically, rates of enzymatic degradation mediated by hyaluronidases are related to the degree of cross-linking of hyaluronan hydrogels, wherein higher degrees of cross-linking are associated with relatively improved stability. Additionally, cross-linking degrees along with polymer concentration greatly influence hydration balance values of specific hydrogels, defining the parameters for environmental water uptake and absorption. As an illustration thereof, the hydration balance in dermal fillers is generally achieved with 5.5 mg HA/mL of water and a typical degree of 4% cross-linking. Considering relatively higher hyaluronan concentrations (i.e. 20–24 mg/mL), the hydration balance values generally lead to important absorption of environing water by the product (Tezel and Fredrickson 2008).

Aqueous modification of hyaluronan hydroxyl groups by ether formation using epoxides was first reported in 1964, in strong alkaline conditions (i.e. pH 13–14) and 50°C (Laurent et al. 1964). To this end, 1,4 butanediol-diglycidyl ether (BDDE) is used for the reaction, mediating the opening of an epoxide ring to form ether bonds with hyaluronan hydroxyl groups. BDDE is a widely used cross-linking agent in the modulation of hyaluronan, due to accessibility of the reaction and absence of degradation product toxicity (Bogdan Allemann and Baumann 2008; Schanté et al. 2011a). Proposed applications consist in mixing and cross-linking HMW and LMW hyaluronan in various proportions using BDDE, thereby obtaining improved stability and resistance to degradation, optimal mechanical properties, and lower cytotoxicity (Xue et al. 2020). Exploiting hyaluronan hydroxyl groups and divinyl sulfone (DVS) as an alternative to BDDE confers the advantage of avoiding heating the reaction, yet intrinsic toxicity of DVS must be taken into account (Lai 2014). Modification of hyaluronan N-acetyl groups is not a first choice, since it requires preliminary modifications with deacetylation, carried out by hydrazinolysis, resulting in exposition of free amine functional groups. Furthermore, the conditions for deacetylation may then cause cleavage in HA backbones (Babasola et al. 2014).

Specifications for Therapeutic Hyaluronan-Based Hydrogel Products: Dermal Fillers and Viscosupplements

Specifications or target parameter ranges for therapeutic products based on hyaluronan strongly depend on the intended application and use, as well as expected clinical outcome, largely influenced by key physical properties of the polymer network. In the specific case of dermal filler hyaluronan hydrogels, specifications playing a preponderant role in design considerations comprise hyaluronan concentration, cross-linking, elastic modulus G', cohesivity, swelling, particle size, and extrusion force. With the introduction of BDDE as a cross-linking agent, the starting hyaluronan MW was rendered non-significant as a parameter, due to the high increase in final product MW after cross-linking (Kablik et al. 2009; Molliard et al. 2018; La Gatta et al. 2019). For viscosupplementation products, considerations encompass the polymer MW distribution, gel rheological behaviour, lubrication capacity, and time of residence (de Rezende and de Campos 2015; Zheng et al. 2019). Raw material properties must be considered in order to fullfil the final product specifications, yet the specific manufacturing technologies and degrees of modification of the hyaluronan polymer may drastically impact product parameters and behaviour. Such processes may therefore be optimised, wherein the use of minimal BDDE reagent amounts to maximally cross-link the polymer with minimal reaction waste (i.e. pendant chains) representing a desirable manufacturing characterisation workflow (Kablik et al. 2009; Edsman et al. 2012; Molliard et al. 2018). Total degrees of modification can thus be defined as the addition of the cross-link percentage and the pendant chain percentage.

In order to enable dermal delivery of the product by injection (e.g. small-bore needles of 27-gauge and 30-gauge), the extrusion force must remain acceptable for human use. This parameter may be modulated by targeting adequate hyaluronan MW, viscosity, and polymer concentration, in order to eventually obtain a gel of appropriate viscosity at high shear rates. The incorporation of linear hyaluronan in a cross-linked formulation is an option to modulate polymer concentration while facilitating the extrusion. In this context, the elastic modulus G' represents gel ability towards volume recovery after deformation, with typical values laying between 10 and 10^3 Pa, most often superior in value to the viscous modulus G" (Pierre et al. 2015). Cohesivity further characterises the capacity to resist dissociation after implantation and may be partly assessed by compression testing.

Considering existing hyaluronan-based viscosupplementation product parameters, the mean values for polymer MW have been described between 0.5 and 10^6 Da. Polymer concentrations range from 0.8 to 30 mg/mL for linear, cross-linked, or hybrid network types in the hydrogel, which is usually conditioned in units of 0.5–6 mL (Bowman et al. 2018). For clinical application in OA patients, 21-gauge to 23-gauge needles are commonly used. Product rheological behaviours in terms of G' and G" (i.e. at constant oscillatory frequencies of 0.5 Hz and 2.5 Hz, simulating walking and running conditions, respectively) should be relatively elevated (e.g. 0.2–91.9 and 1.8–45.9 for G' and G", respectively) to improve joint lubrication (Bhuanantanondh et al. 2010). Improving joint residence time by derivation or addition of antioxidants is a key parameter in improving the clinical response, as short residence times are mediated by respective actions of HYAL enzymes and ROS (Stern et al. 2007; Conrozier et al. 2014; Maudens et al. 2018).

Finally, depending on the specifications of the individual raw material process and the product specifications established in the product master file, optimal storage conditions and instructions for use need to be established, given the relatively unstable nature of hyaluronan and derivatives. In particular, due to high hygroscopicity, materials and products should be stored in sealed original packages in clean, dry, and controlled-temperature environments. Depending on the nature and sensitivity of the additional therapeutic materials incorporated in the product, refrigeration and protection from light or air might be necessary (Huerta-Ángeles et al. 2018).

Sterilisation of Hyaluronan-Based Hydrogel Products

Sterilisation of hyaluronan-based products is complex, due to partial loss of viscosity, structural integrity, and function of the polymeric networks after exposure to high temperatures (i.e. >100°C), which typically occur during terminal sterilisation (e.g. autoclaving). Nevertheless, clinical application requires demonstration of product sterility. Alternative sterilisation methods have been proposed (e.g. ethylene oxide gas, electron beam or gamma radiation), yet each alternative presents potential detrimental effects as well (Huerta-Ángeles et al. 2018). Furthermore, an industrial manufacturing workflow is limited in the choice of validated sterilisation methods, or the task of validating a new method is to be undertaken. Therefore, when dealing with biologics and especially with therapeutic cellular materials or derivatives, aseptic processing may represent an alternative to endpoint sterilisation, in view of maintaining functional structures and related activity. Sterilisation by 0.22 μm filtration represents a good option depending on the concentration and the MW of the polymer. Based on sterilisation impact on the product, steam treatment and plasma radiation were shown to be efficient, in contrast with ethanol treatment or UV irradiation (Shimojo et al. 2015). Post-sterilisation validations of the physico-chemical integrity of the product, such as MW determination, rheological characterisation, and measurement of other key properties, are needed to determine the extent of degradation or depolymerisation and safety of the product. Most hyaluronan-based products used for dermal correction or restoring synovial balance in small joints (e.g. Restylane®, Juvéderm®, and Ostenil®) are terminally sterilised in glass pre-filled syringes with a validated moist heat process in a pressurised autoclave. Specific sterilisation processes for medical devices are described in different normative documents (e.g. ISO 10993-7, ISO 17665-1, ISO 11737-2 and ISO 20857) (Huerta-Ángeles et al. 2018).

Characterisation Workflows for Hyaluronan-Based Hydrogels

Depending on the considered clinical application, claims for regulatory submissions, and related manufacturing and control requirements, many different aspects of HA hydrogels may be characterised and documented. The applied workflow may depend on the contents of the hydrogel or specific behaviours thereof (e.g. pH or temperature sensitivity such as particle thermoformation). The end-product may be analysed using orthogonal methods to determine the polymeric structure (e.g. IR, NMR, MS), degrees of substitution and cross-linking, size and morphology of particles (i.e. scanning electron microscopy, dynamic light scattering, nanoparticle tracking analysis), injectability, rheology (i.e. viscosity curves, elasticity, viscosity), stability (i.e. physical and under enzymatic or oxidative degradation), pH and zeta potential, *in vitro* drug release capacities (i.e. if an active pharmaceutical ingredient is incorporated in the hydrogel), cytocompatibility (i.e. *in vitro*, *ex vivo*, *in vivo*), etc. (Maudens et al. 2018; Miastkowska et al. 2020). Specific norms and reference standards are defined for evaluation of the different classes of products which may comprise hyaluronan-based hydrogels (e.g. ISO 10993: 2009 for medical devices).

Safety Evaluation of Hyaluronan-Based Products

For any type of product registration, adequate preclinical demonstration of safety (i.e. proof that the product is not pyrogenic, mutagenic, toxigenic, genotoxic, hemolytic, or immunogenic) must be provided. For medical devices, biocompatibility (i.e. assessment of cytotoxicity, sensitisation, and irritation/intracutaneous reactivity) and biodistribution must be documented within specific requirements, depending on the device type, category, residence time, and type of use (Huerta-Ángeles et al. 2018). *In vitro* acute cytotoxicity may be studied using adequate cell lines of cell types, within validated models coherent with the intended application (i.e. homologous with implantation site), and adequate methodology-specific sample preparation (e.g. extract-dilution method, test by direct contact, or indirect contact). Accepted readout methodologies for cytocompatibility testing comprise neutral red uptake (NRU), MTT (methyl thiazolyltetrazolium), XTT, resazurin assay, ATP concentration measurement, crystal violet staining, or DNA content measurement. For further preclinical safety evaluations, *in vivo* models (e.g. mice, guinea pigs, rats, rabbits, sheep) may be considered on an application-dependent basis, for evaluation (i.e. macroscopic description and histopathology) of the tissular response after product application. Such assays need to be devised based on projected product degradation rates, types of tissues exposed, and intended clinical exposure times, and may be used as safety demonstration before clinical testing of efficacy (Huerta-Ángeles et al. 2018).

Developing Hyaluronan-Based Hydrogels for Therapeutic Cell Delivery

Notwithstanding the considerations exposed hereabove and set forth for hyaluronan-based hydrogel development, which mainly revolves around the field of medical devices, additional parameters need to be taken into account for the sound development of biological combination products (e.g. cATMPs) for effective and safe delivery of therapeutic cellular materials. Such combination products are of high interest and have the potential to meet clinical needs, as they may act by exerting intrinsic additive or synergistic effects. Specific parameters and dimensions of combination product design and development are presented hereafter.

Therapeutic Cellular Materials for cATMPs

Vast arrays of tissues and cell sources may be considered as starting materials for therapeutic product development, among which platelet-rich plasma (PRP), stem cells, adult autologous stromal cells, or allogeneic foetal progenitor cells (Grognuz et al. 2016). Key considerations when developing combination products are the intended interactions and respective roles of the constituents. In this context, the hyaluronan-based hydrogel may be defined as an inert, cyto- and biocompatible vehicle for cell delivery, or as a functional and synergistic active ingredient. Two major parameters defining the formulation and conditioning of the final product are the extemporaneous mixing of therapeutic components versus the ready-to-use preparation, and incorporation of autologous or allogeneic cellular materials. The various designs may range from autologous and in-clinic preparation of the product (e.g. hyaluronan-suspended PRP) to serially manufactured formulations incorporating devitalised allogeneic cells or derivatives thereof (Abate et al. 2015; Wang et al. 2019). Overall, the key driver in designing the optimal cell therapy combination product is the maintenance or stabilisation of maximal therapeutic effects, which does not presuppose maintenance of cellular viability in all cases. Based on the intended application and therapeutic requirements, the hyaluronan hydrogel component may be tuned as described previously for various indications and modalities of administration, such as infiltration, soft-tissue partial replacement, viscosity enhancement and friction reduction, or complementary delivery of therapeutic materials to subcritical bone or cartilage defects.

In view of artificial tissue development, therapeutic cells are selected and designed to release specific biomolecules, and support matrices are tailored to provide intrinsic structural and regulatory signals to optimise tissue formation (Segura et al. 2005; Schmidt et al. 2008). Indeed, hematopoietic, embryonic, and mesenchymal stem cells physiologically reside in particular hyaluronan-rich microenvironments, remaining in a quiescent form with a low proliferation rate, in particular through interaction with essential hyaluronan receptors such as CD44 and RHAMM. Under control of PDGF, adult MSCs express CD44 and leverage interaction with extracellular hyaluronan to migrate towards regenerating tissues (Dicker et al. 2014).

Relative Proportions of Hyaluronan and Therapeutic Cells

Various parameters of hyaluronan polymeric networks may drastically qualitatively influence the cellular portion of the product. As an example, varying ranges of hyaluronan proportions and MW were used to suspend thermally-challenged murine fibroblasts, demonstrating an absence of clinically significant cytotoxicity in standard conditions, as well as relatively superior performance of hyaluronan versus carbomer and sodium alginate with regard to toxicity (Guo et al. 2015). Typical concentrations of LMW hyaluronan in topical combination treatments revolve around 0.2% (Gariboldi et al. 2008). With regard to the quantity of therapeutic cells *per* individual product dose, *in vivo* studies often report the use of 10^6 to 10^7 cells, yet we have previously shown that for foetal progenitor cells (FPC), clinical doses of 5×10^5 to 2.5×10^6 were sufficient for obtaining potent regenerative stimuli (Hohlfeld et al. 2005; Grognuz et al. 2016; Laurent et al. 2020b).

Cell Encapsulation within Hyaluronan-Based Hydrogels

Encapsulating therapeutic cellular materials within hydrogels may be beneficial for viability maintenance, targeted delivery, and protection from stress and degradation. The rheological properties of cell suspensions prior to hydrogel formation are fundamental to maintaining both cell viability and mutual adhesion during the encapsulation process. Two key process categories may be distinguished, with macro encapsulation (i.e. large quantities of cells in a large volume of product), and microencapsulation (i.e. cells introduced in small volumes of product). Micro-manufacturing techniques have been used to encapsulate cells with hyaluronan before a polymeric cross-linking step leading to hydrogel formation occurred. Excessive viscosity during cell suspension processes and product administration constitutes a major factor for viability loss, due to relatively high shear forces potentially damaging cell plasma membranes and preventing cell-cell contacts (Schmidt et al. 2008). Additionally, it was shown that for cell microencapsulation in hyaluronan hydrogels, high polymer concentrations (i.e. 2%–15%) enable obtention of mechanical integrity of the microstructures, yet said higher amounts prove to be cytotoxic. An optimal concentration of 5% hyaluronan in water or phosphate-buffered saline (PBS) was therefore proposed for cell microencapsulation (Khademhosseini et al. 2006).

In order to improve cell viability in hyaluronan hydrogels, incorporation of dextran microspheres was proposed, due to the biodegradable, hydrophilic, and biocompatible properties of this polysaccharide, as well as its use in tissue engineering matrices. Human embryonic stem cells loaded in such formulations maintained an undifferentiated state, while conserving their full potency. Dextran microspheres provide an interface between the hydrogel and cells, synergistically promoting cell survival, activity, and tissue regeneration stimulation (Kim et al. 2010). It is of prime importance that controlled chemical reactions and hydrogel degradation products are not deleterious to the encapsulated cells, wherein chemical and mechanical properties of hydrogels are generally determined during the cell encapsulation process (Tibbitt and Anseth 2009). From a manufacturing point of view, hyaluronan derivation often requires relatively high temperatures and highly alkaline cross-linking conditions, which prevent the inclusion of living cells during hydrogel preparation. Modern workflows allow circumventing some of these potential pitfalls, with the rapid preparation of ready-to-use hyaluronan-based hydrogels under neutral conditions (Luo et al. 2000). Finally, cell spatial distribution within the hydrogel can be controlled by using bio-printing techniques, thereby accurately depositing the cells and encapsulating them in the hydrogel (Murphy et al. 2013; Madaghiele et al. 2014).

Viscosity Tuning in Hyaluronan-Based Cell Therapies

A precise definition of hydrogel characteristics is necessary to allow optimal product application and persistence (Jones and Vaughan 2005). Modulation of hyaluronan cross-linking degrees directly influences hydrogel dynamic viscosity values, expressed in units of Pa•s. To this end and in a specific tissue engineering application, we had previously compared various hyaluronan hydrogels with a hybrid objective of acceptable rheological properties and cellular viability maintenance (Grognuz et al. 2016). It was shown that for each additional unit of Pa•s, 0.3% of cell survival was lost, but a dynamic viscosity of less than 5 Pa•s was insufficient to appropriately maintain cells in suspension. It was found that an optimal balance between shear force applied on the cells and optimal viscosity for intended applications was obtained with the use of Ostenil® Tendon (i.e. 2% hyaluronan gel, TRB Chemedica SA, Switzerland) in tendon tissue engineering (Grognuz et al. 2016). It has been demonstrated that mechanical properties of cell carriers modulate both behaviour and phenotype of cells in a similar manner to biochemical cues. Particularly, cell migration requires a high degree of cellular spatiotemporal coordination and is therefore closely linked to mechanical properties of the carrier in which cells are distributed. Cells are subjected to high tensile forces in highly viscous environments, wherein the viscosity of the hydrogel may be proportional to the internal adhesion capacity and inversely proportional to migration speed. Therefore, viscosity is a limiting factor, as relatively high values may restrict cellular migration, while increasing adhesion strength. Furthermore, cell proliferation is also proportional to hydrogel viscosity to a certain extent (Ghosh et al. 2007).

Biochemical Signalling and Protein Transport within Hyaluronan-Based Hydrogels

For optimal effectiveness of biological combination products, the cell support matrix must provide an appropriate environment, relay signals directing or modulating cellular processes and tissue formation. Matrix surfaces ensure cell adhesion and migration, potentially influencing their survival or the ability to colonise surrounding recipient tissues. A determinant functional parameter is therefore the facilitation by the hydrogel of both influx and efflux of biological molecules, which may be modulated during polymer and gel synthesis by tuning porosity (Schmidt et al. 2008). A specific example thereof consists in using cryopolymerisation for cryogel formation, with the obtention of three-dimensional networks of interconnected macropores, yielding enhanced mechanical stability as compared to traditional hydrogels. Therefore, cryogels have been proposed for tissue engineering applications, as open pore structures allow the transport of nutrients and oxygen to the cells, and conversely, the removal of cellular waste. Combination of the excellent biocompatibility of hyaluronan with the advantageous properties of cryogels has shown that adequate physiological environments may be created to support long-term cell cultures *in vitro* (Thönes et al. 2017). The advantages of macroporosity in hyaluronan-based matrices are further substantiated by the improved diffusion of encapsulated therapeutic cells towards recipient environing tissues (Burdick and Prestwich 2011).

Complex Hyaluronan-Based Hydrogels for Therapeutic Cell Delivery

Further modulation of therapeutic cell behaviour within hydrogels is attainable by introduction of cell adhesion sites on hyaluronan backbones or by adding adhesion proteins such as collagen, laminin, or fibronectin to the formulation. To this end, soluble hyaluronan is often mixed with collagen gels, in order to mitigate intrinsic mechanical limitations, reduce degradation rates, and restrict the relative quantity of retained water (Segura et al. 2005). A comparative study for vocal cord regeneration investigated the use of hyaluronan hydrogels, hyaluronan-collagen cogels, and hyaluronan-DCM (decellularised ECM). Hyaluronan alone and hyaluronan-collagen cogels were characterised by relatively limited biological activity for the proliferation support of encapsulated adipose stem cells. The incorporation of DCM provided an optimal environment for cell growth and differentiation, indicating that cogels based on hyaluronan and DCM would constitute promising support matrices for standard therapeutic cellular materials (Saddiq et al. 2009; Huang et al. 2016).

Self-Assembling Hyaluronan-Based Hydrogels for Optimal Responsiveness and Behaviour

We have previously reported the novel green chemistry manufacturing of biocompatible and biodegradable hyaluronan-based thermoresponsive hydrogels, cross-linked using a poly(N-isopropylacrylamide) (PNIPAM) polymer, named "HA Nano" or "HA Pearls" (Maudens et al. 2018). An overview of this specific technology is presented, along with advantages and potential therapeutic applications, in Figure 8.3. Appropriately linked to hyaluronan, *via* a cyclooctyne linker and a PEG spacer, PNIPAM enables the spontaneous formation of submicron spherical particles above a defined lower critical solution temperature (LCST). These 200-nm objects result from the self-assembly of PNIPAM moieties grafted onto the hyaluronan backbone which lead, at body temperature, to a unique nanostructured gel containing HA-PNIPAM spheres entangled in the gel network, without the use of chemical crosslinkers (Maudens et al. 2018). We have recently improved the aforementioned chemical production process in order to obtain finely tuned hydrogel biomechanical properties, depending on the chosen application (e.g. dermal fillers, long-lasting viscosupplementation, and vectors for cell therapies). Developed in the context of OA management and cutaneous delivery for improved injectability, persistence, and effectiveness, these materials (i.e. thermoresponsive gels) undergo reversible gelation through spontaneous nanoparticle formation at physiological temperatures (Brown et al. 1991; Schild 1992; Tan et al. 2009; Li and Guan 2011; Cooperstein and Canavan 2013). Major clinical benefits may be deployed by such novel formulations presenting thermoreversible, non-Newtonian pseudoplastic behaviours, enabling a facilitated injection of products and building high viscosity *in situ*, along with improved resistance against degradation (i.e. actions of HYAL and ROS) due to designed steric hindrance (Cilurzo et al. 2011; Maudens et al. 2018).

Hyaluronan-Based Hydrogel Stability and Degradation

In vivo biological activity of hydrogels is closely dependent on both microstructural (i.e. chemical composition, cross-linking density, etc.) and macrostructural (i.e. viscosity, degradation rate, etc.) characteristics, which may directly influence behaviours of encapsulated cells and indirectly promote or restrict integration thereof into recipient tissues. Particularly, cell migratory capabilities may rely on hydrogel degradation, as infiltration is inhibited by narrow dimensions of the polymer mesh (Madaghiele et al. 2014). Hyaluronan degradation in the hydrogel greatly depends on the concentration and exposure time to hyaluronidases (i.e. mainly HYAL-1 and HYAL-2). Due to intrinsic hyaluronidase production by various cell types including fibroblasts, the inclusion of live cells in hyaluronan gels negatively impacts long-term stability, with the introduction of "autophagic" components, adding to the effects exerted by recipient

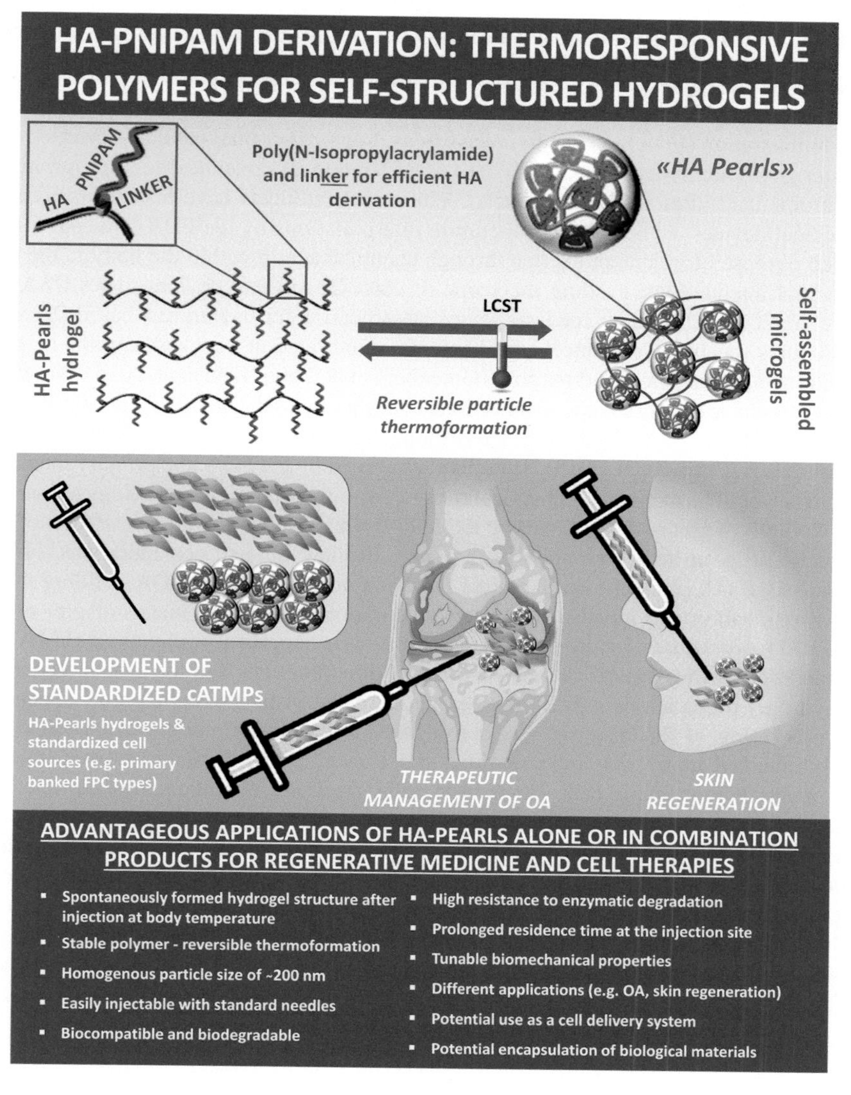

FIGURE 8.3 Overview of thermoresponsive hyaluronan-based hydrogels spontaneously forming particles at body temperature, assorted to potential applications and related benefits thereof. The combined use of appropriate linkers and PNIPAM for modification of the polymeric network confers unique properties to the formulation, namely, thermoresponsive and reversible self-formation of particles in the hydrogel. Chemical optimisation to obtain a specific lower critical solution temperature (LCST) inferior in value to body temperature but superior to ambient temperature enables simple injection of the hydrogel product, which develops specific and differential rheological properties *in situ* after administration. Appropriate conjugation of such thermoresponsive formulations with standardised cellular materials (e.g. FPCs or derivatives thereof) represents highly versatile novel therapeutic product development options for management of diverse conditions (e.g. OA, manifestations of ageing, etc.). Adapted from Maudens et al. 2018.

endogenous hyaluronidases. Nevertheless, it was demonstrated that despite complete and rapid enzymatic degradation of hydrogels with hyaluronidases and some macrophages, encapsulated cells could maintain viability (Khademhosseini et al. 2006; Madaghiele et al. 2014). Considering tissue engineering applications, ideal rates of hydrogel degradation should coincide with the rate of tissue formation. Insufficient degradation would indeed compromise remodelling, while rapid degradation would induce early resorption of the support matrix. It was hypothesised that for cutaneous applications, complete degradation of the hydrogel in seven days would allow optimal migration of angiogenic cells and induce tissue neo-vascularisation (Madaghiele et al. 2014).

Regulatory-Guided Hyaluronan-Based Biological Combination Product Development

Ambivalence in the current regulatory context of hyaluronan-based product development should be taken into account and is highlighted in the example of injectable products for OA, which the USA Food and Drug Administration (FDA) intends to reclassify as drugs rather than medical devices (https://www.federalregister.gov/documents/2018/12/18/2018-27351/intent-to-consider-the-appropriate-classification-of-hyaluronic-acid-intra-articular-products). While such products have historically been classified as drugs, medical devices, or both, recent literature interpretations by the FDA have put forward a primary intended purpose of treatment exerted through chemical action within the body in the case of OA, hence a potential intended use evading the scope of class III medical devices under USA regulations (Braithwait et al. 2016). Until now, medical device classification for hyaluronan-based OA viscosupplementation products was based on a mechanical action (i.e. lubrication, shock absorption by synovial liquid prostheses) as a primary intended purpose (Greenberg et al. 2006; Jahn et al. 2016). Notwithstanding, recent publications have supported chemical actions contributing to the analgesic, anti-inflammatory, and chondroprotective effects of hyaluronan-based orthopaedic products (Moreland 2003; Vasi et al. 2014; Richards et al. 2016; Altman et al. 2018). Therefore, as argued by the FDA, the primary intended purpose of such products would be attained by chemical means, by mitigation of the pathological condition itself. This aspect would be further substantiated by the long-lasting clinical therapeutic effects of hyaluronan on pain reduction (i.e. up to six months) in the knee. In view of such considerations, official advice to hyaluronan-based OA product manufacturers has been put forth by the FDA, guiding them towards preliminary classification and jurisdictional determination of products by means of pre- or request for designation (RFD) procedures, before submission of premarket approval applications (PMA).

Notwithstanding, various classifications of combination products may be considered for hyaluronan-based hydrogel preparations containing therapeutic cells or derivatives thereof, depending on the source and processing of the material (i.e. autologous or allogeneic, culture-expanded or minimally manipulated), intended application and modalities (i.e. topical or invasive), and defined state of the biological raw materials (e.g. living cells, devitalised cells, fractions, conditioned medium, etc.) (Laurent et al. 2020b). Although some specific applications may be defined as medicalised cosmetics, most applications covered by various legal and regulatory frameworks separate the products described herein in specific categories, such as cATMPs and class III medical devices, depending on the mode of action of the product (i.e. principal or ancillary effect of the biologicals) and the claimed effect (Racchi et al. 2016). Due to highly different requirements and possibilities for the aforementioned regulatory classifications, close attention must be paid and risk mitigation must be performed at the beginning of the product development phase, in order to ensure compliance with applicable regulations in the intended markets.

Tissue Engineering Clinical Applications of Hyaluronan-Based Hydrogels for Standardised Cell Therapies

Clinical Hindsight and Advantages of Hyaluronan-Based Therapeutics

Hyaluronan has been extensively investigated as a vehicle for drug delivery, whether considering ophthalmic, nasal, pulmonary, parenteral, or topical applications (Brown and Jones 2005). In topical applications, hyaluronan appears to increase the residence time of other agents, as during an inflammatory state, CD44 and RHAMM receptors are up-regulated and favour hyaluronan binding (Weindl et al. 2004). Due to ECM-mimicking properties of hydrogels, along with optimal biocompatibility, biodegradability, and applicability to support encapsulation of cells or growth factors, such preparations have been widely used in various pharmaceutical fields. Even though hyaluronan was first discovered in 1934, the first clinical application was reported in 1968 for the treatment of burn victims (Fatini et al. 1968). Since then, multiple facets of human and veterinary medicine have benefitted from its application, proven to be effective in various fields such as ophthalmic and aesthetic surgery, pulmonary and rheumatological pathologies, development of cosmetic products, as well as skin regeneration (Juhász et al. 2012). Various hyaluronan-based biomaterials have been developed for the controlled release of growth factors or cytokines which induce cell proliferation. The immunomodulatory action of hyaluronan may be beneficial, through the increases of tissue regeneration capacities in burns, surgical epithelial wounds, or chronic wounds (Fouda et al. 2016).

Topical Management of Burns and Wounds using Hyaluronan-Based Biological Therapies

As a non-adherent material allowing generation of moist and sterile "*in vivo*-like" environments, hyaluronan hydrogels allow for rehydration of tissues and absorption of skin lesion exudates in a reversible manner, depending on environmental stimuli (i.e. temperature, pH). Various hyaluronan-based products have been proposed for the topical management of burn wounds, including cell-free (e.g. Hyalomatrix®, Hyalosafe®, HYAFF®-11, Ialugen®) and cell-laden constructs (e.g. Hyalograft 3D™, Laserskin™) (Turner et al. 2004; Tezel and Fredrickson 2008; Shevchenko et al. 2010; Longinotti 2014; Dalmedico et al. 2016). Clinical studies have confirmed the benefits of such constructs in the treatments of burn victims (Harris et al. 1999; Price et al. 2007; Gravante et al. 2010; Voigt and Driver 2012; Fino et al. 2015). Hyaluronan hydrogels have been characterised as moderately beneficial for topical delivery of therapeutic materials, depending on the extent and gravity of cutaneous lesions, with variable tissue penetration capacities, wherein a MW of 100 kDa enabled optimal passage through the disrupted skin barrier (Mesa et al. 2002; Witting et al. 2015). Qualitatively, hyaluronan hydrogels provide comfort during application, with a refreshing sensation and soothing effect which may contribute to significantly alleviate pain (Jones and Vaughan 2005). Structurally, hyaluronan hydrogels mimic native ECM and promote maintenance of tissue hydration and oxygenation *via* water retention, detritus and pathogenic microorganisms trapping, and cell protection by creating a physical barrier (Madaghiele et al. 2014; Guo et al. 2015). Their use as a supporting matrix for wound treatment may be complemented by the combined use of therapeutic cells, which may be encapsulated within the hydrogel, wherein culture conditions within such three-dimensional environments have been shown to be optimal. In addition to intrinsic biocompatibility and biological activity, natural hyaluronan confers various benefits in tissue engineering workflows, such as ECM remodelling chaperoning or the promotion of cellular functions of both therapeutic transplanted cells and recipient endogenous cells (Khademhosseini et al. 2006; Dicker et al. 2014; Thönes et al. 2017). Additionally, cell encapsulation within a biomaterial may potentially reduce inherent immunogenicity of therapeutic materials (Schmidt et al. 2008).

Various alternative or functionalised hyaluronan-based formulations may be considered for conjugation to therapeutic cells, notably with the inclusion of classical drugs, peptides, or other biological products (e.g. vitronectin) for burn repair and re-epithelialisation stimulation (Brown and Jones 2005; Xie et al. 2011). Storage and controlled release of bioactive factors, zinc, or silver sulfadiazine by hyaluronan gels may present therapeutic interest in complex chronic ulcer or burn cases requiring long-term anti-inflammatory effects and anti-microbial management (Koller 2004; Costagliola and Agrosì 2005; Lee et al. 2008; Juhász et al. 2012; Su et al. 2014; Das and Baker 2016). Intrinsically, hyaluronan was shown to elicit anti-bacterial responses (i.e. stimulation of β-defensin 2 by LMW hyaluronan), immunomodulatory, and anti-inflammatory effects (i.e. Ki-67 proliferation antigen expression reduction and CD44 receptor blocking by HMW hyaluronan) (Mesa et al. 2002; Lataillade et al. 2010; Schlesinger and Rowland Powell 2014). Cell-laden hyaluronan hydrogels were shown to accelerate tissue repair, re-epithelialisation, and neo-vascularisation in diversified applications (Neuman et al. 2015).

Hyaluronan-Based Products for Effective Management of OA using Autologous PRP

Relatively simple and standardised cell therapies, which have recently been democratised for various applications, consist in extemporaneous concentration of patient platelets and injection in appropriate vehicles such as hyaluronan. Despite the widespread use of bedside kits for PRP preparation (i.e. contained blood processing materials and a centrifuge), it should be noted that appropriate GMP manufacturing is required. Such standardised autologous approaches have been proposed and studied for topical rejuvenation of the skin, management of subcritical tendon injuries, or OA (Ulusal 2017; Tan et al. 2021). Encouraging results have been reported for the use of PRP-yielding hyaluronan gels in OA patients which had previously not responded optimally to hyaluronan injections alone (Saturveithan et al. 2016; Renevier et al. 2018; Yu et al. 2018; Pereira et al. 2019). Specifically, PRP-hyaluronan injections were shown to be safe and to significantly delay the need for replacement surgeries and arthroplasties in suffering OA patients (Altman et al. 2015; Honvo et al. 2019; Ong et al. 2019).

Effective Modulation of Cartilage Repair by Hyaluronan-Based Therapies

OA remains a major clinical challenge, often refractory in its degenerative progression, which might synergistically benefit from combination products containing hyaluronan-based vehicles and therapeutic cell sources (Sekiya et al. 2012; Li et al. 2018). Hyaluronan-based formulations (i.e. HA Pearls) were shown to exert significant intrinsic effects on pro-inflammatory cytokine levels in a preclinical OA model (Maudens et al. 2018). For combination products, most preclinical experience has been gathered around the intra-articular use of stem cells for tissue repair stimulation, wherein hyaluronan mediates both the repair process itself and the function of therapeutic cells (Wang et al. 2020). Specifically, control of inflammation and chaperoning of exogenous and endogenous cell repair processes have been attributed to such combination products, wherein the level of hyaline cartilage restoration is dependent on the effective chondrogenesis (Ha et al. 2015). In this context, cartilage FPCs have been shown to present robust advantages for tissue repair or regeneration promotion after local delivery (Darwiche et al. 2012; Choi et al. 2016; Park et al. 2020). Whereas cartilage cell therapies have often been limited by the surgical approach or choice of the scaffold, subcritical defects might largely benefit from designed cell-laden hyaluronan hydrogels. Recent clinical trials for OA have been launched, confirming the safety and probable functional benefits of cells from foetal sources (Lee et al. 2018, 2020).

Importance of Standardised Cell Sources in Regenerative Medicine Products and Therapies

The task of defining novel biological raw materials or active pharmaceutical ingredients (API) is burdened by specific stringent prerequisites and quality considerations, in view of regenerative medicine product development or study of cell therapies (Platt et al. 2020; Yamazaki et al. 2020). Vast arrays of heterogenous tissue or cell sources have been proposed, comprising autologous, allogeneic, or xenogeneic materials of various developmental stages to be further processed by culture expansion, thereby introducing potential inherent multifactorial complications in development processes (De Buys Roessingh et al. 2013). Such proposed sources have non-exhaustively comprised embryonic stem cells (ESC), adult stem cells [i.e. adipose stem cells (ASC), bone marrow-derived mesenchymal stem cells (BM-MSC)], neural stem cells (NSC), limbal stem cells (LSC), hematopoietic stem cells (HSC)], endothelial progenitor cells (EPC), foetal progenitor cells (FPC), umbilical cord cells, neonatal foreskin cells, platelets, placenta, and amniotic fluid cells (Vertelov et al. 2013; Heathman et al. 2015; Mount et al. 2015; Muraca et al. 2017; Li and Maitz 2018; Sacchetti et al. 2018; Jayaraj et al. 2019; Torres-Torrillas et al. 2019). Imperious biological, logistical, sustainability, and clinical considerations synergistically contribute to the optimisation of cell source selection and product design phases. Harnessing of appropriate therapeutic cell sources for bioprocessing and product formulation presupposes safety and consistency of considered materials, traceability and sustainability of the source, high inherent expansion potential, and adaptability to clinical delivery methods often based on bioengineered scaffolds (Monti et al. 2012). Many cell sources are technically demanding and require specific processing or strong biochemical cues, wherein complex requirements often delay or restrict the potential for product development (Heathman et al. 2015). Potential barriers might arise with slow cellular proliferation, rarity of donors, unstable potency, low stability, or high propensity towards communicable disease transmission (Rayment and Williams 2010; Ratcliffe et al. 2011; Abbasalizadeh and Baharvand 2013; Hunsberger et al. 2015).

Facilitated Translational Research using Standardised FPC Sources for Complex Biological Products

Cell source selection is paramount for sound translational development and implementation of cellular therapy products. Iterative optimisation of standardised cell selection workflows has allowed to select allogeneic primary FPC types as potent candidates for cell therapy development, in addition to their recognised role as vaccine production substrates since the 1960s (Hayflick et al. 1962; De Buys Roessingh et al. 2006; Mirmalek-Sani et al. 2006; Larijani et al. 2015; Kim et al. 2018). A detailed overview of primary FPC isolation and biobanking is presented in Figure 8.4. Adequately isolated, such cell sources are advantageous in many aspects, such as a defined and tissue-specific phenotype, low *in vitro* growth requirements (i.e. independence from growth factors), high cytocompatibility and tolerance to oxidative stress, extensive lifespans, low immunogenicity, and high stimulatory potential (Cass et al. 1997; Quintin et al. 2007). Robust multi-tiered banking enables scalable good manufacturing practices (GMP) production, wherein sterility, safety, identity, purity, potency, stability, and efficacy may be demonstrated under stringent quality standards for cATMP development. Additionally, generation of large stocks of allogeneic cellular raw materials allows for drastic minimisation of product clinical availability delays. Foetal tissues or derived FPCs were used in various clinical applications, notably neurology (i.e. Huntington's or Parkinson's disease, strokes, spinal cord injuries), haematological or metabolic disorders, or liver failure (Touraine et al. 1993; Freeman 1997; Rosser and Dunnett 2003; Montanucci et al. 2013). Over the past three decades, we have reported notable clinical advances in the field of allogeneic FPC therapy, notably for managing burns and chronic ulcers, as well as arrays of musculoskeletal potential applications (Hohlfeld et al. 2005; Ramelet et al. 2009; Laurent et al. 2020b).

Swiss Foetal Progenitor Cell Transplantation Program for FPC Sourcing and cATMP Development

Within current regulatory frameworks for cell therapy product development, transplantation programs are well adapted for traceable procurement of safe and effective biological samples serving for standardised cell isolation procedures. Practical and sustainable designs for cell sourcing workflows are primordial for cell therapy or tissue bioengineering product development, guaranteeing homogeneity, consistency, robustness, and efficacy of therapeutic materials (Kent and Pfeffer 2006). Transplantation programs offer solid bases for optimised material procurement and therapeutic cell type establishment, ensuring regulated and traceable processing within. Swiss foetal progenitor cell transplantation programs were initially designed in the 1990s in Lausanne for the controlled cell banking of FPCs in view of tissue engineering development. Such federally registered programs are governed by organ transplantation laws and approved by ethics committees, public health authorities (i.e. FOPH, FDA, TFDA, and PMDA to date), and eventually by the national therapeutic products agency (i.e. Swissmedic). Multidisciplinary collaboration and exhaustive descriptive documentation enable the identification of qualifying donors after voluntary pregnancy terminations, in view of safe and standardised FPC derivation (Laurent et al. 2020d). Repeated serological testing (i.e. EBV, HIV, HTLV, hCMV, HHV, HSV, HBV, HCV, HPV, West Nile virus, syphilis) allows for screening of donors, before differential and simultaneous FPC isolation may take place. To this end, primary diploid cell culture initiation may be performed, after a single organ donation, on arrays of tissues such as skin, cartilage, tendon, bone, muscle, intervertebral disc, or lung, using enzymatic or mechanical methods for rapid, safe, sustainable, and efficient cell bank generation (Quintin et al. 2007; Laurent et al. 2020b). Subsequent multi-tiered cell banking optimised workflows (i.e. sub-tiering cryopreserved cell lots in Parental, Master, Working, and End of Production Cell Banks) may be transposed to industrialised processing, providing starting materials for decades of research and clinical applications (Laurent et al. 2020c). High stability and consistency of such scalable cell sources present optimal fits for modern safety and quality-driven regulations governing therapeutic product development (Hunsberger et al. 2015). A single qualifying organ donation may, therefore, for each considered tissue, simultaneously yield sufficient starting material for eventual GMP manufacturing of $> 10^8$ therapeutic product units of relatively small effective doses (i.e. 5×10^5 to 2×10^6 cell equivalents *per* dose on a

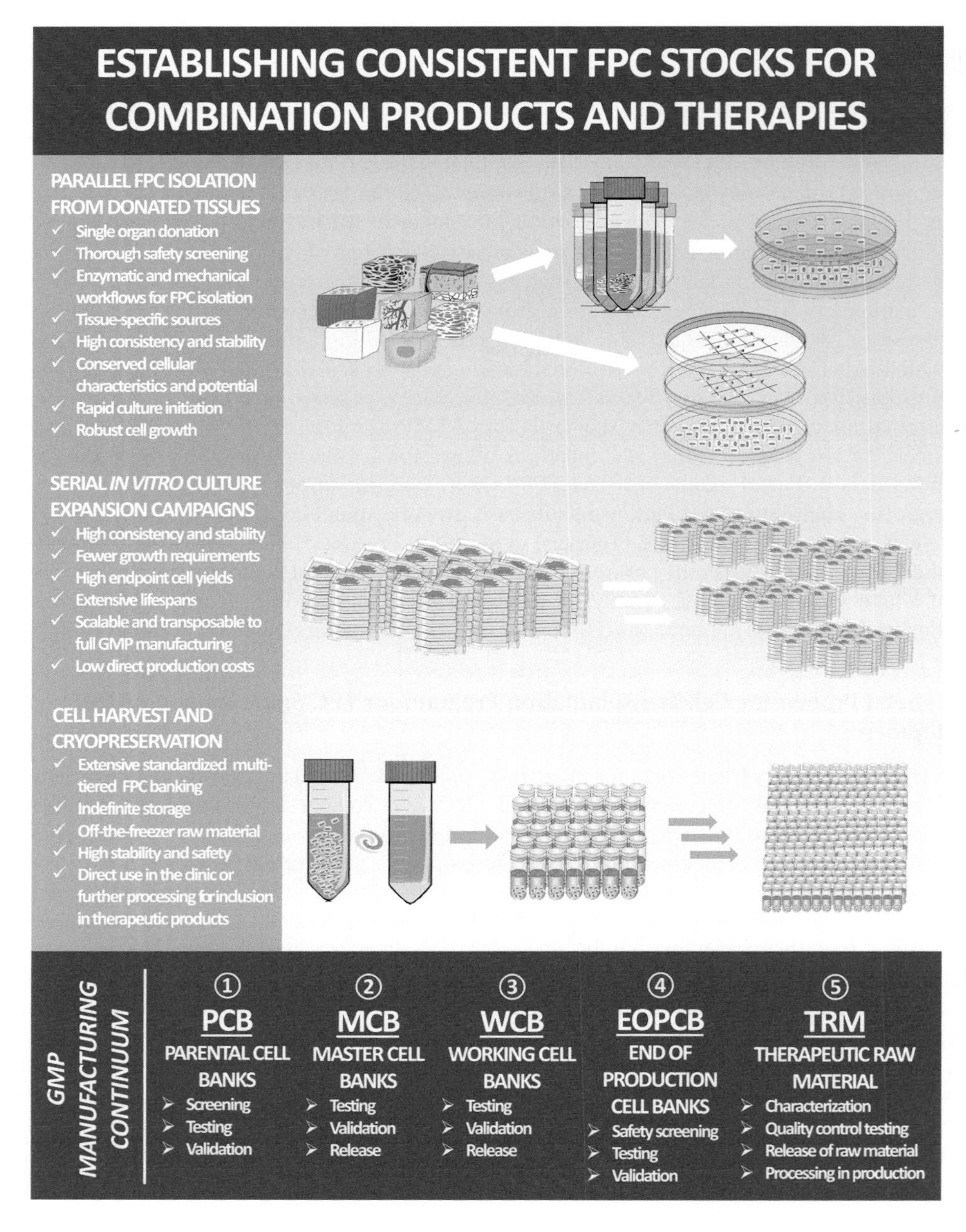

FIGURE 8.4 Schematic overview of standardised FPC isolation and banking methodologies for combined regenerative medicine product formulation. High efficiency may be attained by harvesting several tissues after a single organ donation. Subsequent parallel processing of biopsies (i.e. enzymatic and mechanical culture initiation) enables the rapid establishment of extensive and consistent Parental Cell Banks (PCB), which may be cryopreserved for long-term storage. Simple *in vitro* culture conditions and intrinsic robustness of primary FPC types allow for multi-tiered biobanking and optimal valorisation of therapeutic material stocks. In such GMP workflows, Parental Cell Bank vials serve for successive establishment of Master (MCB), Working (WCB), and End of Production Cell Bank (EOPCB) stocks, allowing for a thorough evaluation of safety and quality of manufactured batches. Selection of adequate lots thereafter enables optimal usage of cell banks for processing of therapeutic raw materials (TRM) to be used in combination products.

cell type-specific basis) (Hohlfeld et al. 2005; Darwiche et al. 2012). Applicable bioprocessing workflows additionally allow for the implementation of safety screening or testing and quality controls (e.g. assays to detect mycoplasma, viruses, prions, endotoxins, virus-like particles, retroviral activity, fungi, yeasts, bacteria, and testing for immunogenicity or tumorigenicity). Overall, FPC transplantation programs as described herein enable optimal efficiency at an industrial level of innovative therapeutic product manufacturing and translation, as well as sustainable and on-demand availability of therapies meeting clinical unmet needs (Laurent et al. 2020b).

Hyaluronan-Based Hydrogels for Cell Delivery of Banked FPCs

Hyaluronan-based vehicles have been investigated for formulation and delivery of equine and human FPC therapeutic cell types (Grognuz et al. 2016; Laurent et al. 2020a). Such formulations present great interest for non-surgical injection delivery of biological therapies to injured tendons, muscles, or other soft tissues (Tezel and Fredrickson 2008). Marketed or authorised commercial injectable products (e.g. Ostenil®, Synovial®, Hyalgan®) were compared in the context of extemporaneous suspension of living FPCs for clinical delivery, allowing for optimal survival of therapeutic materials, as well as acceptable and stable rheological parameters. It was shown that refrigeration storage may be used in case of delays before transplantation (i.e. several days), in view of slowing cell metabolism and significantly prolonging survival. The clear influence of gel viscosity on cell survival rates was outlined, as well as on homogeneity of the preparations. Furthermore, results indicated that small product volumes (i.e. 500 µl) containing several million cells could be extruded through standard-sized needles (e.g. 22-gauge) and stored, while preserving structural integrity, high relative viable cell proportions, metabolic activity, adhesion, migration, and proliferation of therapeutic FPCs. In the specific case of human tendon FPC delivery, it was outlined that Ostenil® Tendon HA (TRB Chemedica SA, Switzerland) presented optimal characteristics for on-demand biological therapeutic combination product development (Grognuz et al. 2016).

Nanoscale Hyaluronan-Based Biological Combination Products for Optimised Regenerative Effects

Both hyaluronan and banked FPCs benefit from extensive historical hindsight and industrial manufacturing applicability evidence, which places such materials in key positions to illustrate modern paradigm shifts towards biological therapeutics development and use. Indeed, emerging evidence lead increasing numbers of researchers and clinicians to investigate alternative treatment options to those proposed by classical pharmacotherapeutic guidelines, focusing on biologically sourced products. Considering the proposed combinations (i.e. FPC-loaded hyaluronan hydrogels), further developments shall shed some light on the optimal processing conditions and target parameters for tailoring and adaptation of raw materials. To this end, a schematic summary of such combination product lifecycles is presented in Figure 8.5. Indeed, for maximised therapeutic benefits to be obtained, hydrogel functionalisation and tuning may necessitate the presence, inherent or acquired, of nanoscale biological particles or aggregates, optimising the delivery or preservation of key therapeutic components. Similarly, standardised cellular materials may be further processed before inclusion in final formulations, with the derivation of nanoscale cellular compartments or components, such as exosomes, microvesicles, or soluble fractions of lysates and conditioned medium. While unfractionated cellular materials may exceed the nanometric scale, native hyaluronan from biological sources has been characterised by means of average root-mean-square radius determination to generally be under 100 nm for polymers around 10^6 Da. Therefore, pragmatic exploitation of high therapeutic value biological raw materials (e.g. hyaluronan and banked primary FPC sources), coupled to sound product development and manufacturing, shall further enable tangible contributions in the wide field of cell therapies and regenerative medicine.

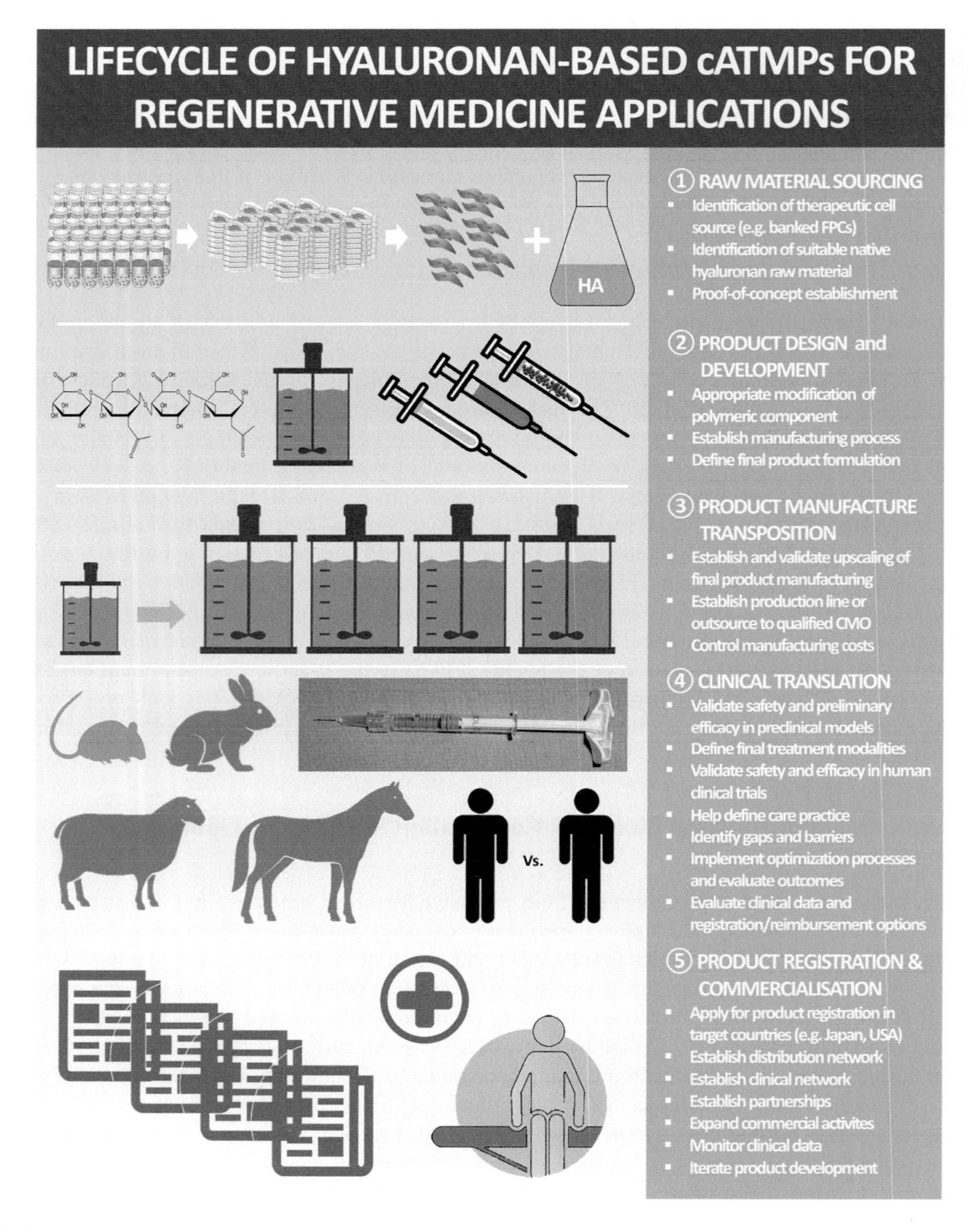

FIGURE 8.5 Overview of developmental activities and steps for creating, registering, and commercialising biological combination products based on hyaluronan and therapeutic cells. The sourcing phase should allow for the identification of optimal therapeutic cell types (e.g. FPCs), polymer structure and properties, and chemistry involved in the formulation. The design and development phase then leads to the formal establishment of specifications, processes, and parameters to obtain working prototypes, or at least a minimal viable product (MVP). Upscaling of manufacturing must then be validated, and production may be outsourced to appropriate and qualified contract manufacturing organisations (CMO). Clinical translation of the product is achieved after appropriate evidence and clinical data has been gathered successfully. The final stage of product registration and commercialisation then marks the beginning of the product-market lifecycle, which must be iteratively optimised and ameliorated, to ensure maximal numbers of patients may benefit from the therapeutic benefits yielded by the product.

Conclusion and Perspectives

The various considerations relative to product development in regenerative medicine, using highly versatile and safe biological raw materials as presented herein, enable us to conclude on the high interest for further industrial development and clinical translation efforts. Hyaluronan-based hydrogels constitute excellent versatile working bases for sound design, development, and manufacturing of innovative cATMPs, related TEPs, and medical devices. Indeed, key physico-chemical properties and parameters of hyaluronan hydrogels may be tailored at will for the optimal support or conjoined effect provision within cATMPs containing standardised cellular substrates or derivatives thereof.

To this end, primary FPCs may be isolated, after a single organ donation, from tissues such as skin, cartilage, tendon, bone, muscle, intervertebral disc, or lung, using enzymatic or mechanical methods for rapid, safe, sustainable, and efficient Parental Cell Bank generation. Subsequent robust and multi-tiered GMP cell banking workflows may be scalably transposed to industrialised processing, providing starting materials for decades of research and clinical applications, potentially addressing the needs of millions of patients. Generation of large stocks of such allogeneic cellular raw materials allows for drastic minimisation of product clinical availability delays.

Furthermore, pragmatic optimisation of combination product developmental workflows enables optimal adequation with modern regulatory and economic constraints relative to cellular therapy products, cell-based products, or cell-based cell-free preparations. The overarching benefits of exploiting standardised substrates such as hyaluronan and banked primary FPCs consist in the demonstrable industrial and clinical experience with such materials, wherein the scalable manufacturing of both components provides a key competitive advantage. Therefore, the critical product design considerations covered in this chapter, directly impacting the potential of proposed formulations in terms of clinical translation and effectiveness, shall contribute to the further development of high clinical value products and therapies for optimised regenerative stimulation and management of wide arrays of cutaneous and musculoskeletal affections globally.

Acknowledgements

Authors thank Dr. Carlota Salgado for review work. Dr. Abdel-Sayed and Prof. Laurent-Applegate acknowledge funding from the Marie Sklodowska-Curie Action (N°833594 - PHAS). Template graphical elements were obtained from www.biorender.com.

REFERENCES

Abate M, Verna S, Schiavone C, Di Gregorio P, Salini V. *Eur J Orthop Surg Traumatol 25*(8) (2015): 1321–6. doi:10.1007/s00590-015-1693-3.

Abbasalizadeh S, Baharvand H. *Biotechnol Adv 31*(8) (2013): 1600–23. doi:10.1016/j.biotechadv.2013.08.009.

Agerup B, Berg P, Akermark C. *Bio Drugs 19*(1) (2005): 23–30. doi:10.2165/00063030-200519010-00003.

Ahmed EM. *J Adv Res 6*(2) (2015): 105–21. doi:10.1016/j.jare.2013.07.006.

Albano GD, Bonanno A, Cavalieri L et al. *Mediators Inflamm 2016* (2016): 8727289. doi:10.1155/2016/8727289.

Altman R, Lim S, Steen RG, Dasa V. *PLoS One 10*(12) (2015): e0145776. doi:10.1371/journal.pone.0145776.

Altman RD, Dasa V, Takeuchi J. *Cartilage 9*(1) (2018): 11–20. doi:10.1177/1947603516684588.

Aruffo A, Stamenkovic I, Melnick M, Underhill CB, Seed B. *Cell 61*(7) (1990): 1303–13. doi:10.1016/0092-8674(90)90694-a.

Aya KL, Stern R. *Wound Repair Regen 22*(5) (2014): 579–93. doi:10.1111/wrr.12214.

Babasola O, Rees-Milton KJ, Bebe S, Wang J, Anastassiades TP. *J Biol Chem 289*(36) (2014): 24779–91. doi:10.1074/jbc.M113.515783.

Baeva LF, Sarkar Das S, Hitchins VM. *J Biomed Mater Res B Appl Biomater 105*(5) (2017): 1210–5. doi:10.1002/jbm.b.33659.

Bertram TA, Tentoff E, Johnson PC, Tawil B, Van Dyke M, Hellman KB. *Tissue Eng Part A 18*(21–22) (2012): 2187–94. doi:10.1089/ten.TEA.2012.0186.
Bhuanantanondh P, Grecov D, Kwok E. *J Med Biol Eng 32*(1) (2010): 12–6. doi:10.5405/jmbe.834.
Bogdan Allemann I, Baumann L. *Clin Interv Aging 3*(4) (2008): 629–34. doi:10.2147/cia.s3118.
Bourguignon LY, Ramez M, Gilad E et al. *J Invest Dermatol 126*(6) (2006): 1356–65. doi:10.1038/sj.jid.5700260.
Bowman S, Awad ME, Hamrick MW, Hunter M, Fulzele S. *Clin Transl Med 7*(1) (2018): 6. doi:10.1186/s40169-017-0180-3.
Braithwaite GJ, Daley MJ, Toledo-Velasquez D. *J Biomater Sci Polym Ed 27*(3) (2016): 235–46. doi:10.1080/09205063.2015.1119035.
Brown MB, Jones SA. *J Eur Acad Dermatol Venereol 19*(3) (2005): 308–18. doi:10.1111/j.1468-3083.2004.01180.x.
Brown TJ, Laurent UB, Fraser JR. *Exp Physiol 76*(1) (1991): 125–34. doi:10.1113/expphysiol.1991.sp003474.
Burdick JA, Prestwich GD. *Adv Mater 23*(12) (2011): 41–56. doi:10.1002/adma.201003963.
Cass DL, Bullard KM, Sylvester KG, Yang EY, Longaker MT, Adzick NS. *J Pediatr Surg 32*(3) (1997): 411–5. doi:10.1016/s0022-3468(97)90593-5.
Chen Y, Peng F, Song X, Wu J, Yao W, Gao X. *Carbohydr Polym 181* (2018): 150–8. doi:10.1016/j.carbpol.2017.09.017.
Choi WH, Kim HR, Lee SJ et al. *Cell Transplant 25*(3) (2016): 449–61. doi:10.3727/096368915X688641.
Cilurzo F, Selmin F, Minghetti P et al. *AAPS Pharm Sci Tech 12*(2) (2011): 604–9. doi:10.1208/s12249-011-9625-y.
Conrozier T, Mathieu P, Rinaudo M. *Rheumatol Ther 1*(1) (2014): 45–54. doi:10.1007/s40744-014-0001-8.
Cooperstein MA, Canavan HE. *Biointerphases 8*(1) (2013): 19. doi:10.1186/1559-4106-8-19.
Costagliola M, Agrosì M. *Curr Med Res Opin 21*(8) (2005): 1235–40. doi:10.1185/030079905X56510.
Dalmedico MM, Meier MJ, Felix JV, Pott FS, Petz FD, Santos MC. *Rev Esc Enferm USP 50*(3) (2016): 522–8. doi:10.1590/S0080-623420160000400020.
Danishefsky I, Siskovic E. *Carbohydrate Res 16*(1) (1971): 199–205. doi:10.1016/S0008-6215(00)86114-5.
Darwiche S, Scaletta C, Raffoul W, Pioletti DP, Applegate LA. *Cell Med 4*(1) (2012): 23–32. doi:10.3727/215517912X639324.
Das S, Baker AB. *Front Bioeng Biotechnol 4* (2016): 82. doi:10.3389/fbioe.2016.00082.
De Buys Roessingh AS, Guerid S, Que Y et al. *Def Manag 3* (2013): 003. doi:10.4172/2167-0374.S3-003.
De Buys Roessingh AS, Hohlfeld J, Scaletta C et al. *Cell Transplant 15*(8–9) (2006): 823–34. doi:10.3727/000000006783981459.
Dicker KT, Gurski LA, Pradhan-Bhatt S, Witt RL, Farach-Carson MC, Jia X. *Acta Biomater 10*(4) (2014): 1558–70. doi:10.1016/j.actbio.2013.12.019.
Dong Y, Arif A, Olsson M et al. *Sci Rep 6* (2016): 36928. doi:10.1038/srep36928.
Edsman K, Nord LI, Ohrlund A, Lärkner H, Kenne AH. *Dermatol Surg 38*(7) (2012): 1170–9. doi:10.1111/j.1524-4725.2012.02472.x.
Fallacara A, Baldini E, Manfredini S, Vertuani S. *Polymers 10*(7) (2018): 701. doi:10.3390/polym10070701.
Fatini G, Gallenga G, Veltroni A. *Osp Ital Chir 19*(3) (1968): 283–7.
Fino P, Spagnoli AM, Ruggieri M, Onesti MG. *G Chir 36*(5) (2015): 214–8. doi:10.11138/gchir/2015.36.5.214.
Fouda MM, Abdel-Mohsen AM, Ebaid H et al. *Int J Biol Macromol 89* (2016): 582–91. doi:10.1016/j.ijbiomac.2016.05.021.
Fraser JR, Kimpton WG, Pierscionek BK, Cahill RN. *Semin Arthritis Rheum 22*(6) (1993): 9–17. doi:10.1016/s0049-0172(10)80015-0.
Fraser JR, Laurent TC, Laurent UB. *J Intern Med 242*(1) (1997): 27–33. doi:10.1046/j.1365-2796.1997.00170.x.
Freeman TB. *Exp Neurol 144*(1) (1997): 47–50. doi:10.1006/exnr.1996.6387.
Gao F, Yang CX, Mo W, Liu YW, He YQ. *Clin Invest Med 31*(3) (2008): E106–16. doi:10.25011/cim.v31i3.3467.
Gariboldi S, Palazzo M, Zanobbio L et al. *J Immunol 181*(3) (2008): 2103–10. doi:10.4049/jimmunol.181.3.2103.
Ghosh K, Pan Z, Guan E et al. *Biomaterials 28*(4) (2007): 671–9. doi:10.1016/j.biomaterials.2006.09.038.
Gravante G, Sorge R, Merone A et al. *Ann Plast Surg 64*(1) (2010): 69–79. doi:10.1097/SAP.0b013e31819b3d59.
Greenberg DD, Stoker A, Kane S, Cockrell M, Cook JL. *Osteoarthritis Cartilage 14*(8) (2006): 814–22. doi:10.1016/j.joca.2006.02.006.

Grognuz A, Scaletta C, Farron A, Pioletti DP, Raffoul W, Applegate LA. *Cell Med 8*(3) (2016): 87–97. doi:10.3727/215517916X690486.

Guo X, Huang S, Sun J, Wang F. *Adv Skin Wound Care 28*(9) (2015): 410–4. doi:10.1097/01.ASW.0000467303.39079.59.

Gurski LA, Jha AK, Zhang C, Jia X, Farach-Carson MC. *Biomaterials 30*(30) (2009): 6076–85. doi:10.1016/j.biomaterials.2009.07.054.

Ha CW, Park YB, Chung JY, Park YG. *Stem Cells Transl Med 4*(9) (2015): 1044–51. doi:10.5966/sctm.2014-0264.

Harris PA, di Francesco F, Barisoni D, Leigh IM, Navsaria HA. *Lancet 353*(9146) (1999): 35–6. doi:10.1016/s0140-6736(05)74873-x.

Hartmann-Fritsch F, Marino D, Reichmann E. *Transfus Med Hemother 43*(5) (2016): 344–52. doi:10.1159/000447645.

Hayflick L, Plotkin SA, Norton TW, Koprowski H. *Am J Hyg 75* (1962): 240–58. doi:10.1093/oxfordjournals.aje.a120247.

Heathman TR, Nienow AW, McCall MJ, Coopman K, Kara B, Hewitt CJ. *Regen Med 10*(1) (2015): 49–64. doi:10.2217/rme.14.73.

Hoffman AS. *Adv Drug Deliv Rev 64* (2012): 18–23. doi:10.1016/j.addr.2012.09.010.

Hohlfeld J, de Buys Roessingh AS, Hirt-Burri N et al. *Lancet 366* (2005): 840–2. doi:10.1016/S0140-6736(05)67107-3.

Honvo G, Reginster JY, Rannou F et al. *Drugs Aging 36*(1) (2019): 101–27. doi:10.1007/s40266-019-00657-w.

Huang D, Wang R, Yang S. *Biomed Res Int 2016* (2016): 6584054. doi:10.1155/2016/6584054.

Huerta-Ángeles G, Nešporová K, Ambrožová G, Kubala L, Velebný V. *Front Bioeng Biotechnol 6* (2018): 62. doi:10.3389/fbioe.2018.00062.

Hunsberger J, Harrysson O, Shirwaiker R et al. *Stem Cells Transl Med 4*(2) (2015): 130–5. doi:10.5966/sctm.2014-0254.

Jahn S, Seror J, Klein J. *Annu Rev Biomed Eng 18* (2016): 235–58. doi:10.1146/annurev-bioeng-081514-123305.

Jayaraj JS, Janapala RN, Qaseem A et al. *Cureus 11*(9) (2019): 5585. doi:10.7759/cureus.5585.

Jeffery AF, Churchward MA, Mushahwar VK, Todd KG, Elias AL. *Biomacromolecules 15*(6) (2014): 2157–65. doi:10.1021/bm500318d.

Jiang D, Liang J, Noble PW. *Annu Rev Cell Dev Biol 23* (2007): 435–61. doi:10.1146/annurev.cellbio.23.090506.123337.

Jones A, Vaughan D. *J Orthop Nurs 9* (2005): 1–11. doi:10.1016/S1361-3111(05)80001-9.

Juhász I, Zoltán P, Erdei I. *Ann Burns Fire Disasters 25*(2) (2012): 82–5.

Kablik J, Monheit GD, Yu L, Chang G, Gershkovich J. *Dermatol Surg 35*(1) (2009): 302–12. doi:10.1111/j.1524-4725.2008.01046.x.

Kang DY, Kim WS, Heo IS, Park YH, Lee S. *J Sep Sci 33*(22) (2010): 3530–6. doi:10.1002/jssc.201000478.

Kavasi RM, Berdiaki A, Spyridaki I et al. *Food Chem Toxicol 101* (2017): 128–38. doi:10.1016/j.fct.2017.01.012.

Kent J, Pfeffer N. *BMJ 332*(7546) (2006): 866–7. doi:10.1136/bmj.332.7546.866.

Khademhosseini A, Eng G, Yeh J et al. *J Biomed Mater Res A 79*(3) (2006): 522–32. doi:10.1002/jbm.a.30821.

Kim BS, Choi JS, Kim JD, Yeo TY, Cho YW. *J Biomater Sci Polym Ed 21*(13) (2010): 1701–11. doi:10.1163/092050609X12548957288848.

Kim HR, Kim J, Park SR, Min BH, Choi BH. *Tissue Eng Regen Med 15*(5) (2018): 649–59. doi:10.1007/s13770-018-0132-z.

Koller J. *Drugs Exp Clin Res 30*(5–6) (2004): 183–90.

Kong JH, Oh EJ, Chae SY, Lee KC, Hahn SK. *Biomaterials 31*(14) (2010): 4121–8. doi:10.1016/j.biomaterials.2010.01.091.

Kuiper NJ, Sharma A. *Osteoarthritis Cartilage 23*(12) (2015): 2233–41. doi:10.1016/j.joca.2015.07.011.

Kulkarni SS, Patil SD, Chavan DG. *J Appl Nat Sci 10*(1) (2018): 313–5. doi:10.31018/jans.v10i1.1623.

La Gatta A, Salzillo R, Catalano C et al. *PLoS One 14*(6) (2019): e0218287. doi:10.1371/journal.pone.0218287.

Lai JY. *Carbohydr Polym 101* (2014): 203–12. doi:10.1016/j.carbpol.2013.09.060.

Larijani B, Ghahari A, Warnock GL et al. *Med Hypotheses 84*(6) (2015): 577–9. doi:10.1016/j.mehy.2015.03.004.

Lataillade JJ, Albanese P, Uzan G. *Ann Dermatol Venereol 137*(2010): 15–22. doi:10.1016/S0151-9638(10)70004-1.

Laurent A, Darwiche SE, Hirt-Burri N et al. *AJBSR 8*(4) (2020a): 252–71. doi:10.34297/AJBSR.2020.08.001284.

Laurent A, Hirt-Burri N, Scaletta C et al. *Front Bioeng Biotechnol 8* (2020b): 557758. doi:10.3389/fbioe.2020.557758.
Laurent A, Lin P, Scaletta C et al. *Front Bioeng Biotechnol 8*(581) (2020c). doi:10.3389/fbioe.2020.00581.
Laurent A, Scaletta C, Hirt-Burri N, Raffoul W, de Buys Roessingh AS, Applegate LA. *Methods Mol Biol* (2020d): 1–24. doi:10.1007/7651_2020_294.
Laurent TC, Hellsing K, Gelotte B. *Acta Chem Scand 18*(1) (1964): 274–5.
Lee H, Choi K, Kim H et al. *Osteoarthrit Cart 26* (2018): 125. doi:10.1016/j.joca.2018.02.272.
Lee H, Kim H, Seo J et al. *Inflammopharmacology 28*(5) (2020): 1237–52. doi:10.1007/s10787-020-00738-y.
Lee JH, Jung JY, Bang D. *J Eur Acad Dermatol Venereol 22*(5) (2008): 590–5. doi:10.1111/j.1468-3083.2007.02564.x.
Li L, Duan X, Fan Z et al. *Sci Rep 8*(1) (2018): 9900. doi:10.1038/s41598-018-27737-y.
Li Z, Guan J. *Expert Opin Drug Deliv 8*(8) (2011): 991–1007. doi:10.1517/17425247.2011.581656.
Li Z, Maitz P. *Burns Trauma 6* (2018): 13. doi:10.1186/s41038-018-0117-0.
Liao YH, Jones SA, Forbes B, Martin GP, Brown MB. *Drug Deliv 12*(6) (2005): 327–42. doi:10.1080/10717540590952555.
Longinotti C. *Burns Trauma 2*(4) (2014): 162–8. doi:10.4103/2321-3868.142398.
López-Ruiz E, Jiménez G, Álvarez de Cienfuegos L et al. *Eur Cell Mater 37* (2019): 186–213. doi:10.22203/eCM.v037a12.
Luo Y, Kirker KR, Prestwich GD. *J Control Release 69*(1) (2000): 169–84. doi:10.1016/s0168-3659(00)00300-x.
Madaghiele M, Demitri C, Sannino A, Ambrosio L. *Burns Trauma 2*(4) (2014): 153–61. doi:10.4103/2321-3868.143616.
Mahedia M, Shah N, Amirlak B. *Plast Reconstr Surg Glob Open 4*(7) (2016): 791. doi:10.1097/GOX.0000000000000747.
Mandal A, Clegg JR, Anselmo AC, Mitragotri S. *Bioeng Transl Med 5*(2) (2020): e10158. doi:10.1002/btm2.10158.
Maudens P, Meyer S, Seemayer CA, Jordan O, Allémann E. *Nanoscale 10*(4) (2018): 1845–54. doi:10.1039/C7NR07614B.
Mesa FL, Aneiros J, Cabrera A et al. *Histol Histopathol 17*(3) (2002): 747–53. doi:10.14670/HH-17.747.
Miastkowska M, Kulawik-Pióro A, Szczurek M. *Processes 8*(11) (2020): 1416. doi:10.3390/pr8111416.
Mirmalek-Sani SH, Tare RS, Morgan SM et al. *Stem Cells 24*(4) (2006): 1042–53. doi:10.1634/stemcells.2005-0368.
Molliard SG, Bétemps JB, Hadjab B, Topchian D, Micheels P, Salomon D. *Plast Aesthet Res 5*(17) (2018): 1–8. doi:10.20517/2347-9264.2018.10.
Montanucci P, Pennoni I, Pescara T, Basta G, Calafiore R. *Biomaterials 34*(16) (2013): 4002–12. doi:10.1016/j.biomaterials.2013.02.026.
Monti M, Perotti C, Del Fante C, Cervio M, Redi CA. *Biol Res 45*(3) (2012): 207–14. doi:10.4067/S0716-97602012000300002.
Moreland LW. *Arthritis Res Ther 5*(2) (2003): 54–67. doi:10.1186/ar623.
Mount NM, Ward SJ, Kefalas P, Hyllner J. *Philos Trans R Soc Lond B Biol Sci 370*(1680) (2015): 20150017. doi:10.1098/rstb.2015.0017.
Muraca M, Piccoli M, Franzin C et al. *Int J Mol Sci 18*(5) (2017): 1021. doi:10.3390/ijms18051021.
Murphy SV, Skardal A, Atala A. *J Biomed Mater Res A 101*(1) (2013): 272–84. doi:10.1002/jbm.a.34326.
Neuman MG, Nanau RM, Oruña-Sanchez L, Coto G. *J Pharm Pharm Sci 18*(1) (2015): 53–60. doi:10.18433/j3k89d.
Nyman E, Huss F, Nyman T, Junker J, Kratz G. *J Plast Surg Hand Surg 47*(2) (2013): 89–92. doi:10.3109/2000656X.2012.733169.
Ong KL, Runa M, Lau E, Altman R. *Cartilage 10*(4) (2019): 423–31. doi:10.1177/1947603518775792.
Pardue EL, Ibrahim S, Ramamurthi A. *Organogenesis 4*(4) (2008): 203–14. doi:10.4161/org.4.4.6926.
Park DY, Min BH, Park SR et al. *Sci Rep 10*(1) (2020): 5722. doi:10.1038/s41598-020-62580-0.
Peppas NA, Bures P, Leobandung W, Ichikawa H. *Eur J Pharm Biopharm 50*(1) (2000): 27–46. doi:10.1016/s0939-6411(00)00090-4.
Peppas NA, Slaughter BV, Kanzelberger MA. Hydrogels. In *Polymer Science: A Comprehensive Reference*, 385–95. Amsterdam: Matyjaszewski, Müller, 2012.
Pereira LC, Schweizer C, Moufarrij S et al. *Pilot Feasibility Stud 5*(56) (2019): 1–8. doi:10.1186/s40814-019-0443-4.

Pierre S, Liew S, Bernardin A. *Dermatol Surg 41* (2015): 120–6. doi:10.1097/DSS.0000000000000334.
Platt A, David BT, Fessler ARG. *Medicines 7*(5) (2020): 27. doi:10.3390/medicines7050027.
Prestwich GD. *J Control Release 155*(2) (2011): 193–9. doi:10.1016/j.jconrel.2011.04.007.
Price RD, Berry MG, Navsaria HA. *J Plast Reconstr Aesthet Surg 60*(10) (2007): 1110–9. doi:10.1016/j.bjps.2007.03.005.
Price RD, Myers S, Leigh IM, Navsaria HA. *Am J Clin Dermatol 6*(6) (2005): 393–402. doi:10.2165/00128071-200506060-00006.
Quintin A, Hirt-Burri N, Scaletta C, Schizas C, Pioletti DP, Applegate LA. *Cell Transplant 16*(7) (2007): 675–84. doi:10.3727/000000007783465127.
Racchi M, Govoni S, Lucchelli A, Capone L, Giovagnoni E. *Expert Rev Med Devices 13*(10) (2016): 907–17. doi:10.1080/17434440.2016.1224644.
Ramelet AA, Hirt-Burri N, Raffoul W et al. *Exp Gerontol 44*(3) (2009): 208–18. doi:10.1016/j.exger.2008.11.004.
Ratcliffe E, Thomas RJ, Williams DJ. *Br Med Bull 100* (2011): 137–55. doi:10.1093/bmb/ldr037.
Rayahin JE, Buhrman JS, Zhang Y, Koh TJ, Gemeinhart RA. *ACS Biomater Sci Eng 1*(7) (2015): 481–93. doi:10.1021/acsbiomaterials.5b00181.
Rayment EA, Williams DJ. *Stem Cells 28*(5) (2010): 996–1004. doi:10.1002/stem.416.
Renevier J-L, Marc J-F, Adam P, Sans N, Le Coz J, Prothoy I. *Int J Clin Rheum 13*(4) (2018): 230–8. doi:10.4172/1758-4272.1000191.
de Rezende MU, de Campos GC. *Rev Bras Ortop 47*(2) (2015): 160–4. doi:10.1016/S2255-4971(15)30080-X.
Richards MM, Maxwell JS, Weng L, Angelos MG, Golzarian J. *Phys Sportsmed 44*(2) (2016): 101–8. doi:10.1080/00913847.2016.1168272.
Rosser AE, Dunnett SB. *CNS Drugs 17*(12) (2003): 853–67. doi:10.2165/00023210-200317120-00001.
Sacchetti M, Rama P, Bruscolini A, Lambiase A. *Stem Cells Int 2018* (2018): 8086269. doi:10.1155/2018/8086269.
Saddiq ZA, Barbenel JC, Grant MH. *J Biomed Mater Res A 89*(3) (2009): 697–706. doi:10.1002/jbm.a.32007.
Šafránková B, Gajdova S, Kubala L. *Mediators Inflamm 2010* (2010): 380948. doi:10.1155/2010/380948.
Šafránková B, Hermannová M, Nešporová K, Velebný V, Kubala L. *Int J Biol Macromol 107*(A) (2018): 1–8. doi:10.1016/j.ijbiomac.2017.08.131.
Saturveithan C, Premganesh G, Fakhrizzaki S et al. *Malays Orthop J 10*(2) (2016): 35–40. doi:10.5704/MOJ.1607.007.
Schanté C, Zuber G, Herlin C, Vandamme TF. *Carbohydrate Polymers 85*(3) (2011a): 469–89. doi:10.1016/j.carbpol.2011.03.019.
Schanté C, Zuber G, Herlin C, Vandamme TF. *Carbohydrate Polymers 86*(2) (2011b): 747–52. doi:10.1016/j.carbpol.2011.05.017.
Schild HG. *Prog Polym Sci 17*(2) (1992): 163–249. doi:10.1016/0079-6700(92)90023-R.
Schlesinger T, Rowland Powell C. *J Clin Aesthet Dermatol 7*(5) (2014): 15–8.
Schmidt JJ, Rowley J, Kong HJ. *J Biomed Mater Res A 87*(4) (2008): 1113–22. doi:10.1002/jbm.a.32287.
Schneider A, Picart C, Senger B, Schaaf P, Voegel JC, Frisch B. *Langmuir 23*(5) (2007): 2655–62. doi:10.1021/la062163s.
Segura T, Anderson BC, Chung PH, Webber RE, Shull KR, Shea LD. *Biomaterials 26*(4) (2005): 359–71. doi:10.1016/j.biomaterials.2004.02.067.
Sekiya I, Ojima M, Suzuki S et al. *J Orthop Res 30*(6) (2012): 943–9. doi:10.1002/jor.22029.
Selyanin MA, Boykov PY, Khabarov VN, Polyak F. The history of hyaluronic acid discovery, foundational research and initial use. In *Hyaluronic Acid: Preparation, Properties, Application in Biology and Medicine*, 1–8. John Wiley & Sons, Ltd., 2015. doi:10.1002/9781118695920.
Shevchenko RV, James SL, James SE. *J R Soc Interface 7*(43) (2010): 229–58. doi:10.1098/rsif.2009.0403.
Shimojo AA, de Souza Brissac IC, Pina LM, Lambert CS, Santana MH. *Biomed Mater Eng 26*(3–4) (2015): 183–91. doi:10.3233/BME-151558.
Stern R, Kogan G, Jedrzejas MJ, Soltés L. *Biotechnol Adv 25*(6) (2007): 537–57. doi:10.1016/j.biotechadv.2007.07.001.
Su Z, Ma H, Wu Z et al. *Mater Sci Eng C Mater Biol Appl 44* (2014): 440–8. doi:10.1016/j.msec.2014.07.039.
Tan H, Ramirez CM, Miljkovic N, Li H, Rubin JP, Marra KG. *Biomaterials 30*(36) (2009): 6844–53. doi:10.1016/j.biomaterials.2009.08.058.
Tan J, Chen H, Zhao L, Huang W. *J Arthrosc Rel Surg 37*(1) (2021): 309–25. doi:10.1016/j.arthro.2020.07.011.
Tezel A, Fredrickson GH. The science of hyaluronic acid dermal fillers. *J Cosmet Laser Ther 10*(1) (2008): 35–42. doi:10.1080/14764170701774901.

Thomas V, Namdeo M, Mohan YM, Bajpai SK, Bajpai M. *J Macromol Sci A 45*(1) (2007): 107–19. doi:10.1080/10601320701683470.

Thönes S, Kutz LM, Oehmichen S et al. *Int J Biol Macromol 94*(A) (2017): 611–20. doi:10.1016/j.ijbiomac.2016.10.065.

Tibbitt MW, Anseth KS. *Biotechnol Bioeng 103*(4) (2009): 655–63. doi:10.1002/bit.22361.

Torres-Torrillas M, Rubio M, Damia E et al. *Int J Mol Sci 20*(12) (2019): 3105. doi:10.3390/ijms20123105.

Touraine JL, Roncarolo MG, Bacchetta R et al. *Bone Marrow Transplant 11* (1993): 119–22.

Turner NJ, Kielty CM, Walker MG, Canfield AE. *Biomaterials 25*(28) (2004): 5955–64. doi:10.1016/j.biomaterials.2004.02.002.

Ulusal BG. *J Cosmet Dermatol 16*(1) (2017): 112–9. doi:10.1111/jocd.12271.

Vasi AM, Popa MI, Butnaru M, Dodi G, Verestiuc L. *Mater Sci Eng C Mater Biol Appl 38* (2014): 177–85. doi:10.1016/j.msec.2014.01.052.

Vertelov G, Kharazi L, Muralidhar MG, Sanati G, Tankovich T, Kharazi A. *Stem Cell Res Ther 4*(1) (2013): 5. doi:10.1186/scrt153.

Voigt J, Driver VR. *Wound Repair Regen 20*(3) (2012): 317–31. doi:10.1111/j.1524-475X.2012.00777.x.

Wang AT, Zhang QF, Wang NX et al. *Front Bioeng Biotechnol 8* (2020): 87. doi:10.3389/fbioe.2020.00087.

Wang C, Liang C, Wang R et al. *Biomater Sci 8*(1) (2019): 313–24. doi:10.1039/c9bm01207a.

Watterson JR, Esdaile JM. *J Am Acad Orthop Surg 8*(5) (2000): 277–84. doi:10.5435/00124635-200009000-00001.

Weindl G, Schaller M, Schäfer-Korting M, Korting HC. *Skin Pharmacol Physiol 17*(5) (2004): 207–13. doi:10.1159/000080213.

Witting M, Boreham A, Brodwolf R et al. *Mol Pharm 12*(5) (2015): 1391–401. doi:10.1021/mp500676e.

Xie Y, Upton Z, Richards S, Rizzi SC, Leavesley DI. *J Control Release 153*(3) (2011): 225–32. doi:10.1016/j.jconrel.2011.03.021.

Xue Y, Chen H, Xu C, Yu D, Xua H, Hu Y. *RSC Advances 10*(12) (2020): 7206–13. doi:10.1039/C9RA09271D.

Yamazaki K, Kawabori M, Seki T, Houkin K. *Int J Mol Sci 21*(11) (2020): 3994. doi:10.3390/ijms21113994.

Yu W, Xu P, Huang G, Liu L. *Exp Ther Med 16*(3) (2018): 2119–25. doi:10.3892/etm.2018.6412.

Zheng Y, Yang J, Liang J et al. *Biomacromolecules 20*(11) (2019): 4135–42. doi:10.1021/acs.biomac.9b00964.

9

Extracellular Matrix: The State of the Art in Regenerative Medicine

Gurpreet Singh
Guru Nanak Dev University, Amritsar, Punjab, India

Pooja A Chawla
ISF College of Pharmacy, Moga, Punjab, India

Abdul Faruk
HNB Garhwal University (A Central University) Srinagar-Garhwal, Uttrakhand, India

Viney Chawla
Baba Farid University of Health Sciences, Faridkot, Punjab, India

Anmoldeep Kaur
Guru Nanak Dev University, Amritsar, Punjab, India

CONTENTS

Introduction

Regenerative medicine gained significant interest in the treatment of life-threatening diseases and disorders, especially in cardiovascular and neurodegenerative diseases (Mao and Mooney 2015). It is a multidisciplinary approach, which restores the normal physiological functions of the human body by replacement or repair of tissues and organs (Christ et al. 2013). Regenerative medicines are innovative therapies that involve various strategies of tissue engineering, stem cell biology, gene, and cellular therapeutics (Lorden et al. 2015). All regenerative medicine approaches depend upon cellular level events and their constituents, which are involved in various developmental or repair processes of human tissues, i.e. replacing damaged cells in the brain and pancreas (Mao and Mooney 2015). These transplanted cells perform all normal functions and functionally participate in the all tissue events (Chen and Liu 2016). Presently, regenerative medicine-based treatment is very expensive and not affordable by all (Mahalatchimy 2016).

DOI: 10.1201/9781003153504-9

Regenerative medicine is defined as a cellular therapeutic approach which "substitutes or repair human cells, various tissues or organ systems, to restore normal physiological function of human body" (Han et al. 2020; Sampogna et al. 2015).

There are a number of regulatory issues that influence the development of regenerative medicine and thus, in this scenario, need additional focus on legislation for regenerative medicine (Kleiderman et al. 2018). Recent research reports suggested that stem cell-based therapy has a promising role in the treatment of deadly human diseases, i.e. leukaemia, breast cancer, and others (Aly 2020). The ultimate objective of regenerative medicine is the isolation of specialised cell constituents and implanted into a patient where it replaces or repairs damage part of tissue or cells through self-repair remodelling (Mao and Mooney 2015). Therefore, it regulates the functioning of native tissues or cells. It offers transformative and effective outcomes for targeting life-threatening acute and chronic conditions and also an alternative for degenerative and genetic disorders (Mahla 2016).

According to the status of the Global Regenerative Medicine Market forecast, the international market of regenerative medicine is continuously growing and expected to reach USD 17.9 billion by 2025 (marketsandmarkets 2020). Food and Drug Administration (FDA, United States) implemented the 21st Century Cures Act in 2016 for the regulation of regenerative medicine therapies under a special section 3033, which describes the term and conditions for designation of drug under Regenerative Medicine Advanced Therapy (RMAT) (Barlas 2018). The Cures Act improves the ability of scientific, technical, and professional experts regarding clinical trial designs for regenerative medicine. It will accelerate the production of regenerative medicine products with safety of patients (FDA 2020).

The Cures Act defines the regenerative medicine as:

> *cell therapy, therapeutic tissue engineering product, human cell and tissue product, or any combination product using such therapies or products intended to treat, modify, reverse, or cure a serious or life threatening disease.*
>
> (FDA 2021b)

There are currently four main categories of stem cells that have the clone ability and differentiate into particular types of cells.

i. **Embryonic stem cells:** Derived from the initial developmental phase of few days old embryos at the blastocyst stage. It has the potential to differentiate into various cells with a distinct biological response. Such cells are known as pluripotent (Romito and Cobellis 2016).
ii. **Foetal stem cells:** Isolated from aborted human foetuses, especially foetal blood, foetal tissues, and also bone marrow. They have the ability to differentiate but not all cells. They are known as multipotent and have been utilised in the regeneration and repair of damaged tissues/organs (Biehl and Russell 2009).
iii. **Cord blood and placental stem cells:** Obtained from umbilical cord blood and placentas. They possess the therapeutic potential and used in bone-marrow replacement therapies. They are not able to differentiate into all types of cells (Weiss and Troyer 2006).
iv. **Adult stem cells:** They are the most abundant cells, which are used for various therapies/conditions. They are isolated from almost all human tissue and organs. They are known as "somatic stem cells" (Liras 2010).

There is no doubt that regenerative medicine products provide a better treatment option than conventional drugs. But still, there are certain limitations and challenges for researchers and pharmaceutical companies that need to be addressed for the improvement of these specialised products (Dodson and Levine 2015). The following are the few noticeable points that should be considered during the design and production of regenerative medicine (Herberts et al. 2011):

i. **Safety:** The derived product should be safe and effective without any tumour formation or production of unwanted cell types.

ii. **Regulatory aspects and standardisation:** Must meet regulatory requirements which ensure product quality, safety, and efficacy as mention by standards (Rosemann et al. 2019).
iii. **Imaging and Monitoring:** Need sophisticated techniques with the features of observing all the changes and variation during cell behaviour (Leahy et al. 2016) and also, monitoring the migration of cells after administration (Naumova et al. 2014).
iv. **Manufacturing:** Manufacturing of viable (living) cells for regenerative medicine must follow through optimised process protocol to avoid cell variability (Martin et al. 2014).
v. Multidisciplinary research involves in regenerative medicine requiring effective communication within all research communities for better outcomes (Shineha et al. 2017).
vi. **Animal Models:** Appropriate animal models are needed for the comparison of animal embryos/human genetic or cellular material information (Ribitsch et al. 2020).
vii. **Scale up/Technology Transfer**: Large-scale production reduces the overall cost of the product. The scalable production processes provide safe and effective products (Pigeau et al. 2018).
viii. **Immunogenicity:** In regenerative therapies, a major issue is the rejection of transplanted cell by the patient. This could be overcome by exploring the research for new generation of immunosuppressant drugs (Charron 2013).
ix. **Cell Viability:** Cell viability and storage conditions (Yu et al. 2018).

A number of regenerative medicine which have already received FDA approval (FDA 2021a) and are commercially available are listed in Figure 9.1. This chapter explores the role of extracellular matrix (ECM) in regenerative medicine.

The Extracellular Matrix

Regenerative strategies mainly focus on stem cell-based or tissue engineering applications for remodelling and regeneration of defective cells, tissues, and organs. Stem cell differentiation is modulated by signals from the extracellular microenvironment including the extracellular matrix (ECM) (Chen and Liu 2016). Cellular migration and differentiation events are the main key factors that are considered for the design of regenerative medicine (Mata et al. 2017). The ECM is composed of several types of collagens, proteoglycans, glycoproteins, and glycosaminoglycans, which are assembled into a complex structure (Yue 2014). The composition of ECM varies from tissue to tissue and organ to organ (Kular et al. 2014). The distinctive functions of the ECM include cell adhesion, the physical barrier for different tissues. It also impacts many cellular functions, including mechanical stimulation from substrates, activation of intracellular signalling by cell adhesion molecules, and availability and action of soluble factor (Muncie and Weaver 2018).

The extracellular matrixes (ECM) define the tissue architecture and biochemical and biophysical features. The main organisational unit of the ECM called core matrisome, which includes different kinds of collagen (divided into several families), glycoproteins, and proteoglycans (Hynes and Naba 2012). Other than ECM, there are numerous non-ECM varying factors, which also participate in different cellular events, i.e. remodelling and cell behaviour. They mainly include proteases, growth factors, cytokines, and cross-linking enzymes (Vaday and Lider 2000). Collagen is the most abundant protein of mammalian ECM and accountable for the structural and functional integrity of the tissue (Frantz et al. 2010). Other structural molecules of ECM belong to the glycosaminoglycans class which includes hyaluronic acid (HA, non-sulphated glycosaminoglycan), chondroitin sulphate (CS, sulphated glycosaminoglycan), and heparin (natural glycosaminoglycan) Figure 9.2 (Pomin and Mulloy 2018). They play a vital role in elasticity, water retention, and resistance to compressive forces, while adhesion proteins play a significant role as molecular glue for a structural network of ECM complex. Examples of adhesion molecules are laminin, fibronectin, and tenascin-C (Walker et al. 2018).

Early in the 20th century, cell biologists worked in a two-dimensional (2D) framework, which includes separating and culturing cells from living tissue for replacing damaged or diseased tissue (O'Brien and Duffy 2015). With the advancement in the field of bioengineering and regenerative medicine, it is

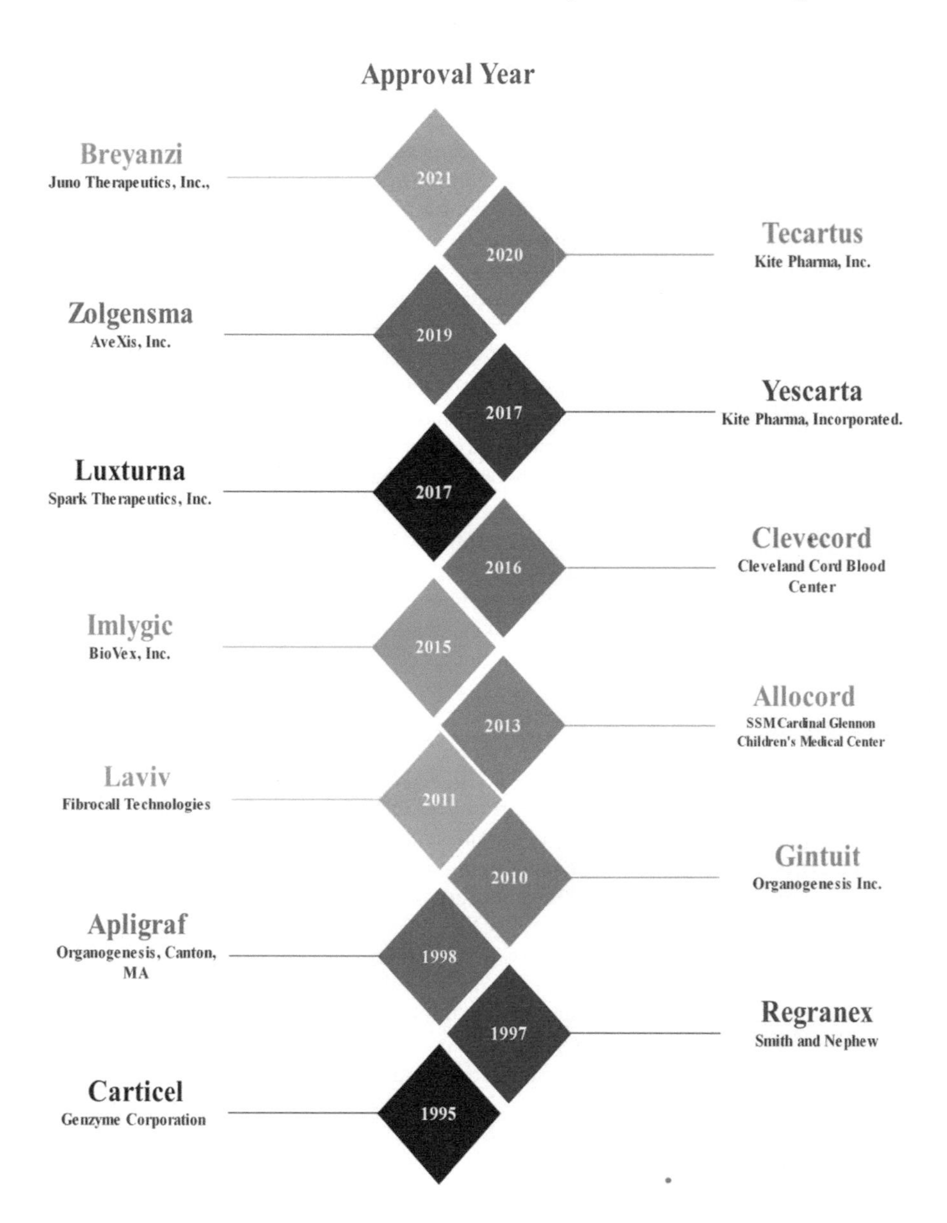

FIGURE 9.1 List of FDA Approved Products.

observed that multicellular organisms require a three-dimensional (3D) framework for structural integrity with specific microenvironments (Chen and Liu 2016). It is required to incorporate the knowledge of cell biology and cell transplantation with the discipline of material science for providing a 3D environment for growing cells and tissues. The evolution in the medical field has opened a new horizon for use of ECM in regenerative medicine. Various ECM analogues have been developed from synthetic scaffolds (Nikolova and Chavali 2019), hydrogels, and ceramic-based scaffolds (Hussey et al. 2018). These scaffolds are commonly made-up of synthetic and biodegradable polymers (Chaudhari et al. 2016). Commonly used polymers include polycaprolactone, polyethylene glycol, polyacrylic acid, hydroxyapatite or tricalcium phosphate, alginate, chitosan, and cellulose derived from plants (Hussey et al. 2018). Biomaterials used in regenerative medicine are broadly classified into two groups, i.e. naturally obtained and synthetic

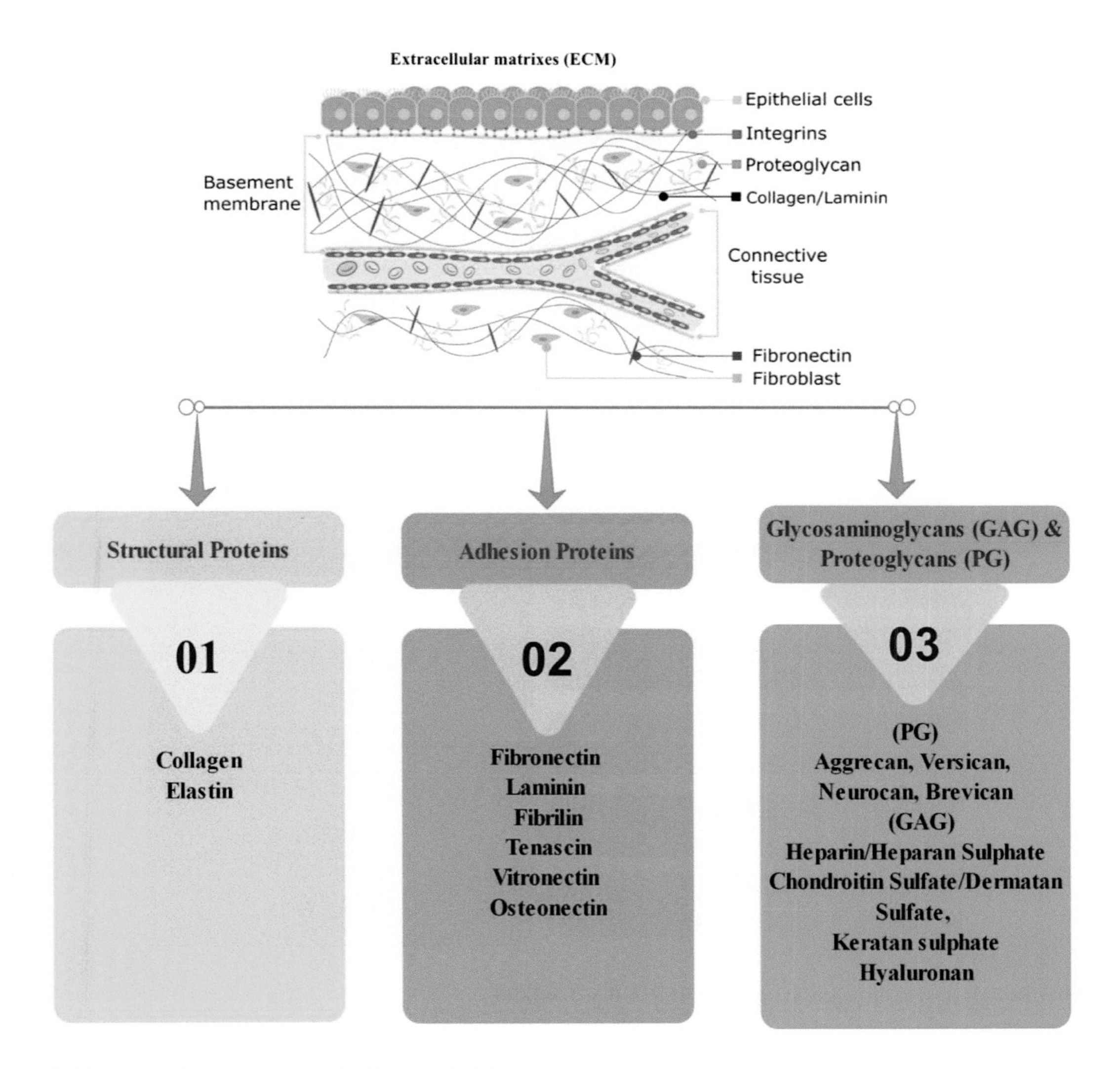

FIGURE 9.2 Composition of ECM (Kular et al. 2014).

materials (Fernandes et al. 2009). Natural materials are generally extracted or purified from ECM or its components such as collagen, laminin, and fibronectin (Frantz et al. 2010). Synthetic materials include polymers, metals or derived substrates. Both synthetic and natural materials have distinct pros and cons in regenerative medicine. Ideally, they are selected on the basis of condition and requirement of treatment. Biomaterials isolated from ECM show more unpredictability than synthetic polymers. In the case of synthetic polymers, immune response and their antigenicity is the major issue (Chen and Liu 2016). New trends and technologies in the bioengineering field reveal the functions of the ECM in regenerative medicine. This enriched the knowledge of ECM signalling in the functions of stem cells. These outcomes revealed the use of synthetic ECM scaffolds, which promote the endogenous stem cell repair and healing of damaged cells/tissues and mimic the native microenvironment (Chan and Leong 2008). A list of potential components of the extracellular matrix which are utilised in regenerative medicine (Traphagen and Yelick 2009) are summarised in Figures 9.3 and 9.4.

ECM-based biomaterials promote tissue remodelling in a precise and controllable manner. The decellularised ECM (DECM), which is water-insoluble matrix obtained after removal of cellular constituents

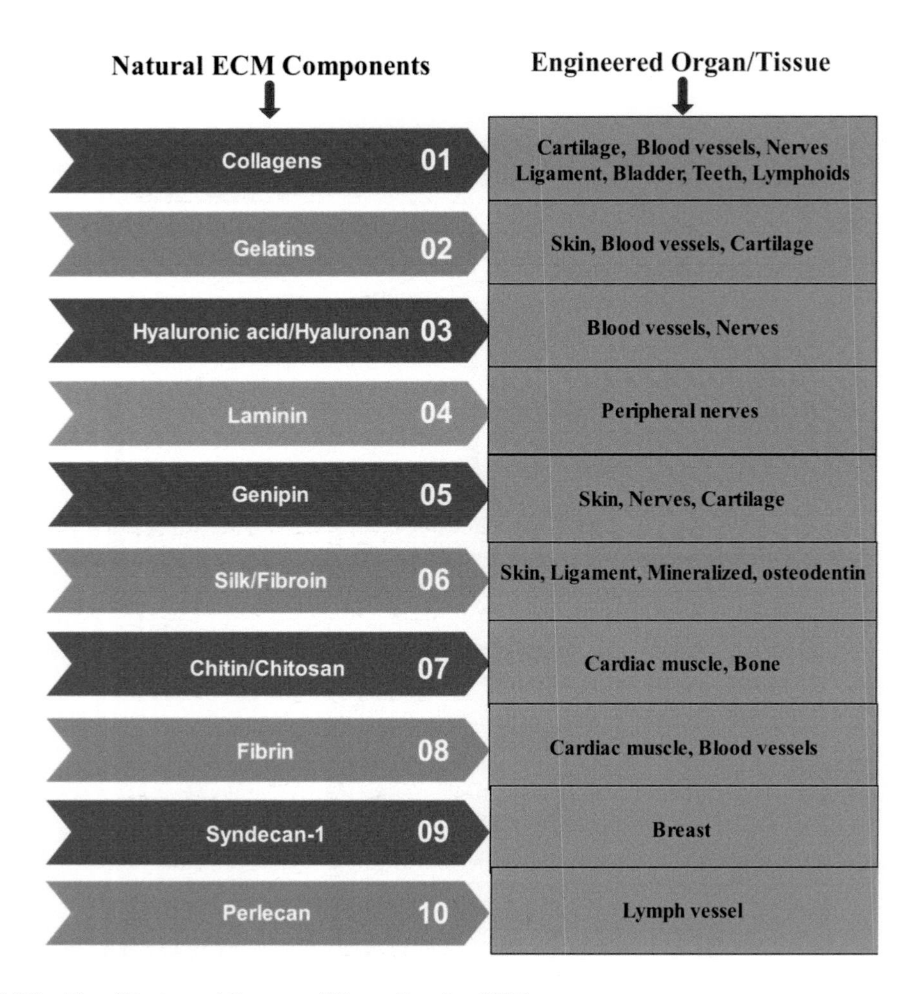

FIGURE 9.3 List of Engineered Organs and Tissues Based on ECM.

from ECM, also plays a significant role in the remodelling and repair process (Chakraborty et al. 2020). Due to biocompatibility and biodegradability, the DECM offers better results than other commonly used biomaterials (Liao et al. 2020). DECM-based tissue/organ, hydrogel, and microparticles have high demand in regenerative medicine (Parmaksiz et al. 2016).

Biomimetic materials can be fabricated using different techniques, i.e. soft lithography (Whitesides et al. 2001) (micro-contact printing), electrospinning (Braghirolli et al. 2014), and 3D printing (Atala and Forgacs 2019). Cellular constituents present within all tissues are required for tissue morphogenesis, differentiation, and the homeostasis process. Fundamentally, ECM can resolve various syndromes, physiological conditions, and defects in the body (Theocharis et al. 2019). In recent years, many studies indicate the role of native ECMs/DECM in regenerative medicine (Ramos and Moroni 2020). The main applications of ECMs include 3D tissue culturing (Edmondson et al. 2014), stimulate the wound healing process (Agren and Werthen 2007), activate stem cell differentiation (Gattazzo et al. 2014), and drug screening assays (Langhans 2018). It's also applied in cell repair pathways and functional recovery of kidney (Bulow and Boor 2019), adrenal glands (Ruiz-Babot et al. 2015), and reproductive organs (Yalcinkaya et al. 2014). ECMs have many applications due to their biocompatibility and *in vivo* replicate ability (Aamodt and Grainger 2016). This chapter summarises some research investigations based on EMCs in regenerative medicine.

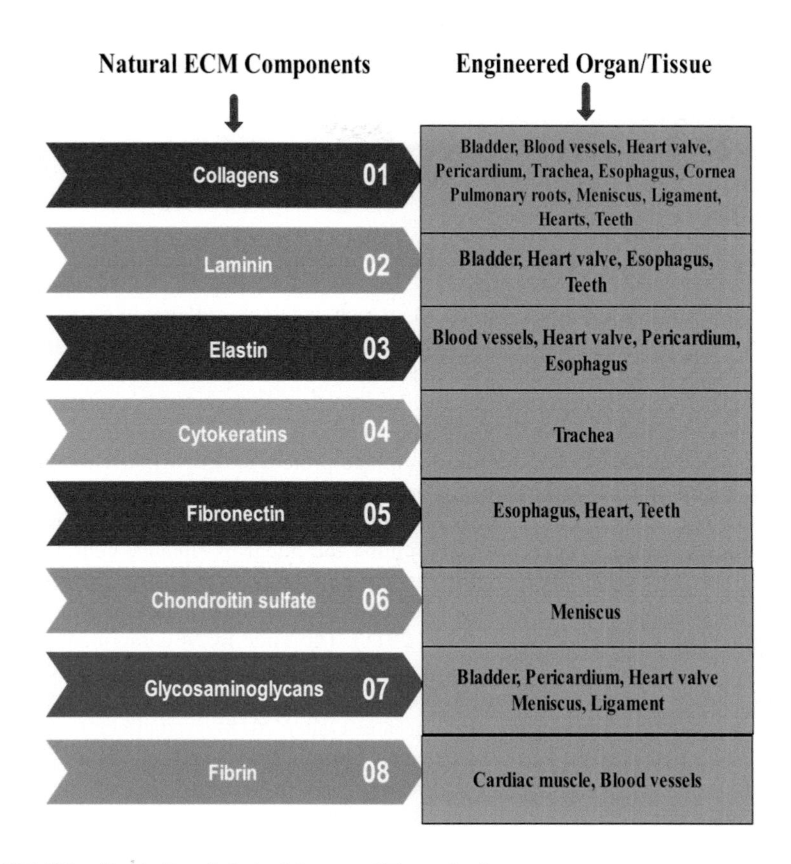

FIGURE 9.4 ECM Identified in Decellularised Organ and Tissue Studies.

Application of Extracellular Matrix

The chief proteins of the ECM are collagens and elastin. They are considered for biomedical applications because of their tensile strength and viscoelasticity to tissues. Other proteins include fibronectin, laminin, and nidogen, which act as connectors or linking proteins in the matrix network. Glycosaminoglycans (GAGs), proteoglycans, and growth factors are also promoting the *in vivo* construction of functional tissue (Mouw et al. 2014). Overall it is a challenging task due to limited knowledge and tissue to tissue variability. Ultimate goals of regenerative medicine can be achieved only if biomaterials maintain desired morphology, differentiation, proliferation, and metabolism of the cell.

Cardiac Extracellular Matrix

The heart is a vital and delicate organ of our body that requires sophisticated tissue architecture for normal functioning. It acts as a circulatory motor for our body that supplies the blood and fulfils the variable demands during the rest and exercise phase (Lee and Walsh 2016). This excitation-contraction, the cycle of the heart, developed a physical force at the cellular level of the myocardial structure. Overall, this process is regulated by the delicate organisation of the cardiac extracellular matrix. In each excitation and contraction cycle, a number of mechanical events are involved in myocardial elements (Stoppel et al. 2016).

Recent investigations suggest that ECM is found in all the segments of the heart. However, it is particularly present in mesenchyme structures and plays a role in valvuloseptal morphogenesis (Lockhart et al. 2011). Any impairment in the composition of ECM in mesenchyme structures often leads to congenital heart disease. Several reported animal studies describe the involvement of ECM in congenital heart diseases (Hacker 2018). Studies indicate the involvement of aggrecan, hyaluronan, versican, collagen type I-V, fibulin1, and fibronectin (Wight 2018). The common complications are vascular defects, blood vessel rupture, and cardiomyopathy. ECM involves in the regulation of cell differentiation and proliferation, which serve the cell survival (Ponticos and Smith 2014).

Extracellular Matrix in Brain

A major part of the brain is occupied by ECM, which contains collagens, fibronectin, vitronectin, laminin, and perlecan especially in amyloid deposits of the brain (Bonneh-Barkay and Wiley 2009). These ECM components play the main role in the development of nervous tissue and also regulate cell adhesion (Barros et al. 2011). Matrix proteins are almost absent in the adult brain (Ruoslahti 1996). Any change that occurs in the composition of ECM after neural injury may result in drastic consequences. Brain injury may induce changes in chondroitin sulphate proteoglycans, which influence myelin repair (Rhodes and Fawcett 2004). During the early stage of neural growth, the ECM provides structural support and stimulates signalling pathways of proliferation, especially by proteoglycans, laminins, and integrins. Proteoglycans provide structural support while laminins and integrins enhance neural progenitor proliferation (Bonnans et al. 2014). They also modify the shape of neural progenitors and neurons. In addition to this, ECM components affect the migration of newborn neurons during cortical growth. The role of the ECM in the brain is highly complicated (Lu et al. 2011). The same ECM component performs multiple roles during neural development and also influences the functioning of neighbouring cells (Rozario and DeSimone 2010). Recently reported evidence indicates the involvement of the ECM in several disease conditions, such as traumatic brain injury (George and Geller 2018), Alzheimer's disease (Lepelletier et al. 2017), age-associated cognitive deficits (Richard and Lu 2019), and schizophrenia (Lubbers et al. 2014). ECM-based regenerative approaches are widely used in the repair of peripheral soft tissue but not in the case of the brain due to the invasive route of administration. It requires a very specific narrow needle-guided administration approach for specific targeting. Current research efforts in regenerative medicine suggest that ECM-based biomaterials could serve as regenerative therapies in the brain (Hwang et al. 2020). A variety of underlying factors and mechanisms are still under observation and site-specific administration of ECM-based biomaterials is another issue in development of regenerative medicine (Chen and Liu 2016).

Pulmonary Extracellular Matrix

Pulmonary ECM is a structural complex system of protein molecules, which participate in various biochemical processes (Burgstaller et al. 2017). The remodelling mechanism is important for tissue homeostasis and any change in it may result in conditions like chronic obstructive pulmonary disease (COPD). Impaired expression of ECM proteins seen in COPD leads to the degradation and disruption of alveolar walls and stiffening of minor airways, which result in obstruction of airways (Ito et al. 2019). Alterations in ECM composition also influence the immune cell movement and their maintenance in the lung (Bonnans et al. 2014). Any abnormal functioning of ECM and response of inflammatory cell surface receptors may modify the collagen microstructure of the lung (Hussell et al. 2018). It is observed that there is a change in collagen organisation in COPD lung as compared to normal lung. The imbalance of enzymes like lysyloxidase and transglutaminase2 may involve structural changes of ECM during COPD (Burgess et al. 2016). ECM regulates normal interstitial fluid dynamics and strength and elasticity, tissue repair, and remodelling in the lungs. Versican and perlecan participate in the balancing of tissue fluid homeostasis (Pelosi et al. 2007). In the area of regenerative medicine, several studies reported lung scaffolds from small and large animals as an alternative to lung transplantation (Ohata and Ott 2020). These lung scaffolds were decellularised and reseed with lung perfusion culture in bioreactors. The resulting bioartificial lungs are probable to solve the problem of donor organ shortage and also reduced the immunogenicity (Panoskaltsis-Mortari 2015).

Extracellular Matrix in Inflammatory Bowel Disease

Inflammatory Bowel Disease (IBD) is a global health issue and the specific aetiology of IBD is unknown. Ulcerative colitis and Crohn's disease are the two main forms of IBD and they are characterised by an unusual immune response linked with defects functioning in the intestinal epithelial cell barrier (Zhang and Li 2014). Macroscopic tissue injury and clinical features of IBD are developed by changes in the ECM. Any change in ECM constituents may result in intestinal inflammation and progression of IBD (Petrey and de la Motte 2017). The conventional treatments merely target treatment of inflammation not repair/recovery of damaged tissue. Recently published work reports the use of hematopoietic or mesenchymal stem cells (HSCs or MSCs) for the management of IBD (Martinez-Montiel Mdel et al. 2014). It may help to establish an effective regenerative medicine for IBD patients. The development of decellularisation techniques in biomedical engineering greatly assisted the site-specific applications of ECM bio-scaffolds in the gastrointestinal tract (Almeida-Porada et al. 2013).

Conclusion

In conclusion, it is clear that the manipulation of ECM may serve as natural mimicking scaffolds in the arena of regenerative medicine. Regenerative medicine will change the traditional methods of management of various life-threatening diseases and conditions. Moreover, there is no doubt that all classes of stem cells (embryonic, adult, and induced pluripotent stem cells) have the potential to control the variety of diseases. ECM and ECM-like materials are biocompatible and having integration with the physiological microenvironment and mimic the ECM structure of the target tissues. ECM supports various biological functions and preserves the structures of entire organs. Ideally, they are preferred over synthetic polymers for biomedical engineering because of their immune tolerance. ECM regulation can play a significant role in several body conditions such as COPD, spinal cord injury, and neurodegenerative disorder. Furthermore, innovative interdisciplinary approaches and advancements in methodologies may lead to the improvement and discovery of new treatments for human disease.

REFERENCES

Aamodt J.M., Grainger D.W. 2016. *Biomaterials 86*: 68–82. doi:10.1016/j.biomaterials.2016.02.003.

Agren M.S., Werthen M. 2007. *The International Journal of Lower Extremity Wounds 6* (2): 82–97. doi:10.1177/1534734607301394.

Almeida-Porada G., Soland M., Boura J., Porada C.D. 2013. *Regenerative Medicine 8* (5): 631–644. doi:10.2217/rme.13.52.

Aly R.M. 2020. *Stem Cell Investigation 7*: 8. doi:10.21037/sci-2020-001.

Atala A., Forgacs G. 2019. *Stem Cells Translational Medicine 8* (8): 744–745. doi:10.1002/sctm.19-0089.

Barlas S. 2018. *P & T: A Peer-Reviewed Journal for Formulary Management 43* (3): 149–179.

Barros C.S., Franco S.J., Muller U. 2011. *Cold Spring Harbor Perspectives in Biology 3* (1): a005108. doi:10.1101/cshperspect.a005108.

Biehl J.K., Russell B. 2009. *The Journal of Cardiovascular Nursing 24* (2): 98–103; quiz 104–105. doi:10.1097/JCN.0b013e318197a6a5.

Bonnans C., Chou J., Werb Z. 2014. *Nature Reviews. Molecular Cell Biology 15* (12): 786–801. doi:10.1038/nrm3904.

Bonneh-Barkay D., Wiley C.A. 2009. *Brain Pathology 19* (4): 573–585. doi:10.1111/j.1750-3639.2008.00195.x.

Braghirolli D.I., Steffens D., Pranke P. 2014. *Drug Discovery Today 19* (6): 743–753. doi:10.1016/j.drudis.2014.03.024.

Bulow R.D., Boor P. 2019. *The Journal of Histochemistry and Cytochemistry: Official Journal of the Histochemistry Society 67* (9): 643–661. doi:10.1369/0022155419849388.

Burgess J.K., Mauad T., Tjin G., Karlsson J.C. et al. 2016. *The Journal of Pathology 240* (4): 397–409. doi:10.1002/path.4808.

Burgstaller G., Oehrle B., Gerckens M., White E.S. et al. 2017. *The European Respiratory Journal 50* (1). doi:10.1183/13993003.01805-2016.

Chakraborty J., Roy S., Ghosh S. 2020. *Biomaterials Science 8* (5): 1194–1215. doi:10.1039/C9BM01780A.

Chan B.P., Leong K.W. 2008. *European Spine Journal 17* Suppl 4: 467–479. doi:10.1007/s00586-008-0745-3.

Charron D. 2013. *The Indian Journal of Medical Research 138* (5): 749–754.

Chaudhari A.A., Vig K., Baganizi D.R., Sahu R. et al. 2016. *International Journal of Molecular Sciences 17* (12): 1974.

Chen F.M., Liu X. 2016. *Progress in Polymer Science 53*: 86–168. doi:10.1016/j.progpolymsci.2015.02.004.

Christ G.J., Saul J.M., Furth M.E., Andersson K.E. 2013. *Pharmacological Reviews 65* (3): 1091–1133. doi:10.1124/pr.112.007393.

Dodson B.P., Levine A.D. 2015. *BMC Biotechnology 15*: 70. doi:10.1186/s12896-015-0190-4.

Edmondson R., Broglie J.J., Adcock A.F., Yang L. 2014. *Assay and Drug Development Technologies 12* (4): 207–218. doi:10.1089/adt.2014.573.

FDA. 2020. 21st Century Cures Act. Retrieved 5/2/2021, 2021, from https://www.fda.gov/regulatory-information/selected-amendments-fdc-act/21st-century-cures-act

FDA. 2021a. Approved Cellular and Gene Therapy Products. Retrieved 3 March 2021, 2021, from https://www.fda.gov/vaccines-blood-biologics/cellular-gene-therapy-products/approved-cellular-and-gene-therapy-products

FDA. 2021b. Regenerative Medicine Advanced Therapy Designation. Retrieved 20/2/2021, from https://www.fda.gov/vaccines-blood-biologics/cellular-gene-therapy-products/regenerative-medicine-advanced-therapy-designation

Fernandes H., Moroni L., van Blitterswijk C., de Boer J. 2009. *Journal of Materials Chemistry 19* (31): 5474–5484. doi:10.1039/B822177D.

Frantz C., Stewart K.M., Weaver V.M. 2010. *Journal of Cell Science 123* (Pt 24): 4195–4200. doi:10.1242/jcs.023820.

Gattazzo F., Urciuolo A., Bonaldo P. 2014. *Biochimica et biophysica acta 1840* (8): 2506–2519. doi:10.1016/j.bbagen.2014.01.010.

George N., Geller H.M. 2018. *Journal of Neuroscience Research 96* (4): 573–588. doi:10.1002/jnr.24151.

Hacker T.A. 2018. *Advances in Experimental Medicine and Biology 1098*: 45–58. doi:10.1007/978-3-319-97421-7_3.

Han F., Wang J., Ding L., Hu Y. et al. 2020. *Frontiers in Bioengineering and Biotechnology 8*: 83. doi:10.3389/fbioe.2020.00083.

Herberts C.A., Kwa M.S., Hermsen H.P. 2011. *Journal of Translational Medicine 9*: 29. doi:10.1186/1479-5876-9-29.

Hussell T., Lui S., Jagger C., Morgan D. et al. 2018. *European Respiratory Review: An Official Journal of the European Respiratory Society 27* (148). doi:10.1183/16000617.0032-2018.

Hussey G.S., Dziki J.L., Badylak S.F. 2018. *Nature Reviews Materials 3* (7): 159–173. doi:10.1038/s41578-018-0023-x.

Hwang J., Sullivan M.O., Kiick K.L. 2020. *Frontiers in Bioengineering and Biotechnology 8*: 69. doi:10.3389/fbioe.2020.00069.

Hynes R.O., Naba A. 2012. *Cold Spring Harbor Perspectives in Biology 4* (1): a004903. doi:10.1101/cshperspect.a004903.

Ito J.T., Lourenco J.D., Righetti R.F., Tiberio I. et al. 2019. *Cells 8* (4). doi:10.3390/cells8040342.

Kleiderman E., Boily A., Hasilo C., Knoppers B.M. 2018. *Stem Cell Research & Therapy 9* (1): 307. doi:10.1186/s13287-018-1055-2.

Kular J.K., Basu S., Sharma R.I. 2014. *Journal of Tissue Engineering 5*: 2041731414557112. doi:10.1177/2041731414557112.

Langhans S.A. 2018. *Frontiers in Pharmacology 9*: 6. doi:10.3389/fphar.2018.00006.

Leahy M., Thompson K., Zafar H., Alexandrov S. et al. 2016. *Stem Cell Research & Therapy 7* (1): 57. doi:10.1186/s13287-016-0315-2.

Lee R.T., Walsh K. 2016. *Circulation 133* (25): 2618–2625. doi:10.1161/CIRCULATIONAHA.115.019214.

Lepelletier F.X., Mann D.M., Robinson A.C., Pinteaux E. et al. 2017. *Neuropathology and Applied Neurobiology 43* (2): 167–182. doi:10.1111/nan.12295.

Liao J., Xu B., Zhang R., Fan Y. et al. 2020. *Journal of Materials Chemistry B 8* (44): 10023–10049. doi:10.1039/D0TB01534B.

Liras A. 2010. *Journal of Translational Medicine 8*: 131. doi:10.1186/1479-5876-8-131.
Lockhart M., Wirrig E., Phelps A., Wessels A. 2011. *Birth Defects Research. Part A, Clinical and Molecular Teratology 91* (6): 535–550. doi:10.1002/bdra.20810.
Lorden E.R., Levinson H.M., Leong K.W. 2015. *Drug Delivery and Translational Research 5* (2): 168–186. doi:10.1007/s13346-013-0165-8.
Lu P., Takai K., Weaver V.M., Werb Z. 2011. *Cold Spring Harbor Perspectives in Biology 3* (12). doi:10.1101/cshperspect.a005058.
Lubbers B.R., Smit A.B., Spijker S., van den Oever M.C. 2014. *Progress in Brain Research 214*: 263–284. doi:10.1016/B978-0-444-63486-3.00012-8.
Mahalatchimy A. 2016. *Medical Law Review 24* (2): 234–258. doi:10.1093/medlaw/fww009.
Mahla R.S. 2016. *International Journal of Cell Biology 2016*: 6940283. doi:10.1155/2016/6940283.
Mao A.S., Mooney D.J. 2015. *Proceedings of the National Academy of Sciences of the United States of America 112* (47): 14452–14459. doi:10.1073/pnas.1508520112.
marketsandmarkets. 2020. Regenerative Medicine Market by Product. Retrieved 14 Feb, 2021, from https://www.marketsandmarkets.com/Market-Reports/regenerative-medicine-market-65442579.html
Martin I., Simmons P.J., Williams D.F. 2014. *Science Translational Medicine 6* (232): 232fs216. doi:10.1126/scitranslmed.3008558.
Martinez-Montiel Mdel P., Gomez-Gomez G.J., Flores A.I. 2014. *World Journal of Gastroenterology 20* (5): 1211–1227. doi:10.3748/wjg.v20.i5.1211.
Mata A., Azevedo H.S., Botto L., Gavara N. et al. 2017. *Current Stem Cell Reports 3* (2): 83–97. doi:10.1007/s40778-017-0081-9.
Mouw J.K., Ou G., Weaver V.M. 2014. *Nature Reviews. Molecular Cell Biology 15* (12): 771–785. doi:10.1038/nrm3902.
Muncie J.M., Weaver V.M. 2018. *Current Topics in Developmental Biology 130*: 1–37. doi:10.1016/bs.ctdb.2018.02.002.
Naumova A.V., Modo M., Moore A., Murry C.E. et al. 2014. *Nature Biotechnology 32* (8): 804–818. doi:10.1038/nbt.2993.
Nikolova M.P., Chavali M.S. 2019. *Bioactive Materials 4*: 271–292. doi:10.1016/j.bioactmat.2019.10.005.
O'Brien F.J., Duffy G.P. 2015. *Journal of Anatomy 227* (6): 705–706. doi:10.1111/joa.12401.
Ohata K., Ott H.C. 2020. *Surgery Today 50* (7): 633–643. doi:10.1007/s00595-020-02000-y.
Panoskaltsis-Mortari, A. 2015. *Current Transplantation Reports 2* (1): 90–97. doi:10.1007/s40472-014-0048-z.
Parmaksiz M., Dogan A., Odabas S., Elcin A.E. et al. 2016. *Biomedical Materials 11* (2): 022003. doi:10.1088/1748-6041/11/2/022003.
Pelosi P., Rocco P.R., Negrini D., Passi A. 2007. *Anais da Academia Brasileira de Ciencias 79* (2): 285–297. doi:10.1590/s0001-37652007000200010.
Petrey A.C., de la Motte C.A. 2017. *Current Opinion in Gastroenterology 33* (4): 234–238. doi:10.1097/MOG.0000000000000368.
Pigeau G.M., Csaszar E., Dulgar-Tulloch A. 2018. *Frontiers in Medicine 5*: 233. doi:10.3389/fmed.2018.00233.
Pomin V.H., Mulloy B. 2018. *Pharmaceuticals 11* (1): 27.
Ponticos M., Smith B.D. 2014. *Journal of Biomedical Research 28* (1): 25–39. doi:10.7555/JBR.27.20130064.
Ramos T., Moroni L. 2020. *Tissue Engineering. Part C, Methods 26* (2): 91–106. doi:10.1089/ten.TEC.2019.0344.
Rhodes K.E., Fawcett J.W. 2004. *Journal of Anatomy 204* (1): 33–48. doi:10.1111/j.1469-7580.2004.00261.x.
Ribitsch I., Baptista P.M., Lange-Consiglio A., Melotti L. et al. 2020. *Frontiers in Bioengineering and Biotechnology 8*: 972. doi:10.3389/fbioe.2020.00972.
Richard A.D., Lu X.H. 2019. *Neural Regeneration Research 14* (4): 578–581. doi:10.4103/1673-5374.247459.
Romito A., Cobellis G. 2016. *Stem Cells International 2016*: 9451492. doi:10.1155/2016/9451492.
Rosemann A., Vasen F., Bortz G. 2019. *Science as Culture 28* (2): 223–249. doi:10.1080/09505431.2018.1556253.
Rozario T., DeSimone D.W. 2010. *Developmental Biology 341* (1): 126–140. doi:10.1016/j.ydbio.2009.10.026.
Ruiz-Babot G., Hadjidemetriou I., King P.J., Guasti L. 2015. *Frontiers in Endocrinology 6*: 70. doi:10.3389/fendo.2015.00070.
Ruoslahti E. 1996. *Glycobiology 6* (5): 489–492. doi:10.1093/glycob/6.5.489.
Sampogna G., Guraya S.Y., Forgione A. 2015. *Journal of Microscopy and Ultrastructure 3* (3): 101–107. doi:10.1016/j.jmau.2015.05.002.

Shineha R., Inoue Y., Ikka T., Kishimoto A. et al. 2017. *Regenerative Therapy 7*: 89–97. doi:10.1016/j.reth.2017.11.001.

Stoppel W.L., Kaplan D.L., Black L.D., 3rd 2016. *Advanced Drug Delivery Reviews 96*: 135–155. doi:10.1016/j.addr.2015.07.009.

Theocharis A.D., Manou D., Karamanos N.K. 2019. *The FEBS Journal 286* (15): 2830–2869. doi:10.1111/febs.14818.

Traphagen S., Yelick P.C. 2009. *Regenerative Medicine 4* (5): 747–758. doi:10.2217/rme.09.38.

Vaday G.G., Lider O. 2000. *Journal of Leukocyte Biology 67* (2): 149–159. doi:10.1002/jlb.67.2.149.

Walker C., Mojares E., Del Rio Hernandez A. 2018. *International Journal of Molecular Sciences 19* (10). doi:10.3390/ijms19103028.

Weiss M.L., Troyer D.L. 2006. *Stem Cell Reviews 2* (2): 155–162. doi:10.1007/s12015-006-0022-y.

Whitesides G.M., Ostuni E., Takayama S., Jiang X. et al. 2001. *Annual Review of Biomedical Engineering 3*: 335–373. doi:10.1146/annurev.bioeng.3.1.335.

Wight T.N. 2018. *Matrix Biology: Journal of the International Society for Matrix Biology 71–72*: 396–420. doi:10.1016/j.matbio.2018.02.019.

Yalcinkaya T.M., Sittadjody S., Opara E.C. 2014. *Maturitas 77* (1): 12–19. doi:10.1016/j.maturitas.2013.10.007.

Yu N.H., Chun S.Y., Ha Y.S., Kim H.T. et al. 2018. *Tissue Engineering and Regenerative Medicine 15* (5): 639–647. doi:10.1007/s13770-018-0133-y.

Yue B. 2014. *Journal of Glaucoma 23* (8 Suppl 1): S20–23. doi:10.1097/IJG.0000000000000108.

Zhang Y.Z., Li Y.Y. 2014. *World Journal of Gastroenterology 20* (1): 91–99. doi:10.3748/wjg.v20.i1.91.

10

Hydrogels with Ubiquitous Roles in Biomedicine and Tissue Regeneration

Priyanka
Punjabi University, Patiala, Punjab, India

Pooja A Chawla
ISF College of Pharmacy, Moga, Punjab, India

Aakriti
Medical College, Baroda, Vadodara, Gujarat, India

Viney Chawla
Baba Farid University of Health Sciences, Faridkot, Punjab, India

Durgesh Nandini Chauhan
Columbia Institute of Pharmacy, Raipur, Chhattisgarh, India

Bharti Sapra
Punjabi University, Patiala, Punjab, India

CONTENTS

DOI: 10.1201/9781003153504-10

Introduction

The term tissue engineering (TE) was conceived in 1980. Every year, a huge number of patients endure from loss or failure of an organ or a tissue. Organ or tissue grafting is the most satisfactory remedy; however, its use is abridged due to a limited number of donors. The global market size of TE is likely to index an annual average growth rate (AAGR) of 17.84% to make USD 53,424.00 million by 2024. Escalating elderly people is, in turn, escalate the prevalence rate of chronic/infectious diseases due to which TE approach is budding as a novel technology in biomedicine (Martin et al., 2019). The TE market is divided into various segments based on its applications (Figure 10.1) such as gastrointestinal, gynaecology, neurology, cardiology, urology, dental, skin/integumentary, orthopaedics, cancer, musculoskeletal, and spine. Out of these, orthopaedics, musculoskeletal, and spine will have the maximum share in the market in the future according to estimated years (Jansen and Peyton, 2018). The key players in the TE field are Stryker, Allergan, Medtronic, Zimmer, Baxter International, Integra Life Sciences, Organovo Holdings Inc., Cook Medical, DePuy Synthes, and Acelity.

TE is a biomedical engineering area that uses an amalgamation of cells, material, methods, and appropriate components to restore, maintain, or improve different types of cells and tissues (Griffith and Naughton, 2002; Lysaght and Reyes, 2001). Shortage of organs and substandard prosthetic substances used for repairing or replacing diseased or damaged organs is the main reason for augmented research in the field of TE (Stock and Vacanti, 2001). Its unique characteristic is regenerating a patient's tissues and organs that show poor biocompatibility and have low biofunctionality as well as dreadful immune rejection. Because of these advantages, TE is frequently considered as an ideal medical treatment for repairing tissues (Ikada, 2006).

Biomaterials are the components that play a very important role in TE of new functional tissues for repairing or replacing lost or damaged tissues or organs. They provide a transitory scaffolding to direct

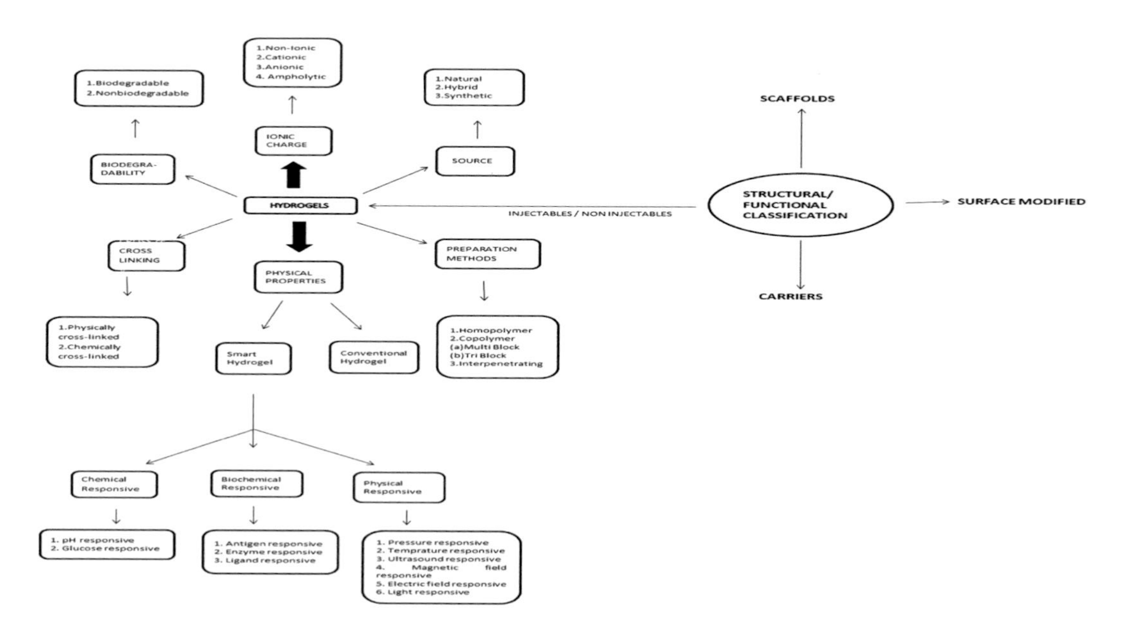

FIGURE 10.1 Classification of hydrogels.

fresh tissue development and may convey biologically active signals critical for the detection of tissue-specific gene expression (Kim et al., 2000).

This chapter explores different biomaterials that can be used for TE. The stress has been laid upon polymers that can be used for the preparation of hydrogels; classification, methods of preparation, and characterization of hydrogels; and patents pertaining to this field and applications in various fields of TE.

Biomaterials

Natural polymers (e.g. collagen and alginate), acellular tissue matrices (e.g. decellularised tissues or organs), and synthetic biodegradable polymers (e.g. polyglycolic acid (PGA), polylactic acid (PLA), poly (lactic-co-glycolic acid) and their copolymers are frequently used for TE application (Kim et al., 2000; Lee et al., 2018). Biomaterials can be classified on the basis of structure as well as function (Dolcimascolo et al., 2019) (Figure10.1 summarises this classification), e.g. whether biomaterials are applied directly in the tissue in the form of injectable (Gutowska et al., 2001) or non-injectable hydrogels; carriers or scaffolds and as surface modification, etc. (Baroli, 2007; Kretlow et al., 2007).

Hydrogels have the properties, e.g. they can hold high-water content, demonstrate excellent biocompatibility, low interfacial tension, least frictional and mechanical irritation, and have various other advantages (Baroli, 2007; Domb et al., 1998). Their tissue-engineered scaffolds can be used in gene therapy and may achieve cellular function (Lee et al., 2018) due to which they have attained remarkable attention among various biomaterials that are used in TE. Table 10.1 summarises different biomaterials with their advantages and disadvantages.

Hydrogels

Hydrogels can be defined as polymeric structures that are represented as water-swollen gels. Because of the presence of water and the cross-linking mechanisms, hydrogels exhibit flexibility. Various mechanisms alone or in combination can lead to crosslinking (Peppas and Hoffman, 2020) such as primary covalent cross-linking, ionic forces, hydrogen bonding, hydrophobic interactions, affinity interactions, polymer crystallites, or physical entanglements of individual polymer chains.

TABLE 10.1

Advantages, Disadvantages, and Use of Various Biomaterials

	Natural Polymer	Synthetic Polymer	Composites	Hydrogels
Advantages	1. Biocompatibility 2. Bioactivity	1. Possibility of modulating porosity and mechanical properties during synthesis process	1. Biocompatibility 2. Good mechanical strength	1. Biocompatibility 2. Controlled *in vivo* biodegradation 3. Possibility to modulate the parameters (e.g. cross-linking, density, porosity, and pore size)
Disadvantages	1. Poor mechanical properties 2. Fast biodegradation	1. Low biocompatibility 2. Low mechanical strength	1. Processing difficulties	1. At all times do not offer the requisite physiochemical and biological properties for most favourable cell development
Use in TE	1. Bone and cartilage 2. Tendon and ligament	1. Cardiovascular prosthesis 2. Bone cements	1. Hard and soft tissue	1. Hard and soft tissue

Hydrogels have been of immense concern to researchers for many years because of their hydrophilic nature and potential to be biocompatible. In addition, polymeric hydrogels are a striking topic of interest in the field of TE as matrices for repairing and regenerating a variety of tissues and organs (Zhang and Khademhosseini, 2017).

Classification of Hydrogels

They can be categorised based on the ionic strength, method of preparation, physical structures, sources, biodegradability, nature of swelling with changes in the environment, nature of crosslinking, etc. (Figure 10.1) (Peppas and Hoffman, 2020).

Nature of Crosslinking

One of the key classifications of hydrogels is based on their cross-linking nature. Hydrogels can hold the network strength in their swollen condition due to the episode of chemical or physical crosslinking. The use of chemically cross-linked hydrogels is commonly inadequate due to dearth of fabricability and post-process modifications. Nevertheless, physically cross-linked hydrogels uphold their physical permanence due to the presence of amendable physical junction domains coupled with hydrogen bonding, chain entanglements, crystallinity hydrophobic interactions, and ionic complexation (Park and Bae, 2002).

Smart Hydrogels

They are named smart hydrogels because of their reaction to the outside milieu. These polymers demonstrate substantial changes in their swelling performance, network arrangement, sol–gel changeover, permeability potency, and mechanical strength in response to small changes in the outside environment, i.e. ionic strength, pH, temperature, and salt type. Smart hydrogels can be additionally classified as chemical, physical, and biochemical responsive hydrogels (Mantha et al., 2019).

(a) pH-responsive hydrogel: They accept or donate protons in response to a small change in pH (Xu et al., 2018)
(b) Temperature-responsive hydrogels: They swell or shrink in response to even minute changes in the proximate temperature (Klouda, 2015; Huang et al., 2019)
(c) Glucose-responsive hydrogels: These are in fact insulin carriers and/or glucose oxidase mixture. Glucose oxidase is the enzyme that oxidises glucose into gluconic acid, which changes the pH of the system, thereby facilitating controlled delivery of insulin from pH-susceptible hydrogels (Gu et al., 2013; Dong et al., 2016)
(d) Antigen-responsive hydrogels: They are used for targeted delivery. Antigens are grafted on hydrophilic polymers to deliver biomolecules at an unambiguous site (Lu et al., 2003)
(e) Light-responsive hydrogels: These polymeric networks have a photoreceptive functional moiety that responds to light (Qiu and Park, 2001; Ter Schiphorst et al., 2014)
(f) Magnetic-responsive hydrogels: These hydrogels are able to amend their properties even in the presence of a minute change in electric stimuli or outside fields (Araujo-Custodio et al., 2019; Ghadban et al., 2016)

Based on Source

Hydrogels can also be classified on the basis of their origin into natural and synthetic. Natural hydrogels are obtained from a non-manmade source; hence, they are biocompatible and bioactive, whereas synthetic hydrogels are produced using chemical polymerisation techniques and they are preferred due to high durability, strength, and capacity for water absorption. Hybrid hydrogels are used in response to maintain the chemical and mechanical properties without affecting biocompatibility (Khansari et al., 2017).

Based on Methods of Preparation

Hydrogels can be categorised as homopolymers (Jeong et al., 2012), copolymers (Mathur et al., 1996), and interpenetrating network polymers (Miyata, 2010). This cataloguing is based on the nature, number, and ratio of monomers. Homopolymers comprise a single type of monomer units, whereas copolymer hydrogels contain two dissimilar monomers, out of which one monomer should be hydrophilic. Interpenetrating polymeric network hydrogels are composed of two intertwined polymer networks and there should be no chemical linkage between these polymers.

Based on Ionic Charge

Hydrogels can be classified as non-ionic (do not have any charge on the backbone or side groups), cationic (have a positive charge), and anionic hydrogels (have a negative charge) on the basis of the presence or absence of an electrical charge in crosslinked chains (Mun and Khutoryanskiy, 2015).

Methods of Preparation for Hydrogel

In general, three main components of the hydrogel preparation are monomer, initiator, and cross-linker (Figure 10.2). Any technique that is used to put in order a cross-linked polymer is used to prepare a hydrogel also. Preparation methods of both natural and synthetic origin hydrogels are described in Figure 10.3 (Ahmed, 2015).

Different techniques that can be used for the preparation of hydrogels are illustrated in Figure 10.4. Small multifunctional molecules (monomers and oligomers) are brought together to prepare covalently cross-linked hydrogels (Lee and Mooney, 2001; Zhu and Marchant, 2011).

Physical hydrogels are considered to be most suitable for cell and sensitive molecules as their preparation method is devoid of any chemical reaction (Lee and Mooney, 2001). In fact, the method of preparation is considered obligatory amid the application of hydrogel, i.e. type of hydrogel as well as the site where it has to be used. Based on these impediments, components are sorted and synthesised (e.g. mechanical strength of gel can be fortified by varying the mechanical properties) (Ghobril and Grinstaff, 2015).

The predominantly considered structural elements of hydrogel are the total quantity of water that hydrogel can absorb and binding of polymer chains in the gel network (Mantha et al., 2019).

Hydrogels in TE must meet the requirements of design criteria for proper functioning. The mechanism of gelling is one of the criteria that should be considered (Lee and Mooney, 2001). Polysaccharides (9, 63 Baier et al., 2003, Kim et al., 2008) (hyaluronic acid, chitosan), fibrous proteins (Glowacki and Mizuno, 2008; Sakai et al., 2009) (collagen, gelatin), protein/polysaccharide hybrid polymers (chitosan/gelatin), and DNA (Xing et al., 2011) are four main classes of natural materials used in hydrogel preparation (Zhu and Marchant, 2011).

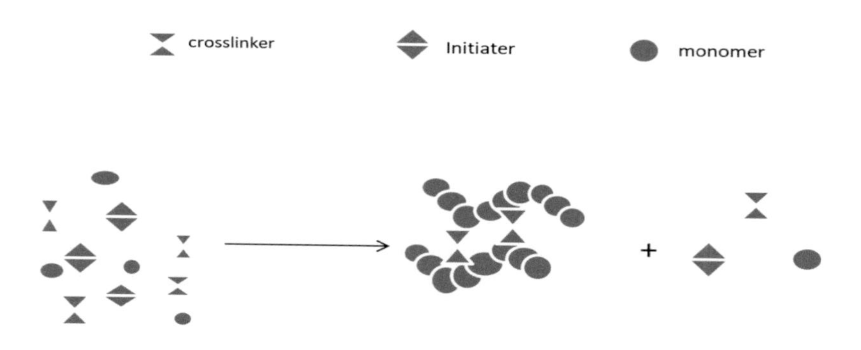

FIGURE 10.2 General Mechanism of Hydrogel Preparation.

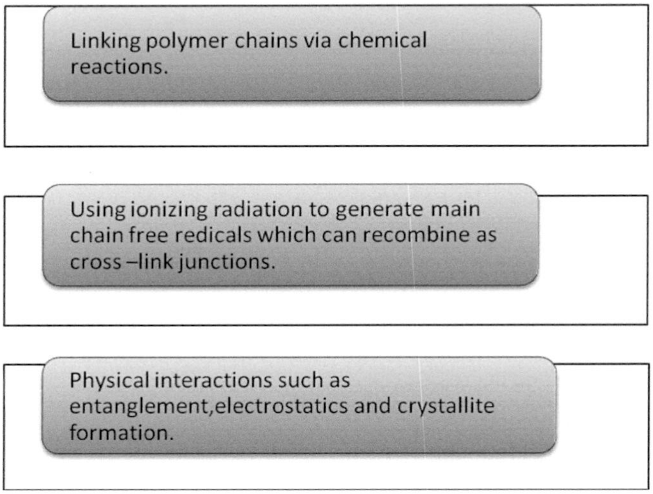

Linking polymer chains via chemical reactions.

Using ionizing radiation to generate main chain free redicals which can recombine as cross –link junctions.

Physical interactions such as entanglement,electrostatics and crystallite formation.

FIGURE 10.3 Basics of Preparation Methods of Hydrogels.

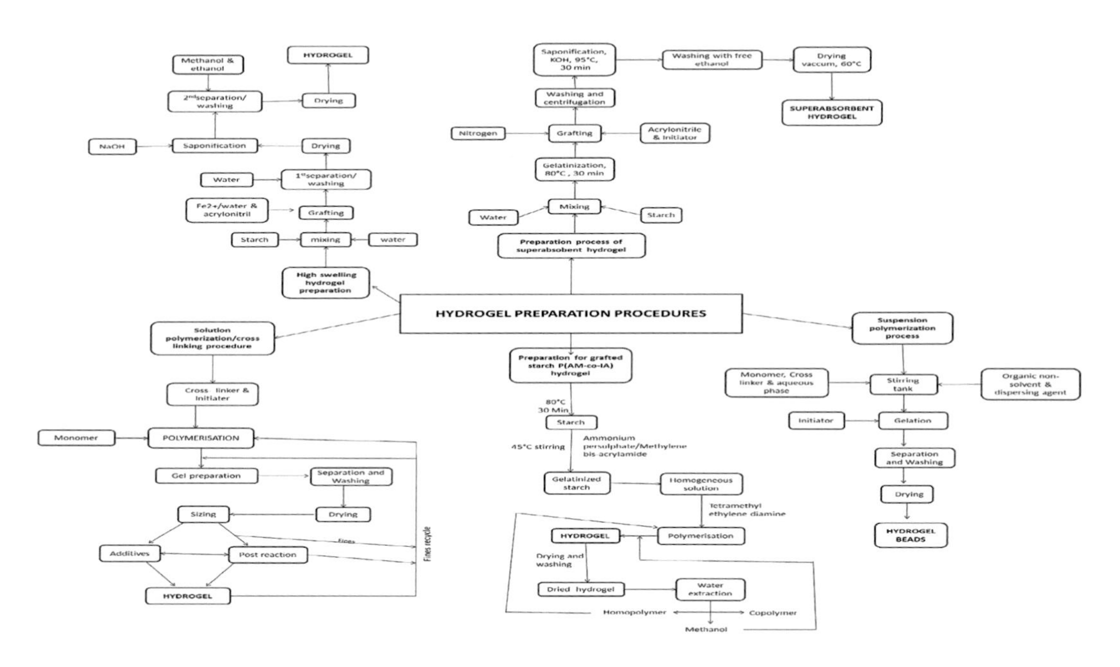

FIGURE 10.4 Methods of Preparation of Hydrogels.

Characterisation of Hydrogels

The characterisation (Mantha et al., 2019; Onaciu et al., 2019) of hydrogels include parameters such as biocompatibility (Luo et al., 2000), degree of crosslinking, swelling properties (Larrañeta et al., 2018), rheology (Fanesi et al., 2018), degradation profile (Pradhan et al., 2017), mechanical properties (Glowacki and Mizuno, 2008), porosity, permeation (Chirani et al., 2015), and thermal behaviour (Wang et al., 2018) (Figure 10.5).

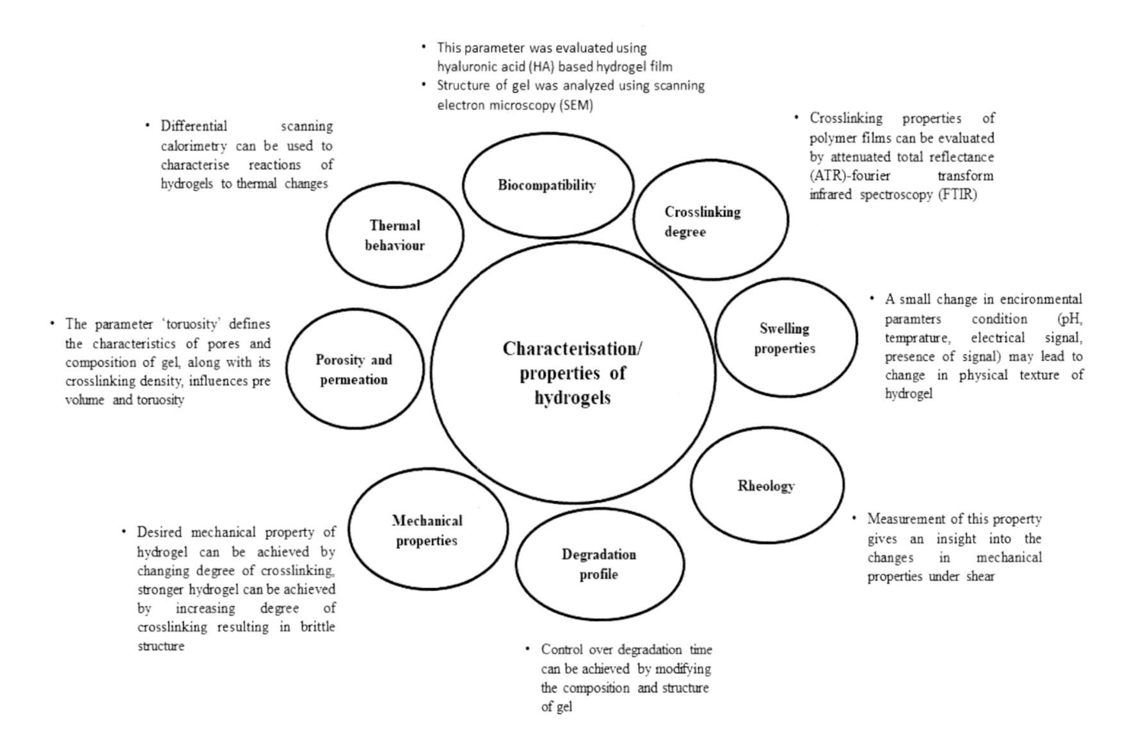

FIGURE 10.5 Characterisation of Hydrogels.

Types of Tissue Engineering

TE aims to investigate the intensification and remodelling of tissues using different approaches from an array of fields. Recent developments in collaborative discipline of TE have yielded many tissue replacements. Systematic and methodological advances in stem cell research and bioinspired environments have shaped exceptional opportunities to engineer tissues in the laboratory.

Amongst different biomaterials, the market contribution of hydrogel is likely to witness a sturdy escalation for the duration 2017–2022. The global market size of hydrogels summed up to $16.5 billion in 2017 and is likely to be $22.3 billion by 2022, growing at an AAGR of 6.3% from 2017–2022 (researchandmarkets.com). The key players running the global hydrogel market are Johnson & Johnson Pvt Ltd. (Nu-Gel Hydrogel; WC983, Acuvue® Oasys silicone gel-based contact lense), Axelgaard Manufacturing Co. Ltd. (AmGel hydrogels; AG500, AG600, A9500, AG2500), Cardinal Health (Cardinal Health™; Kendall™), The Cooper Companies, Inc. (Biofinity™; Silicone Hydrogel Contact Lens), Dow Corning Corporation (Dow Corning®; 9509 Silicone Elastomer Suspension), R&D Medical Products Inc. (Comfort™ Gel), and BSN Medical (Cutimed®, Sorbact® gel), among others (marketsandmarkets.com).

Cornea TE

TE approaches have been useful in generating viable cornea tissue equivalents from engineering of epithelium, stroma, and endothelium layers to rebuild the inhabitants. Due to shortage of donor cornea, tissue engineered human corneal endothelial cell monolayer can be implanted on a carrier, which is used as only a solution for corneal transplantation in corneal blindness (Rizwan et al., 2017) (Figure 10.6).

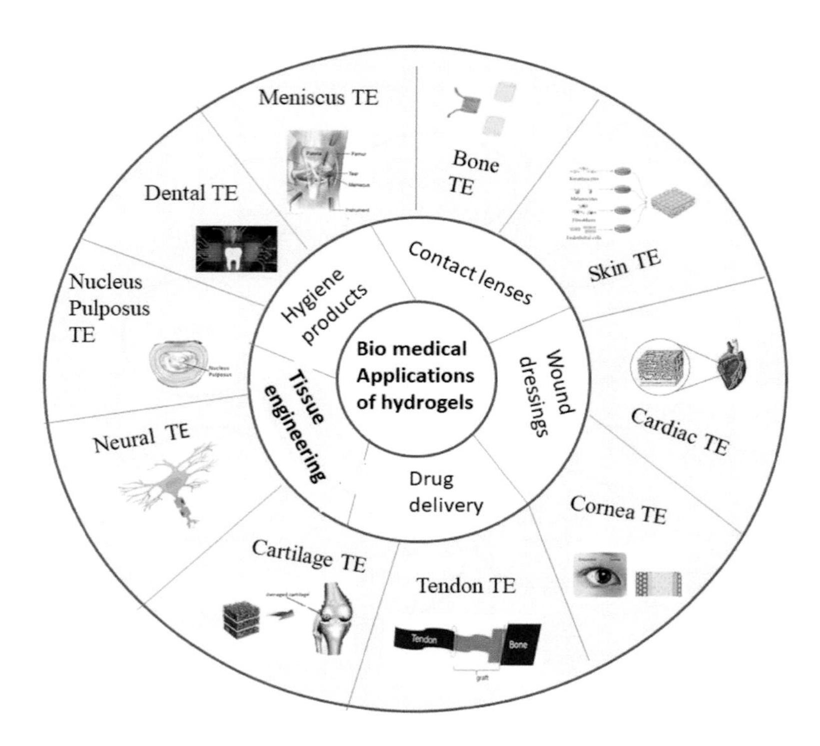

FIGURE 10.6 Applications of Hydrogels for Tissue Engineering.

Tendon TE

Tendon-related diseases or injuries are very common worldwide, which include processes such as tendon suturing and reattaching, tendon transplantation of healthy tissue to a distorted, and lost or malfunctioning part of the tendon. Pluronic is a thermosensitive synthetic polymer, which can be injected *in vivo* to fill a tissue defect and undergoes gelation at body temperature with minimum invasion. Multilayered scaffolds have also been used for tendon TE, which are based on electrospun substrates coated with cell-laden hydrogel layers (Rinoldi et al., 2019).

Skin TE

Tissue-engineered skin has been used as an encouraging projection to restore various skin defects (Dhandayuthapani et al., 2010; Priya et al., 2008). Several types of tissue engineered scaffolds, e.g. gelatin methacrylamide hydrogels have been used in skin restoration by the development of stratified epidermis with a definite barrier role (Narayanan et al., 2019). Bioprinting has been used for the fabrication of multilayered skin, cartilage, and liver constructs due to its great potential. Bioprinting of skin constructs involves the fabrication of hydrogel constructs containing different skin cells (keratinocytes and fibroblasts) and *in situ* printing of skin cells and biomaterials over a wound site (Ng et al., 2016).

Neural TE

Damage to the central nervous system (CNS) may lead to rigorous neurological dysfunction because of neuronal cell death and axonal degeneration. The ability of neurons to restore following damage is very stumpy in adults. Neural TE approaches focus on developing scaffolds that synthetically create complimentary cellular microenvironments that support the renewal of neural tissue (Willerth and

Sakiyama-Elbert, 2007). Various hydrogels have been successfully used for neural TE (Nisbet et al., 2008), e.g. silk scaffolds (Kaplan, 2012; Kim et al., 2020). Because of the slow degradation rate, silk hydrogels are capable of giving adequate time for restoring nerve damage *in vivo* (Hopkins et al., 2013). Another example is the use of photo cross linkable chitosan (Valmikinathan et al., 2012). This polymer possesses a number of advantages, such as minimal quantities of cytotoxic chemicals are used in the production process and possibility of encapsulation of cells or bioactive agents within the gel (Schmidt and Leach, 2003). Laminin-1-functionalised methylcellulose (a temperature-responsive hydrogel), when injected in the CNS wound cavity, go through sol–gel transitions, and negligible invasive delivery becomes possible because of *in situ* gelling (McIntosh et al., 1996; Stabenfeldt et al., 2006).

Meniscus TE

Injury to or degeneration of the meniscus leads to compromised meniscus function, which is a reason of progression of knee osteoarthritis. A number of TE approaches have been used to enhance replacement and repair of damaged meniscus such as meniscus allograft transplantation to replace lost meniscal tissue and direct replacement of meniscal tissue by natural or synthetic biomaterial scaffolds (An et al., 2018). Polylactic acid (PLA) has great prospective in meniscal repair devices because of its sufficiently long half-life (6 months) as the time period required for meniscal repair is 6–12 weeks. Along with this, PLA also have other advantages such as biocompatibility, suitable degradation time, and strength due to which it may take off the role of indigenous extracellular matrix proteins and provides support to normalise tissue development and restoration (Baek et al., 2015; Katz and Scott, 2009)]. Another approach used in meniscal TE involves injectable photo crosslinked cell-based collagen scaffold, in which photo crosslinking has been induced by riboflavin, has great potential for utilization in meniscus regeneration (Heo et al., 2016).

Nucleus Pulposus TE

Nucleus pulposus (NP) degeneration results in lower back pain, which can be treated by various TE approaches. Tissue-engineered construct or injectable implants can be used in cases of healthy annulus fibrosus. They can reduce pain while concurrently restoring spinal mobility and impeding disc deterioration. Alginate and agarose are capable of maintaining cell phenotype and hence are frequently used scaffolds for NP cell-seeding (Cloyd et al., 2007). Notochordal cells can also be used either as the primary source of stem cells or as an organiser cell in this approach (Hunter et al., 2003).

Dental TE

In dentistry, root canal therapy (RCT) is commonly performed. During RCT, the removal of tender or necrotic dental pulp is carried out followed by replacement with a synthetic material. On the contrary, current investigations provide substantial data on the possibility of dental pulp and dentin engineering. As TE approach holds the promise to override conventional RCT, there is a need for customised scaffolds for the same. Various investigations in recent times have demonstrated the capability of dental pulp stem cells (DPSCs) along with a scaffold material to produce soft connective tissue, which is analogous to dental pulp (Cordeiro et al., 2008; Huang et al., 2010; Iohara et al., 2009; Nakashima and Iohara, 2011; Prescott et al., 2008; Sakai et al., 2010).

The proliferation rates of DPSCs were found to improve in peptide gels containing the RGD motif, where cell adhesion is considered as a condition of proliferation (Galler et al., 2010). At the same time, DPSCs was found to have adequate spreading and formation of a collagenous matrix. For significant clinical applications, two parameters are of utmost importance, i.e. viscoelastic properties and sheer recovery. By changing the sequence, crosslinking and stiffness can be modified. Viscoelastic properties are modified especially in cases where the hydrogel has to be injected into the root canal using a syringe (Lymperi et al., 2013).

Bone TE

A range of hydrogels have been used for the restoration of imperfect osteochondral interface or articular cartilage tissues; even then it is very challenging to refurbish these interfaces and defects regardless of incredible growth in the discipline of regenerative medicine. This is mainly due to the lack of availability of suitable as well as correct tissue-engineered biopolymers that have the capability of restoring the injured site and stimulating tissue regeneration (Alsberg et al., 2001; Park, 2011).

Cartilage TE

The formation of cartilage takes place in the embryonic stage. It is avascular viscoelastic connective tissue, of low cell density as well as low proliferative activity. Due to these properties, the cartilage is almost incapable to regenerate after injury or degeneration, e.g. after osteoarthritis. Cartilage engineering involves *in vitro* techniques that involve neocartilage engineering (Camarero-Espinosa et al., 2016).

TE of the cartilage has been progressively chased for about 20 years; however, hardly any of the research studies have reached the stage of clinical trials. In the early 1990s, Brittberg and Peterson used analogous chondrocyte implantation for restoring the cartilage (Brittberg et al., 1994).

In the 1950s, Pridie and Ficat reported a technique of removing a dead or infected tissue for wound healing (i.e. debridement) (Ficat et al., 1979). For the treatment of microfractures, debridement of the cartilage was done followed by grooving into the subchondral bone. However, this was not an adequate method. Hydrogels, porous foams, or fibrous matrices have been used for the preparation of these types of scaffolds. The hydrogels in cartilage TE permit the flow of nutrients and waste, encapsulation of cells, and the transduction of mechanical loads.

Cardiac TE

Heart failure is one of the critical diseases that causes maximum deaths globally. Heart transplantation is considered to be the unsurpassed treatment for the last-stage heart failure patients. Because of the restricted and inadequate repository of donor hearts (Barr and Taylor, 2015; Kikuchi and Poss, 2012; Macdonald et al., 2015; Prabhu and Frangogiannis, 2016) and the indigent regenerative ability of the myocardium, researchers have moved towards the therapeutic approaches which ameliorate myocardial function.

Stem cell therapy and TE approaches have been used by various researchers and clinicians for myocardial support as well as for regeneration (Vizzardi et al., 2012). Amongst various approaches, a few popular TE techniques are the implantation of scaffold-free cell sheets, (Masuda et al., 2008) *in situ* engineering, (Tokunaga et al., 2010), and biomaterial implants (Wang et al., 2018). In Table 10.2, the research work of a few researchers pertaining to TE has been compiled.

Patents on Hydrogels

A considerable number of patents have been published involving applications of hydrogels. Recently, hydrogels have been used as delivery vehicles for biomolecules or as space-filling agents in TE. Space-filling agents are a group of scaffolds that are used commonly for bulking and to prevent adhesion (Drury and Mooney, 2003). The patents provide a detailed description of applications of hydrogels in TE, from the time when the pioneering work was done by Wichterle to most recent hydrogel-based inventions and products in the market. They also depict the progress in the use of hydrogels (originally defined in 1988). Various alternatives such as metals, ceramics, and synthetic polymers have been tried as scaffolds for many years (Li et al., 2009).

Various patents describing hydrogel inventions have been published since 1998 until today. However, recent patents pertaining the formation of new tissues, along with drug delivery, gene delivery, regeneration of tissues, organ transplantation, and wound healing, possess many other applications in TE that can explain great developments in the field of hydrogels. Various patents describing hydrogel applications are summarised in Table 10.2.

TABLE 10.2
Research Work on Hydrogels in TE

Sr. No.	Polymer Used	Methods Used	Remarks	Applications	Reference
1	Poly(2-hydroxyethyl methacrylate), uncrosslinked polymethyl methacrylate microspheres	Sphere-templating method; controlled pore structure, size, and interconnectivity; mixture with a new photo-patterning procedure; autonomous control over the macro architecture of scaffolds	New and straightforward technique; designing of porous, degradable hydrogel scaffolds	Supplementary means to modify the three-dimensional scaffold structure of a cell or tissue	(Bryant et al., 2007)
2	Hyaluronic acid (HA), chitosan, gelatin, conventional alginate, and agarose	Formulation A (1% HA solution + 7% polyethylene glycol-g-chitosan solution): Gelation technique; formulation B (2.6% HA solution + 7% polyethylene glycol-g-chitosan solution): Gelation technique; formulation C (moderately oxidised hyaluronan and gelatin): Swift mixing of equal parts of both solutions followed by direct injection into a mould	Formulation A exhibited 48% grafting; injectable hydrogels; choice of implants must be done on the basis of similarity of mechanics of the implanted material to that of indigenous tissue to imitate *in vivo* performance .	For nucleus pulposus replacement	(Cloyd et al., 2007)
3	Poly(2-hydroxyethyl methacrylate) (PHEMA)	Polymerisation by free-radical mechanism; triethylene glycol dimethacrylate: cross-linker; ammonium persulfate, and sodium bisulphate: Initiator systems	Bioresorbable pHEMA hydrogel	Cardiac TE	(Atzet et al., 2008)
4	Polyethylene glycol (PEG), chitosan, collagen, carbodiimide, 1- ethyl-3-(3-dimethyl amino propyl) carbodiimide (EDC), and N-hydroxy succinimide (NHS)	Simple carbodiimide cross-linker Hybrid cross-linking system comprised bi-functional cross-linkers; used either long-range or short-range bi-functional cross-linkers	Biocompatible hydrogels; improvement in mechanical strength and elasticity; good optical properties; good suturability or surgical seaming and better permeability	Cornea TE	(Rafat et al., 2008)
5	PEG, HA	Three-dimensional cell encapsulation; *in situ* free radical polymerization	Maximum cytocompatibility; minimum number of processing steps;	In general TE	(Khetan and Burdick, 2009)

(Continued)

TABLE 10.2 (Continued)

Sr. No.	Polymer Used	Methods Used	Remarks	Applications	Reference
6	Alginate	Combination of cell and ionic cross-linking techniques	Cross-linking improved the viscoelastic properties; no inflammation in regenerated tissues; amplification in total reconstructed cell volume	Cartilage TE	(Park et al., 2009)
7	Agarose	Peptide scaffold encapsulation; peptide sequence: Ac-RADARADARADARADA-$CONH_2$	Shape and cell viability were retained in peptide hydrogels; stable in physiological environment; capable of producing extra cellular matrix similar to the cartilage	Cartilage engineering	(Liu et al., 2010)
8	Fibrinogen	Crosslinking of PEGylated fibrinogen using photo initiation	Biological viability because of the existence of fibrinogen backbone; regulation of cellular features would be possible via structural modifications of the PEG–fibrinogen scaffold	For TE	(Dikovsky et al., 2008; Frisman et al., 2012)
9	Polyvinylpyrrolidone, PEG	Crosslinking by means of Michael-type reaction	Increase in the firmness of PEGylated hydrogels after the incorporation of carbon nanotubes in polymeric milieu; cytocompatible nano composites; PEG–carbon nanotubes resulted in comparatively softer and more swellable hydrogels	In general TE	(Van den Broeck et al., 2019)
10.	Chitosan, HA, sodium alginate	Coagulation and wet-spinning method	Hybridisation of HA improved the tensile strength of chitosan; hybridization of chitosan did not improve the tensile strength of alginate; type I collagen was found to increase with chitosan–HA scaffold; increased type II collagen in the pericellular matrix	Musculoskeletal TE; feasibility in cartilage regeneration	(Majima et al., 2007)
11	Dextran, hyaluronic acid sodium salt, hydroxyethyl methacrylate	Functionalising dextran followed by photoinitiation	Elasticity increased with increasing HA concentration; formulated hydrogels could deform effortlessly; illustrated similar viscoelastic behaviour as that of natural tissues	Appropriate for TE	(Pescosolido et al., 2011)

12	Collagen	Injectable hydrogels; meniscus ECM (mECM) hydrogel	Shielded pathologic calcification along with tissue regeneration; vehicle to retain mesenchymal stem cells (MSCs) within the intra-articular space for repairing meniscus; repository of indigenous tissue signals to normalise performance and development to fibrocartilage	Tissue-specific decellularised mECM hydrogel	(Yuan et al., 2017)
13	Chitosan powder, graphite powder	Casting of films was done using homogeneous dispersion of lactic acid and mixture of graphene and chitosan; graphene oxide and lactic acid are used as cross-linkers	Composites showed good mechanical properties and reserved the swellability of the polymer matrix	Conductive, processable, biocompatible hydrogels with promising use in TE	(Sayyar et al., 2015)
14	Methacrylated kappa – carrageenan (k–CA)	Combination of chemical and physical cross-linking; photo cross-linkable hydrogels	Cross-linking was linked to the degree of hydration, mechanical and rheological properties	Can be used for customised TE	(Mihaila et al., 2013)
15	Cellulose	Carboxylated cellulose nanofibrils; hydrogelation using metal salt solution	Biocompatible and biodegradable nanofibrils	Potential for biomedical applications in TE	(Benselfelt and Wågberg, 2019; Dong et al., 2013; Zander et al., 2014)
16	Poly (N-isopropylacrylamide) (PNIPAAm)–gelatin	Free radical polymerization	Injectable hydrogel; gelatin provided adequate cell adhesion, better biological activity, and influenced water retention	Cardiac TE	(Navaei et al., 2016)]
17	Gelatin with methacrylamide (GelMA)	Photocross-linking	Concentration of GelMA prepolymer solution affected the physical and biological properties; modifiable mechanical and degradation characteristics	Skin TE and as wound dressings	(Zhao et al., 2016)
18	Poly-ε-caprolactone (PCL), methacrylated gelatin (mGLT)	Dual electro spinning; free radicals generation lead to cross-linking during the processing	Structural similarity with that of the indigenous structure of tendon; biocompatible composite	For tendon TE	(Yang et al., 2016)

(Continued)

TABLE 10.2 (Continued)

Sr. No.	Polymer Used	Methods Used	Remarks	Applications	Reference
19	Xylan, chitosan	Xylan–chitosan hydrogel (1:3); chitosan hydrogel (thermogelling properties)	The composite hydrogel ameliorated the curing of traumatic tibia and femur fractures; no exogenous proteins, growth factors; ease in the application because of injectable form	Bone TE (healing of bone fracture)	(Bush et al., 2016)
20	Reduced graphene oxide, gelatin methacryloyl	Ascorbic acid was used as a reducing agent to overcome previous limitations of toxicity	Conductive hydrogel; synchronous beating activity observed	Potential in cardiac TE	(Shin et al., 2016)
21	PEG	Michael-type addition	Improved porosity; controlled mechanical and swelling properties	As TE scaffolds	(Frisman et al., 2012; Yu et al., 2016)
22	Collagen	Photo-oxidation using riboflavin as a photosensitiser	Improved mechanical properties; delayed degradation rate due riboflavin; no cytotoxicity; lesser gelation time	Meniscus TE	(Heo et al., 2016)
23	Lignin, PVA (polyvinyl alcohol)	Crosslinking method; for production of lignin-based hydrogels, quantitative design of experiments, i.e. three-level-two-factorial design in combination with response surface methodology has been used	Concentration of lignin influenced swelling capacity as well as the porosity of hydrogels	Promising role of lignin for hydrogel-based TE	(Morales et al., 2020)
24	Chitin nanocrystals (ChNCs), Methylcellulose (MC)	Mechanical mixing followed by physical cross-linking	Amalgamation of MC with β-ChNCs lead to hydrogel formation; no hydrogel formation with α-ChNCs; improved mechanical strength and gelling rate; chitin improved the compression strength of composite	Potential of chitin for TE scaffolds	(Jung et al., 2019)
25	Bentonite nanoparticles, gelatin methacryloyl	Sonication followed by polymerisation;	Porosity and cell viability of hydrogels was found to be bentonite-dependent	Promising role of bentonite for TE	(Sakr et al., 2020)

TABLE 10.3
Summary of Patents

Sr. No.	Patent No.	Polymers/Components	Application in TE	Patent Date	Remarks	Reference
1	5,836,313	Cross-linkable polymer; polyethylene oxide or hydroxyl methyl methacrylate	Corneal epithelial cell growth	17 Nov, 1998	Applications in keratoprosthesis	(Perez et al., 1998)
2	6,027,744	Calcium-alginate; Pluronic™:	Permeable, biocompatible hydrogel for replacement TE; better tissue strength	22 Feb, 2000	Programmed or pre-decided shape can be achieved; tissue precursor cells are incorporated in hydrogel	(Vacanti and Vacanti, 2000)
3	US 6,268,405 B1	Polyvinyl alcohol	Used as a substitute and for the expansion of facial, head, and cranial bones	31 July, 2001	Novel porous and solid hydrogels	(Yao and Swords, 2001)
4	US 7,396,537 B1	Polytetrafluoro ethylene	Application in enlargement of the heart ventricle; cell delivery patch for engineering of myocardial tissue	8 Jul, 2008	Role of allograft parenchymal cells in direct allorecognition	(Krupnick et al., 2008)
5	US20080193536A1	Alginate, collagen, HA, agarose, chondroitin sulphate	Microscale cell-laden hydrogels with capability to self-assemble	14 Aug, 2008	Three-dimensional scaffolds of cell-laden hydrogels; for organ/tissue formation in biomedicine	(Khademhosseini et al., 2008)
6	US20100179659A1	Self-assemble peptide, fibrin, alginate, agarose, PEO (polyethyleneoxide), PEG, collagen (type 1 and type 2,) gelatin, alginic acid, chitin, chitosan	Natural tissue composition; generated from a construct made up of subject's own cells; combination with scaffold for replacement of defective tissue	15 Jul, 2010	Tissue-engineered intervertebral disc comprising at least one inner layer (hydrogel) and an exterior layer (nanofibrous)	(Li et al., 2010)
7	US 7,790,194 B2	Combination of acrylamide and methylene bis-acrylamide	For soft tissue augmentation, e.g. facial cosmetic surgery, lip augmentation, and soft tissue correction of body	7 Sep, 2010	Methods for filling a soft tissue in mammal using endoprosthesis, a prosthetic device	(Petersen, 2010)

(Continued)

TABLE 10.2 (Continued)

Sr. No.	Patent No.	Polymers/Components	Application in TE	Patent Date	Remarks	Reference
8	US 8,039,258 B2	Polyamide, methylcellulose, collagen, extracellular matrix, polysaccharides such as alginates, polyacrylates that are cross-linked to polyethyleneoxide–polypropyleneglycol block copolymers	As a template for cells to grow and produce new tissue	18 Oct, 2011	A microporous scaffold and a non-fibrous, non-porous hydrogel formed from a self-assembling peptide	(Harris et al., 2011)
9	US 20110288199A1	Polyvinyl alcohol (PVA)	Suitable for repairing or replacing musculoskeletal tissues and/or fibrocartilage such as ligaments and tendons	24 Nov, 2011	Tailored fibre-reinforced hydrogel composites for implantation into a subject	(Lowman et al., 2011)
10	US 20130236971A1	One or more PVA polymers and one or more phenylboronate containing copolymers which is composed of phenylboronate ligands (e.g. 4-vinylphenylboronic acid, N-acryloyl-3-aminophenylboronic acid), acrylic monomers (acrylamide, methacrylamide, methacrylic acid), alkaline tertiary amines (e.g. N,N-dimethylaminoethylmethacrylate, N,N-dimethylaminopropylacrylamide)	Can be used for tissue reconstruction such as cell and tissue grafting, skin grafting, wound healing grafts, skin replacement, cardiac patching, cartilage tissue repair, and reconstruction	12 Sep, 2013	Tissues emanate from one or more various cell types (fibroblasts, keratinocytes), method for obtaining a sheet of cells separated from hydrogel, administering the regenerated cells	(Kumar, 2013)
11	US 8,664,202 B2	High molecular weight components such as polyglycans (dextran, alginate) and polypeptides (gelatin, collagen), enhancing agents such as amino acids (e.g. cysteine, arginine, glutamic acid, or their combination)	Used as cell attachment scaffold that promotes wound healing, tissue repair, tissue regeneration, wound adhesive, for tissue bulking, for site-specific tissue regeneration including vasculogenesis	4 Mar, 2014	Cross-linked hydrogel matrix stabilised by enhancing the stabilising agent, hydrogel is mixed with other materials to form castable structures	(Lamberti et al., 2014)

12	US 20140178344A1	Derivatives of (tetrahydropyranyl) methyl and (tetrahydrofuranyl) methyl-acrylamide	Acrylamide hydrogel as scaffold material may be used for wound healing, tissue regeneration, artificial organ growth, vascularisation and creation of new tissues, repair of wounded tissue, and to replace non-functional tissues	26 Jun, 2014	Drugs entrained within hydrogel, invention provides methods of wound healing or tissue regeneration	(Carlson and Phelan, 2014)
13	US 20140186460A1	Acellular tissue matrices	As implants for the face or neck; as tissue fillers for tissue regeneration after loss of bulk soft tissue, for tissue repair or treatment	3 July, 2014	Method comprised harvesting an arterial tissue, decellularising the tissue and treating the arterial tissue with an elastase, method of treating a tissue after the removal of bulk soft tissue	(Wenquan et al., 2014)
14	US 8,883,503 B2	Synthetic terpolymers complexed with PVA polymer, phenylboronate-containing copolymers composed of one or more phenylboronate ligands (4-vinylphenylboronic acid, N-acyloyl-3-aminophenylboronic acid), one or more acrylic monomers (e.g. acrylamide, N-isopropylacrylamide) and one or more alkaline tertiary amines (e.g. N,N-dimethylaminoethyl methacrylate)	One or more cell layers are suitable for cell and tissue grafting, skin grafting, skin replacement, liver tissue reconstruction, bone tissue repair, and reconstruction, thyroid tissue reconstruction or any combination thereof	11 Nov, 2014	Hydrogel scaffolds produced, provided methods for culturing cells on a PCC-PVA hydrogel, forming one or more cell layers; the intact hydrogel with regenerated cells or tissues administered to the subject	(Kumar, 2014)

(Continued)

TABLE 10.2 (Continued)

Sr. No.	Patent No.	Polymers/Components	Application in TE	Patent Date	Remarks	Reference
15	US 20150147397A1	Chitosan, collagen, elastin, fibrin, alginate, gelatin.	Used as a bone filler, useful for treating bone oedema, useful for wound healing, useful for enhancing bone formation, cartilage formation	28 May, 2015	Hydrogel biomatrix comprised a marine skeletal derivative biomatrix; organism skeletal derivative is aragonite, used for wound healing, orthopaedic application, dental applications, carniofacial skeletal surgery, three-dimensional printing methods	(Altschuler, 2015)
16	US 9,782, 517 B2	Hyaluronic acid component cross-linked to collagen component	Augmented soft tissue of a human being; used to enhance, promote, or support cell proliferation	10 Oct, 2017	Methods of augmenting soft tissue comprise injecting or implanting a hydrogel	(Pollock et al., 2017)
17	US 9,610,353 B2	Fibrous network of a plurality of peptides (e.g. phenylalanine-phenylalanine dipeptide, napthylalanine–napthylalanine dipeptide) and biocompatible polymer (hyaluronic acid, chitosan)	Tissue regeneration	4 Apr, 2017	Method of repairing a damaged tissue, each peptide in the plurality of peptide is an end-capping modified peptide, method comprising contacting the damaged tissue with hydrogel	(Aviv et al., 2017)
18	US 20180346902A1	Polymer comprising PEG (polyethyleneglycol), agarose, collagen, fibrin, polyacrylamide	For tissue-specific materials	6 Dec, 2018	The integrin heterodimers bound to peptides represented binding motifs in two or more of full-length proteins, the hydrogel comprised cells from the tissue of origin; method for making hydrogel tissue specific, allowing for substrates	(Jansen and Peyton, 2018)

19	US 20180207319A1	Thermosensitive nanocomposite composed of cross-linked polymeric matrix of protein (keratin, collagen, gelatin) and a conjugated copolymer (e.g. poloxamer, PEG)	Cartilage TE	26 Jul, 2018	Injecting a nanocomposite into a defect site of cartilage, forming the protein-conjugated copolymer by cross-linking, forming hydrogel via sol–gel transition technique by increasing the temperature of nanocomposite	(Eslahi and Simchi, 2018)
20	US 9,968,708 B2	Collagen, gelatin, elastin, polylactic-co-glycolic acid (PLGA), polyglycolic acid(PGA), polylactic acid (PLA)	Used for vascularisation of wound healing and tissue regeneration	15 May, 2018	Exhibit improved ability to facilitate cellular invasion and vascularisation for wound healing and tissue regeneration	(Fishman et al., 2018)
21	US20180372725A1	Gelatin, collagen, arginine, fibrin, fibronectin, glycoprotein, and glucose or their combination	Polymeric fibre-scaffolded engineered tissues for identifying compounds that modulate a contractile function	27 Dec, 2018	Devices, constructs, and methods of use of polymeric fibre-scaffolded engineered tissues	(Parker et al., 2018)
22	US 10,260,039 B2	Acrylate-PEG20k-acrylate, conjugated acrylate-PEG-streptavidin	Generation of three-dimensional micro tissue-based liver on a chip; invention of micro gels and micro tissues for TE	16 Apr, 2019	Micro gels or micro tissues assembled into larger ordered constructs that mimic *in vivo* tissue architecture, encoded microtissues are assembled onto DNA-patterned templates	(Bhatia and Li, 2019)
23	US 10,300,169 B2	Hyaluronic acid component cross-linked to silk fibroin	Used to augment soft tissue of human being, such as the skin, promote/support tissue viability, or proliferation, create space in tissue; used in soft tissue aesthetic product	28 May, 2019	Methods of promoting or supporting cell proliferation or survival, methods provided for preparing space in human/animal tissue	(Yu et al., 2019)

(*Continued*)

TABLE 10.2 (Continued)

Sr. No.	Patent No.	Polymers/Components	Application in TE	Patent Date	Remarks	Reference
24	US 20190192739A1	Fibrin hydrogel conjugated to one or more peptides of laminin-111; natural polymers (polypeptides, polysaccharides, alginate, collagen, fibrin), synthetic polymers (PEG, PVA, PLA)	Can be used as carriers for cell transplantations, scaffolds, barriers against restenosis, salivary TE	27 Jun, 2019	Composition comprising a fibrin hydrogel conjugated to one or more peptides of laminin-111; method of generating and repairing damaged salivary tissue	(Baker et al., 2019)
25	US 10,471,181 B2	Polymeric fibre included a biocompatible biodegradable polyester, polycaprolactone, poly(lactic acid), hydrogel material includes hyaluronic acid, poly(ethylene glycol), collagen, dextran, elastin, alginate, fibrin, etc.	A surgical scaffold device comprising laminar composite for purpose of reducing foreign body response, as an implant for tissue or cartilage repair; for reducing or reversing tissue defect (tissue defect include pleural tissue, muscle tissue, skin etc.)	12 Nov, 2019	Laminar sheet scaffold, sheet composed of surgical mesh sheet and hydrogel composite, invention provided an implantable material that included a scaffold complex; method for reducing or reversing a tissue defect; method for manufacturing a composite material for use in soft tissue healing; method for preparing an implant for tissue or cartilage repair	(Martin et al., 2019)
26	US 10,647,755 B2	Light-responsive protein hydrogels made of recombinant protein, extracellular matrix protein (e.g. collagen, gelatin, elastin), two or more cross-linkable proteins (e.g. SdyTag and SdyCatcher)	Used for cell encapsulation, culturing; selective release under appropriate light conditions; used to treat skin disorder	12 May, 2020	Method of encapsulating cells in a photo-responsive hydrogel matrix; provided evidence for culturing cells *in vitro* using light-sensitive protein hydrogels; composition of this invention for use in the treatment of skin disorders	(Sun and Wang, 2020)

27	US 20200397953A1	Shear thinning hydrogel comprising layer silicate (e.g. laponite) and an anionic polysaccharide (e.g. alginate)	Used in forming a submucosal cushion, which is used for removing protrusions (e.g. lesions such as polyps or tumours)	24 Dec, 2020	Methods of forming a submucosal cushion, separating portions of a tissues; methods of removing lesions, preventing or treating cancer	(Langer et al., 2020)
28	US 10,744,229 B2	Comprised functionalised tissue particles (e.g. devitalised tissue particles having amine group functionalised with thiol groups) and functionalised hyaluronic acid (e.g. chemically modified hyaluronic acid molecules that have been functionalised with 4-pentonate groups)	Used for bone regeneration in craniofacial bones; used for regeneration and repair of cartilage, tendon, and meniscus; to repair osteochondral defects	18 Aug, 2020	Methods of forming functionalised tissue particles and functionalised hyaluronic acid combined to form a devitalised tissue particle—functionalised hyaluronic acid hydrogel precursor composition that could be placed in an implant site	(Townsend and Detamore, 2020)

Summary

The triumph of many space-filling agents, use of bioactive molecules for drug release, and tissue constructs is exceedingly reliant on the blueprint of the scaffold. However, the blueprint of the same depends chiefly on the tissue as well as on its environment, e.g. if a scientist wishes to engineer a bone or a cartilage, the major area of concern is its load-bearing capacity of the engineered tissue. In addition, many materials (polymers for hydrogels) used for the TE until date have been used because they have the ability to mix with the cells, i.e. biocompatibility, irrespective of their mechanical properties or the rate of degradation. These properties are equally important as they also play a role in deciding the life of engineered tissues. It would be of great interest to users as well as consumers if these engineered tissues are associated with some electrical signals that dictate the properties such as degradation or change in mechanical strength.

REFERENCES

Ahmed E.M., 2015. *Journal of Advanced Research 6*(2): 105–121. doi:10.1016/j.jare.2013.07.006.

Alsberg E., Anderson K.W., Albeiruti A., Franceschi R.T., Mooney. et al. 2001. *Journal of Dental Research 80*(11): 2025–2029. DOI:10.1177/00220345010800111501.

Altschuler N. Biomatrix hydrogels and methods of use thereof. U.S. Patent 10342897, May 28, 2015.

An Y.H., Kim H.D., Kim K., Lee S.H., Yim et al. 2018. *International Journal of Biological Macromolecules 110*: 479–487. doi:10.1016/j.ijbiomac.2017.12.053.

Araujo-Custodio S., Gomez-Florit M., Tomas A.R., Mendes B.B., Babo et al. 2019. *ACS Biomaterials Science & Engineering 5*(3): 1392–1404. doi:10.1021/acsbiomaterials.9b00416.

Atzet S., Curtin S., Trinh P., Bryant S., Ratner et al. 2008. *Biomacromolecules 9*(12): 3370–3377. doi:10.1021/bm800686h.

Aviv M., Buzhansky L., Einav S., Nevo Z., Gazit E., Adler-Abramovich L. Malleable hydrogel hybrids made of self-assembled peptides and biocompatible polymers and uses thereof. U.S. Patent 9610353, Apr 4, 2017.

Baek J., Chen X., Sovani S., Jin S., Grogan et al. 2015. *Journal of Orthopaedic Research 33*(4): 572–583. doi:10.1002/jor.22802.

Baier Leach J., Bivens K.A., Patrick JrC.W., Schmidt C.E. 2003. *Biotechnology and Bioengineering 82*(5): 578–589. doi:10.1002/bit.10605.

Baker O., Nam K., Lei P., Andreadis S. Salivary tissue regeneration using laminin peptide-modified hydrogels. U.S. patent application 16/326326, Jun. 27, 2019.

Baroli B. 2007. *Journal of Pharmaceutical Sciences 96*(9): 2197–2223. doi:10.1002/jps.20873.

Barr M.L., Taylor D.O. 2015. *Am Journal Transplant 15*: 7–9. doi:10.1111/ajt.13032.

Benselfelt T., Wågberg L. 2019. *Biomacromolecules 20*(6): 2406–2412. doi:10.1021/acs.biomac.9b00401.

Bhatia SN, Li C.Y. Microgels and Microtissues for use in Tissue Engineering. U.S. Patent 10260039 B2, Apr 16, 2019.

Brittberg M., Lindahl A., Nilsson A., Ohlsson C., Isaksson et al. 1994. *New England Journal of Medicine 331* (14): 889–895. doi:10.1056/NEJM199410063311401.

Bryant S.J., Cuy J.L., Hauch K.D., Ratner B.D. 2007. *Biomaterials 28*(19): 2978–2986. doi:10.1016/j.biomaterials.2006.11.033.

Bush J.R., Liang H., Dickinson M., Botchwey E.A. 2016. *Polymers for Advanced Technologies 27*(8): 1050–1055. doi:10.1002/pat.3767.

Camarero-Espinosa S., Rothen-Rutishauser B., Foster E.J., Weder C. 2016. *Biomaterials Science 4* (5): 734–767. doi:10.1039/c6bm00068a.

Carlson W.B., Phelan G.D. Acrylamide hydrogels for tissue engineering. U.S. Patent application US 13/879976, Jun 26, 2014.

Chirani N., Yahia L., Gritsch L., Motta F.L., Chirani et al. 2015. *Journal of Biomedical Science 4*: 13. doi:10.4172/2254-609X.100013.

Cloyd J.M., Malhotra N.R., Weng L., Chen W., Mauck et al. 2007. *European Spine Journal 16*(11): 1892–1898. doi:10.1007/s00586-007-0443-6.

Cordeiro M.M., Dong Z., Kaneko T., Zhang Z., Miyazawa et al. 2008. *Journal of Endodontics 34*(8): 962–969. doi:10.1016/j.joen.2008.04.009.

Dhandayuthapani B., Krishnan U.M., Sethuraman S. 2010. *Journal of Biomedical Materials Research Part B: Applied Biomaterials 94*(1): 264–272. doi:10.1002/jbm.b.31651.

Dikovsky D., Bianco-Peled H., Seliktar D. 2008. *Biomaterials 27*(8): 1496–1506. doi:10.1016/j.biomaterials.2005.09.038.

Dolcimascolo A., Calabrese G., Conoci S., Parenti R. 2019. Innovative biomaterials for tissue engineering. In: Barbeck M., Jung O., Smeets R., Koržinskas T. [eds.] *Biomaterial-supported Tissue Reconstruction or Regeneration*. IntechOpen, doi:10.5772/intechopen.83839.

Domb A.J., Kost J., Wiseman D. 1998. *Handbook of biodegradable polymers*, CRC press, United States.

Dong H., Snyder J.F., Williams K.S., Andzelm J.W. 2013. *Biomacromolecules 14*(9): 3338–3345. doi:10.1021/bm400993f.

Dong Y., Wang W., Veiseh O., Appel E.A., Xue et al. 2016. *Langmuir 32*(34): 8743–8747. doi:10.1021/acs.langmuir.5b04755.

Drury J.L., Mooney D.J. 2003. *Biomaterials 24*(24): 4337–4351. doi:10.1016/s0142-9612(03)00340-5.

Eslahi N., Simchi A. Hydrogel for cartilage tissue regeneration. U.S. Patent application 15/935014, Jul 26, 2018.

Fanesi G., Abrami M., Zecchin F., Giassi I., Dal Ferro et al. 2018. *Pharmaceutical Research 35*(9), 171. doi:10.1007/s11095-018-2427-0.

Ficat R.P., Ficat C., Gedeon P., Toussaint J.B. 1979. *Clinical Orthopaedics and Related Research, 144*, 74–83.

Fishman R., Havener R., Fattah I.A., Abdelazim A., Newell S., Bishop T.H., Khayal T., Kyi S., Taylor R., Harriott D., De Remer M. Systems and methods for ex vivo lung care. U.S. Patent 10750738, May 15, 2018.

Frisman I., Seliktar D., Bianco-Peled H. 2012. *Acta Biomaterialia 8*(1): 51–60. doi:10.1016/j.actbio.2011.07.030.

Galler K.M., Aulisa L., Regan K.R., D'Souza R.N., Hartgerink et al. 2010. *Journal of the American Chemical Society 132*(9): 3217–3223. doi:10.1021/ja910481t.

Ghadban A., Ahmed A.S., Ping Y., Ramos R., Arfin et al. 2016. *Chemical Communications 52*: 697–700. doi:10.1039/C5CC08617E.

Ghobril C., Grinstaff M.W. 2015. *Chemical Society Reviews 44*(7): 1820–1835. doi:10.1039/C4CS00332B.

Glowacki J., Mizuno S. 2008. *Biopolymers 89*(5): 338–344. doi:10.1002/bip.20871.

Griffith L.G., Naughton G. 2002. *Science 295*(5557): 1009–1014. doi:10.1126/science.1069210.

Gu Z., Dang T.T., Ma M., Tang B.C., Cheng et al. 2013. *ACS Nano 7*: 6758–6766. doi:10.1021/nn401617u.

Gutowska A., Jeong B., Jasionowski M. 2001. *The Anatomical Record 263*(4): 342–349. doi:10.1002/ar.1115.

Harris I.R., Harmon A.M., Brown L.J. Tissue-engineering scaffolds containing self-assembled-peptide hydrogels. U.S. Patent 8039258, Oct 18, 2011.

Heo J., Koh R.H., Shim W., Kim H.D., Yim et al. 2016. *Drug Delivery & Translational Research 6*(2): 148–158. doi:10.1007/s13346-015-0224-4.

Hopkins A.M., De Laporte L., Tortelli F., Spedden E., Staii et al. 2013. *Advanced Functional Materials 23*(41): 5140–5149. doi:10.1002/adfm.201300435.

Huang G.T.J., Yamaza T., Shea L.D., Djouad F., Kuhn et al. 2010. *Tissue Engineering Part A 16*(2): 605–615. doi:10.1089/ten.TEA.2009.0518.

Huang H., Qi X., Chen Y., Wu Z. 2019. *Saudi Pharmaceutical Journal 27*(7): 990–999. doi:10.1016/j.jsps.2019.08.001.

Hunter C.J., Matyas J.R., Duncan N.A. 2003. *Tissue Engineering 9*(4): 667–677. doi:10.1089/107632703768247368.

Hydrogel market. Hydrogel Market by Raw Material Type (Natural, Synthetic, Hybrid), Composition (Polyacrylate, Polyacrylamide, Silicon), Form (Amorphous, Semi-crystalline, Crystalline), Application (Contact Lens, Personal Care & Hygiene), Region - Global Forecast to 2022. https://www.marketsandmarkets.com/Market-Reports/hydrogel-market-181614457.html assessed on 18 November, 2019.

Ikada Y. 2006. *Journal of the Royal Society Interface 3*(10): 589–601. doi:10.1098/rsif.2006.0124.

Iohara K., Zheng L., Ito M., Ishizaka R., Nakamura et al. 2009. *Regenerative Medicine 4*(3): 377–385. doi:10.2217/rme.09.5.

Izzo G.M. Market Analysis on Biomaterials & Tissue Engineering. *Der Pharmacia Sinica*, 2019, *10*(2).

Jansen L., Peyton S. 3d synthetic tissue hydrogels. U.S. Patent application 15/895710, Dec 6, 2018.

Jeong B., Kim S.W., Bae Y.H. 2012. *Advanced Drug Delivery Reviews 64*: 154–162. doi:10.1016/s0169-409x(01)00242-3.

Jung H.S., Kim H.C., Park W.H. 2019. *Carbohydrate Polymers 213*: 311–319. doi:10.1016/j.carbpol.2019.03.009.

Kaplan D.L. 2012. *Acta Biomaterialia 8*(7):2628–2638. doi:10.1016/j.actbio.2012.03.033.

Katz Jeffrey N., Scott D.M. 2009. *Arthritis and Rheumatism 60*(3): 633. doi:10.1002/art.24363.

Khademhosseini A., Karp J.M., Farokhzad O.C., Langer R.S. Cell-Laden Hydrogels. U.S. Patent 11/838,752, Aug. 14, 2008.

Khansari M.M., Sorokina L.V., Mukherjee P., Mukhtar F., Shirdar et al. 2017. *JOM 69*(8): 1340–1347. doi:10.1007/s11837-017-2412-9.

Khetan S., Burdick J. 2009. *Journal of Visualized Experiments*(32): e1590. doi:10.3791/1590.

Kikuchi K., Poss K.D. 2012. *Annual Review of Cell and Developmental Biology 28*: 719–741. doi:10.1146/annurev-cellbio-101011-155739.

Kim B.S., Baez C.E., Atala A. 2000. *World Journal of Urology 18*(1): 2–9. doi:10.1007/s003450050002.

Kim I.Y., Seo S.J., Moon H.S., Yoo M.K., Park et al. 2008. *Biotechnology Advances 26*(1): 1–21. doi:10.1016/j.biotechadv.2007.07.009.

Kim S.H., Seo Y.B., Yeon Y.K., Lee Y.J., Park et al. 2020. *Biomaterials 260*: 120281. doi:10.1016/j.biomaterials.2020.120281.

Klouda L. 2015. *European Journal of Pharmaceutics & Biopharmaceutics 97*: 338–349. doi:10.1016/j.ejpb.2015.05.017.

Kretlow J.D., Klouda L., Mikos A.G. 2007. *Advanced Drug Delivery Reviews 59*(4–5): 263–273. doi:10.1016/j.addr.2007.03.013.

Krupnick A., Kreisel D., Rosengard B.R. Cell delivery patch for myocardial tissue engineering. U.S. Patent 7396537, Jul. 8, 2008.

Kumar A. Hydrogel scaffolds for tissue engineering. U.S. Patent 8883503, Sep. 12, 2013.

Kumar A. Hydrogel scaffolds for tissue engineering. U.S. Patent 8883503, Nov. 11, 2014.

Lamberti F.V., Klann R.C., Hill R.S. Cross-linked bioactive hydrogel matrices. U.S. Patent 8664202, Mar. 4, 2014.

Langer R.S., Traverso C.G., Liu J., Pang Y. Injectable shear-thinning hydrogels and uses thereof. U.S. Patent 16/857279, Dec. 24, 2020.

Larrañeta E., Henry M., Irwin N.J., Trotter J., Perminova et al. 2018. *Carbohydrate Polymers 181*: 1194–1205. doi:10.1016/j.carbpol.2017.12.015.

Lee K.Y., Mooney D.J. 2001. *Chemical Reviews 101*(7): 1869–1880. doi:10.1021/cr000108x.

Lee S.J., Yoo J.J., Atala A. 2018. Biomaterials and tissue engineering, pp 17–51. In: Lim B.(Eds.) *Clinical regenerative medicine in urology*, Springer, Singapore.

Li W.J., Nesti L.J., Tuan R.S. Cell-nanofiber composite and cell-nanofiber-hydrogel composite amalgam based engineered intervertebral disc. U.S. Patent 12/443393, Jul 15, 2010.

Li X., Liu X., Yu Y., Qu X., Feng Q., Cui F., Watari F. et al. 2009. *Recent Patents on Biomedical Engineering 2*(1):65–72.

Liu J., Song H., Zhang L., Xu H., Zhao et al. 2010. *Macromolecular Bioscience 10*(10): 1164–1170. doi:10.1002/mabi.200900450.

Lowman A.M., Palmese G.R., Maher S.A., Warren R.F., Wright T.M., Holloway J.L. Fiber-Hydrogel Composite for Tissue Replacement. U.S. Patent12/783393, Nov 24, 2011.

Lu Z.R., Kopečková P., Kopeček J. 2003. *Macromolecular Bioscience 3*(6):296–300. doi:10.1002/mabi.200390039.

Luo Y., Kirker K.R., Prestwich G. D. 2000. *Journal of Controlled Release 69*(1): 169–184. doi:10.1016/s0168-3659(00)00300-x.

Lymperi S., Ligoudistianou C., Taraslia V., Kontakiotis E., Anastasiadou et al. 2013. *The Open Dentistry Journal 7*: 76. doi:10.2174%2F1874210601307010076.

Lysaght M.J., Reyes J. 2001. *Tissue Engineering 7* (5): 485–493. doi:10.1089/107632701753213110.

Macdonald P., Verran D., O'Leary M., Cavazzoni E., Dhital et al. 2015. *Transplantation 99*(6): 1101–1102. doi:10.1097/TP.0000000000000791.

Majima T., Irie T., Sawaguchi N., Funakoshi T., Iwasaki et al. 2007. *Journal of Engineering in Medicine 221*(5): 537–546. doi:10.1243%2F09544119JEIM203.

Mantha S., Pillai S., Khayambashi P., Upadhyay A., Zhang et al. 2019. *Materials 12*(20): 3323. doi:10.3390/ma12203323.

Market research future. Global Tissue Engineering Market: Information by Material (Nano-fibrous Material, Biomimetic Material, Composite Material and Nano-composite Material) by Application (Orthopedics, Musculoskeletal and Spine, Skin/Integumentary, Cancer, Dental, Cardiology, Urology, Neurology, Cord Blood & Cell Banking and GI & Gynecology) and Region - Forecast till2024.https://www.marketresearchfuture.com/reports/tissue-engineering-market-2134 (accessed Dec 15, 2020).

Martin R., Reddy S., Sacks J., Li X., Cho B.H., Mao H.Q. Fiber-hydrogel composite surgical meshes for tissue repair. U.S. Patent 10471181, Nov 12, 2019.

Masuda S., Shimizu T., Yamato M., Okano T. 2008. *Advanced Drug Delivery Reviews 60*: 277–285. doi:10.1016/j.addr.2007.08.031.

Mathur A.M., Moorjani S.K., Scranton A.B. 1996. *Journal of Macromolecular Science, Part C: Polymer Reviews 36*(2): 405–430. doi:10.1080/15321799608015226.

McIntosh T.K., Smith D.H., Meaney D.F., Kotapka M.J. 1996. *Laboratory Investigation 74*(2): 315. doi:10.1002/adhm.201200317.

Mihaila S.M., Gaharwar A.K., Reis R.L., Marques A.P., Gomes et al. 2013. *Advanced Healthcare Materials 2*(6): 895–907. doi:10.1002/adhm.201200317.

Miyata T. 2010. *Polymer Journal 42*(4): 277–289. doi:10.1038/pj.2010.12.

Morales A., Labidi J., Gullón P. 2020. *Journal of Industrial & Engineering Chemistry 81*: 475–487. doi:10.1016/j.jiec.2019.09.037.

Mun G., Khutoryanskiy V. (2015). *Synthesis and characterization of polymer-protected gold nanoparticles*, Elmira Nurgaziyeva, Doctoral dissertation, University of Reading, UK.

Nakashima M., Iohara K. 2011. *Advances in Dental Research 23*(3): 313–319. doi:10.1177%2F0022034511405323.

Narayanan K.B., Choi S.M., Han S.S. 2019. *Colloids and Surfaces B: Biointerfaces 181*: 539–548. doi:10.1016/j.colsurfb.2019.06.007.

Navaei A., Truong D., Heffernan J., Cutts J., Brafman et al. 2016. *Acta Biomaterialia 32*: 10–23. doi:10.1016/j.actbio.2015.12.019.

Ng W.L., Yeong W.Y., Naing M.W. 2016. *Procedia Cirp 49*: 105–112. doi:10.1016/j.procir.2015.09.002.

Nisbet D.R., Crompton K.E., Horne M.K.Finkelstein D.I., Forsythe et al. 2008. *Journal of Biomedical Materials Research Part B: Applied Biomaterials 87*(1): 251–263. doi:10.1002/jbm.b.31207.

Onaciu A., Munteanu R.A., Moldovan A.I., Moldovan C.S. 2019. *Pharmaceutics 11*(9): 432. doi:10.3390/pharmaceutics11090432.

Park H., Kang S.W., Kim B.S., Mooney D.J., Lee et al. 2009. *Macromolecular Bioscience 9*(9):895–901. doi:10.1002/mabi.200800376.

Park J.B. 2011. *Med Oral Patol Oral Cir Bucal 16*(1): e115–8. doi:10.4317/medoral.16.e115.

Park J.H., Bae Y.H. 2002. *Biomaterials 23*(8): 1797–1808. doi:10.1016/s0142-9612(01)00306-4.

Parker K.K., Badrossamay M.R., Agarwal A. Polymeric fiber-scaffolded engineered tissues and uses thereof. U.S. Patent 15/869228, Dec 27, 2018.

Peppas N.A. and Hoffman A.S. 2020. *Biomaterials science*. Academic Press, US.

Perez E., Miller D., Merrill E.W. Methods for making composite hydrogels for corneal prostheses. U.S. Patent 5836313, Nov 17, 1998.

Pescosolido L., Schuurman W., Malda J., Matricardi P., Alhaique et al. 2011. *Biomacromolecules 12*(5): 1831–1838. doi:10.1021/bm200178w.

Petersen J. Polyacrylamide hydrogel as a soft tissue filler endoprosthesis. U.S. Patent 7790194, Sep. 7, 2010.

Pollock J.F., Kokai L.E., Cui C., Yu X., Epps D.E.V., Messina D.J. Crosslinked Hyaluronic acid-Collagen Gels for improving Tissue graft viability and Soft Tissue augmentation. U.S. Patent 9782517 B2, Oct. 10, 2017.

Prabhu S.D., Frangogiannis N.G. 2016. *Circulation Research 119*(1): 91–112. doi:10.1161/CIRCRESAHA.116.303577.

Pradhan S., Keller K.A., Sperduto J.L., Slater J.H. 2017. *Advanced Healthcare Materials 6*: 1700681. doi:10.1002/adhm.201700681.

Prescott R.S., Alsanea R., Fayad M.I., Johnson B.R., Wenckus et al. 2008. *Journal of Endodontics 34*(4): 421–426. doi:10.1016/j.joen.2008.02.005.

Priya S.G., Jungvid H., Kumar A. 2008. *Tissue Engineering Part B: Reviews 14*(1): 105–118. doi:10.1089/teb.2007.0318.

Qiu Y., Park K. 2001. *Advanced Drug Delivery Reviews 53*: 321–339. doi:10.1016/S0169-409X(01)00203-4.

Rafat M., Li F., Fagerholm P., Lagali N.S., Watsky et al. 2008. *Biomaterials 29*(29): 3960–3972. doi:10.1016/j.biomaterials.2008.06.017.

Research and markets. Hydrogels: Applications and global markets to 2022. https://www.researchandmarkets.com/reports/4425798/hydrogels-applications-and-global-markets-to-2022 (accessed on Dec 23).

Rinoldi C., Costantini M., Gawrońska E., Testa S., Fornetti et al. 2019a. *Advanced Healthcare Materials 8*(7): 1801218. doi:10.1002/adhm.201801218.

Rinoldi C., Fallahi A., Yazdi I.K., Campos Paras J., Kijeńska-Gawrońska et al. 2019b. *ACS Biomaterials Science & Engineering 5*(6): 2953–2964. doi:10.1021/acsbiomaterials.8b01647.

Rizwan M., Peh G.S., Ang H.P., Lwin N.C., Adnan et al. 2017. *Biomaterials 120*: 139–154. doi:10.1016/j.biomaterials.2016.12.026.

Sakai S., Hirose K., Taguchi K., Ogushi Y., Kawakami et al. 2009. *Biomaterials 30*(20): 3371–3377. doi:10.1016/j.biomaterials.2009.03.030.

Sakai V.T., Zhang Z., Dong Z., Neiva K.G., Machado et al. 2010. *Journal of Dental Research 89*(8): 791–796. doi:10.1177%2F0022034510368647.

Sakr M.A., Mohamed M.G., Wu R., Shin S.R., Kim et al. 2020. *Applied Clay Science, 199*: 105860. doi:10.1016/j.clay.2020.105860.

Sayyar S., Murray E., Thompson B.C., Chung J., Officer et al. 2015. *Journal of Materials Chemistry B 3*(3): 481–490. doi:10.1039/C4TB01636J.

Schmidt C.E., Leach J.B. 2003. *Annual Review of Biomedical Engineering 5*(1), 293–347. doi:10.1146/annurev.bioeng.5.011303.120731.

Shin S.R., Zihlmann C., Akbari M., Assawes P., Cheung et al. 2016. *Small 12*(27): 3677–3689. doi:10.1002/smll.201600178.

Stabenfeldt S.E., García A.J., LaPlaca M.C. 2006. *Journal of Biomedical Materials Research Part A 77*(4): 718–725. doi:10.1002/jbm.a.30638.

Stock U.A., Vacanti J.P. 2001. *Annual Review of Medicine 52*(1): 443–451. doi:10.1146/annurev.med.52.1.443.

Sun F., Wang R. Photoresponsive protein hydrogels and methods and uses thereof. U.S. Patent 10647755, May 12, 2020.

Ter Schiphorst J., Coleman S., Stumpel J.E., Ben Azouz A., Diamond et al. 2014. *Chemistry of Materials 27*(17): 5925–5931. doi:10.1021/acs.chemmater.5b01860.

Tokunaga M., Liu M.L., Nagai T., Iwanaga K., Matsuura et al. 2010. *Journal of Molecular and Cellular Cardiology 49*(6): 972–983. doi:10.1016/j.yjmcc.2010.09.015.

Townsend J.M., Detamore M.S. Thiolated Devitalized Tissue Particle Implantable Hydrogel. U.S. Patent 10744229 B1, Aug. 18, 2020.

Vacanti C.A., Vacanti J.P. Guided development and support of hydrogel-cell compositions. U.S. Patent 6027744, Feb. 22, 2000.

Valmikinathan C.M., Mukhatyar V.J., Jain A., Karumbaiah L., Dasari et al. 2012. *Soft Matter 8*(6): 1964–1976. doi:10.1039/C1SM06629C.

Van den Broeck L., Piluso S., Soultan A.H., De Volder M., Patterson et al. 2019. *Materials Science and Engineering: C 98*: 1133–1144. doi:10.1016/j.msec.2019.01.020.

Vizzardi E., Lorusso R., De Cicco G., Zanini G., D'Aloia et al. 2012. *The Journal of Cardiovascular Surgery 53*(5): 685.

Wang W.Y., Pearson A.T., Kutys M.L., Choi C.K., Wozniak et al. 2018. *APL Bioengineering 2*(4): 046107. doi:10.1063/1.5052239.

Wenquan S.U., Xu H., Wan H. Elastic tissue matrix derived hydrogel. U.S. Patent 10,092,602, July 3, 2014.

Willerth S.M., Sakiyama-Elbert S.E. 2007. *Advanced Drug Delivery Reviews 59*(4–5): 325–338. doi:10.1016/j.addr.2007.03.014.

Xing Y., Cheng E., Yang Y., Chen P., Zhang et al. 2011. *Advanced Materials 23*(9): 1117–1121. doi:10.1002/adma.201003343.

Xu L., Qiu L., Sheng Y., Sun Y., Deng et al. 2018. *Journal of Materials Chemistry B 6*(3): 510–517. doi:10.1039/C7TB01851G.

Yang G., Lin H., Rothrauff B.B., Yu S., Tuan et al. 2016. *Acta Biomaterialia 35*: 68–76. doi:10.1016/j.actbio.2016.03.004.

Yao L., Swords G.A. Hydrogels and methods of making and using same. U.S. Patent 6268405, Jul. 31, 2001.

Yu J., Chen F., Wang X., Dong N., Lu et al. 2016. *Polymer Degradation and Stability 133*: 312–320. doi:10.1016/j.polymdegradstab.2016.09.008.

Yu X., Messina D.J., Pavlovic E., Cui C., Smither K.M. Co-crosslinked hyaluronic acid-silk fibroin hydrogels for improving tissue graft viability and for soft tissue augmentation. U.S. Patent 10,300,169, May 28, 2019.

Yuan X., Wei Y., Villasante A., Ng J.J., Arkonac et al. 2017. *Biomaterials 132*: 59–71. doi:10.1016/j.biomaterials.2017.04.004.

Zander N.E., Dong H., Steele J., Grant J.T. 2014. *ACS Applied Materials & Interfaces 6*(21): 18502–18510. doi:10.1021/am506007z.

Zhang Y.S., Khademhosseini A. 2017. *Material Science 356*: 500. doi:10.1126/science.aaf3627.

Zhao X., Lang Q., Yildirimer L., Lin Z.Y., Cui et al. 2016. *Advanced Healthcare Materials 5*(1): 108–118. doi:10.1002/adhm.201500005.

Zhu J., Marchant R.E. 2011. *Expert Review of Medical Devices 8*(5): 607–626. doi:10.1586/erd.11.27.

11

Lutein: A Nutraceutical Nanoconjugate for Human Health

Ishani Bhat and Bangera Sheshappa Mamatha
Nitte (Deemed to be University), Nitte University Center for Science Education and Research (NUCSER), Mangaluru, Karnataka, India

CONTENTS

Introduction

Changes in the 20th-century lifestyle have contributed to many lifestyle-oriented diseases, the prevalence of which is growing with each day. The alarming consequences of these lifestyle-oriented diseases and adverse effects of their medication have revoked the use of natural substances. Today, natural substances are not only consumed for therapeutic purposes but also for preventing the occurrence of such diseases. The rise in the health-conscious population has driven the emergence of new technologies to manufacture food that provides certain health benefits. This has contributed to the augmented use of natural ingredients for nutraceutical foods with unique health benefits. Natural pigments can reduce oxidative stress

DOI: 10.1201/9781003153504-11

caused by various reasons due to their chemical structure are natural antioxidants. Carotenoids are a class of natural pigments that have inspired and changed the course of research because of their biological activity. Carotenoids include two subclasses, namely, carotenes and xanthophyll carotenoids. Although carotenes such as α and β-carotenes are involved in provitamin-A activity, xanthophyll carotenoids such as lutein and zeaxanthin exhibit anti-oxidative activity. Lutein has been recognised in the human body as a predominant carotenoid that has a protective function in various parts of the body, including ocular tissues, the central nervous system, and the cardiovascular system. Despite being a compound of utmost importance in the human body, it is not endogenously synthesised by the human body. Although lutein can be included in the daily diet, its bioavailability is impaired during oral intake by extrinsic (food matrix, conutrients, etc.) and intrinsic factors (nature of the host, genetic constitution, etc.). This has drawn the attention of researchers to further explore plants and animals for lutein extraction, purification, and integration into an effective delivery medium for enhanced bioavailability. Recent studies on lutein activity and the complexities of bioavailability of intact lutein have contributed to the growth of novel strategies for delivering lutein into the systemic circulation. The bioavailability of lutein can be significantly amplified by conjugating it with nanosized particles (10–400 nm). Also, nanoconjugates are more soluble, stable, and potent than their counterparts of greater size (>1 nm). Currently, attempts are being made to develop nanoconjugates of lutein with adjuvants that support their targeted delivery. Besides, lutein is a photolabile, lipophilic compound requiring care for its pairing with adjuvants that stabilise it during storage. This chapter is a compilation of recent advances in formulating nanoconjugated delivery systems of lutein. Innovative techniques for the preparation of lutein nanoconjugates using materials that not only boost lutein bioavailability but also deliver a safe and controlled lutein release are presented.

The Biological Value of Lutein

Lutein is an oxygenated xanthophyll carotenoid consisting of nine conjugated double bonds in a C_{40} isoprenoid with two ionone rings, each bound to a hydroxyl group (OH) at 3 and 3' positions, at either end of the isoprenoid (Figure 11.1). Zeaxanthin and *meso*-zeaxanthin are isomers of lutein and all the three

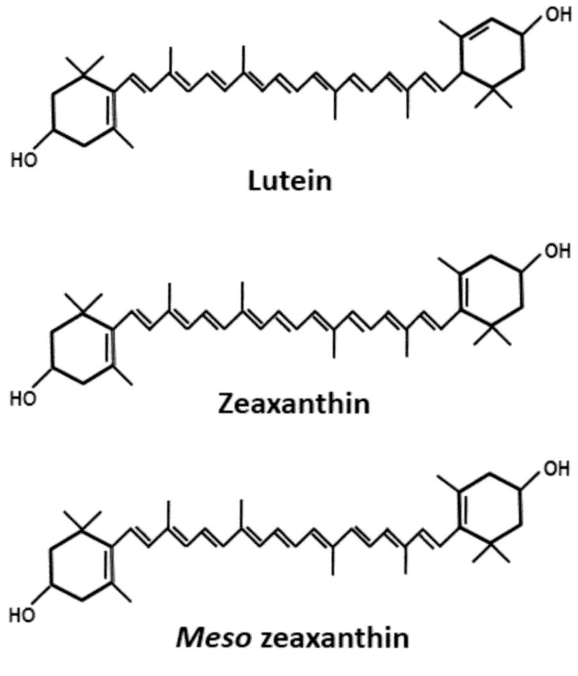

FIGURE 11.1 Chemical Structure of Macular Pigments.

compounds share a common molecular formula $C_{40}H_{56}O_2$. Because the three xanthophyll carotenoids are concentrated in the human eye macula (lutein: 36%, zeaxanthin: 18%, and *meso*-zeaxanthin: 18%), they are collectively called macular pigments (Bone et al., 1993). The chemical configuration of lutein is not only responsible for manifesting major biological activities such as antioxidant and light-absorbing properties but also defines its polarity and solubility (Woodall et al. 1997). These pigments have their peak absorption at 460 nm, which corresponds to the wavelength of "blue light hazard" (400–500 nm). The incident high-energy, short-wavelength visible blue light causes oxidative stress in the eyes. The MP absorbs 40–90% of blue light, which is incidental to the retina. However, this filtering is Macular Pigment concentration-dependent. Thus, macular pigments function as blue light filtering anti-oxidants to protect the retinal pigment epithelial cells in the eyes from the consequences of light-induced oxidative stress (Krinsky et al. 2003). Henceforth, they are positively associated with preventing age-related macular degeneration.

However, lutein is synthesised only by photosynthetic organisms such as plants, algae, bacteria, yeast, and moulds. In 1988, the first research on the correlation between food intake and age-related macular degeneration was published (Goldberg et al. 1988). Additionally, a diet that is rich in lutein could also reduce the risk factors associated with AMD (Eisenhauer et al. 2017). Macular pigments display good antioxidant effects on reactive oxygen species such as singlet oxygen (Vershinin 1999). This property is attributed to its typically long polyene arrangement comprising conjugated double bonds. In particular, present-day intervention studies using randomised controlled trials have shown lutein intake to beneficially affect cognitive functions (Hammond et al. 2017; Power et al. 2018). Lutein extends its protective action against oxidation of docosahexaenoic acid (DHA), in all active areas of the brain namely the prefrontal cortex, cerebellum, striatum, and hippocampus (Mohn et al. 2017), to overcome oxidative stress in age-related cognitive degeneration (Feeney et al. 2017; Johnson 2014; Vishwanathan et al. 2014) and Alzheimer's disease (AD) (Nolan et al. 2014). In addition to this, lutein supplementation can improve memory, attentiveness, and reasoning ability (Renzi-Hammond et al. 2017), cause a reduction in the levels of serum cholesterol, hepatic cholesterol, and triglycerides (Murillo et al. 2016; Qiu et al. 2015), protect low-density lipoproteins (LDL) from lipid peroxidation (Kishimoto et al. 2017), reduce diabetes-induced oxidative-testicular damage (Fatani et al. 2015), provide angioprotective efficacy in diabetic retinopathy (Sharavana & Baskaran 2017), provide chemo-protectivity against atherosclerosis (Han et al. 2015) and improve skin conditions by causing skin lightening (Juturu et al. 2016).

Bioavailability of Lutein

The amount of lutein consumed in the diet is the foremost factor that influences its oral bioavailability (Figure 11.2). However, the processing conditions dietary sources undergo majorly influence the changes in the source matrix and molecular linkages. The reduction in lutein concentration in plant sources starts right from the harvest. Green-leafy and yellow vegetables have been observed to contain the highest lutein concentration at maturity, followed by a decrease at hypermaturity (Lefsrud et al. 2007; Ma et al. 2015). Lutein is prone to degradation by light and heat. Exposure to cooking, frying, baking, and other extreme conditions can lower lutein content in the food (Gutiérrez-Uribe et al. 2014; Shen et al. 2015). Simple processing techniques such as peeling have slight detrimental effects on carrot lutein (Ma et al. 2015). When maize undergoes nixtamalisation, lutein content is enhanced. However, when it is baked and converted to tortillas, lutein level drops down to the initial quantity (Gutiérrez-Uribe et al. 2014). Cooking foxtail millets also results in paling of the millets indicating a loss of yellow pigments (Shen et al. 2015). However, reducing the cooking time and increasing the pressure can subsequently lower the carotenoid losses in vegetables (Sánchez et al. 2014). Nevertheless, blanching and enzyme liquefaction can improve lutein and zeaxanthin levels in vegetables (Ma et al. 2015; Mamatha et al. 2012).

On consumption, the bioactivity of lutein depends on its release during the digestion of food and its delivery to the target sites. These two concepts are known as bioaccessibility and bioavailability, respectively. To make lutein bioaccessible, gastrointestinal enzymes digest the food matrix and liberate the embedded lutein (Figure 11.2). Lutein in vegetables is better bioaccessible (80%) than that in fruits (35%) (Goni et al. 2006). Lutein absorption is highly influenced by its association with macro and micronutrients

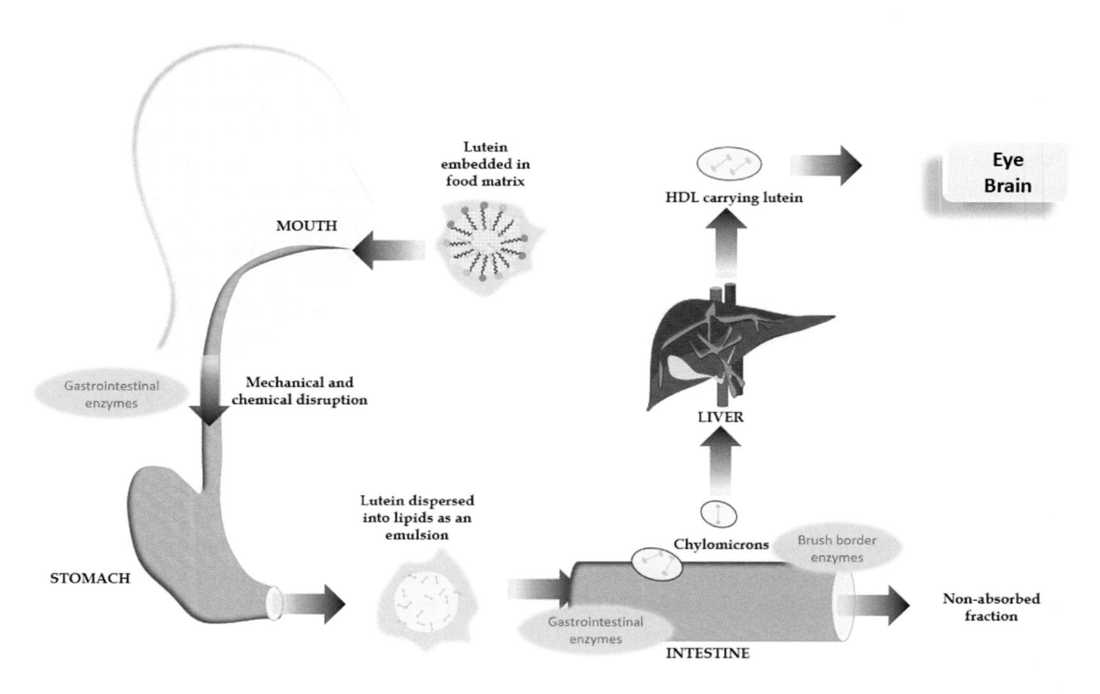

FIGURE 11.2 Schematic Representation of Gastrointestinal Absorption of Lutein.

(Korobelnik et al. 2017). Supplementation of lutein with dietary fibres and other carotenoids is unfavourable to lutein absorption (Mamatha & Baskaran, 2011; Kostic et al. 1995; Reboul et al. 2007; Riedl et al. 1999). The presence of β-carotene along with lutein restricts its intestinal absorption (Mamatha & Baskaran, 2011). The cleavage enzymes BCO1 and BCO2 along with vitamin A status in the host may play an important role in this phenomenon (Section 2). However, supplementation of lutein with lipids, oils, and lipoproteins has positively influenced lutein absorption (Mamatha & Baskaran, 2011; Marriage et al. 2017; Kijlstra et al. 2017; Roodenburg et al. 2000). Although poly-unsaturated fatty acids (PUFA) has been claimed to reduce the risk of AMD, in combination with lutein in the diet, the bioavailability of lutein is reduced (Wolf-Schnurrbusch et al. 2015). Voiding fat from products suppressed lutein bioavailability (Xavier et al. 2014), whereas food prepared with a higher fat content improved the bioavailability of lutein (Read et al. 2015; Xavier et al. 2018). Also, variability in the genes coding for proteins that participate in lutein absorption affects the oral bioavailability of lutein. Modulation in the chylomicron triacylglycerol response has been associated with single nucleotide polymorphisms (SNPs) in 15 genes, which in turn modulate the oral bioavailability of lutein (Borel et al. 2014). In recent years, human apolipoprotein E genotype studies suggest that this genotype can modulate the metabolism of carotenoids (Huebbe et al. 2016). However, more research is needed to clarify the variability in genes affecting lutein absorption.

Modulation of lutein bioavailability can be achieved by modifying variables such as bioaccessibility, absorption, and transformation (Salvia-Trujillo et al. 2016). Poor solubility, stability, and oral bioavailability are the key obstacles to the integration of lutein into commercial food products. In response to these issues as well as to promote the incorporation of lutein in functional foods and beverages, the use of colloidal encapsulation systems can be made. Encapsulation of lutein ideally involves covering it with a wall material (Madene et al. 2006), which thereby allows the formation of a colloidal suspension. The wall material acts as a barrier against adverse environmental conditions, enhance its viability, and promote optimized delivery. Besides, the use of wall material improves the shelf-life of encapsulated lutein by preventing evaporation, migration, crystallisation, and precipitation of lutein. Henceforth, embedding lutein into an encapsulation system is a promising approach to improve its oral bioavailability. However, encapsulating a nutraceutical component for oral consumption requires the use of permitted pharmaceutical or food-grade materials marked safe on the basis of toxicity (McClements et al. 2007). Additionally, the encapsulation system needs to be built using economical materials that can withstand physicochemical degradation and release lutein at a controlled rate (Joseph & Bunjes2013; McClements 2018).

Mostly, encapsulation systems are emulsion-based, which have core-shelled spheroid lipid droplets distributed in an aqueous medium. Such emulsions generally contain a hydrophobic core (surfactant tails) consisting of lutein and a hydrophilic shell (surfactant head), which makes the emulsion thermodynamically stable. The suspended particles are commonly sized below 100 nm and are called microemulsions. Because hydrophobic drugs are well incorporated into such emulsions within the hydrophobic core, they are commonly used to enhance their oral bioavailability in the pharmaceutical industry. As lutein is also a hydrophobic molecule of biological significance, incorporating it into microemulsions could help improve its oral bioavailability. However, the extent of encapsulation and release from the microemulsion depends on the surfactants, co-surfactants, and oil phases used (Setya et al. 2014; Lo et al. 2016). Tween 80 is a non-ionic food-grade surfactant, which has been successfully used to prepare microemulsions of lutein and zeaxanthin and incorporated into beverages (Amar et al. 2004). A combination of Tween 80, capryol, and transcutol, used to prepare the microemulsion of *Rhinacanthusnasutus* carotenoid extract by sonication, encapsulated 98.6% of carotenoids (Ho et al. 2016). Besides, the microemulsion exhibited enhanced oral bioavailability (6.25%) in comparison to distilled water suspension.

The process of electrostatic deposition enables designing multilayer emulsions consisting of emulsifier-covered lipid droplets, which is layered with single or multiple biopolymer (polysaccharides and proteins) layers (Bortnowska 2015; Guzey & McClements, 2006). In case of protein-coated emulsions, exposure to isoelectric pH values, raised temperatures, high ionic strengths, and other environmental factors can lower its stability. Thus, coating a layer of polysaccharides is suggested to improve the stability of such emulsions (Burgos-Díaz et al. 2016). The chemical stability of a lutein-loaded emulsion containing lipid droplets layered with whey protein isolate was improved by coating with multilayers of flaxseed gum and chitosan (Xu et al. 2016). Further, linking the polysaccharides covalently to proteins by Maillard reaction can improve the lipid droplet size and emulsification (Hou et al. 2017). In addition, Maillard conjugates allow the emulsion to be stable during changes in pH, temperature, and ionic strength. However, in an emulsion, reducing the droplet size further (<50 nm) causes changes in physicochemical and functional properties of emulsions. In contrast to emulsions, the resulting nanosized emulsions are more stable for flocculation, creaming, more easily digested, and less opaque (Kale & Deore, 2017).

Lutein Nanoconjugates

Nanotechnology is an evolving discipline aimed at enhancing the bioaccessibility and bioavailability of a compound by size reduction. Currently, research focus lies on designing particles of nanosize (10^{-9} m or 1 nm) and conjugating them with the bioactive component (mostly hydrophobic) to attain targeted delivery in the human body. Although the possible uses of nanotechnology in medical science were recognised early, it was only recently that practical applications in food processing were implemented. At present, however, principles of nanotechnology are being applied in food science to deliver nutritional supplements in a targeted way. Lutein incorporation can be made in solid or liquid nanoformulations. Although lutein-loaded lipid nanoparticles, lutein nanocrystals, and lutein-loaded polymeric nanoparticles are solid nanoconjugates, liquid nanoconjugates include lutein-loaded nanoemulsions, lutein-loaded nanoliposomes, and lutein-loaded nanopolymersomes (Figure 11.3). Lutein is a hydrophobic molecule, which has a high affinity for lipids and low affinity for aqueous medium. Henceforth, researchers have proposed using lipid-based organic solvents for designing nanoconjugates of lutein. The nanoconjugates of lutein designed using specific methods, in an attempt of improving its bioavailability and optimal delivery is listed in Table 11.1.

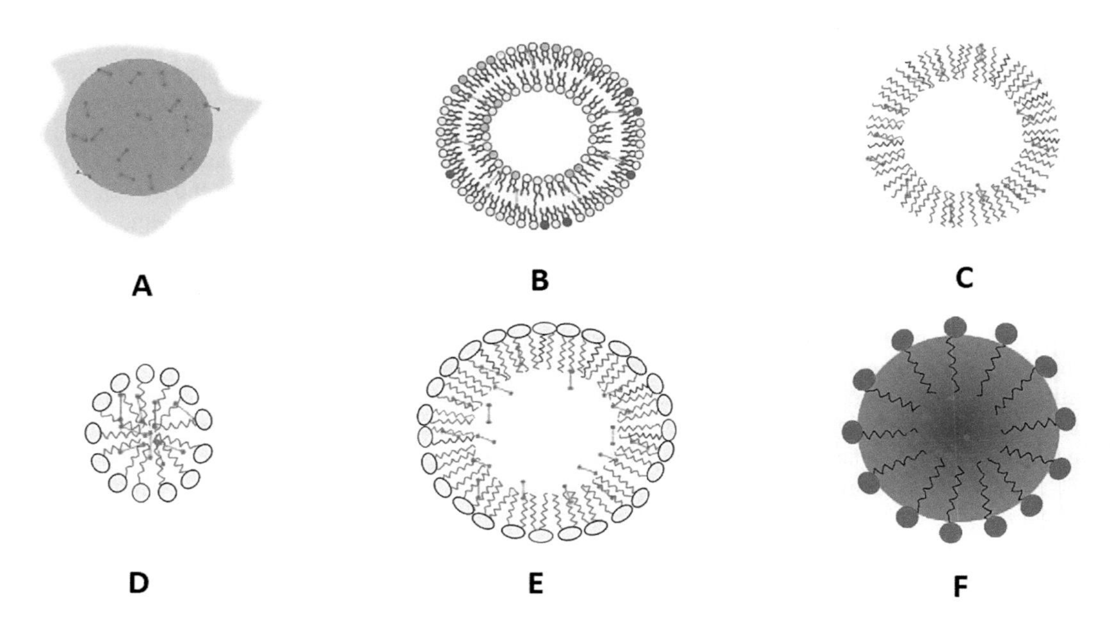

FIGURE 11.3 Lutein-Loaded Nanoconjugates Represented Diagrammatically. A- Lutein-loaded nanoemulsion; B- Lutein-loaded nanoiposomes; C- Lutein-loaded nanopolymersomes; D- Lutein-loaded nanocrystals; E- Lutein-loaded polymeric nanoencapsulate; F- Lutein-loaded nanoparticles.

Lutein-Loaded Liquid Nanoconjugates

Lutein-Loaded Nanoemulsions

When two immiscible liquids are combined, which includes a lipid phase (containing lutein) and an aqueous phase (solvent), it often results in lutein-loaded nanoemulsions (Figure 11.3A). However, mechanical methods such as microfluidizers, homogenisers, or sonicators that provide shearing action or pressure are often applied to combine the two phases (Kim et al. 2006; Vishwanathan et al. 2009; Liu et al. 2015; Tan et al. 2016a; Tan et al. 2016b; Poorani et al. 2017; Teo et al. 2017). The extent of mechanical pressure applied, directly influences the droplet size in the emulsion. For example, specific energy vibrations during ultrasonication improve the solubility of lutein and disperse the particles of favourable size into the dispersal medium (Chen and Yao 2017). Further, the use of surfactants, protein fractions, and saccharides can stabilise the nanoemulsion formed but also improve the lutein holding capacity of the emulsion formed (Liu et al. 2015; Tan et al. 2016a; Chen & Yao, 2017; Teo et al. 2017). Encapsulation efficiency of 95% was achieved in a lutein-loaded nanoemulsion stabilised with cyclodextrin prepared by hot sonication (Liu et al. 2015). Although this method could reduce the size of the droplets to 91.7 nm, high pressure homogenisation at 500 bar and 80 MPa (four cycles) could reduce the lutein-corn oil droplet size to 150 nm and 147 nm, respectively (Mitri et al. 2011; Teo et al. 2017). However, although electrostatic stabilisation was provided by incorporating whey protein isolate into the lutein-corn oil conjugate, it could further reduce the droplet size to 68 nm (Teo et al. 2017). Conversely, this combination reduced the efficiency to encapsulate lutein due to free radical formation that resulted from the unsaturated part of corn oil.

Lutein-Loaded Nanoliposomes

Unlike nanoemulsions, nanoliposomes consist of a phospholipid bilayer that can entrap lutein, provide protection, and serve targeted delivery (Figure 11.3B). Generally, encapsulation of lutein into these bilayers involves injection methods or supercritical fluids (Tan et al. 2013; Zhao et al. 2017). When the

TABLE 11.1

Methods and Adjuvants used to Prepare Lutein-Loaded Nanoconjugates and Their Resulting Characteristics

Lutein Nanoconjugates	Method Employed	Adjuvant	Particle Size (nm)	Entrapment Efficiency (%)	Release (%)	References
Lutein-loaded nanoemulsion	High pressure homogenisation	Corn oil, whey protein isolate	68.0	80.0	—	Teo et al. (2017)
	Hot sonication	Cyclodextrin	91.7	95.0	—	Liu et al. (2015)
	Solvent displacement	Tween 80	123.0	93.0	—	Tan et al. (2016)
	High-pressure valve homogenisation	Tween 80	136.0	91.0	—	Tan et al. (2016)
	High-pressure homogenisation	Corn oil, Miglyol 812 Plantacare 810	150.0	—	19 (24h)	Mitri et al. (2011)
	Solvent displacement	Tween 80	74.0	83.0	—	Tan et al. (2016)
		SDS	96.0	82.2		
		Sodium caseinate	125.0	83.4		
		SDS-Tween 80	66.0	82.1		
	Nano emulsification	Phospholipon, soybean oil, Vit E oil	155–157	—	—	Vishwanathan et al. (2009)
	Nano emulsification	TPGS, MCT (60:40 caprylic and capric acid)	254.0	—	—	Murillo et al. (2016)
Lutein-loaded nanoliposomes	Liposome formation using supercritical carbon dioxide	Soy lecithin, phospholipids	160–170	96.0	—	Zhao et al. (2017)
	Liposome formation using ethanol injection	EYPC, Tween 20	76–134	82–91	—	Tan et al. (2013)
	Liquid–liquid dispersion	Zein	156.0	69.0	—	Chuacharoen et al. (2016)
		Lecithin, plutonic F127	216.0	83.0		
	Liquid–liquid dispersion	Zein	398.3	85.4	—	Jiao et al. (2018)
		Zein derived peptides	297.7	90.3		
	Ethanol injection and ultra-high pressure homogenisation	Casein-dextran conjugates	118.5	97.2	70 (1h)	Muhoza et al. (2018)
Lutein-loaded nanopolymersomes	Self-assembly by injection method	Octenyl succinic anhydride modified short glucan chains	187–249	85.0	30 (8h) 80 (25h) 68 (75h)	Chang et al. (2017)

(*Continued*)

TABLE 11.1 (Continued)

Lutein Nanoconjugates	Method Employed	Adjuvant	Particle Size (nm)	Entrapment Efficiency (%)	Release (%)	References
Lutein-loaded nanocrystals	Anti-solvent precipitation for oral fast-dissolving films	THF, HPMC, PEG400, sorbitol, citric acid, aspartame	360–750	—	80 (1h)	Liu et al. (2017)
	Nanoprecipitation by solvent displacement	Ethyl cellulose, PVP, Tween 80	490.0	84–92	92 (4h)	Bhadkariya et al. (2017)
	Nanoprecipitation	Zein, sodium caseinate	242.0	95.9	—	do Prado Silva et al. (2017)
	Nanoprecipitation for nanosuspension	PVA	152.4	—	—	Mishra et al. (2015)
	Interfacial deposition	Poly-ε-caprolactone, sorbitan monostearate, medium chain triglycerides	191.0	99.5	—	Brum et al. (2017)
Lutein-loaded nanoparticles	Sonication	Myverol 18-04K, Pluronic F68, Estasan 3575	199.0	—	35 (55h)	Liu and Wu (2010)
	Nanocarrier	Myverol 18-04K, Pluronic F68, Precirol ATO5,	134.0	90.0	90 (55h)	Liu and Wu (2010)
	High-pressure homogenisation for SLN	Corn oil, cetylpalmitate, glyceryltripalmitate, carnauba wax, Plantacare 810	220–240	—	0.4 (24h)	Mitri et al. (2011)
	High-pressure homogenisation for NLC	Corn oil, glyceryltripalmitate, carnauba wax, Miglyol 812, Plantacare 810	160–190	—	7–12 (24h)	Mitri et al. (2011)
	Hot sonication	Corn oil, capric acid, span 60, Tween 80 HPβCD	229.0 336.0	59.0 68.0	—	Liu et al. (2014)
	High-shear homogenisation for NLC	Corn oil, glycerol stearate, carnauba wax, fish oil	167.5	88.5	90 (27h)	Lacatusu et al. (2013)

Lutein-loaded polymeric nanoencapsulate	Nanoencapsulation	Low molecular weight chitosan	80–600	85.0	—	Arunkumar et al. (2013)
	Enhanced dispersion by supercritical fluids	Zein, DMSO	198–244	75.0	85 (5h)	Hu et al. (2012)
	Supercritical anti-solvent precipitation	Hydroxypropyl methylcellulose phthalate	163–314	67–95	—	Heyang et al. (2009)
	Dissolution in common solvent	PVP, Tween 80	200.0	—	—	do Prado Sailva et al. (2017)
	Rotary evaporation	Tween 80	<450000	93.0	—	Zhao et al. (2014)
	Nanoencapsulation	Soybean seed, ferritin, chitosan	12.0	17.0	5 (4°C) 27 (20°C) 69 (37°C) (168h)	Yang et al. (2016)
	Emulsion evaporation	PLGA	124.0	52.0	—	Kamil et al. (2016)
	Nano encapsulation	PLGA-PEG	80–500	—	66 (72h)	Arunkumar et al. (2013)

Note: SDS – sodium dodecyl sulphate; TPGS – d-α-tocopherol polyethylene glycol succinate; MCT – medium chain triglyceride; EYPC – egg yolk phosphatidylcholine; THF – tetrahydrofuran; HPMC – hydroxypropyl methylcellulose; PEG400 – polyethylene glycol 400; PVP – polyvinyl pyrrolidone; PVA – polyvinyl alcohol; HPβCD – 2-hydroxy-beta-cyclodextrin; DMSO – dimethyl sulfoxide; PLGA – polylactic co glycolic acid; PLGA-PEG – polylactic co glycolic acid-polyethylene glycol; SLN – solid lipid nanoparticles; NLC – nanostructured lipid carriers.

above-mentioned methods are performed after applying shearing pressure, it aids self-conjugation of the phospholipid bilayer. Henceforth, lutein can be encapsulated into the bilayer without the use of organic solvents. This is an added advantage in producing a commercial nutraceutical to be incorporated into functional foods and beverages. Egg yolk phosphatidylcholine and soy lecithin have been used to make such nanoconjugates. In addition, 82–91% of lutein was entrapped into egg yolk phosphatidylcholine vesicles of 76–134 nm (Tan et al. 2013), whereas 96% of lutein was entrapped into soy lecithin vesicles of 160–170 nm (Zhao et al. 2017). Further, these conjugates can be stabilised by using protein structures and surfactants. Both zein protein and zein-derived peptides (alcalase hydrolysis) can stabilise lutein-loaded nanoliposomes, improve roundness, and encapsulation efficiency of lutein (Jiao et al. 2018). Similarly, using a surfactant (Pluronic F12) can also improve the encapsulation efficiency of lutein-loaded nanoliposomes (Jiao et al. 2018). Further, positively charged adjuvants can be used to interact with the electronegative regions in lutein to improve the stability of such nanoconjugates. Besides, a steric layer can be coated onto the protein lutein-loaded nanoliposomes by conjugating the protein with polysaccharides. For example, glycosylating casein with dextrin moieties by Maillard reaction keeps the positive charge of the protein away from the outer aqueous dispersion medium (Muhoza et al. 2018). However, although the addition of an extra layer could improve the stability of the nanoconjugate, it can also lead to particle size enlargement (Jiao et al. 2018). The proportion of lipid to lutein along with conditions during processing such as temperature and pressure affects the rate of liposome formation. For example, lutein entrapment can be enhanced by high temperature treatment, which can break down hydrophobic bonds in the phospholipids and ease interaction with lutein (Zhao et al. 2017). For targeted delivery, permeability of these bilayers is also crucial.

Lutein-Loaded Nanopolymersomes

Lutein-loaded nanopolymersomes, similar to nanoliposomes, contain amphiphilic polymers as a substitute for phospholipid bilayers (Figure 11.3C). The self-assembly of polymersomes can be initiated by an injection method similar to liposomes to obtain an optimum particle size and entrapment efficiency. Self-assembled, lutein-loaded nanopolymersomes were designed using a polymer of octenyl succinic anhydride-modified short glucan chains (Chang et al. 2017). Despite the ultra-small size of 10–20 nm, the nanoconjugate was able to entrap 85% of lutein within the polymersome. Similarly, methoxypoly(-ethylene glycol)-co-polyaspartic acid-imidazole and methoxy poly(ethylene glycol)-co-poly β-benzyl L-aspartate was used for such nanoconjugates (Zhang et al. 2018).

Lutein-Loaded Solid Nanoconjugates

Lutein-Loaded Nanocrystals

Lutein-loaded nanocrystals (Figure 11.3D) are prepared by nanoprecipitation or antisolvent precipitation method, which involves injecting lutein into a surfactant dispersion medium followed by shearing action (Liu et al. 2017). Nanocrystals when surrounded by surfactants can entrap up to 99% of lutein due to increased surface area (Poorani et al. 2017). Ethyl cellulose and lutein nanoprecipitation was done with a surfactant (Tween 80) and a stabiliser (polyvinylpyrrolidone) by homogenising the combination at 10,240 rpm. This resulted in the formation of 490 nm lutein-loaded nanocrystals, effectively trapping 84%–92% of lutein (Bhadkariya et al. 2017). Although ethyl acetate and sodium caseinate precipitation could entrap 95.9% lutein into a nanocystal of 242 nm (do Prado Silva et al. 2017). Further, particle size of the lutein-loaded nanocrystal was reduced to 152.38 nm by using polyvinyl alcohol (Mishra et al. 2015). In addition, interfacial deposition using medium-chain triglycerides on polymers such as poly(ε-caprolactone) sorbitanmonostearate and polysorbate 80 could entrap 99.5% of lutein into a nanocrystal of 191 nm (Brum et al. 2017). In contrast, high-energy sonication (20kHz, 2 cycles, 30 seconds) using coconut oil, which innately is composed of medium-chain fatty acids, could entrap 99% of lutein into a nanocrystal of 1370 nm (Poorani et al. 2017). Henceforth, solvent displacement method can efficiently entrap lutein into small-sized nanocrystals (Tan et al. 2013).

Lutein-Loaded Lipid Nanoparticles

The enhanced lipid affinity of lutein enables it to be loaded into solid lipid nanoparticles and nanostructured lipid carriers (Figure 11.3E). Efficient nanostructured lipid carriers of as low as 130 nm size can be developed using an ultrasonic probe (Liu & Wu, 2010). The same technique can also be used to design solid lipid nanoparticles of 220–240 nm (Mitri et al. 2011). Similarly, nanoscaled lipid carriers using decanoic acid and span 60 resulted in a size of 229 nm and entrapping up to 59% of lutein (Liu et al. 2014). Further, 9% increase in lutein entrapment (68%) was achieved by incorporating a polysaccharide (cyclodextrin) with lipid carriers. However, a linear increment in particle size (336 nm) due to enhanced lutein accommodation by the adjuvant (cyclodextrin) was reported. Conversely, an increase in the particle size (with increasing lutein concentration) was observed with a decrease in lutein loading into the lipid carriers made of glycerol stearate, carnauba wax, corn oil, and fish oil when prepared by high-pressure homogenisation at 25,000 rpm (Lacatusu et al. 2013). Henceforth, lutein overloading can result in the failure of lutein adsorption on surfactant surface, which leads to reduced entrapment efficiency.

Lutein-Loaded Nanopolymeric Encapsulates

Lutein-loaded polymeric nanoparticles contain a polymer core where lutein is entrapped (Figure 11.3F). Methods such as nanoencapsulation and enhanced dispersion by supercritical fluids are adopted to prepare such nanoconjugates. Low molecular weight chitosan used as a polymer encapsulated 85% of lutein into an 80–600 nm polymeric encapsulate (Arunkumar et al. 2013). However, only a simplified shearing action at 500 rpm was provided. Incorporating zein by enhanced dispersion by supercritical CO_2 at 32°C and 1 mL/min flow rate, encapsulated 75% of lutein into the 198–244 nm nanoencapsulate (Hu et al. 2012). Particle size and encapsulation efficiency of the lutein-loaded nanopolymeric encapsulate is negatively affected by increase in the temperature. A similar sized (163–314 nm) lutein nanoencapsulate developed using atomisation by supercritical CO_2 showed better encapsulation efficiency (67–95%) (Heyang et al. 2009). Pulse sonication (30 seconds) for 15 min of PVP enabled the production of 200 nm sized lutein nanospheres (do Prado Silva et al. 2017). An association between the hydroxyl part of lutein and carbonyl part of PVP results in formation of soluble complexes and improves lutein solubility by 43 times. However, the use of Tween 80 lowers the encapsulating capacity of PVP due to the development of hydrogen between molecules that can transport lutein from the centre to the periphery (Zhao et al. 2014). Encapsulating lutein into soybean seed ferritin resulted in reduced size (12 nm) lutein–chitosan nanoconjugate (Yang et al. 2016). However, due to the small size, the conjugate was not able to encapsulate more than 16.8% of lutein. Emulsion evaporation technique adopted to develop lutein-loaded PLGA nanoencapsulates resulted in nanoconjugates of 124 nm size and 52% encapsulation efficiency (Kamil et al. 2016). Besides, lutein-loaded PLGA-PEG nanospheres exhibited 735-fold increase in aqueous solubility (Arunkumar et al. 2013).

Lutein Nanoconjugates as Nutraceutical Supplements

Owing to reduced size and other related properties including stability, protectiveness, and shelf life, nanoconjugates of lutein offer the potential for delivering lutein to its targeted sites. However, efficient encapsulation of lutein into the nanoconjugates and controlled release during and after gastrointestinal digestion into the blood is crucial for optimal delivery. The *in vitro* release behaviour of the nanoconjugate is therefore essential to review.

Release of Lutein from Nanoconjugates

Although the release pattern of lutein-loaded nanoconjugates is important to predict its stability, shelf, and oral bioavailability, only a few reports have attempted to study it (Arunkumar et al. 2013; Bhadkariya et al. 2017; Chang et al. 2017; Poorani et al.). *In vitro* release study can be performed in sophisticated setups such as Franz diffusion cell, USP (United States Pharmacopeia) dissolution apparatus, dialysis chamber, and release tester; otherwise by simple centrifugal method (Arunkumar et al. 2013). The amount of lutein released from the nanoconjugates needs to be measured at regular time intervals. Self-conjugated lutein nanoliposomes showed faster lutein release in the pH of blood (80% within 25h) than in the pH of the intestine (68% within 75h) (Chang et al. 2017). However, a release spurt of 30% was documented within the first 8h in either conditions. When lutein-loaded nanocrystals of PVP were subjected to release studies (2h) in gastric and intestinal pH using the USP dissolution apparatus, lutein release of 21% at gastric pH was followed by 92% at intestinal pH (Bhadkariya et al. 2017). Strong influence of lutein–polymer ratio affects the release profile of lutein-loaded nanocrystals. A micro-emulsion of lutein with coconut oil using Tween 80, when subjected to release study at blood pH by Franz diffusion cell, lutein release of 8.7% was reported within 2.5h (Poorani et al. 2017). The lutein release profile of lutein-loaded nanostructured lipid carriers developed with precirolAT05, Myverol18-04K, and PluronicF68 was compared to a nanoemulsion, which revealed that within 55h, nanoemulsion was able to release 90% of lutein and solid lipid carriers could release only 35% in simulated gastrointestinal conditions (Liu & Wu 2010). Similarly, lutein release patterns of lutein-loaded nanoemulsion, lutein-loaded nanostructured lipid carriers, lutein-loaded solid lipid nanoparticles were studied for 24h at 5.5 pH, where lutein release of 19.54%, 7–12% and 0.4% was observed, respectively (Mitri et al. 2011). The potential of fish oil to be used as a lipid nanocarrier for lutein was estimated by observing lutein release patterns of the product made with different proportions of fish oil to lutein. After 27h, a combination of 30% oil with 0.12% and 0.08% lutein, released 99% and 90% lutein, respectively (Lacatusu et al. 2013). The increased lutein release was due to the surface adsorption phenomena supported by lowered encapsulation efficiency (12% decrease). Although lutein–zeinnano encapsulates released 60% of lutein at blood pH within 40 min, only 85% of lutein was released at the end of 5h (Hu et al. 2012). Storage temperature, along with the type and quantity of adjuvants used also influences the product's stability and shelf life. A seven-day study of ferritin–chitosan-lutein shell core nanocomposites at 4, 20, and 37°C, reported release of 5%, 27%, and 69% of lutein, respectively (Yang et al. 2016). Besides, the intake of other nutrients has also been found to influence the release profile of nanocomposites. Although the co-existence of antioxidants such as epigalocatechingallate (EGCG), proanthocyanidins, and polyphenols lowered lutein release, consumption of milk promoted it (Yang et al. 2016). The effect of nutrients on the release of lutein nanoparticles is therefore important to be understood. The release pattern of lutein nanocrystals in combination with oral fast-dissolving films, when investigated in dissolution apparatus was able to release 17% more lutein within 3 min than conventional dissolving films (Liu etal.2017). On intake, the authors claimed better absorption of lutein-loaded nanocrystals.

During gastrointestinal digestion, lutein released from its carrier can further be depleted in the gut. It is therefore important to add stabilising agents. A recent study reported pH-dependent lutein release from polymersomes of mPEG-PAsp-IM cross linked with Fe^{+3} at pH 5 than pH 7.4 (Zhang et al. 2018). At pH 5, the breakage of a crosslink between the polymer and Fe^{+3} allows the polymer to show the hydrophilic nature in the same pH condition. This allows the release of lutein by micelle disassembly. In the case of nanoconjugates stabilised by protein, an association of protein with lutein can be disrupted on enzymatic hydrolysis during gastrointestinal digestion. However, the degradation of lutein in such nanoconjugates has been successfully reduced in comparison to free lutein degradation (Jiao et al. 2018). Besides, a controlled delivery of lutein was attained using glycosylated-casein (Millard reaction between casein and dextran). The bioaccessibility of lutein from this nanoconjugate increased by 47.82% (Muhoza et al. 2018). Dextran moieties protect the protein core by forming a steric cover over the micelle and keeping their charges away from adsorptive digestive enzymes. Therefore, incorporating high molecular weight dextran moieties by layering an interfacial layer of desired thickness offers steric hindrance and enables controlled delivery of lutein from nanoconjugates. It has been shown that in lutein entrapped polymersomes and polymeric encapsulates, lutein release is greatly affected by the nature of the chosen polymer, its amphiphilicity, and cross-linking with ions (Zhang et al. 2018). Although polymer is a major component of nanoconjugates, the medium in which it is dispersed is also equally important in lutein release. Besides, during the gastrointestinal digestion, intrinsic factors of the host such as bile salts significantly influence the absorption of molecules that are lipophilic through the mucosal lining in the gut (Maldonado-Valderrama et al. 2011). Because lutein absorption also follows the pattern of lipid absorption by chylomicrons, an association between bile salts and the absorption of lutein must definitely be deciphered. A recent research on glycosylated casein lutein nanomicelles indicates a favourable release and increase in the bioacessibility through the dissolution effect of bile salts under replicated gut conditions (Muhoza et al. 2018).

Cellular Uptake of Lutein Released from Nanoconjugates

Cellular uptake profile of lutein in *in vitro* conditions is usually studied in intestinal cells (Caco-2 cell line), which offers valuable advantage in recognising and interpreting the oral bioavailability of lutein. However, only a few scientists reported such studies. Increases of 543.4 pmol in the cellular uptake of lutein/mg cell protein were noted when comparing a traditional emulsion to a nanoemulsion (Teo et al. 2017). Phytoglycogen incorporation also increased lutein uptake by Caco-2 cells (Hua et al. 2017). Uptake and secretion of lutein was studied in Caco-2 cells on introduction to polylactic co glycolic acid nanoparticles (PLGA-NP). In comparison to free lutein, lutein uptake and secretion was tenfold and 50.5-fold higher in PLGA-NP (Kamil et al. 2016).

Lutein Permeation Through Simulated Gut Models

Ex vivo accumulation studies can deduce a better conception of lutein uptake, where a functioning tissue is tested for the permeation of the previously exposed quantity of lutein. Lutein released and permeated through the porcine cornea from lutein-loaded nanostructured lipid carriers associated with HPβCD was 4.9 times that of control particles (Liu et al. 2014). The preparation, however, did not penetrate the examined tissues fully. Similarly, an increase (10%) in goat eye lutein penetration was observed when micro-sized lutein emulsion with coconut oil was added (Poorani et al., 2017). A 9.2-fold increase in lutein accumulation in porcine sclera compared to free lutein was observed with cyclodextrin-associated lutein nanoemulsion (Liu et al., 2015). In a comparative study between lutein-loaded nanoemulsion, nanolipid carriers and solid lipid nanoparticles, 60, 19, and 7% of lutein was found to be penetrating the nitrocellulose membrane, respectively (Mitri et al., 2011). This suggested rapid release of lutein by lutein-loaded nanoemulsion. However, with dermal tissue from pigs, the findings were not reproducible.

Oral Bioavailability of Lutein Nanoconjugates in an Animal Model

The maximum clarification of the oral bioavailability and functionality of the lutein-loaded nanoconjugate be obtained only by conducting an *in vivo* analysis. Lutein encapsulated in low molecular weight chitosan (LMWC) administered orally (single dose) to mice after 8h manifested 62.8% accumulation in the eyes than in control (Arunkumar et al. 2013). Oral fast-dissolving films (OFDF) incorporated with lutein-loaded nanocrystals had a C_{max} of 828 ng/mL and $AUC_{0\text{-}24h}$ of 4021 ng.h/mL within 2 h of administration to male rats. The conventional OFDFs presented in the same study had lower C_{max} 295 ng/mL and $AUC_{0\text{-}24h}$ 2551 ng.h/mL (Liu et al. 2017). Lutein-loaded nanocrystals precipitated by polyvinyl alcohol (PVA) when administered to diabetic rats (50 mg/kg body weight) lowered the blood glucose to 105 mg/dL. The reduction was comparable to the positive control used (96.6 mg/dL), which was metformin (Mishra et al. 2015). The administration of lutein nanospheres in PVP to Swiss albino mice showed a dissimilarity in memory improvement, which was insignificant in comparison to free lutein (do Prado Silva, et al. 2017). However, increased synthesis of N-acetyl-beta-D-glycosaminidase and myeloperoxidase enhanced neuroprotective effects. Several studies have shown that, relative to free lutein, nanoconjugates of lutein have greater bioavailability. When administered to rats, lutein-loaded PLGA-NPs showed 15.6-fold increased plasma availability compared to free lutein (Kamil et al. 2016). Polylactic co glycolic acid polyethylene glycol (PLGA-PEG) nanoencapsulates loaded with lutein also demonstrated a 5.4-fold increase in postprandial plasma lutein concentration in mice relative to free lutein (Arunkumar et al. 2013). The antiproliferative activity of the nanoconjugate was also stated to be increased by 43.6%. An increase (twofold) in lutein concentration in the plasma in comparison to free lutein was recorded with long-term administration of lutein nanoemulsion to guinea pigs for 6 weeks. There was also a decrease (24%) in the steatosis score and a decrease (31%) in the accumulation of hepatic cholesterol (Murillo et al. 2016).

Oral Bioavailability of Lutein Nanoconjugates in Humans

With respect to functionality and safety, conclusions can only be taken on the basis of clinical trials. In humans over the age of 18 years, a five-week analysis (avoiding other sources of carotenoids) was performed by administering 6 mg of lutein nanoemulsion supplemented with orange juice. First week measurements were considered as baseline. Lutein supplements were given in the second week, followed by a washout phase (third and fourth week). Post treatment, measurements were taken during the fifth week and 62% increase in bioavailability was reported compared to the supplementation period (Vishwanathan et al. 2009). Henceforth, lutein-loaded nanoconjugates can be utilized for delivering lutein to targeted sites.

Concerns Related to Nanoconjugates of Lutein

Although lutein is a nutraceutical component of health benefit due to differential conditioning and distribution whose exposure can be harmful to the environment, risk assessment of nanomaterials has to be understood. As nanoconjugates have improved the bioaccessibility, altered absorption kinetics and improved bioavailability due to nanometre scale, a particular concern about the safety of nanonutraceutical supplements is needed. The behaviour of supporting materials used for engineering nanoscale supplements must also be included in the safety evaluation. During the development of lutein-loaded nano conjugates, the use of different stabilising agents and media necessitates studies on their toxicological aspects. This has not been presented satisfactorily by researchers.

Toxicity of Lutein Nanoconjugates

Cytotoxicity posed by pH-responsive lutein-mPEG-PAsp-IM (Fe^{+3}) micelles was examined in HeLa cells by the MTT assay in comparison with lutein-mPEG-PLBA micelles and free lutein (Zhang et al. 2018). Cytotoxicity caused by free lutein ($IC_{50} \sim 33$ μM) was higher than the polymersomes ($IC_{50} \sim 58$ μM). Further, cell viability studies were performed on HeLa cells by exposing them to polymersomes without lutein. Also, 40 μg/mL was the maximum concentration of polymersomes, which was found to be safe for cells. However, the application of the same in an *in-vivo* model was not reported. Acute and subacute toxicities of lutein-loaded PLGA-phospholipid nanoencapsulates were studied by administering series of single dose (0.1, 1, 10, and 100 mg/kg body weight) and daily dose (1 and 10 mg/kg body weight) for 4 weeks, respectively, to mice (Ranganathan et al. 2016). No signs of severe toxicity were identified upon examination of clinical and ophthalmic factors. However, the authors recommend further confirmation by clinical trials.

Regulatory Concerns

Regulatory authorities are currently concerned about the consideration of relevant technical questions (Justo-Hanani & Dayan 2016). Regardless of an encouraging societal outlook towards nutraceuticals and nanotechnology, a public perception of risk and hypersensitivity exists. This is due to the absence of a consistent description of nanofoods and a supervisory structure for the ingredients used in the formulation of nanoconjugates. This has additionally hampered the industrial development of nanoscience and the selling of nanoproducts (Chen & Yada 2011). A new team was formed in August 2006 to regulate supervisory theories relating to the use of nanomaterials in food, according to the Food and Drug Administration (FDA 2007) newsletter entitled "Nanotechnology Task Force Report." Although a regulatory definition and designated nanofood policy has not been achieved, guidelines for the classification of foods as nanofoods have been given. The primary aim of this was to continuously promote the production of novel and innovative nanosized food that is healthy to the environment. Apart from consumer awareness and pros-cons of nanosupplements, governing bodies should consider ethical issues and provide researchers with a forum to achieve legitimate market approval for nanosupplements that have been proven to be secure (Coles & Frewer 2013). Ultimately, the eco-friendliness, protection, corroboration of community benefits, and population health security are the considerations that play a role when launching a novel lutein nanoconjugate (Chen & Yada 2011).

Conclusion

Biological viability, biochemical affinity and stability, molecular relationships with the food matrix, and economical aspects are to be considered when selecting nanosized formulation for lutein. Colloidal particles can be employed to overcome the physicochemical deterioration of lutein nanoconjugates during gastrointestinal digestion by reducing the interaction with reactive constituents in food as well as gastrointestinal fluids. As the industry moves to customer-friendly labels, upcoming developments will persist to explore the replacement of synthetic ingredients with natural ingredients, making it increasingly necessary to identify successful nanoconjugates of lutein. However, it is still necessary to establish the acceptance of lutein nanoconjugates as marketable goods. Also, the validation for oral consumption of nanofoods is questionable. To direct the development of well-defined regulatory steps, the safety of nanodelivery systems for lutein needs further research. When assessing the safety aspects of nanoconjugates for food supplementation, good judgement must be made by governing bodies that regulate food processing and labelling. However, there is a need to conduct systematic investigations on the toxicological profiles of nanoconjugates. It is also important to develop improved *in vitro* methods that replicate *in vivo* approach. Because lutein has considerably higher oral bioavailability in nanoconjugates, it is important that the biosafety of nanoconjugates be defined. In addition, the societal understanding of nanodelivery systems needs to change for nanofoods to be more generally accepted.

REFERENCES

Amar I., Abraham A., and Nissim G. 2004. *Colloids and Surfaces B-Biointerfaces 33*(3–4): 143–150. doi:10.1016/j.colsurfb.2003.08.009.

Arunkumar R., Keelara V.H.P., and Vallikannan B. 2013. *Food Chemistry 141* (1): 327–337. doi:10.1016/j.foodchem.2013.02.108.

Bhadkariya N., Garud A., Tailang M., and Garud N. 2017. *World Journal of Pharmaceutical Research 6*: 467–477. doi:10.20959/wjpr201714-9903.

Bone R.A., Landrum J.T., Hime G.W., Cains A., et al. 1993. Stereochemistry of the human macular carotenoids. *Investigative Ophthalmology and Visual Science 34*(6): 2033–2040.

Borel P., Charles D., Marion N., Romain B., et al. 2014. *The American Journal of Clinical Nutrition 100*(1): 168–175. doi:10.3945/ajcn.114.085720.

Bortnowska G., 2015. *Polish Journal of Food and Nutrition Sciences 65*(3): 157–166. doi:10.2478/v10222-012-0094-0.

Brum A.A.S., dos Santos P.P., da Silva M.M., Paese K., et al. 2017. *Colloids and Surfaces A: Physicochemical and Engineering Aspects 522*(2017): 477–484. doi:10.1016/j.colsurfa.2017.03.041.

Burgos-Díaz C., Wandersleben T., Marqués A.M., Rubilar M., et al. 2016. *Current Opinion in Colloid & Interface Science 25*(2016): 51–57. doi:10.1016/j.cocis.2016.06.014.

Chang R., Yang J., Ge S., and Zhao M., 2017. *Food Hydrocolloids 67*: 14–26. doi:10.1016/j.foodhyd.2016.12.023.

Chen H., and Yada M. 2011. *Trends in Food Science & Technology 22* (11): 585–594. doi:10.1016/j.tifs.2011.09.004.

Chen H., and Yao Y. 2017. *Food Research International 97*: 258–264. doi:10.1016/j.foodres.2017.04.021.

ColesD., and Frewer L.N. 2013. *Trends in Food Science & Technology 34*(1): 32–43. doi:10.1016/j.tifs.2013.08.006.

Eisenhauer B., Natoli S., Liew G., and Flood V.M. 2017. *Nutrients 9*(2): 120. 10.3390/nu9020120.

Fatani A.J., Al-Rejaie S.S., Abuohashish H.M., Al-Assaf A., et al. 2015. *BMC Complementary and Alternative Medicine 15*(1): 204. doi:10.1186/s12906-015-0693-5.

Feeney J., O'Leary N., Moran R., O'Halloran M.O., et al. 2017. *Journals of Gerontology Series A: Biomedical Sciences and Medical Sciences 72*(10): 1431–1436. doi:10.1093/gerona/glw330.

Goldberg J., Gordon F., Ellen S., Brody J.A., et al. 1988. *American Journal of Epidemiology 128*(4): 700–710. doi:10.1093/oxfordjournals.aje.a115023.

Gutiérrez-Uribe J.A., Rojas-García C., García-Lara S., and Serna-Saldivar S.O. 2014. *Cereal Chemistry 91*(5): 508–512. doi:10.1094/CCHEM-07-13-0145-R.

Guzey D., and McClements J.D. 2006. *Advances in Colloid and Interface Science 128*: 227–248. doi:10.1016/j.cis.2006.11.021.

Hammond Jr B.R., Miller L.S., Bello M.O., Lindbergh C.A., et al. 2017. *Frontiers in Aging Neuroscience 9*: 254. doi:10.3389/fnagi.2017.00254.

Han H., Cui W., Wang L., Xiong Y., et al. 2015. *Lipids 50*(3): 261–273. doi:10.1007/s11745-015-3992-1.

Heyang J.I.N., Fei X.I.A., Jiang C., Yaping Z.H.A.O., et al. 2009. *Chinese Journal of Chemical Engineering 17*(4): 672–677. doi:10.1016/S1004-9541(08)60262-1.

Ho N.H., Inbaraj B.S., and Chen B.H. 2016. *Scientific Reports 6*: 25426. doi:10.1038/srep25426.

Hou C., Wu S., Xia Y., Phillips G.O., 2017. *Food Hydrocolloids 69*: 236–241. doi:10.1016/j.foodhyd.2017.01.038.

Hu D., Lin C., Liu L., Li S., et al. 2012. *Journal of Food Engineering 109* (3): 545–552. doi:10.1016/j.jfoodeng.2011.10.025.

Huebbe P., Lange J., Lietz G., and Rimbach G. 2016. *Biofactors 42* (4): 388–396. doi:10.1002/biof.1284.

Jiao Y., Zheng X., Chang Y., Li D., et al. 2018 *Food & Function 9*(1): 117–123. doi:10.1039/C7FO01652B.

Johnson E.J. 2014. *Nutrition Reviews 72*(9): 605–612. doi:10.1111/nure.12133.

Joseph S., and Bunjes H. 2013. *Drug Delivery Strategies for Poorly Water-Soluble Drugs 2013*: 103–149. doi:10.1002/9781118444726.

Justo-Hanani R., and Dayan T. 2016. *Global Environmental Politics 16*(1): 79–98. doi:10.1162/GLEP_a_00337.

Juturu V., Bowman J.P., and Deshpande J. 2016. *Clinical, Cosmetic and Investigational Dermatology 9*: 325. doi:10.2147/CCID.S115519.

Kale S.N., and Deore S.L. 2017. *Systematic Reviews in Pharmacy 8*(1): 39. doi:10.5530/srp.2017.1.8.

Kamil A., Smith D.E., Blumberg J.B., Astete C., et al. 2016. *Food Chemistry 192*: 915–923. doi:10.1016/j.foodchem.2015.07.106.

Kijlstra A., van der Made S.M., Plat J., and Berendschot T.T.J.M. 2017. *Journal of Food and Nutrition Research 5*(6): 362–369. doi:10.12691/jfnr-5-6-2.

Kishimoto Y., Taguchi C., Saita E., Suzuki-Sugihara N., et al. 2017. *Food Research International 99*: 944–949. doi:10.1016/j.foodres.2017.03.003.

Korobelnik J.F., Rougier M.B., Delyfer M.N., Bron A., et al. 2017. *JAMA Ophthalmology 135*(11): 1259–1266. doi:10.1001/jamaophthalmol.2017.3398.

Kostic D., White W.S., and Olson J.A. 1995. *The American Journal of Clinical Nutrition 62*(3): 604–610. doi:10.1093/ajcn/62.3.604.

Krinsky N.I., Landrum J.T., and Bone R.A. 2003. *Annual Review of Nutrition 23*(1): 171–201. doi:10.1146/annurev.nutr.23.011702.073307.

Lacatusu I., Mitrea E., Badea N., Stan R, et al. 2013. *Journal of Functional Foods 5*(3): 1260–1269. doi:10.1016/j.jff.2013.04.010.

Lefsrud M., Kopsell D., Wenzel A., and Sheehan J. 2007. *Scientia Horticulturae 112*(2): 136–141. doi:10.1016/j.scienta.2006.12.026.

Liu C., Chang D., Zhang X, Sui H., et al. 2017. *Aaps Pharmscitech 18*(8): 2957–2964. doi:10.1208/s12249-017-0777-2.

Liu C.H., Chiu H.C., Wu W.C., Sahoo S.L., et al. 2014. *Journal of Ophthalmology* 2014. doi:10.1208/s12249-017-0777-2.

Liu C.H., Lai K.Y., Wu W.C., Chen Y.J., et al. 2015. *Chemical and Pharmaceutical Bulletin 63*(2): 59–67. doi:10.1016/j.colsurfa.2009.11.006.

Liu C.H., and Wu C.T. 2010. *Colloids and Surfaces A: Physicochemical and Engineering Aspects 353*(2–3): 149–156. doi:10.1016/j.colsurfa.2009.11.006.

Lo J.T., Lee T.M., and Chen B.H. 2016. *Materials 9*(9): 13. doi:10.3390/ma9090761.

Ma T., Tian C., Luo J., Sun X., et al. 2015. *Journal of Functional Foods 16*: 104–113. doi:10.1016/j.jff.2015.04.020.

Madene A., Jacquot M., Scher J., and Desobry S. 2006. *International Journal of Food Science & Technology 41*(1): 1–21. doi:10.1111/j.1365-2621.2005.00980.x.

Maldonado-Valderrama J., Wilde P., Macierzanka A., and Mackie A. 2011. *Advances in Colloid and Interface Science 165* (1): 36–46. doi:10.1016/j.cis.2010.12.002.

Mamatha B.S., Arunkumar R., and Baskaran V. 2012. *Food and Bioprocess Technology 5*(4): 1355–1363. doi:10.1007/s11947-010-0403-8.

Mamatha B.S., and Baskaran V. 2011. *Nutrition 27*(9): 960–966. doi:10.1016/j.nut.2010.10.011.

Marriage B.J., Williams J.A., Choe Y.S., Maki K.C., et al. 2017. *British Journal of Nutrition 118*(10): 813–821. doi:10.1017/S0007114517002963.

McClements D.J. 2018. *Comprehensive Reviews in Food Science and Food Safety 17*(1): 200–219. doi:10.1111/1541-4337.12313.

McClements D.J., Decker E.A., and Weiss J. 2007. *Journal of Food Science 72*(8): R109–R124. doi:10.1111/j.1750-3841.2007.00507.x.

Mishra S.B., Malaviya J., and Mukerjee A. 2015. *Journal of Pharmaceutical Sciences and Pharmacology 2*(3): 242–249. doi:10.1166/jpsp.2015.1067.

Mitri K., Shegokar R., Gohla S., Anselmi C., et al. 2011. *International Journal of Pharmaceutics 414*(1–2): 267–275. doi:10.1016/j.ijpharm.2011.05.008.

Mohn E.S., Erdman Jr J.W., Kuchan M.J., Neuringer M., et al. 2017 *PLoS One 12*(10): e0186767. doi:10.1371/journal.pone.0186767.

Muhoza B., Zhang Y., Xia S., Cai J., et al. 2018. *Journal of Functional Foods 45*: 1–9. doi:10.1016/j.jff.2018.03.035.

Murillo A.G., Norris G.H., DiMarco D.M., Hu S., et al. 2016. *The FASEB Journal 30*: 913–2. doi:10.1096/fasebj.30.1_supplement.913.2.

Nolan J.M., Loskutova E., Howard A.N., Moran R., et al. 2014. *Journal of Alzheimer's Disease 42*(4): 1191–1202. doi:10.3233/JAD-140507.

Poorani T.R., Vellingiria V., Deepa V.S., and Abdula A.N. 2017. *International Journal of Pharmaceutical Science and Research 8*(10): 4159–4171. doi:10.13040/IJPSR.0975-8232.

Power R., Coen R.F., Beatty S., Mulcahy R., et al. 2018. *Journal of Alzheimer's Disease 61*(3): 947–961. doi:10.3233/JAD-170713.

Qiu X., Gao D.H., Xiang X., Xiong Y.F., et al. 2015. *World Journal of Gastroenterology: WJG 21*(26): 8061. doi:10.3748/wjg.v21.i26.8061.

Ranganathan A., Hindupur R., and Baskaran V. 2016. *Materials Science and Engineering: C 69*: 1318–1327. doi:10.1016/j.msec.2016.08.029.

Read A., Wright A., and Abdel-Aal E.S.M. 2015. *Food Chemistry 174*: 263–269. doi:10.1016/j.foodchem.2014.11.074.

Reboul E., Thap S., Tourniaire F., André M., et al. 2007. *British Journal of Nutrition 97*(3): 440–446. doi:10.1017/S0007114507352604.

Renzi H., Lisa M., Bovier E.R., Fletcher L.M.L., et al. 2017. *Nutrients 9*(11): 1246. doi:10.3390/nu9111246.

Riedl J., Linseisen J., Hoffmann J, and Wolfram G. 1999. *The Journal of Nutrition 129*(12): 2170–2176. doi:10.1093/jn/129.12.2170.

Roodenburg A.J.C., Rianne L., van het Hof K.H., WeststrateJ.A., et al. 2000. *The American Journal of Clinical Nutrition 71* (5): 1187–1193. doi:10.1093/ajcn/71.5.1187.

Salvia T., Belloso L.O.M., and McClements D.H. 2016. *Nanomaterials 6*(1): 17. doi:10.3390/nano6010017.

Sánchez C., Baranda A.B., and de Marañón I.M. 2014. *Food Chemistry 163*: 37–45. doi:10.1016/j.foodchem.2014.04.041.

Setya S., Talegaonkar S., and Razdan B.Z. 2014. *World Journal of Pharmacy and Pharmaceutical Sciences 3*(2): 2214–2228.

Sharavana G., and Baskaran V. 2017. *Journal of Functional Foods 31*: 97–103. doi:10.1016/j.jff.2017.01.023.

Shen R., Yang S., Zhao G., Shen Q., et al. 2015. *Journal of Cereal Science 61*: 86–93. doi:10.1016/j.jcs.2014.10.009.

Tan C., Xia S., Xue J., Xie J., et al. 2013. *Journal of Agricultural and Food Chemistry 61*(34): 8175–8184. doi:10.1021/jf402085f.

Tan T.B., Yussof N.S., Abas F., Mirhosseini H., et al. 2016a. *Journal of Food Engineering 177*: 65–71. doi:10.1016/j.jfoodeng.2015.12.020.

Tan T.B., Yussof N.S., Abas F., Mirhosseini H., et al. 2016b. *Food Chemistry 194*: 416–423. doi:10.1016/j.foodchem.2015.08.045.

Teo A., Lee S.J., Goh K.K.T., and Wolber F.M. 2017. *Food Chemistry 221*: 1269–1276. doi:10.1016/j.foodchem.2016.11.030.

Vershinin A. 1999. *Biofactors 10*(2–3): 99–104. doi:10.1002/biof.5520100203.

Vishwanathan R., Iannaccone A., Scott T.M., Kritchevsky S.B., et al. 2014. *Age and Ageing 43*(2): 271–275. doi:10.1093/ageing/aft210.

Vishwanathan R., Wilson T.A., and Nicolosi R.J. 2009. *Nano Biomedicine and Engineering 1*(1): 38–49. doi:10.5101/nbe.v1i1.p38-49.

Wolf-Schnurrbusch U.E.K., Zinkernagel M.S., Munk M.R., Ebneter A., et al. 2015. *Investigative Ophthalmology & Visual Science 56*(13): 8069–8074. doi:10.1167/iovs.15-17586.

Woodall A.A., Britton G., and Jackson M.J. 1997. *Biochimicaet BiophysicaActa (BBA)-General Subjects 1336*(3): 575–586. doi:10.1016/S0304-4165(97)00007-X.

Xavier A.A.O., Carvajal-Lérida I., Garrido-Fernández J., and Pérez-Gálvez A. 2018. *Journal of Food Composition and Analysis 68*: 60–64. doi:10.1016/j.jfca.2017.01.015.

Xavier A.A.O., Mercadante A.Z., Garrido-Fernández J., and Pérez-Gálvez A. 2014. *Food Research International 65*: 171–176. doi:10.1016/j.foodres.2014.06.016.

Xu D., Aihemaiti Z., Cao Y., Teng C., et al. 2016. *Food Chemistry 202*: 156–164. doi:10.1016/j.foodchem.2016.01.052.

Yang R., Gao Y., Zhou Z., Strappe P., et al. 2016. *RSC Advances 6*(42): 35267–35279. doi:10.1016/j.foodchem.2019.125097.

Zhang D., Wang L., Zhang X., Bao D., et al. 2018. *Journal of Drug Delivery Science and Technology 45*: 281–286. doi:10.1016/j.jddst.2018.03.023.

Zhao C., Cheng H., Jiang P., Yao Y., et al. 2014. *Food Chemistry 156*: 123–128. doi:10.1016/j.foodchem.2014.01.086.

Zhao L., Temelli F, Curtis J.M., and Chen L. 2017. *Food Research International 100*: 168–179. doi:10.1016/j.foodres.2017.06.055.

12

Advances in Nanonutraceuticals: Indian Scenario

Amthul Azeez, Mubeen Sultana, Lucky, and Noorjahan
Justice Basheer Ahmed Sayeed College for Women, Chennai, Tamil Nadu, India

CONTENTS

Introduction

Since time immemorial, man has looked towards a wide variety of raw materials procured from both naturally occurring substances as well as synthetic origins, to produce medicines. Over the past few years, the interest of researchers and scientists has been focused on food products of natural origin alternatives to pharmaceuticals that can be used as tools to prevent and in some cases cure or delay the onset of a health issue. It was the breakthrough in comprehending the correlation between health and nutrition that

DOI: 10.1201/9781003153504-12

gave an impetus to the conception of nutraceuticals as the most trustworthy approach to attain indefectible health as well as decrease the risks associated with certain diseases. Nutraceutical is a term given by Stephen De Felice. Nutraceuticals are phytocomplexes derived from foods of vegetal origin and as a pool of secondary metabolites. They are obtained from the sources of living organisms given in the form of drugs. It is a non-toxic food if supplement that aids in the prevention and treatment of health defects (Santini et al. 2017). Nutraceuticals are complementary natural compounds that are usually configured as tablets, capsules, and palatable syrups comprising broad-spectrum vitamins, naturally found herbs, other plant-derived components that can be consumed or used as dietary supplements, which can help to enhance the health of individuals. In recent times, the nutraceutical industry has expanded greatly according to the needs and conscious consumer interest and demands (Keservani et al. 2010). Most of the nutraceuticals, which are in great demand, are antioxidants, natural vitamins, polyunsaturated fatty acids (PUFA), dietary fibres. Apart from these nutraceuticals, which fall under the category of prebiotics and probiotics, also have a great consumer appeal. There is a fragile line of distinction in classifying a nutraceutical as a food or drug. The success of use or consumption of nutraceuticals depends on the safety and efficacy of a specific compound used and ideally supported by clinical data. The majority of the nutraceuticals differ from the regular health-protecting and disease-preventing measures, as they are often used as dietary supplements to improve the overall health condition of an individual (Josef and Katarina, 2019).

Nutraceuticals in India

It is a known fact that India is a highly populated country and its population is known to increase by 2024. In this scenario, with the growing population, the obvious concern is about health and well-being, which can be made better and achieved by focusing on nutrition, which essentially can prevent major health issues. Therefore, healthcare and nutrition are in for a mega revolutionary change. The use of natural compounds, termed nutraceuticals, for better health is not a new concept to India. The Indian ancient traditional medicinal system, known as Ayurveda, has long been involved in identifying and using traditional herbs and other plant components for the benefit of human health. An Ayurvedic preparation, such as Chavanprash, which is a mixture of various natural compounds, is frequently used in almost every household for improving overall health. Certain extracts from plants, such as aswagandha, triphala, and dasamkantichurna, are some of the ingredients used frequently to improve oral/dental health. In addition, certain commonly available plants or their components, such as turmeric, chilies, garlic and ginger are frequently included in the everyday diet because of their nutritional value and as immunity boosters. In recent times, compounds such as Vanaspati fortified with Vitamin A and iodinated salt are also freely available in the market (Potential for functional foods in the Indian market 2018). It is quite interesting, to note that, nutraceuticals have now superseded the pharma supremacy, specifically in the domain of nutritional supplements, which is quite evident by the 67% share taken by nutraceuticals. This paradigm shift from pharmaceuticals to nutraceuticals can be attributed to increased awareness of the population about the health hazards and an overall better understanding of healthcare. In India, nutraceuticals are frequently used in southern parts, followed by the eastern part of the country and most rural population is gaining more awareness about nutraceuticals (India Infoline News Service 2018).

Nutraceuticals: A Natural Approach

Regularly used foodstuffs and beverages are considered to contain substantial amounts of natural ingredients that qualify them to regard as nutraceuticals, which are believed to promote robust growth and overall good health. Nutraceuticals are not some exotic compounds but naturally occurring phenols, abundantly found in food items, such as green tea, red berries and in different types of spices. Most of these active phenolic compounds are known to treat amyloid neurodegenerative disorders. Apart from these regular foods, a number of other routinely used in cuisine, such as flax seed oil (tocotrienols), carrots (carotenoids), turmeric (curcumin) all contain one or other form of bioactive compounds, which may be recognised as nutraceuticals. Fruits, vegetables, and seafood also contain abundant quantiles of omega fatty acids; allyl-sulphides found in garlic, beta-glucans of mushrooms/cereals are some of the examples of natural compounds that can be considered for the synthesis of nutraceuticals. Nuts contain fatty acids and are also rich in polyunsaturated fatty acids (PUFA) that can help limit cholesterol. Egg shells are a rich source of calcium carbonate and calcium powder that is derived from these shells can be used as a calcium supplement to treat calcium deficiency (Josef & Katarina 2019). There are quite a number of studies that show convincing results, illustrating that certain bioactive compounds present in soya bean, garlic, ginger, green tea and honey are known to have an apoptotic (programmed cell death) effect on cancer cells and hence these natural substances may aid in preventing chemotherapy for the treatment of cancer and therefore be recommended as an ancillary remedy along with standard cancer treatment protocols. As there are quite a number of valuable, active molecules being identified as anticancer agents in the commonly used food, these food-based nutraceuticals have caught the imagination of scientific researchers. The reason for this widespread clamor for nutraceuticals for cancer therapy is due to is their easy accessibility, high degree of absorption and transformation, which leads to the easy availability of the compound in the biological system. One such example is worth mentioning, wherein neem is used via cinnamon oil, which acts as a lipophilic agent to convey vitamin D to the lungs, where it has been identified to treat alveolar carcinoma cells in humans and hence vitamin D qualifies to be used in the food industry as well as processed to be an anticancer nutraceutical.

Nanotechnology

A new field has emerged in recent times, known as nanotechnology, wherein the particles are scaled down to nanosize and these nanosized particles have unique properties and have advanced capacities that can be passed on to a broad spectrum of operations. Nanomaterials are natural or synthetically developed compounds that are present as individual entities or in a combination with other components. These nanosized particles are of great importance as they can be efficiently used in diagnostics for disease detection, tissue repair, and regeneration and also effectively used in drug delivery systems. In certain diseases, such as cancer, it is often necessary to convey the drug exactly to the target site, which can be efficiently done by nanoparticles. The advantages of using nanoparticles in the drug delivery system over the regular system are that nanoparticles increase the half-life of the drug, when injected into the system, minimise the risk of a drug-triggered immune response and enhance the solubility of many hydrophobic drugs and also discharge the medication in a continuous or uninterpreted manner, thereby reducing the frequency of drug administration and thereby minimising the drug-related side effects. This evidently suggests that nanotechnology has certainly revolutionised the way pharmaceutical concentrations are synthesised and administered. With the advent of nanotechnology, drug dosage concentration as minute as mini, micro or nano can be administered and hence this technology can be a resourceful means to regulate treatment guidelines for the administration of biomolecules.

From Nanopharmaceuticals to Nanonutraceuticals

Recent advancements in the field of nanotechnology have totally revolutionised the field of medicine and the pharmaceutical industry. In the present day, there are two common terms that are constantly being referred are nanopharmaceuticals and nanonutraceuticals. Nanopharmaceuticals are a combination of manufactured or synthesised drugs amalgamated with nanoparticles and similarly a nanonutraceutical is a composite of a natural plant constituent merged with nanoparticles. The entire motive in developing these nano-based compounds is to promote and enhance the pharmacodynamics and pharmacokinetics of the drug administered, which can completely revolutionise the treatment protocols and methodology. As per the standards of nanotechnology, nanosized particles must be prepared so that these are less than 100 nm, atmost in one dimension. These nanosized particles have exceptional abilities, all-round functionality that make them supreme wonder material, which finds an application in almost all fields of everyday life such as agriculture, cosmetic industry and medical diagnostics to name a few (Farokhzad and Langer 2006; Davis et al. 2008). In the pharmaceutical industry, these nanomaterial/nanoparticles are big players, where pharma products are combined with nanoparticles and are used in the treatment of various diseases such as cancer. Several neurodegenerative diseases and infectious diseases are also successfully treated using nanopharmaceuticals. Natural compounds, such as polysaccharides and lipid molecules, can be used to process nanoparticles, in addition to this synthetic nanoparticles can also be prepared using polymers. Nanomaterials or nanoparticles, which can be used in the pharma industry, must be biocompatible, easily degradable, and also certified as generally recognised as safe (GRAS), to be qualified for usage in the pharma industry and in the manufacture of nutraceuticals (Muller et al. 2006; Martins et al. 2009; Souto and Muller 2010; Doktorovova et al. 2016).

Perhaps the best suitable candidate for orally administering nanonutraceuticals would be lipid molecules. This is because lipids can easily penetrate the mucosal lining and can rapidly enhance the possible availability of the injected drug or molecule. Lipid nanoparticles are frequently used in drug delivery systems for drugs that have poor solubility; these are packaged into suitable lipid-based nanoparticle formulations for easy entry into the biological systems. Nanoparticles of lipid origin can be synthesised using lipid entities such as cholesterol, triglycerides, fatty acids and also phospholipids, as these are more biocompatible and biodegradable and are abundantly available in the human body or food items (Severino et al. 2011). Apart from lipids, carbohydrates such as chitosan, alginates are also used as source material for nanoparticle synthesis (Severino et al. 2015, 2019; Andreani et al. 2019). Among these polysaccharides, chitosan has adhesive properties and is easily permeable through membranes and therefore can increase the drug availability to the cells if the nutraceuticals are administered orally as chitosan-derived nanoparticles. In addition to being a good source of nanoparticles, chitosan is a natural polymer, least toxic and by itself exhibits antimicrobial properties and has a functional group that can be altered as per requirements and can be used for target-specific drug delivery. Another natural polymer is alginate that also exhibits adhesive characteristics also known to be less lethal and so can be used safely *in vivo*. Alginate-derived nanoparticles are more hydrophilic and hence can hold large concentrations of water-soluble drugs in a small volume and are used for targeted drug delivery. Apart from being used for drug delivery, alginate nanoparticles may also be used as a stimulant or as a synergist for vaccinations. Alginate can also be conjugated with dextran to produce nanoparticles, which can be used to adjust the concentrations of proteins or any other biomolecule that needs to be given orally (Sarei et al. 2013). Nutraceuticals that are processed as nanonutraceuticals have evolved to be encouraging contenders with the potential to be effective in treatment without compromising the safety of the patients especially for those who cannot be qualified for regular or routine therapeutic procedures. A more comprehensive study and detailed experimentation must be conducted for conclusive results and these data may be analysed to expand this new field of nanonutraceuticals for extensive and broader applications to be implemented into further categories of medical therapeutics.

Materials Used in Nanoencapsulation

A wide array of food components is sometimes coated or enclosed in special packaging material to serve a specific cause. This method is known as encapsulation. This encapsulation may be done using any state of matter such as solid, liquid or gas. At times nanoscale particles are also used in the packaging or encapsulation. This encapsulation is deliberately done in food packaging such that certain minute components of bioactive components of nutraceuticals can possibly enter the food source (De Felice 1992; Santini et al. 2017). All nutraceuticals cannot be encapsulated in the same way, depending upon their different molecular and physiological functionalities; the delivery system for each nutraceutical varies. The methods used for encapsulation are spray drying or emulsification (Rekha et al. 2009).

Encapsulation is a process, wherein a material is coated with a biocompound. In this context, proteins are frequently used for encapsulation. It is interesting to note that proteins that are used for encapsulation possess certain special characteristics which aid them to develop into gels, protein and also colloids. Recent research suggests protein encapsulated lipid molecules have an enhanced bioavailability when they are consumed orally and hence may prove to be effective.

Fat molecules are yet another class of biological compounds that can be processed to encapsulate bioactive ingredients. Liposomes are a class of lipids molecules that are used to synthesise vesicular carriers which are drug delivery vehicles, with a phospholipid backbone that operates as a semi-permeable membrane. A subclass of lipids, known as phospholipids, which are bilayered are frequently used as vesicular carriers and are known as nanoliposomes. These are the most encouraging contenders for the nanonutraceutical industry. They have the unique capacity to encapsulate both lipophilic and hydrophilic components concurrently resulting in an effective synergistic reaction. This synergism has been extensively exploited to retain the sensitivity of bioactive compounds, boost their accessibility, secure, uninterrupted release and also promote better stability when stored. These exclusive properties of nanoliposomes qualify them to be availed as dietary supplements (DISs) for the maintenance of good health and disease avoidance (Josef and Katrina 2019).

Another nano-based vesicular carrier is the nanophytosome, which in recent times has gained importance in drug delivery systems especially to deliver plant-based bioactive components. These are nanoparticles that contain water-soluble plant contents at one end bonded with phospholipids to cause easy passage through the biomembranes. These are frequently used in the manufacture of food commodities and food beverages.

In addition to proteins and lipids acting as nano-based drug delivery vehicles, carbohydrates are also equally competent molecules, qualified to be processed into drug delivery vehicles. They are very ideal for the synthesis of nano-based compounds as they are biocompatible and also can be degraded easily. They have a profound capacity to interact with a wide range of functional groups and so establish them to be accomplished molecules, capable of binding to an array of hydrophilic and hydrophobic compounds (Jose Roberto 2019).

Nanoparticles in Tissue Engineering and Regenerative Medicine

Regeneration of lost tissues involves a number of calculated preprogrammed phases identified as cellular adhesion, cell migration and cellular regeneration. An advanced branch of biology, identified as regenerative medicine, involves the development of new methods and techniques to achieve tissue repair, dead tissue replacement, and also restore normal shape and function of dying or lost tissue or organ. With the advent of nanotechnology and nanomedicine, a whole new era has dawned in the field of regenerative medicine. This technology has substantially revolutionised the developmental dynamics of cells/tissues. The internal microenvironment of a cell and its biochemical and mechanical functions can be easily manipulated using efficient nano-based drug delivery systems. The advantages of nanoparticles, over other regular materials, are that these nanoparticles have completely different physical, chemical and magnetic properties which make them more suitable and efficient vehicles or vectors for drug delivery and therefore change the entire treatment regime of a disease. Apart from this, nanoparticles can also be shaped to form scaffolds that can be successfully used for tissue engineering, where lost or damaged parts can be replaced. The advantages of employing nanoparticle-based tissue scaffolds are that they considerably resemble the original cellular matrix and cause the release of certain developmental molecules and also cause the expression of certain specific spatiotemporal genes, which can direct behavioural changes ultimately governing the establishment of an original tissue that can be implanted to replace the damaged one. Apart from these applications, the role of nanotechnology specifically with nano-based drug delivery systems in stem cell-based tissue regeneration is worth knowing. As stem cells have the potential to be differentiated into any cell type triggering their capacity to differentiate into custom-made cell types may be induced, by delivering certain biologically active components, using a nano-based drug delivery system, which can revolutionise regenerative medicine. Nanotechnology-based solutions, for the treatment of diseases and tissue regeneration therapy and techniques are more affordable and easily available.

Gold Nanoparticles in Tissue Engineering and Regenerative Medicine

Nanoparticles can be processed using quite a number of metals and also using precious metals, such as gold. Gold nanoparticles, AuNPs (gold nanoparticles), are processed as sub-microscopic particles, entangled or suspended in a gel-like matrix. These nanoparticles have entirely different physical, chemical properties, unlike solid gold (Daniel and Astruc 2004). These matrix entangled gold nanoparticles, can be affixed to a variety of ligand molecules such as polypeptides, antibodies and also protein complexes.

Not only biomolecules but also other compounds such as phosphines, thiols, amines, etc., all the metals, which have a greater affliction for gold can be conjoined with gold nanoparticles (Alivisatos et al. 1996). Gold nanoparticles find a greater application in regenerative medicine for replacing tissue, as it is least reactive and safe to implant when cancerous tissue is removed.

Literature survey reveals that a number of compounds have been used for the preparation of nanoparticles that can be successfully used as scaffolds for tissue regeneration, repair, and drug delivery system. The majority of tissue regeneration studies have shown that gold is being used as the most preferable scaffolding material for bone growth as they have a greater capacity to cause cell differentiation (Zhang et al. 2014; Ko et al. 2015). Recent reports suggest that there was augmented bone growth, during regeneration of a bone when a composite of gold nanoparticles and gelatin were used for the preparation of scaffold (Heo et al. 2014). A combination of gold and gelatin is known to cause bone differentiation from the stem cells derived from adipose tissue, both *in vivo* and *in vitro*. A compound, called 2,2 66-tetramethylpiperidine-N-oxyl (TEMPO) if when conjugated with gold nanoparticles (AuNPs) is known to be absorbed easily into mesenchymal stem cells of humans (MSCs) capable of lowering the concentration of ROS and cause the heightening of bone differentiation from stem cells while suppressing the differentiation of adipose tissue (Li et al. 2017). This property of the combined compound has been since then used to achieve two functions, namely reduction of ROS and also to cause differentiation of mesenchymal stem cells into desired cell types. In addition to bone regeneration, gold nanoparticles have also been used successfully in skin wound healing experiments. Gold nanoparticles, in combination with electrospun silk fibre used in wound healing, revealed that there was a better vascularization, (blood vessel formation) and improved tissue granulation when compared to the untreated skin control group, illustrating the effectiveness of gold nanoparticles (Akturk et al. 2016).

Silver Nanoparticles in Tissue Engineering and Regenerative Medicine

In addition to gold, another precious metal that has promising medical applications is silver. This metal is known to have extensive antimicrobial properties against a broad spectrum of microorganisms. The destruction of microorganisms by silver ions is because it is able to obstruct the respiratory mechanism of the bacterial cells and precipitate the cellular protein, thereby causing their death (Abbasi et al. 2016). Silver nanoparticles; have also been effectively used in wound dressing and also to heal wounds caused by burns that are highly prone to microbial infections. In the context of microbial infections, it is important to learn about multidrug-resistant microorganisms that are of major health concern in recent times. To treat these multidrug-resistant infections in wound healing scaffolds are prepared using an assortment of materials consisting of PVA (Poly Vinyl Alcohol), gelatin, chitosan in combination with alginic acid (alginate), and also cellulose acetate, along with silver nanoparticles and these preparations proved to be quite effective in the treatment against multidrug-resistant infections. Silver nanoparticles that are used in the preparation of scaffold can be prepared separately and incorporated into the scaffold, or they can be directly prepared at the very site of infection (*in situ*), by exposure to heat, UV radiation or by a chemical. Using these composite materials, including silver nanoparticles used in scaffold preparation it can be molded into electrospun fibres, fibre mats and porous scaffolds (Son et al. 2004; Hong et al. 2006; Mehrabani et al. 2018; Yahyaci et al. 2018). The roles of silver nanoparticles in wound healing and their antimicrobial properties were extensively studied using animal models. These experiments revealed that silver nanoparticles had an increased rate of healing and also displayed effective antibacterial action (Tian et al. 2007). It was also observed that silver nanoparticles are known to regulate cytokines which are present in the wounds caused by burns. Wounds, when treated with silver nanoparticles, showed a decline in neutrophil content indicating that there was a substantial reduction in regional inflammation followed by formation and deposition of new skin layers, which were evident after 14 days of treatment, consolidated with histological analysis (Wang et al. 2018).

Bioactive compounds such as protein, carbohydrates, lipids and vitamins are sensitive to the heat and acidic content of the stomach. Encapsulation of these compounds helps them to resist and also to assimilate into food products. Nanoemulsification, nanocomposite and nanofabrication techniques are involved to encapsulate the substances which deliver the nutrients such as protein and antioxidants for targeted nutritional and health benefits.

Bioactive compounds present in functional foods are often degraded due to the hostile environment leading to degradation. Nanoencapsulation of active components extends the shelf life and prevents degradation; moreover, the nanocoatings on food materials prevent moisture and gas exchange thus preserving the colour and flavour of the food. Curcumin, the bioactive component of turmeric is the least stable and upon encapsulation, it is found to be stable to pasteurization. Encapsulating the functional component within droplets decreases the chemical degradation process by engineering the properties of the interfacial layer surrounding them. Nanotechnology shows a huge prospective in the field of tissue engineering to mimic the porous structure of the natural extracellular matrix. Scaffolds from natural sources have protein or carbohydrate with specific biochemical properties, low-level toxicity and found to be biodegradable and also shows compatibility. Aloe vera is rich in nutrients and polysaccharides. The use of aloe vera gel for scaffold fabricating is done by electrospinning and freeze-drying techniques and a thin coating of gellan gum (GG) was applied to Aloe vera gel to increase structural stability (Silva et al. 2014). Aloe vera-based sponges showed heterogeneous porous formation, great porosity and interconnected pores. Antibacterial potency was also displayed by blended membranes. *In vitro* assays have shown that Aloe vera blended membranes have good compatibility with primary human dermal fibroblasts (Silva et al. 2013). Aloe Vera based sponges have potential uses in biomedical applications and might be a promising wound dressing material.

Curcumin reported pharmacological effects such as antioxidant, anti-inflammatory, and anti-tumour activities; however, its clinical applicability is limited due to its low availability during oral administration. Aloe vera and curcumin loaded oxidised pectin-gelatin (OP-Gel) matrices used as antimicrobial finishes on nonwoven cotton fabrics produce composite wound care devices (Tummalapalli et al. 2016). OP-Gel-Aloe treated wounds exhibited 80% of wound healing in just 8 days, further aloe vera exerted anti-inflammatory and scar prevention effects. Histological analysis revealed orderly collagen and neovascularization along with nuclei migration. OP-Gel-Aloe bio composite dressings are proposed as variable materials for effective wound management. The development of a composite material for dressing wounds with nanosilver nanohydrogel along with aloe vera and curcumin promotes antimicrobial activity with wound healing and infection control (Anjum et al. 2016).

Role of Herbal Plants in Regenerative Medicine

In ancient times, man has used different herbal plants to cure certain diseases. In India, more than 2,500 plant species have been identified to have medicinal values. The Indian Pharmacopoeia (1966) recognised 85 drug plants whose ingredients are utilised for the preparation of pharmaceuticals. Herbal medicines have the capacity to affect the body systems and possess curative properties due to the presence of various chemical substances present in the plant. However, the popularity of the usage of herbal medicine is declining due to the spread of modern medicine and surgery. Herbal plants not only provide nutrients but also improve the activity of the digestive system, enhance the rate of processing food including the absorption of nutrients. It also improves the capacity of the body system to remove toxins. Once the toxins are decreased, the body system is able to obtain high resources repairing and strengthening damaged tissues and weakened organs. The factors that influence wound healing and regeneration of new tissues include vascularity, nutritional deficiency or imbalance such as deficiency of Vitamin A, C, E, K, and proteins. The medicinal plants contain all ingredients/factors needed for proper wound healing and new tissue regeneration. They act as antiseptics preventing the growth of bacteria and fungi such as onion (*Bulbus alliicepa*) in Tamil Vengayam and Garlic (*Allium sativum*) in Tamil Acanam, Neem (*Azadirachta indica*) in Tamil Veppai. Sengumaru possesses antibiotic activity. Turmeric (*Curcumin longa*) L, in Tamil Manjal, has proteins, fats, vitamins (A, B, C), etc., which act as an antiseptic agent and play a significant role in the healing of wounds and regeneration of new tissues. It possesses anti-inflammatory properties and contains Vitamin A and proteins that help in the early synthesis of collagen fibres by initiating fibroblastic activity. Several herbal plant extracts have the ability to initiate adult stem cell proliferation and thus aids in the regeneration of damaged or disease tissues.

When wounds are treated with honey it leads to fast and early healing of wounds as honey contains a large quantity of glycine, methionine, and proline that play a significant role in the formation of collagen.

A large content of sugar in honey may also be responsible for the early growth of granulation tissues in the wounds. Moreover, honey produces hydrogen peroxide by oxidation of glucose and the presence of inhibin, which contributes to its strong antimicrobial property. It is also an excellent source of energy in the catabolic environment which is one of the factors that enhances the process of healing, with the capacity to fight against infections.

Spices such as anise, basil, bay leaf, caraway, celery, citrus, clove, coriander, and horse radish have neurogenic components that have an ability for growth of nerves or regeneration of nerves. Compounds such as daidzein and genistein found in most edible legumes exhibit cytoprotective activity that aids in Parkinson's disease, N-acetyl serotonin is a natural chemical intermediate compound that helps in the synthesis of melatonin which is a potent neurogenic. The herbal plants that contain melatonin are cabbage, ginger, radish and mustard greens which also possess the precursor N-acetyl serotonin.

Basil is a potential herb with flavonoid compounds and also acts to restrict the growth of harmful bacteria being a strong antibacterial agent. Flavonoid blocks the molecule that stimulates inflammation. Oregano is one of the delicious herbs rich in antioxidants that help fight the damage that occurs due to harmful free radicals in the body. This has also enriched folate compounds which help to form RNA and DNA building blocks for the regeneration of cells. Thyme is a herb that contains apigenin, a flaxonoid that reduces cytokine release from mast cells and helps to decrease inflammation. Rosemary has the ability to block the proliferation of certain cancer cells particularly ovarian cancer cells. It has flavonoids which are essential for strengthening blood capillaries which are important for providing nutrients to the nerves. The polyphenols and terpenes in rosemary are great weapons to fight against inflammation. They have the power to neutralise free radicals in the blood which shuts down the inflammatory responses. As a result, veins and arteries are less likely to become narrowed and reduce pain, swelling and stiffness associated with peripheral neuropathy. Dill seed is deemed a chemoprotective agent that protects healthy tissues from the toxic effects of anticancer drugs and has the capacity to neutralise toxins that arise due to air pollution and automobile exhaust. Parsley is rich in apigenin flavonoid that reduces inflammation. It contains eugenol that inhibits the enzymes that cause inflammation. Ashwagandha is a popular herb that helps in memory and neurodegenerative diseases such as Parkinson's disease and AD and improves the energy level. Herbs such as dandelion and artichoke protect liver cells from being damaged and help in the normal functioning of the liver.

Sofaniim incanum - Bitter apple and thorn apple have wound healing properties. In a burn wound, the damage occurring to the epidermis is restored by the process of reepithelization in which keratinocytes move from the lower layer of the skin and form mature cells that cover the wound bed. The signs of proper wound healing are healthy granulation tissue, sebaceous glands and new blood vessels. *Sofaniim incanum* promotes movement and differentiation of keratinocytes, new connective tissue and blood vessel formation which are the important features of healthy granulated tissues proving the best healing property. It also contains tannins and flavonoids which are responsible for wound healing activity and stimulates wound healing at cellular and tissue level.

Sauromatum guttatum extract (Schott) helps in wound healing and regeneration of tissues in burn wounds and is traditionally utilised for the treatment of wounds. It has the capacity to control the infections associated with burn wounds, initiates wound healing by the upregulation of growth factors, cell division, maturation, and migration of various cells involved in the process of healing. Keratinocytes, epidermal cells and fibroblasts are also stimulated. Moreover, the presence of phenolic compounds inhibits the growth of microbes. Mushrooms such as oyster mushrooms (*Pleurotus giganteus*) have bioactive substances that are responsible for neurite stimulation.

Medicinal Plants

Since ancient times, people are exploring plant species in search of new drugs that have resulted in the exploitation of a large number of medicinal plants with curative properties to treat various ailments. The importance of medicinal plants becomes more potent in developing countries. In India, it is estimated that 80% of the population depends on plants, about 60% of the population uses medicinal plants habitually to battle certain ailments and almost 40% of humans use such plants in pharmaceutical industries.

Phytochemicals and Medicinal Plants

The active ingredients of the majority of nutraceuticals are phytochemical compounds, which are naturally occurring chemicals of plants (Dillard and German 2000). These are active biological compounds, which promote growth and plant defence mechanisms. Their properties have been long studied, for their medicinal value and extensive biological effects. The majority of nutraceuticals contain plant phytochemicals such as flavanoids, aglycones, glycosides and methylated derivatives which are of biological importance (Harborne and Mabry 1998). Plant phenolic compounds have potent antioxidant properties. Other phytochemical components such as metal chelators are endowed with anti-inflammatory, antimicrobial, properties. Metal chelators are also known to have anti-allergic, hepatoprotective, antiviral, and anticarcinogenic properties (Tapas et al. 2008; Clark and Lee 2016). The anticancerous effect of certain phytochemical compounds is attributed to their antioxidant properties, which can successfully repress the DNA damage induced by oxidative stress, thereby preventing cancer (Dandawate et al. 2016; Chikara et al. 2018). In this context, it is worth mentioning anticancerous compounds such as indole, found in cabbage which is known to reduce the risk of breast cancer, and similarly, capsaicin present in chilli peppers is known to protect DNA from damage by carcinogens (Biersack and Shschobert 2012). A number of phytochemical compounds, which are known to possess anticancerous properties, can be frequently processed and used as effective nutraceuticals, but seem to have certain limitations, such as solubility, stability and permeability. Compounds such as polyphenols, polysterols and carotenoids can be successfully used as effective anti-cancerous nutraceuticals, if these limitations can be defeated and this can be achieved, by using a nanoassembly of chitosan-derived fatty acids (Lavinia et al. 2020).

Examples of Medicinal Plants

***Jasrninum grandiflorum* L.** commonly known as Jasmine is a twining shrub widely grown throughout India. The leaves are used in the treatment of leprosy, skin diseases, ulcers, wounds, etc. Its flowers and leaves are used in folk medicine largely to prevent and treat breast cancer and stop uterine bleeding. It is widely used in Ayurveda as an antileprotic, wound-healing substance (Praveen et al. 2016).

Hibiscus rosasinensis commonly called shoe flower – in Tami semparuthi poo – enriched with polyphenolic phytochemicals similar to tannins, phenols, proteins, etc., These compounds have excellent antimicrobial, anti-oxidative and proliferative activity and therefore used to treat cancer, especially lung cancer, cardiovascular diseases, asthma and pulmonary dysfunctions (Rajendran et al. 2018).

Azadirachta indica Neem is a tree wherein each part of the tree is bitter and contains compounds with verified antiviral, anti-inflammatory, anti-ulcer, antifungal, antiseptic, antipyretic and antidiabetic activities. The bioactive components include alkaloids, flavonoids, phenolic compounds, steroids, etc. Its leaves are used among the tribes of India to remedy cuts, wounds, and minor dermis (skin-related) illnesses (Sushree Priyanka et al. 2017).

Mimosa pudica known as chuemue is a shrub by plant with compound leaves that are sensitive on touching and grows as a weed in almost all parts of the country. The leaves have alkaloids, mimosine and mucilage, whereas the root contains tannins. It is used for its anti-hyperglycemic, anti-diarrhoeal, anticonvulsant and cytotoxic properties. Leaves and roots of the plant are utilised to treat piles and fistula. The plant is also used in the treatment of sore gum, as a blood purifier. *M. pudica* is also used in ayurvedic and the Unani system of medicine for treating diseases such as piles, jaundice, leprosy, ulcers, bilious fever and small pox (Rekha et al. 2009).

Musa paradisiac L. Banana is one of the largest herb groups in the world and the earliest crop cultivated in the history of human agriculture. It contains compounds such as saponins, terpenoids and proteins, which have nutraceutical properties. It has potent anti-microbial, antioxidant and anti-inflammatory activities.

Ocimum sanctum Tulsi plant is an aromatic plant, commonly used in Ayurveda. It contains bioactive components which include oleic acid, orientin, eugenol and gallic acid. It possesses anti-inflammatory, analgesic, antipyretic and immunomodulatory activities. It also prevents skin, liver, oral and lung cancers.

Moringa oleifera Drumstick tree, in Tamil – Murugai – the miracle tree, is a fast-growing and drought-resistant tree. It has been used for centuries due to its medicinal properties and health benefits. It is enriched with bioactive components such as flavonoids, tocopherols, phenolic compounds. It possesses antifungal, antiviral, antidepressant, anti-inflammatory, hepatoprotective, and anti-diabetic properties (Eflong et al. 2013).

Prebiotics, Probiotics, and Synbiotics

Certain fibrous, dietary foods are considered to be probiotic, as these fibrous foods initiate the multiplication of certain, host microbes, which are beneficial to the host, as they help in the digestion of fibrous food and aid in the good maintenance of the digestive tract. These are abundantly found in soybeans, raw oats, breast milk, etc. Prebiotics, apart from maintaining the gut health is known to aid in the clearance of cholesterol, prevention of constipation and minimise the risk of obesity, these prebiotics, also known to be antioxidant, anti-cancerous agents, thereby protecting colon and other organs, from cancerous damage and reduce the risk of cardiovascular diseases. Another group of nonpathogenic microorganisms is known as probiotics. These promote host health if used appropriately in a regular diet or also as dietary supplements (Ravindran et al. 2015). They occur in natural environments, in various foods, and also in gut microbiota. The majority of probiotics are present as gram-positive bacteria such as *Bacillus* sp, *Lactobacillus sp.*, yeast species such as *Saccharomyces cerevisiae*. These probiotics can efficiently attach to the gut epithelium and eliminate pathogenic microbes, thereby ensuring healthy immune progression. Beneficial effects of probiotics upon human health include that they are quite efficient as anticancer (Lidia et al. 2019), antiallergic (Chen et al. 2018) and anti-diabetic (Koh et al. 2018) agents. They are also known for their antiobesity (Choi et al. 2019), antipathogenic (Hsu et al. 2018) and anti-inflammatory effects (Rocha-Ramirez et al. 2020).

Synbiotics

An additional class of food components known as synbiotics are known to impact human health in a positive way. A combination of prebiotics and probiotics creates a synergistic effect or synergism on the human health and is therefore referred to as synbiotics. These compounds selectively promote the growth of microorganisms, advantageous for the host digestive system, and help in promoting the overall well-being of the humans.

Role of Nanotechnology in Pre, Pro, and synbiotics

Probiotics are indeed gaining attention as an alternative to antibiotics to overcome microbial resistance. Researchers now focus on the impact of nanomedicine together with the practices of prebiotics, probiotics and symbionts that represent an approach to creating an ideal environment within the gut. Using nanonisation approaches of probiotics, the utility of nanoprobiotics in the delivery of encapsulated bacteria is carried out. Selenium and gold particles of size (10–1000 nm) have been mainly used for the encapsulation of probiotic bacteria along with nanolayers. The principal option for encapsulating probiotic bacteria is for protecting the cells against the hostile environment of the gastrointestinal tract and also to allow their discharge in a metabolically active state in the intestine. The research was done to evaluate the cellular effects of the nanoprobiotics and it was found that nanoprobiotics were found with improved anti-microbial effect compared to the probiotics alone. Nanocapsulation of probiotic bacteria represents a promising approach in enhancing their viability and existence against gastrointestinal conditions.

Prebiotics have been reported to enhance the immune system, inhibition of colitis, reduction of cardiovascular diseases, antioxidant activity and maintenance of gut health, and generation of bacteriocins.

Supplementing inulin with *L. plantarum*STl6 had increased the growth rate. Some studies on patients with irritable bowel syndrome revealed that when they were given *M. longum*BB536 and *L. rliamnosus*HN00l along with vitamin B6 which resulted in the refurbishment of gut microbiota, restored via intestinal permeability as well as amelioration of disease symptoms. Fermented beverages with symbiotic properties of nanoparticles incorporation with specific bioactivity have opened a new horizon in the food sector with improved absorption, nutrient delivery, and high organoleptic characteristics. Extensive studies are carried on with the formulation of protein-based, inulin-incorporated nanoemulsion for enhanced stability of probiotics. The symbiotic AMFTM 15+ manuka honey yogurt showed antibacterial properties and also increased probiotic bacteria and produced lactic and propionic acids (Mohan et al. 2020). Seaweed-based synbiotic of *Gracilaria coronopifolia* caused the reduction of inflammation, generation of reactive oxygen species (ROS) and diminution of the oxidative stress-induced cell damage (Li et al. 2019).

Traditional Medicine System

The use of herbal preparations dates back to centuries, when humans used a number of plant products and their related compounds, both as food and also to treat and cure a number of pathological conditions. This scenario of using herbal medications, changed in the early time of 20th century, particularly in the western part of the world. The traditional system of herbal medicines was recouped by allopathic medicines. The possible logical explanation for this deviation lies in the very fact that most of the plant constituents, which are flavonoids, terpenoids, tannins have biological activity but are water-soluble. Their water solubility leads to low absorption across the lipid membranes of the cell. The large molecular size of these molecules leads to a low absorption rate and tends to cause diminished bioavailability and efficacy. This deficiency of the herbal extracts can be defeated by the use of the nanoparticulate system. The use of herbal medicines, in combination with nanoparticles, can intensify the efficacy of the extract, and there by a low dosage of the compound can be used and minimise the side effects. Another advantage of using nanosystem for herbal drug delivery is that the desired dosage can be accurately administered at the exact location in the human system, where this cannot be achieved by the regular method of treatment. In the context of the herbal medical system, India is long known to be the core of traditional medicine. The practice of using metals such as gold, silver and copper, is not a new concept in the preparation of medical concoctions in the traditional system. In Ayurveda, Bhasma, a kind of ash, is the safest method of medicine. This bhasma ideally is a medicated metal-based ash prepared by the method known as bhasmikaran. Based upon the material used, bhasmikaran, can be distinct and involves various methods such as shodhana (purification), Bhavana, Jarana, Putapak, and Marana (calcination). Depending upon the symptoms and ailment, the bhasma is prepared according to the methods, previously mentioned using different metals such as gold, silver, iron and lead. Ghee-based formulations are well scripted in the Ayurvedic system of medicines and are used for wound healing purposes.

Physicians following ayurvedic medical practices have been using ghee as enemas for centuries to decrease inflammation. Ghee is also known to stimulate the secretion of gastric juices which inturn aids in the digestive process. In Ayurveda, ghee is placed under most sattvic foods and is considered to promote positivity, growth and expansion of consciousness. Ghee is used as a suitable carrier for many herbs and spices with different medicinal properties, which are to be absorbed and transported to targeted areas of the body. Ayurveda uses ghee in multiple different herbal preparations for treating various diseases and ailments.

In India, there are three major branches of traditional medicines which are in use:

- Ayurveda
- Unani
- Siddha

Ayurveda

In India, the ancient system of traditional medicine, known as Ayurveda was in practice almost 5000 years ago. This system originated and evolved as a standardised most commonly acknowledged ancient method of the health management system. The ancient application of nanomedicine in the form of Ayurvedic bhasmas gives an insight into the safer use of present-day nanomedicine for humans (Kapoor 2010; Sarkar and Chaudhry 2010; Chaudhry 2011). The use of metal nanoparticles, in therapeutics has been a common practice in Ayurveda. Bhasmas, literally meaning ash, is unique. In Ayurveda, it is of common practice to use, mineral or metallic compounds, of very minute size, per se, of nanoproportion (5–50 nm), which is very much in accordance with the size standards, suggested by modern-day microscopic and spectroscopic methods (Rohit and Prajapati 2016). Bhasmas, which are special types of Ayurvedic concoctions are prepared with a rightful combination of mineral and herbal juices and are subjected to heat exposure as according to the puta system of Ayurveda to cure various ailments. This method is in vogue in the Indian continent since the 7th century AD. These so-called ingredients called bhasmas ideally contain nanosized biologically active compounds in combination with other medical formulations, synthesised following either the Putapaka method or the Kupipkwa method.

Ayurveda in Regenerative Medicine

In Ayurveda, human physiology is explained as a dynamic exchange between regeneration and destruction of tissues. According to this principle, tissues are known to continuously undergo destruction and regeneration. This balance is appropriately maintained by fluid-like constituents known as doshas or humors. These doshas are known to coordinate the overall metabolic activities of the body. These doshas are of three types and classified as Vata, a humor that can balance the catabolic activity of the tissues, Kapha, a humor, known to vitalise the formation of new tissues and Pitta, which controls digestion. In the Ayurvedic system, to maintain a proper balance between dosha and nourishment of Dhatu, several dietary and lifestyle changes are recommended, which are known to promote good health and prolong longevity. There are certain traditionally practiced methods in Ayurveda, known as Panachakarma, and Rasayana, which are known to rejuvenate the body system and it has been scientifically proven that with the unique herbal treatment methods of Ayurveda, diseases such as Parkinson's arthritis and Alzheimer's disease can be cured (Chopra et al. 2013; Sehgal et al. 2012). A few herbal plants used in Ayurvedic medicine and their documented regenerative properties are tabulated in Table 12.1.

TABLE 12.1

Herbal Plants Used in Ayurvedic Medicine and Their Regenerative Functions

S. No:	Name of the Herbal Plant	Functions
1.	*Phyllanthus emblica*	Chondroprotective activity by inhibiting the activities of hyaluronidase and collagenase type 2 *in vitro* (Sumantran et al. 2008)
2.	*Phyllanthus emblica, Shorea robusta sin and Yashada bhasma*	Wound healing, fracture, anaemia, corneal ulcers, and repair of DNA damage in experimental models (Datta et al. 2011)
3.	*Amalaki rasayana*	Reduction in DNA damage in brain cells demonstrating its genomic stability in neurons and astrocytes. (Swain et al. 2012)
4.	*Dhanvantar kashaya* (decoction of herbs having regenerative property	The decoction increases the proliferation rate, decreases the turnover time, and also delays the senescence of WJMSC's Wharton Jelly mesenchymal cells (Warrier et al. 2013).
5.	*Curcumin*	Stimulates developmental and adult hippocampal neurogenesis, and biological activity and that may enhance neural plasticity and repair (Kim et al. 2008).

Ayurveda and Nanodrug Delivery System

The main drawbacks in the worldwide usage of ayurvedic drugs or their formulations are:

- Very high dosage
- Poor bioavailability
- Repeated administration
- Stability issues

Because of the above-stated reasons, nanocarriers can be used as a viable drug delivery system. The main advantage of the nanodrug delivery system is:

- They can withstand extreme conditions such as the acidic pH of the stomach
- Reducing the dose of repeated administration
- Enhances the solubility
- Increases bioavailability
- Provides protection against toxicity
- Enhances the pharmacological activity

Hence, the use of ayurvedic drugs and their formulations in the nanocarrier drug delivery system will increase its potential to treat various chronic diseases and ailments and one of the examples of such enhancement is seen in Haridra (*Curcuma longa*), where the bioavailability was enhanced credibly when it was incorporated with nanocarriers and modified into nanoform.

Ayurveda and Rejuvenation

In Ayurveda, there is the traditional method, known as Rasayana, also called Jarachikitsa, is one of the eight branches in Ayurveda. It is known for rejuvenating the body and effectively strengthening the body's immune system. Rasayanas are prepared with a specific combination of plants, herbs, and spices, which are processed into exclusive composition, aimed at definitive target points or organs, to cure the diverse ailments (Shukla et al. 2012). In this context, approximately 200 different rasayanas, aimed at exclusive target organs, are known to be registered in Ayurveda. Rasayana is a generalised term, which can have a vast inclusive action upon human body, but there are rasayanas, which can be tissue-specific or organ-specific. The success of treatment with rasayana is determined by its ability to reach the target organ or tissue, which in Ayurveda is termed as Gamitra (reaching the target). The efficacy of the prescribed rasayana depends upon certain other factors, such as a method of extraction, composition of the ingredients used, dosage strategies, age of an individual and also the quality of health of the individual. It is interesting to note that, apart from being used as treatment, pills, rasayans are also known to provide necessary nourishment (minerals and vitamins) which are lost naturally during the course of life. Rasyanas may not simply be drugs but can also be antioxidants or antitumor agents based upon the ingredients used in their preparation (Datta and Paramesh 2010). There are certain rasayanas which are known to possess defensive properties, to shield the liver and kidneys against damage. There is a certain amount of data available, suggesting that certain herbs are known to act as immune modulators, i.e., regulate the immune system. However, there is no concrete scientific evidence to support that Ayurveda formulations have the ability to modulate the immune system by causing the stem cells (mesenchymal stem cells) to differentiate. In this scenario, there is certainly an urgent need to assess the ability of certain rasayanas for their capacity to act as immune modulators and thereby differentiate the stem cells that can be used therapeutically (Rege et al. 1999; Menon and Nair 2013). Several different types of rasayanas are listed in Table 12.2

TABLE 12.2

Types of Rasayanas and Their Target Organs

S. No:	Rasayana	Target Organ
1.	*Medhaya Rasayana*	Brain
2.	*Hridya Rasayana*	Heart
3.	*Twachya Rasayana*	Skin
4.	*Chakshurya Rasayana*	Eyes

There are quite a number of bone and tissue degenerative diseases, such as osteoarthritis and macular degenerative disease, which are difficult to be treated. Additionally, AD, heart attacks and other injuries need repair and regeneration of lost and dying tissue. In this context, it would be of great value, if certain tissue-specific rasayanas are used for regeneration or to trigger cells for stem cell differentiation, thereby aiding in the treatment of many diseases (Kalpana and Ramesh 2014). Therefore, a holistic approach by combining the knowledge of Ayurveda about repair and regeneration, along with the deeper insights about the development of adult and embryonic stem cells, might pave way for better management and treatment for many disorders.

Unani

The belief that natural medicines are safer than synthetic drugs has gained momentum over the past two decades, which has led to the increased usage of phytopharmaceuticals. The Unani system is one such ancient system that is well-structured and documented. The ancient literature on the Unani system is known as "Qarabadeen."

The major difference between modern medicine and indigenous medicine is that modem medicine deals with a particular constituent/component/moiety, whereas indigenous medicine relies on a holistic approach. The Unani system of medicine is broadly known as Tibb in the Middle East. It is a healthcare system that incorporates theoretical understanding and applies it. Tibb is recognised by the World Health Organization (WHO) as one of the alternative systems of medicine (www.who.int/medicine/en: accessed 10 December, 2012). Unani formulations having metals are formed after many purification methods such as soaking, filtration, washing, powdering, heating, trituration, coating and detoxification. After this processing, the finished product is devoid of toxicity (Talha and Abdul 2020).

In Unani medicine, Kushta (Calx) is the most imperative dosage form of medicine. Kushta is derived from the word "Kushtan," a Persian word that means "Killed or Conquered." Kushta is the nanofine particle of Unani preparation acquired by the process of Taklees (Calcination) of elements. Unani scholars state that the calcination techniques used in the preparation of calx are a specific process in which herbal juices are added. This procedure increases the effectiveness of the end product and also removes the toxicity of the element. Thus, in Unani medicine, Kushta is noted to be nanomaterial or substance that has been prepared by burning of metals/minerals at high temperature up to 700–800°C. Kushta or calx is a dosage form that is very effective in a very minimal dosage but has a very swift action because of its instant absorbing ability.

Kushtas (Calx) undergo a series of classical standardisation techniques before its use. The techniques are summed up below:

- Floating test: A small quantity of Kushtas (Calx) when sprinkled on the surface of the water should float
- Fineness test: The Kushtas should be so fine that when it is rubbed between the fingers, it should enter the creases of the fingertips
- Fuming test: When the Kushtas is put in the fire, it should not produce flames
- Tasteless: Kushtas should not have any taste.

TABLE 12.3
List of Some Kushtas (Calx) Used in Traditional Medicine (Talha and Abdul 2020)

			Ingredients		
S. No:	**Elements**	**Unani Preparation**	**Mineral**	**Herb**	**Uses**
1.	Tin	Kushta Qalai	Tin	*Aloe barbadensis*	Impotence
2.	Iron	Kushta Kubsul – Hadid	Iron oxide	*Aloe barbadensis* *Emblica officinalis* *Teminelia belerica* *Terminalis chebula*	Anaemia and general debility
3.	Zinc	Kushta Jast	Zinc oxide	*Aloe barbadensis*	Spermatorrhea, leucorrhoea, and impotence
4.	Calcium	Kushta Busund Kushtasadaf Kushta Marvareed	Coral roots Oyster shell Pearls	*Aloe barbadensis* *Rosa damascene* *Rosa damascene*	Bleeding disorder and leucorrhoea. It also serves as cardio tonic

However, in recent years, some new modern techniques of standardisation are observed for Kushta (Calz). They are:

- Microbial evaluation
- Heavy metal estimation
- Atomic absorption spectroscopy
- X-ray diffraction studies
- X-ray crystallography studies

In spite of the fact that Kushta has been used in bygone eras without any evident side effects and is well-known in the Unani system of medicine, very few scientific studies have been carried out to date. Unani medications available in the market have a conventional dosage structure; however, in recent times, different rational methodologies of nanotechnology are being used to deliver natural medicine due to their poor rate of absorption and target explicit nature. Kushtas used in the traditional Unani medicine are listed in Table 12.3 (Talha and Abdul 2020).

Nanoemulsion in Unani Medicine

Nanonisation has multiple advantages such as it increases the solubility of the compound, reducing the dose and also improving the absorbency of herbal medicines when compared with their respective crude drug preparations. One such nanopreparation is nanoemulsion. Nanoemulsion has the potential to carry the hydrophilic and lipophilic components of crude drugs. It's a type of colloidal system with particles varying in size from 10 nm–1000 nm (Ratnam et al. 2006). Moreover, it is quite predictable that nanoemulsion may play an increasingly important role in the commercial market, as they can typically be formulated using significantly less surfactant than what is required for nanostructured lyotropic micro emulsion phases (Mason et al. 2006).

Nanoemulsion technique can be explored for both types of liquid dosage forms (Sayyal), internal and external. Dispensing single drugs (Mufaradat) into Nanoemulsion is not a big challenge whereas compound drugs (Murakkbat) pose a bigger challenge. Almost all different forms of drugs can be converted into a nanoemulsion-like solid (Jamid), semi–solid (Neemjamid) and liquid (Sayyal), and also various oils (Roghanbadaam).

In 2013, the Unani system was taken to next level and amalgamated into two systems of medicines of different origins (Mirza et al. 2013). A synergistic Nanoemulsion was formulated using Tea tree oil and Itraconazole for vaginal candidiasis. The dual loading of Itraconazole and tea tree oil in a single formulation seemed promising as it elaborated the microbial coverage. Despite the fact that itraconazole has low solubility in tea tree oil, a homogenous, transparent and stable solution of both was created by co-solvency using chloroform. Complete removal of chloroform was authenticated by gas chromatography–mass spectroscopy and oil solution which were used in making the Nanoemulsion which was further translated into a gel bearing thermosensitive property.

Unani and Regenerative Medicine

Unani is yet another traditional method of treatment, where there is documented evidence, that regeneration was successful, specifically in the treatment of cirrhosis of the liver (Mohammed and Shabnam 2015). Liver damage, attributed to cirrhosis liver has caused 1 million cases all over the globe in the year 2010. Healers who practice Unani, a mode of treatment, have been treating liver cirrhosis, for a quite long time. There is documented evidence, in the entity "Warm-e-JigarBaridSaudawi"/"Taleef-u1-Kabid" of Unani literature, which mimics liver cirrhosis and its associated symptoms, that have been recorded and also information about the drug or a combination of drugs to cure this disease is also available in this ancient text. Bioactive components, used as medications, in Unani, are proven to have antioxidant, antiviral and also immunomodulant properties. The text contains detailed data, regarding plant species, specific bioactive plant components, which can be successfully applied to treat liver cirrhosis, and also information regarding the plant-based drugs that can be used to prevent liver damage. There are instances, where patients suffering from severe liver damage, caused due to cirrhosis, who have been advised for liver transplants had opted for an unani form of treatment. Unani remedies and therapies have been a source of hope for patients with liver cirrhosis, especially in third world countries where liver transplantation is an expensive affair and also there is often a shortage of donors for liver transplantation. In the Unani system of treatment, the herbal medications, used are a combination of multiple herbal concoctions prepared as accordingly detailed in the ancient text.

In Table 12.3, certain drugs have been mentioned, which are known for the effective treatment of persistent liver ailment. Liver cirrhosis, is often, accompanied by fibrosis, which damages the hepatic cells. It is seen that both the innate and adaptive immune responses play a very significant aspect in the progression of fibrosis. In this context, there are pieces of evidence, where Unani-based compounds, when used in animal models, have caused a decline in fibrous tissue formation, which seems to prove that the Unani components could act as immunomodulators of biological systems (Amirghofrana et al. 2000; Kumar et al. 2003; Bafna and Mishra 2004).

Persistent inflammation, following liver damage, often results in the formation of fibrous tissue. Therefore, often, any medication which is given to reduce the inflammation ideally has an anti-fibrotic effect. The sequence of steps, which lead to fibrosis are outlined as oxidative stress caused by the generation of ROS followed by scar formation, leading to a reduced inflammatory response. This inflammation causes the crippling of hepatic stellate cell (HSC) activation causing fibrogenesis as seen in cirrhosis of the liver. In this context, it can be said, that certain natural compounds or Unani preparations which are identified as antioxidants, can be given as a therapy to reduce the ROS and in a way to prevent fibrosis of hepatic tissue. The effectiveness of this treatment depends upon the agility of the antioxidant reaching the target site i.e., liver (Kawada et al. 1998).

Hepatocyte injury can be prevented many times by employing a number of antioxidants, which can directly impact the ROS formation, thereby minimising the fibrosis formation, and salvage liver from major damage. A number of antioxidants have been identified which are often used in Unani preparations, as antioxidants. Some examples include *Fumaria officinalis, Tephrosia purpurea, Swertia chirata, Sphaeranthus indices, Pterocarpus santalinus, Solanum nigrum* and *Cichorium intybus* (Shirwaikar et al. 2006; Arulmozhi et al. 2010; Alam et al. 2012; Gupta et al. 2012; Kshirsagar et al. 2015; Mohammed and Shabnam 2015). There are instances in scientific literature, where liver damage was reduced by remission in fibrosis formation, because of the use of Unani preparations, which are identified as antioxidants and also have revived and rejuvenated the damaged liver tissue. A number of studies have proven that

Unani preparations have a remedial and positive impact upon treating cirrhosis of liver (Ali et al. 2011; Mohammed and Shabnam 2015). Unani components, certainly are known to have reviving and regenerative effects upon damaged liver caused by cirrhosis. *Swertia chirata* was proven to have anti-hepatitis B properties (Saxena et al. 2012).

Siddha

In 2015, the WHO had estimated that 4 billion people which is 80% of the world's population presently uses herbal medicines for primary health care (Rajamaheshwari et al. 2015). Substances derived from the plants remain the basis for a large proportion of commercial medicines produced for the treatment of some chronic diseases. Siddha system of medicine is a pioneer among other traditional therapies and belongs to South East Asia especially in treating dreadful infectious diseases such as tuberculosis. Formulations have been prepared following the available Vedic literature and processed for improving its efficacy before administering the same for clinical use. Metals and minerals play a major role in Siddha preparations (Wang and Brett 2002; Winston 2005). According to the Siddha practitioners, the metallic preparations are detoxified with a specific technique and the levels of metals are so low that they come under the acceptance range as recommended by the WHO guidelines.

Siddha medicines are always formidable because there are so many miraculous medicinal preparations. Siddhars, the practitioners of Siddha medicine have made a specific regime in the field of pharmacology to prove that there is strong literature evidence, Agasthiyar gunavakadam is one such literature. Siddha medical system is based on 3 humours such as Vatham, Pitham, and Kapham, the Siddha drug preparations have high potency and therapeutic value.

Siddha and Nanotechnology

Nanotechnology is an expensive technology, but Siddhars have practised this type of medicine from time immortal at a very low cost and through the programmed method. Though the concept of nanomedicine is very young in modern medicine, the traditional Siddha medicine has many drugs at nanosize with varied chemical characteristics. Some scientific studies do reveal that nanoparticles may produce toxic effects but this has not been encountered in Siddha nanomedicine which has been administered orally for many centuries and may be attributed to the method of preparation and dosage of medicine.

In one standardization study, a Siddha drug named Rasa Mezhugu is used in the treatment of rheumatoid arthritis, hemiplegia, and spondylitis. Ingredients of the drug are tabulated in Table 12.4.

The Fourier transform infrared (FTIR) spectroscopy and inductively coupled plasma atomic emission spectroscopy (ICP-OES) studies of this nanodrug revealed that the drug is absolutely safe as the heavy metals are below detectable levels. The active compound Rasasm is also within 3 ppm as per admissible limits of medicine. The SEM studies revealed that the drug was within 1–100 nm size. Siddha medicine has a lot of hidden treasures that can be applied in the treatment of different life-threatening diseases (Kanniyan and Muthu 2020). Hence scientific validation and interpretation of herbomineral medicines with knowledge of certain modern nanotechnology will help build the gap between the areas of failure in the medical system.

TABLE 12.4

Constituents of *RASA MEZHUGU* (Kannaiyan and Kanniyakumari 2020)

S. No:	Constituents of Drug	Common Name
1.	Vaalai rasam	Purified mercury
2.	China root	*Smilax china*
3.	Long pepper	*Piper longum*
4.	Clove	*Syzigium aromaticum*
5.	Jaggery (sugarcane)	*Saccharum officinarum*

REFERENCES

Abbasi E., Milani M., Fekri A.S., Kouhi M. et al. 2016. *Critical Reviews in Microbiology 42*: 173–180. doi:10.3109/1040841X.2014.912200.

Akturk O., Kismet K., Yasti A.C., Kuru S. et al. 2016. *RSC Advances 6*(16): 13234–13250. doi:10.1039/c5ra24225h.

Alam M.N., Roy S., Anisuzzaman S.M. and Rafiquzzaman M. 2012. *Pharmacognosy Communications 2*(3): 67–71. doi:10.5530/pc.2012.3.14.

Ali A., Shyum Naqvi S.B., Gauhar S. and Saeed R. 2011. *Pakistan Journal of Pharmaceutical Sciences 24*(3): 405–409.

Alivisatos A.P., Johnsson K.P., Peng X., Wilson T.E. et al. 1996. *Nature 382*(6592): 609–611. doi:10.1038/382609a0.

Amirghofrana A., Azadbakhtb M. and Karimi M.H. 2000. *Journal of Ethnopharmacology 72*(1–2): 167–172. doi:10.1016/s0378-8741(00)00234-8.

Andreani T., Fangueiro J.F., Severino P., De Souza A.L.R. et al. 2019. *Nanomaterials 9*(8): 1081. doi:10.3390/nano9081081.

Anjum S., Gupta A., Sharma D., Gautam D. et al. 2016. *Materials Science and Engineering: C 64*(1): 157–166. doi:10.1016/j.msec.2016.03.069.

Arulmozhi V., Krishnaveni M., Karthishwaran K., Dhamodharan G. et al. 2010. *Pharmacognosy Magazine 6*(21): 42–50. doi:10.4103/0973-1296.59965.

Bafna A.R. and Mishra S.H. 2004. *ARS Pharmaceutical 45*(3): 281–291.

Biersack B. and Schobert R. 2012. *Current Drug Targets 13*(14): 1705–1719. doi:10.2174/138945012804545551.

Chaudhry A. 2011. *Ayurvedic Bhasma: Journal of Biomedical Nanotechnology 7*(1): 68–69. doi:10.1166/jbn.2011.1205.

Chen J.C., Tsai C., Hsieh C.C., Lan A. et al. 2018. *Allergologia et Immunopathologia 46*(4): 354–360. doi:10.1016/j.aller.2018.02.001.

Chikara S., Nagaprashantha L.D., Singha J., Horne D. et al. 2018. *Cancer Letters 413*: 122–134. doi:10.1016/j.canlet.2017.11.002.

Choi W.J., Dong H.J., Jeong H.U., Jung et al. 2019. *Journal of Medicinal Food 22*(6): 560–565. doi:10.1089/jmf.2018.4329.

Chopra A., Saluja M., Tillu G., Sarmukkaddam S. et al. 2013. *Rheumatology (Oxford) 52*(8): 1408–1417. doi:10.1093/rheumatology/kes414.

Clark R. and Lee S.H. 2016. *Anticancer Research 36*(3): 837–843.

Dandawate P.R., Subramaniam D., Jensen R.A. and Anant S. 2016. *Seminars in Cancer Biology 40–41*: 192–208. doi:10.1016/j.semcancer.2016.09.001.

Daniel M.-C. and Astruc D. 2004. *Chemical Reviews 104*(1): 293–346. doi:10.1021/cr030698+

Datta H.S., Mitra S.K. and Patwardhan B. 2011. Article ID 134378, 10 pages doi:10.1093/ecam/nep015.

Datta H.S. and Paramesh R. 2010. *Journal of Ayurveda and Integrative Medicine 1*(2): 110–113. doi:10.4103/0975-9476.65081.

Davis M.E., Chen Z.G. and Shin D.M. 2008. *Nature Reviews Drug Discovery 7*(9): 771–782. doi:10.1038/nrd2614.

De Felice S.L. 1992. *Scrip. Mag 9*: 14–15.

Dillard C.J. and German J.B. 2000. *Journal of the Science of Food and Agriculture 80*(12): 1744–1756. doi:10.1002/1097-0010(20000915)80:12<1744:AID-JSFA725>3.0.CO;2-W.

Doktorovova S., Kovacevic A.B., Garcia M.L. and Souto E.B. 2016. *European Journal of Pharmaceutics and Biopharmaceutics 108*: 235–252. doi:10.1016/j.ejpb.2016.08.001. Epub 2016 Aug 9.

Eflong E.E., Lgile G.O., Mgbeje B.I.A. and Ebong P.E. 2013. *Journal of Diabetes and Endocrinology 4*(4): 45–50. doi:10.5897/JDE2013.0069.

Farokhzad O.C. and Langer R. 2006. *Advanced Drug Delivery Reviews 58*(14): 1456–1459. doi:10.1016/j.addr.2006.09.011.

Gupta P.C., Sharma N. and Rao C.V. 2012. *Asian Pacific Journal of Tropical Biomedicine 2*(8): 665–669. doi:10.1016/S2221-1691(12)60117-8.

Harborne J.B. and Mabry T.J. 1998. *Phytochemical methods*. Ed 1. London: Chapman and Hall.

Heo D.N., Ko W.K., Bae M.S., Lee J.B. et al. 2014. *Journal of Materials Chemistry B 2*(11): 1584–1593. doi:10.1039/C3TB21246G.

Hong K.H., Park J.L., Sul I.H., Youk J.H. et al. 2006. *Journal of Polymer Science. Part B 44*(17): 2468–2474. doi:10.1002/POLB.20913.

Hsu T.C., Yi P.J., Lee T.Y. and Liu J.R. 2018. *PLoSOne 13*(4): e0194866. doi:10.1371/journal.pone.0194866.

India Infoline News Service. 2013. Indian nutraceutical market: Poised from neutral gear to top gear. (https://www.indiainfoline.com/article/news-top-story/indian-nutraceuticals-market-poised-from-neutral-to-top-gear-113110816609_1.html) Accessed on 29 Dec 2021

Jampilek J., Kos J. and Kralova K. 2019. *Nanomaterials 9*(2): 296. doi:10.3390/nano9020296.

Jose Roberto V.R. 2019. *Nutraceuticals: International Journal of Biosensors and Bioelectronics 5*(3): 56–61. doi:10.15406/ijbsbe.2019.05.00154.

Kannaiyan N. and Kanniyakumari M. 2020. *International Journal of Ayurveda and Pharma Research 8*(2): 58–63.

Kapoor R.C. 2010. *Indian Journal of Traditional Knowledge 9*(3): 562–575.

Kawada N., Seki S., Inoue M. and Kuroki T. 1998. *Hepatology 27*(5): 1265–1274. doi:10.1002/hep.510270512.

Keservani R.K., Kesharwani R.K., Vyas N., Jain S. et al. 2010. *Der Pharmacia Lettre 2*(1): 106–116.

Kim S.J., Son T.G., Park H.R., Park M. et al. 2008. *Journal of Biological Chemistry 283*(21): 14497–14505. doi:10.1074/jbc.M708373200.

Ko W.K., Heo D.N., Moon H.-J., Lee S.J. et al. 2015. *Journal of Colloid and Interface Science 438*: 68–76. doi:10.1016/j.jcis.2014.08.058.

Koh W.Y., Utra U., Ahmed R., Rather I.A. et al. 2018. *Food Science and Biotechnology 27*: 1369–1376. doi:10.1007/s10068-018-0360-y.

Kshirsagar P.I., Chavan J., Nimbalkar M., Yadav S. et al. 2015. *Natural Product Research 29*(8): 780–784. doi:10.1080/14786419.2014.986124.

Kumar I.V.M.L.R.S., Paul B.N., Asthana R., Saxena A. et al. 2003. *Immunopharmacology and Immunotoxicology 25*(4): 573–583. doi:10.1081/iph-120026442.

Li J., Zhang J., Chen Y., Kawazoe N. et al. 2017. *ACS Applied Materials & Interfaces 9*(41): 35683–35692. doi: 10.1021/acsami.7b12486.

Li P.-H., Lu W.-C., Chan Y.-J., Zhao Y.-P. et al. 2019. *Food 8*(12): 623. doi:10.3390/foods8120623.

Lidia A.-V., Carlos Z.-M., Alicea R.-M., Amalia V. et al. 2019. *International Journal of Biosensors and Bioelectronics 5*(3): 56–61. doi:10.15406/ijbsbe.2019.05.00154.

Martins S., Silva A.C., Ferreira D.C. and Souto E.B. 2009. *Journal of Biomedical Nanotechnology 5*(1): 76–83. doi:10.1166/jbn.2009.443.

Mason T., Graves S.M., Wilking J. and Lin M.Y. 2006. *Condensed Matter Physics 9*(1): 193–199. doi:10.5488/CMP.9.1.193.

Mehrabani M.G., Karimian R., Mehramouz B., Rahimi M. et al. 2018. *International Journal of Biological Macromolecules 114*: 961–971. doi:10.1016/j.ijbiomac.2018.03.128.

Menon A. and Nair C.K. 2013. *Journal of Cancer Research and Therapeutics 9*: 230–234. doi:10.4103/0973-1482.113363.

Mirza M.A., Ahmad S., Mallick M.N., Manzoor N. et al. 2013. *Colloids and Surfaces B: Biointerfaces 103*: 275–282. doi:10.1016/j.colsurfb.2012.10.038.

Mohammed A. Siddique and Shabnam Ansari. 2015. *Asian Journal of Pharmaceutical and Clinical Research 8*(5): 77–79.

Mohan A., Hadi J., Gutierrez-Maddox N., Li Y. et al. 2020. *Food 9*(1): 106. doi:10.3390/foods9010106.

Muller R.H., Runge S., Ravelli V., Mehnert W. et al. 2006. *International Journal of Pharmaceutics 317*: 82–89. doi:10.1016/j.ijpharm.2006.02.045.

Potential for functional foods in the Indian market. Retrieved February 1, 2019, from https://www.figlobal.com/india/visit/news-and-updates/potential-functional-foods-indian-market.

Praveen Chandran R., Bhuvaneshwari V., Amsaveni R. and Ragavendran P. 2016. *Asian Journal of Pharmaceuticals and Clinical Research 9*(1): 171–174.

Rajamaheshwari K., Kumar M.P., Banumathi V. and Sudarshan S. 2015. *Der Pharmacia Lettre 7*(11): 295–300.

Rajendran A., Siva E., Dhanraj C. and Senthilkumar S. 2018. *Journal of Bioprocessing and Biotechniques 8*(3). doi:10.4172/2155-9821.1000324.

Ratnam D.V., Ankola D., Bhardwaj V., Sahana D.K. et al. 2006. *Journal of Controlled Release 113*(3): 189–207. doi:10.1016/j.jconrel.2006.04.015.

Ravindran A., Balasubramanian S. and Toru M. 2015. *Journal of Food Science 5*: 4. doi:10.4172/2155-9600.S1.017.

Rege N.N., Thatte U.M. and Dahanukar S.A. 1999. Adaptogenic properties of six rasayana herbs used in Ayurvedic medicine. *Phytotherapy Research 13*(4): 275–291. doi:10.1002/(SICI)1099-1573(199906)13:4<275::AID-PTR510>3.0.CO;2-S.

Rekha R., Hemalatha S., Akasakalai K., Madhukrishna C.H., Bavan S., Vittal and Meenakshi S.R. 2009. *Journal of Natural Products 2*: 116–122.

Rocha-Ramirez M.L., Hernandez-Ochoa B., Goemez-Manzo S., Marcia-Quino J. et al. 2020. Evaluation of Immunomodulatory activities of the heat killed probiotic strain, *Lactobacillus casei* IMAU 60214 on macrophages *in vitro*. *Microorganisms 8*: 79.

Rohit S. and Prajapati P.K. 2016. *Journal of Pharmacy & Bioallied Sciences 8*(1): 8–81. doi:10.4103/0975-7406.171730.

Santini A., Tenore G.C. and Novellino E. 2017. *European Journal of Pharmaceutical Sciences 96*: 53–61. doi:10.1016/j.ejps.2016.09.003.

Sarei F., Dounighi N.M., Zolfagharian H., Khaki P. et al. 2013. *Indian Journal of Pharmaceutical Sciences 75*(4): 442–449. doi:10.4103/0250-474X.119829.

Sarkar P.K. and Chaudhry A.K. 2010. *Journal of Scientific & Industrial Research 69*: 901–905.

Saxena M.I., Shakya A.K., Sharma N., Shrivastava S. et al. 2012. *Journal of Environmental Pathology, Toxicology and Oncology 31*: 193–201. doi:10.1615/jenvironpatholtoxicoloncol.v31.i3.10.

Sehgal N., Gupta A., Valli R.K., Joshi S.D. et al. 2012. *Proceedings of the National Academy of Sciences of the United States of America 109*: 3510–3515. doi:10.1073/pnas.1112209109.

Severino P., Chaud M.V., Shimojo A., Antonini D., Lancelloti M. et al. 2015. *Colloids and Surfaces. B, Biointerfaces 129*, 191–197. doi:10.1016/j.colsurfb.2015.03.049.

Severino P., Da Silva C.F., Andrade L.N., de Lima Oliveira D. et al. 2019. *Current Pharmaceutical Design 25*(11): 1312–1334. doi:10.2174/1381612825666190425163424.

Severino P., Pinho S.C., Souto E.B. and Santana M.H. 2011. *Colloids and Surfaces B: Biointerfaces 86*(1): 125–130. doi:10.1016/j.colsurfb.2011.03.029.

Shirwaikar A., Prabhu K.S. and Punitha I.S., 2006. *Indian Journal of Experimental Biology 44*: 993–996.

Shukla S.D., Bhatnagar M. and Khurana S. 2012. *Frontiers in Neuroscience 6*: 112. doi:10.3389/fnins.2012.00112.

Silva S.S., Oliveira M.B., Mano J.F. and Reis R.L. 2014. *Carbohydrate Polymers 112*, 264–270. doi:10.1016/j.carbpol.2014.05.042

Silva S.S., Popa E.G., Gomes M.E., Cerqueira M. et al. 2013. *Acta Biomaterialia 9*, 6790–6797. doi:10.1016/j.actbio.2013.02.027.

Son W.K., Youk J.H., Lee T.S. and Park W.H. 2004. *Macromolecular Rapid Communications. Wiley online library 25*(18): 1632–1637. doi:10.1002/marc.200490036.

Souto E.B. and Muller R.H. 2010. *Handbook of Experimental Pharmacology 197*, 115–141. doi:10.1007/978-3-642-00477-3_4.

Sumantran V.N., Kulkarni A., Chandwaskar R. and Harsulkar A. 2008. *Evidence-based Complementary and Alternative Medicine 5*: 329–335. doi:10.1093/ecam/nem030.

Sushree Priyanka D., Sangita D. and Soubhagyalaxmi S. 2017. *Biochemistry and Analytical Biochemistry 6*: 323. doi:10.4172/2161-1009.1000323.

Swain U., Sindhu K.K., Boda U., Pothani S. et al. 2012. *Mechanisms of Ageing and Development 133*: 112–117. doi:10.1016/j.mad.2011.10.006.

Talha M. and Abdul H. 2020. *World Journal of Pharmaceutical and Medical Research 6*(8): 32–37.

Tapas A.R., Sakarkar D.M. and Kakde R.B. 2008. *Tropical Journal of Pharmaceutical Research 7*(3): 1089–1099. doi:10.4314/tjpr.v7i3.14693.

Tian J., Wong K.K., Ho C.M., Lok C.N. et al. 2007. *Chem Med Chem 2*(1): 129–136. doi:10.1002/cmdc.200600171.

Tummalapalli M., Berthet M., Verier B., Alam M.S. et al. 2016. *International Journal of Biological Macromolecules 82*: 104–113. doi:10.1016/j.ijbiomac.2015.10.087.

Wang M.Y. and Brett J.W. 2002. *Acta Parmacologica Sinica 23*(12): 1127–1141.

Wang Y., Dou C., He G., Ban L., Huang L., Li Z., Gong J., Zhang J. and Yu P. 2018. *Nanomaterials 8*: 324. doi:10.3390/nano8050324.

Warrier S.R., Haridas N., Balasubramanian S. and Jalisatgi A. 2013. *Wiley Online Library 46*: 283–290. doi:10.1111/cpr.12026.

Winston D. 2005. Herbalist. *The American Extra Pharmacopoeia 1*: 1–6.

Yahyaei B., Manafi S., Fahimi B., Arabzadeh S. et al. 2018. *Applied Nanoscience 8*: 417–426. doi:10.1007%2Fs13204-018-0711-2.

Zhang D., Liu D., Zhang J., Fong C. et al. 2014. *Materials Science and Engineering: C 42*: 70–77. doi:10.1016/j.msec.2014.04.042.

13

Synthetic Nanoparticles for Anticancer Drugs

Yusnita Rifai
Hasanuddin University, Makassar, Indonesia

CONTENTS

Introduction

Nanoparticles exhibit advanced characteristic features based on size (1–100 nm). They also include solid colloidal particles with a size range of 1–1000 nm. The preferred size of nanoparticles is $\leq$100 nm to about 0.2 nm, which is atomic level, because within this range, materials can have different properties and be improved compared with the same but larger material (Cai et al. 2008; Jain & Das 2011).

Several methods can be used in the manufacturing of nanoparticles, such as solid lipid nanoparticles, precipitated ultrasonication, precipitation, homogenisation, emulsification solvent evaporation, salting out, emulsion solvent diffusion, and polymerization. All methods have advantages and disadvantages in terms of cost and size distribution. If the nanoparticle formula is unstable due to aggregation, the specific surface area will lower its antimicrobial and catalytic activities. Residues of toxic metals in the synthesis process may create some adverse effects in biomedical applications (Lekshmi et al. 2012; Kim et al. 2012). Thus, the biosynthesis of nanoparticles has emerged as an effective alternative to overcome the disadvantages of classical methods.

Anticancer Activity of Nanoparticles

The surfaces of nanoparticles are unique because they can be exploited by reactive terminal groups, with particular proteins or monoclonal antibodies, that are capable of particularly binding to a site of action without interacting with other cells (Nie 2010; Podila & Brown 2013; Son et al. 2007). Therefore, nanomedicine in cancer treatment is capable of destroying cancer cells without affecting other healthy tissues.

Here are some synthetic nanoparticles used in anticancer treatment and their synthesis processes:

Synthesis of Magnetic NPs of Cathelicidin LL-37

Nanoparticles (NPs) are having a considerable impact on modern medical applications. They are used for drug delivery systems and targeted therapy. Nanoparticles are composed of metal, nonmetal atoms, or a

DOI: 10.1201/9781003153504-13

mixture of both. CSA-13 is an example of a synthetic NP, which is composed of an iron oxide anion, an aminosilane group, and the antibacterial peptide LL-37. Propylamine groups, which are placed in the core of the magnetic NPs structure, were synthesized using the modified Stöber method (Geetha et al. 2013).

The nanoparticle surface is immobilised by the LL-37 peptide via an oxidation reaction between the primary amine and the carboxyl peptide group of magnetic NPs. To obtain magnetic NPs that function with ceragenin, ethanol and a 25% glutaricdialdehyde solution were added to the core part of the structure. Finally, magnetic NPs with the terminal aldehyde group were resuspended in ethanol containing the synthetic ceragenin analogue to synthesise CSA-13. After synthesis, the precipitated solvent was washed three times with C_2H_5OH salt solution and a phosphate bufferand then placed at 37°C. To prepare fluorescent nanosystems, magnetic NPs with aldehydes were mixed with propidium iodide (Shukla et al. 2005; Geetha et al. 2013). Magnetic NPs were coated with cathelicidin LL-37. Finally, they were labelled with fluorescein isothiocyanate (Leszczynska et al. 2013; Sorensen et al. 2001).

Nanoparticle Synthesis Gold-Plated Fucoidan Mimetic Glycopolymer

The synthesis of gold NPs coated with fucoidan mimetic glycopolymers demonstrates the development of nanotechnology research. The synthesised NPs showed anticancer activity because they were selectively cytotoxic against human cancer cells. Fucoidan mimetic glycopolymers were synthesised through free-radical-chain-polymerization reactions. The mechanism is similar to that of anti-viral properties possessed by natural polysaccharides (Cai et al. 2008).

Gold NPs with glycopolymers have been employed as a platform for drug delivery systems because they are able to produce anionic polymers such as glycosaminoglycans and nucleic acids for efficient cellular delivery. In addition, gold NPs show synergistic anticancer properties and act as a contrast agent in photothermal therapy and radiosensitisation (Shankar et al. 2004).

To synthesise gold NPs, fucoside monomer-glycopolymer is first prepared using fucoside monomers (2-methacrylamidoethyl-2,3,4-tri-O-acetyl-a-L-fucopyranoside). The monomer is polymerised in dioxane via the classical method with 2-trimethylsilyl-ethanethiol as a chain transfer compound. Glycopolymer is produced by precipitating the per-acetylation polymer with ether. Pellets are collected through filtration and deprotected in $NaOCH_3/CH_3OH$ (Jain & Das 2011). The second glycopolymer will partially sulphate in the presence of sulphur trioxide in the pyridine complex and dimethylformamide. They are converted to sodium bicarbonate to complete the sodium–sodium salt product (Shukla et al. 2005). A dialysis process using deionised water is performed to purify the obtained product. After synthesising the FM-glycopolymer, gold NPs are synthesise dusing NaBH4, which acts as a reducing agent.

Green Synthesised Gold NPs from *Marsdenia tenacissima*

Gold NPs are widely used in antimicrobial, anticancer, anti-inflammatory, and various biomedical treatments. The characteristics of gold NPs enable their use in diagnosing and treating cancer (Jiao et al. 2018). The synthesis of gold NPs usually involves chemical agents, which cause environmental pollution due to their toxic nature; therefore, the synthesis of gold NPs using plant extracts is more effective (Shankar et al. 2004). Plants act as a stabilising and reducing agent in the synthesis of gold NPs (Leszczynska et al. 2013). One of the plants that has been widely used in medical therapy to treat pneumonia and synthesise gold NPs is *Marsdenia tenacissima*.

For the synthesis, the *M. tenacissima* extract is dissolved with aquadest and then $HAuCl_4$ (4 mM) solution is added dropwise under sonication until the volume reaches 20mL. The solvent is then incubated at 25°C. Synthesis results were observed using ultraviolet–visible (UV–Vis) spectrophotometry at wavelengths of 200–1000 nm (Jiao et al. 2018).

Polybutyl Cyanoacrylic NPs Coated with Hyaluronic Acid as Anticancer Drug Carriers

Polybutylalkyl cyanoacrylate nanoparticles were first developed by Couvreur team in year 1979 through ionic emulsion polymerization. In recent years, NPs have been coated with polysaccharides that have specific receptors so they can reach the target. One of the polysaccharides is hyaluronic acid (HA), which is a negatively charged linear carbohydrate, containing two replacement units: d-uronic acid and N-acetyl-d-glucosamine. HA has good wound healing and anticancer activities (Entwistle et al. 1996). HA-coated polybutyl cyanoacrylate NPs were prepared as a drug delivery system that triggers radical emulsion polymerization of the monomer n-butyl cyanoacrylate, which is mediated by ceriumion in the presence of HA.

Synthetics are divided into hyaluronic acid (HA) poly(butyl cyanoacrylate) (PBCA) (HA-PBCA) and hyaluronic acid–Phenyl boronic acid (HA-PBA) NPs loaded by Paclitaxel (PTX). HA-PBCA nanoparticles are synthesised through polymerization of HA radicals and n-butyl cyanoacrylate monomers by dissolving 9 mLof HA in 0.2 mol/L nitric acid while stirring slowly until nitrogen gas bubbles are formed (10 minutes). The solvent is then added tocerium ammonium nitrate. Afterwards, the solvent is added with n-butyl cyanoacrylate monoomers mixed with nitric acid (1 mL). With strong agitation, the polymerization reaction is maintained at about 0.7 and the reaction is stopped after 3 hours. The NP suspension is dialysed with water for 48 hours. Subsequently, unmodified nanoparticles are synthesised by dissolving 50 L of n-butyl cyanoacrylate monomer into 5mL of pH 2.5 hydrochloric acid consisting of 0.5% poloxamer 188 for 4 hours with vigorous stirring. The NP suspension is obtained, which is then dialysed against water for 48 hours. HA-PBCA nanoparticles are synthesised by loading them in PTX in ethanol while stirring. The suspension is sonicated for 240 seconds, then lipolised against water to clean traced ethanol and filtered through a 0.80 mm membrane to withdraw the PTX residue (Entwistle et al. 1996).

Synthesis of Green Silver Nanoparticles using *Cleome viscosa*

Silver nanoparticles (AgNPs) are widely applied in various applications including as an antibacterial agent, a vector for drug delivery, and in anticancer treatment. Physical properties of AgNPs include high conductivity and plasmonic properties (Shankar et al. 2004).

When synthesising AgNPs, various aspects need to be characterised such as size, shape, and morphology. The shape and size of AgNPs are critical in determining their optical, electrical, magnetic, catalytic, and antimicrobial properties. The factors that affect the particle size in the synthesis are temperature, concentration, reducing agent, and reaction time. AgNPs can be synthesised using chemical reduction methods in which a reducing agent is needed to reduce Ag^+ to Ag. In addition, it is important to have a stabilising agent to prevent agglomeration (Siavash 2011).

In the synthesis of AgNPs, 10 grams of *Cleome viscosa* fruit is weighed, washed, and then boiled with 100 mL of water for 30 minutes at 60°C. The solution is then filtered using Whatman filter paper no. 1 to obtain *C. viscose* extract (Siavash 2011).

Cleome viscosa extract was synthesised by placing 10 mL of extract and then adding 1 mM silver nitrate ($AgNO_3$) solution until the volume reached 100 mL. The solution was then incubated in the dark for 24 hours at room temperature (25 ± 2°C) until the colour changed from green to brown, which indicated the formation of AgNPs.

Synthesis of Silver Nanoparticles using *Melia dubia*

Melia dubia leaf extract acts as a reducing and stabilising agent. Before synthesis, an extract of *M. dubia* leaves is first created by washing the fresh leaves using running water and then rinsing using distilled water. A total of 20 g of leaves are weighed for 20 grams and then boiled with 100 mL of aquadest for 15 minutes. The synthesis process is performed by pipetting 10 mL of a 0.01 M $AgNO_3$ solution into an Erlenmeyer flask, then adding the *M. dubia* extract in various concentrations of 1.0, 2.0, 3.0, and 4.0 mL at 28°C. Place the extract for 15 minutes until the colour changes from yellow to dark brown, which indicates the formation of $AgNO_3$. The results of the synthesis of NPs were characterised (Shankar et al., 2004).

REFERENCES

Cai W, Gao T, Hong H. 2008. *Nanotechnol. Sci. Appl. 1*: 17–32. doi:10.2147/NSA.S3788.
Entwistle J, Hall CL, Turley EA. 1996. *J. Cell. Biochem. 61*: 569–577. doi:10.1002/(sici)1097-4644(19960616).
Geetha R, Ashokkumar T, Tamilselvan S. 2013. *Cancer Nano. 4*: 91–98. doi:10.1007/s12645.
Jain S, Das M. 2011. *Nanomedicine. 6*: 599–603. doi:10.2217/nnm.11.60.

Jiao YN, Wu LN, Xue D. 2018. *Cancer Cell. Int. 18*: 149. doi:10.1186/s12935-018-0646-4.

Kim BYS, Rutka JT, Chan WCW. 2012. *NanomedicineN. Engl. J. Med. 362*: 2434–2443. doi:10.1056/NEJMra0912273.

Lekshmi JP, Sumi SB, Viveka S, Jeeva S, Brindha JR. 2012. *J. Microbiol. Biotechnol. Res. 2*: 115–119. doi:10.1016/j.jgeb.2016.12.002.

Leszczynska K, Namiot D, Byfield FJ. 2013. *J. Antimicrob. Chemother. 68*: 610–618. doi:10.1093/jac/dks434.

Nie S. 2010. *Nanomedicine. 5*: 523–528. doi:10.2217/nnm.10.23.

Podila R, Brown JM. 2013. *J. Biochem. Mol. Toxicol. 27*: 50–55. doi:10.1002/jbt.21442.

Shankar SS, Rai A, Ahmad A. 2004. *J. Colloid Interface Sci. 275*: 496–502. doi:10.1016/j.jcis.2004.03.003.

Shukla R, Bansal V, Chaudhary M. 2005. *Langmuir. 21*: 10644–10654. doi:10.1021/la0513712.

Siavash I. 2011. *Green Chem. 13*: 2638–2650. doi:10.1039/C1GC15386B.

Son SJ, Bai X, Lee SB. 2007. *Drug Discov. Today. 12*: 650–656. doi:10.1016/j.drudis.2007.06.002.

Sørensen OE, Follin P, Johnsen AH. 2001. *Blood. 97*: 3951–3959. doi:10.1186/1556-276X-9-248.

14

A 'Biomaterial Cookbook': Biochemically Patterned Substrate to Promote and Control Vascularisation in Vitro and in Vivo

Katie M. Kilgour
North Carolina State University, Raleigh, United States

Brendan L. Turner
North Carolina State University – University of North Carolina Chapel Hill, United States

Augustus Adams
North Carolina State University, Raleigh, United States

Stefano Menegatti
North Carolina State University, Raleigh, United States
Biomanufacturing Training and Education Center, North Carolina State University, Raleigh, United States

Michael A. Daniele
North Carolina State University – University of North Carolina Chapel Hill, United States
North Carolina State University, Raleigh, United States

CONTENTS

DOI: 10.1201/9781003153504-14

Introduction

In vivo, cells are dispersed in an extracellular matrix (ECM) where they form complex tissues with functional and structural characteristics necessary to develop, support, and protect physiological performance (Sawkins, Saldin, Badylak, & White, 2018). The biochemical and mechanical properties of the ECM are both instructional and vital to the resident cells, as they determine their biological identity (*i.e.*, differentiation (Smith, Cho, & Discher, 2018), survival and density (*i.e.*, proliferation (Yue, 2014)), secretion of proteins and biomolecules (Manabe et al., 2008), and organisation into complex structures (Sawkins et al., 2018). Accordingly, interest in developing ECM mimetics for cell engineering and culture, and ultimately tissue construction and repair (*e.g.*, tissue grafting or wound healing), has increased exponentially in the last two decades (Aisenbrey & Murphy, 2020).

A successful engineered tissue construct (ETC) requires a scaffold with instructive functionalisation with biophysical and chemical cues and appropriate resident cell population(s) (Hasan, 2016). The materials employed in tissue scaffolds are typically hydrogels of either natural or synthetic origin. Natural materials comprise polysaccharides (*e.g.*, chitosan, alginate, hyaluronic acid, etc.), proteins (*e.g.*, collagen, gelatin, vitronectin, etc. (J. Zhu & Marchant, 2011)) or self-assembling peptides (*e.g.*, α-helix, β-sheet, collagen-like peptides, elastin-like peptides, or peptide amphiphiles (Hellmund & Koksch, 2019; Rivas, Del Valle, Alemán, & Puiggalí, 2019)). These materials are naturally bioadhesive and exhibit tunable mechanical properties; on the other hand, they are highly variable and often require physicochemical or biochemical modifications that are often challenging to achieve (Cushing & Anseth, 2007; Nguyen & Murphy, 2018; Smithmyer, Sawicki, & Kloxin, 2014). Synthetic materials, instead, are easily tunable both chemically and mechanically and are available in large scale and affordable, but they provide limited biointegration and can trigger adverse foreign-body response in host organisms (Zant & Grijpma, 2016). To date, extensive research has been conducted to fine-tune the mechanical, chemical, and biomolecular properties of synthetic and natural biomaterials and better understand the cell ingredients needed to successfully achieve vascularisation *in vitro* and *in vivo* (*e.g.*, at an injury site) (Berthiaume, Maguire, & Yarmush, 2011; Eberli, Filho, Atala, & Yoo, 2009; Vacanti & Vacanti, 2000; Viola, Lal, & Grad, 2003). Current research revolves around developing 'smart biomaterials' that can direct cell functions and enhance cell performance (Furth & Atala, 2013). Tissue scaffolds can be applied to various tissues (*e.g.*, heart, kidneys, liver, lungs, etc.) and manipulated based on the native complexity (Eberli et al., 2009).

Numerous biomaterials have been fabricated whose (bio)chemical composition is accurately controlled to achieve cell adhesion, proliferation, or differentiation, thus culminating in angiogenesis, vasculogenesis, or anastomosis (Liu et al., 2010; Mhanna et al., 2014; Qiong Liu et al., 2010; Salinas & Anseth, 2008). While vasculogenesis, angiogenesis, and anastomosis are collected under the 'vascularisation' label, they require *ad hoc* natural or engineered environments. Angiogenesis is the formation of new blood vessels from existing vasculature (Adair & Montani, 2010); vasculogenesis is the differentiation of stem cells into endothelial cells (ECs) and the *de novo* formation of a vascular network (Kolte, McClung, & Aronow, 2016); lastly, anastomosis is the formation of a connection between existing blood vessels (Figure 14.1) (van Hinsbergh, 2016). These processes are deeply implicated in tissue regeneration, making it important to develop ETCs that support these processes. The body of work conducted *in vivo*, in both animal models and humans, has clearly shown that scaffolds functionalised with biomolecular patterns and seeded with specific cell populations provide a far superior vascularisation performance compared to biomaterials alone (Atala, 2009). Yet, a major challenge in engineering scaffolds for tissue grafts or implants is ensuring that the substrate promotes the concurrent formation of a tissue (*e.g.*, muscle, bone, liver) and an embedded vascular network that supports the cells therein. The formation and distribution of blood vessels is critical to support large ETCs, as cells that are further than 100–200 μm from a vessel die due

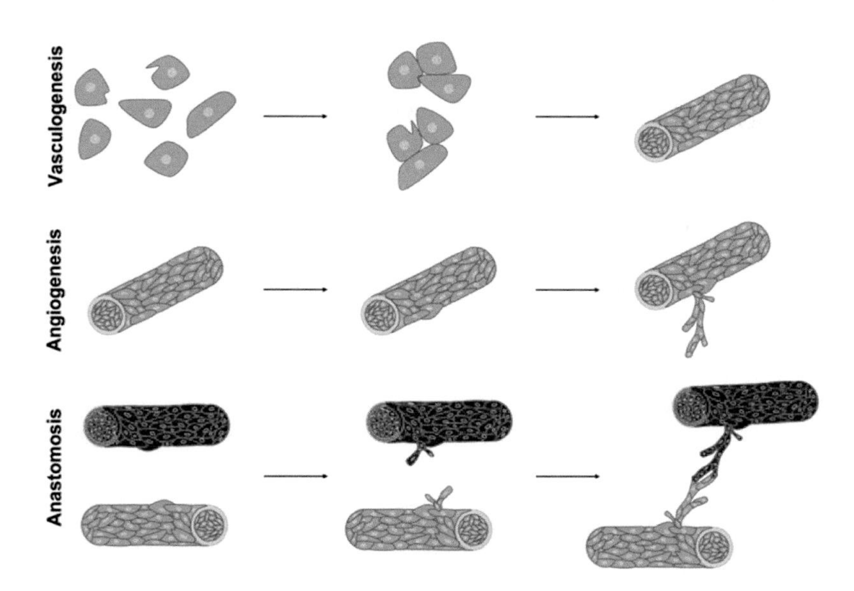

FIGURE 14.1 A schematic representing blood vessel formation via vasculogenesis, angiogenesis, and anastomosis. Vasculogenesis is the *de novo* formation of vessels from endothelial progenitor cells (EPCs). Angiogenesis is the formation of vessels from previously existing ones and anastomosis is the connection of two existing blood vessels.

to a lack of oxygen and nutrients (Grimes, Kelly, Bloch, & Partridge, 2014; Lee, Parthiban, Jin, Knowles, & Kim, 2020; Thomlinson & Gray, 1955).

The biomolecular and mechanical patterning of biomaterials is key to develop scaffolds that induce angiogenesis and vasculogenesis, ultimately enabling successful integration of the construct with the host (Mastrullo, Cathery, Velliou, Madeddu, & Campagnolo, 2020). A variety of techniques are now available to achieve physical patterning, such as Langmuir Blodgett method, immobilisation, electrophoretic deposition, laser deposition, ion beam deposition, plasma treatment, photo-chemical modification, and acid etching (Rana et al., 2017); biomolecular functionalisation can be performed via 3D printing, physisorption and/or chemisorption, and covalent conjugation (Rana et al., 2017). A secondary layer of biomolecular patterning can be provided by the resident or feeder cells, in particular stem and progenitor cells, which respond to local mechanical and biomolecular features by secreting protein factors, cytokines, and chemokines (Murugan, Molnar, Rao, & Hickman, 2009). The combination of engineered solid-phase cues and cell-driven liquid-phase cues forms a complex spatiotemporal pattern of biomolecular signals, which direct downstream cellular fate and tissue organisation (Rana et al., 2017). The angiogenic process is a paradigmatic example of these processes resulting from the crosstalk between biomaterial-regulated and cell-regulated biopatterning (Lovett, Lee, Edwards, & Kaplan, 2009). The engineering of pro-angiogenic tissue scaffolds strives to recapitulate these processes by merging biomolecular engineering, material science, chemical engineering, advanced analytics, and imaging techniques into ECM-mimetic substrates. This review will delve into a detailed discussion of various biomaterials (*e.g.*, natural or synthetic), their properties (*e.g.*, mechanical and biomolecular functionalisation), fabrication methods (*e.g.*, bio-printing, lithography), and the inclusion of cells (*e.g.*, endothelial and stem) to overcome the existing challenges in promoting vascularisation and biomaterial–host integration.

A Bit of History …

First-generation biomaterials for use in humans were developed in the 1960s and 1970s. A pioneer of biomaterials for tissue engineering, Bill Bonfield, contributed to understanding the interactions between implants and living tissues by outlining the rules for designing materials that mimic and replace tissue with minimal-to-no adverse response (Hench & Thompson, 2010). By 1980, documented use of over

50 different medical implants composed of more than 40 different materials had been reported (Hench, 1980). Emphasis then shifted towards the functions at the biomaterial–tissue interface, introducing second-generation biomaterials to control action and reaction (*e.g.*, bioresorption) in a physiological environment. As Larry Hench wrote, 'a second method of manipulating the biomaterials-tissue interface is controlled chemical breakdown, that is resorption of the material ... [namely] the foreign material is ultimately replaced by regenerating tissues' (Hench, 1980). By the 1990s, resorbable polymer materials were routinely used as sutures and began to be used for drug delivery and tissue regeneration scaffolds with the ability to promote cell adhesion and proliferation (Guo & Ma, 2014; Park & Bronzino, 2002). By the dawn of 21st century, biomaterial research focused on regenerative medicine. This led to a new generation of biomaterials capable of stimulating cellular responses at the molecular level (Narayan, 2010). The ground breaking novelty provided by these biomaterials was their ability to stimulate specific interactions, guide cell proliferation and differentiation, and ultimately promote ECM production and organisation. This was achieved by integrating proteins, peptides, and other biomolecules into structured composite biomaterials, resulting in close ECM mimetics (Hench & Polak, 2002; Kim & Mooney, 1998).

Today, the focus is on biomaterials that stimulate the formation of large (up to 1 in^3) microvascular networks to sustain long-term cell differentiation and stability post-implantation (J. H. Lee et al., 2020; Song et al., 2019). Meeting this need relies on the convergence of advanced chemistry and biochemistry, material and surface sciences, and biomedical and biochemical engineering. The latest advances in high-resolution micro- and nanofabrication techniques provide tissue scaffolds that closely resemble the ECM microenvironment by fine-tuning the 3D distribution of mechanical and biomolecular features stimulating cell adhesion and desired functions (von der Mark & Park, 2013). Recent advancements in building blocks and fabrication techniques have led to an improved availability, affordability, and safety of biomaterials, as demonstrated by the growing number of animal trials and human trials of cardiovascular, dermal, and hepatic implants and transplants (van Hinsbergh, 2016).

The Role of ECM in Angiogenesis Inspires the Design of ECM Mimetics

While arteries and veins convey the flow of blood driven by the heart, the distribution and exchange of oxygen and nutrients occur through the microcirculation present within the organs, which fulfils their metabolic needs by tissue perfusion (Williams, Gilchrist, Strain, Fraser, & Shore, 2020). In case of tissue injury or ischemic pathologies, the metabolic demand cannot be met, impacting significant tissue portions and impairing their function. Damaged tissues cannot fully regenerate on their own and require some level of intervention (Krafts, 2010). Implantable scaffolds represent a promising strategy to restore tissue functions (Dhandayuthapani, Yoshida, Maekawa, & Kumar, 2011; Pina et al., 2019). Critical to this end is the ability of tissue scaffolds to promote angiogenesis, the proliferation of new blood vessels from pre-existing, and supply the oxygen and nutrients needed for cell recruitment and integration with the host tissue (Pina et al., 2019).

The Biomolecular Fundamentals Angiogenesis

Angiogenesis is the prime method for promoting vascularisation in most ETCs because it harnesses the instructive properties of the existing vascular network. Angiogenesis mainly occurs during tissue development, healing, or inflammation, and can be categorised into sprouting and intussusceptive angiogenesis (van Hinsbergh, 2016). Sprouting angiogenesis is a process where new blood vessels grow from pre-existing ones, whereas intussusceptive angiogenesis is the process of new blood vessels forming by splitting existing vessel into two vessels (van Hinsbergh, 2016), while sprouting angiogenesis is well understood, intussusceptive angiogenesis occurs in the vessel lumen, making it difficult to study.

Both angiogenetic mechanisms are controlled by growth factors, their receptors, and inhibitors. Chief among them is the family of vascular endothelial growth factors (VEGF), which comprises the four subclasses VEGF-A, VEGF-B, VEGF-C, and VEGF-D, and their receptors VEGFR1, VEGFR2, and VEGFR3 (Pandey et al., 2018). In particular, the VEGF-A/VEGFR2 signalling pair regulates most of the interactions involved in angiogenesis (Pandey et al., 2018). Upon binding VEGF-A, the VEGFR2 receptor activates the pathways (Akt and MAPK) that regulate metabolic, proliferative, and growth processes of endothelial cells resulting in angiogenesis (Pandey et al., 2018); concurrently, it triggers the production of nitric oxide, which enhances vascular permeability by vasodilation and lowers the blood pressure (Pandey et al., 2018). Other proteins involved in endothelial cell growth and migration are angiogenin, epidermal growth factor (EGF), oestrogen, interleukin 8 (IL-8), prostaglandins E1 and E2, and tissue necrotic factor α (TNF-α) (Barnabas, Wang, & Gao, 2013; Harada et al., 1994; Montrucchio et al., 1994; Semino, Kamm, & Lauffenburger, 2006; Shi & Wei, 2016). These actors form the biomolecular core of the processes of stem cell differentiation into endothelial cells and regulate their adhesion, migration, and proliferation, which collectively contribute to angiogenesis. *In vivo*, the cell-surrounding ECM comprises a variety of structural (*e.g.*, collagen, laminin, and fibronectin (FN) (Wang & Borenstein, 2017) and signalling proteins (*e.g.*, cytokines and growth factors (Wang & Borenstein, 2017)), as well as complex topological features (*e.g.*, ports, ridges, and fibres), which cooperatively guide the activity and the motion of cells via chemotaxis, haptotaxis, and mechanotaxis (Nelson et al., 1996).

Most ECM proteins contain peptide motifs that target cell surface receptors, thus promoting cell adhesion, selectively bind growth factors in solution, thus adsorbing them onto the matrix and presenting them to the resident cells (Carson & Barker, 2009). The affinity adsorption drives the formation of combined gradients of growth factors, which directs cell signalling with high spatiotemporal precision, ultimately resulting in a meaningful spatial distribution of different cell phenotypes (Yue, 2014). The structural features and protein gradients in the ECM determine cell activity not only biologically, but also spatially, by directing cellular motion and congregation into tubular constructs that evolve into blood vessels (Yue, 2014).

The second main contribution of ECM to angiogenesis stems from its topological and morphological properties, namely, the combined patterns of structural features (*e.g.*, geometry, topology, and morphology) and stiffness (*i.e.*, Young's modulus) that contribute to determine the morphology, proliferation, migration, and differentiation of cells (Ghasemi-Mobarakeh, 2015). Specifically, substrates with higher stiffness provide increased adhesion, proliferation, and expression of pro-angiogenic factors as the endothelial cells migrate into the substrate and align on fibronectin fibres of ECM (Canver, Ngo, Urbano, & Clyne, 2016; Goli-Malekabadi, Tafazzoli-Shadpour, Tamayol, & Seyedjafari, 2017; Zhao et al., 2018). The arrangement and alignment of nanofibres/tubes and matrix porosity guides vascular smooth muscle cells (VSMCs) and endothelial cell into forming functional microvessels (Lee, Cerchiari, & Desai, 2014; Lin et al., 2018; Nakayama et al., 2015; Peng et al., 2010); it has also been recently shown that the alignment of endothelial cells inhibits inflammation, thus setting an appropriate environment for angiogenesis (Huang et al., 2013; Nakayama et al., 2016).

While the aim of many synthetic and natural scaffolds is to induce angiogenesis, the majority of ETCs are formed without the benefit of being adjacent to or containing their own vascular network. The biomaterials developed in recent years for tissue engineering have been designed to recapitulate the physical, chemical, and biomolecular features of the ECM (Ma, Kotaki, Inai, & Ramakrishna, 2005). Since cell proliferation in 3D scaffolds requires a steady influx of oxygen and nutrients, the substrate biomaterials (*e.g.*, biopolymers, self-assembling peptides, peptide mimetics, etc.) must provide an environment that enables rapid transport of nutrients, bioactive factors, and delivery of oxygen to ensure long-term cell survival (Edelman, Mathiowitz, Langer, & Klagsbrun, 1991); furthermore, their mechanical properties can be fined tuned via physical or chemical modification of native materials to induce desired physiological responses (*e.g.*, proliferation, angiogenesis, vasculogenesis) (Klimek & Ginalska, 2020; Zhu & Marchant, 2011). Finally, biomolecular patterning can be introduced by integrating soluble factors, hormones, growth factors, or small chemicals in the matrix with spatial precision (Martin, Caliari, Williford, Harley, & Bailey, 2011; Mitchell, Briquez, Hubbell, & Cochran, 2016). To these ends, myriad fabrication and modification techniques to impart structural and biomolecular patterning are now available, as discussed below.

Biopatterning of Substrates for Angiogenesis

Patterning synthetic or natural biomaterials to fine-tune properties such as stiffness, pore size, surface morphology, and biomolecular cues for optimising cellular function has recently become mainstream in ECM mimicry (Geckil, Xu, Zhang, Moon, & Demirci, 2010). The toolbox for biopatterning substrates for

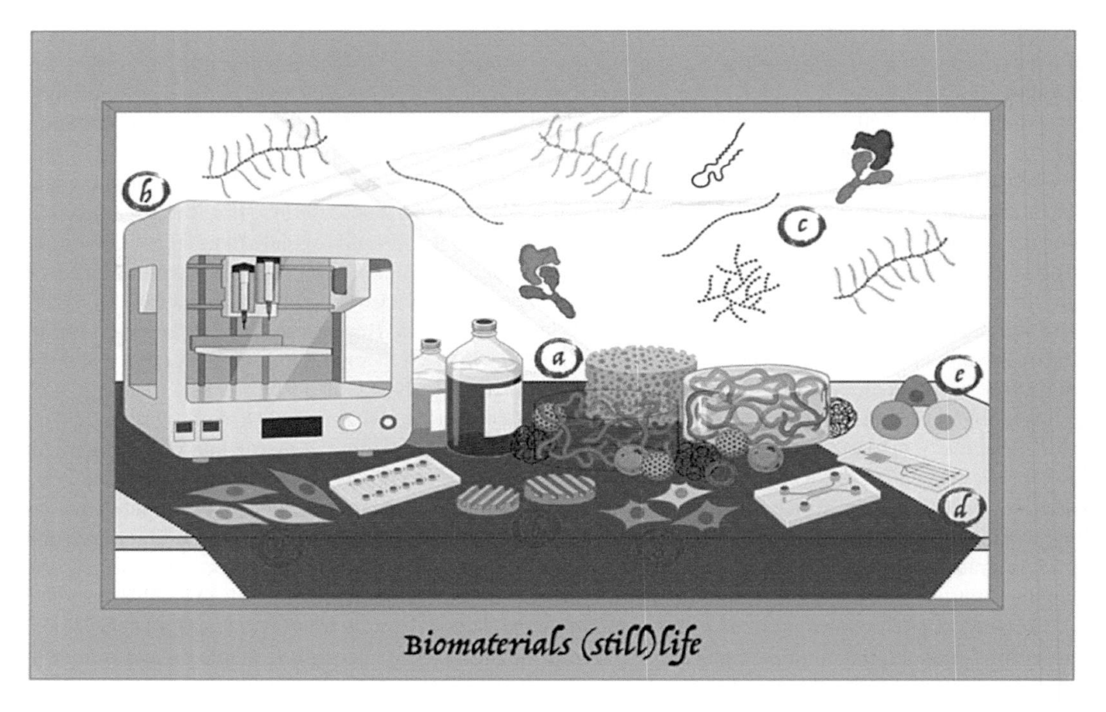

FIGURE 14.2 A still life of different techniques including 3D printing and microfluidics with the necessary cells, reagents, and materials to achieve a biopatterned ETC.

angiogenesis includes bottom-up, post-gelation, and microfluidic techniques (Figure 14.2). Some natural biomaterials possess the inherent ability to promote cell adhesion and ECM modelling and degradation through biological recognition; however, they suffer from batch-to-batch variability in both their biomolecular and mechanical attributes (Chen & Liu, 2016).

Bottom-up approaches employ layer-by-layer deposition to form a stack up to ~ 100 μm of biopolymer enriched with chemical and biochemical cues (Chan & Leong, 2008; Sekiya, Shimizu, Yamato, Kikuchi, & Okano, 2006; Shimizu et al., 2006). Available techniques for bottom-up biopatterning include photolithography, stereolithography, additive crosslinking, and 3D printing (Khetan & Burdick, 2011). The latter is at the forefront owing to its affordability and precision in spatial control of biomaterials and biomolecular cues (Bishop et al., 2017). By constructing each layer individually, 3D printing affords substrates with spatially accurate and reproducible biopatterns and enables encapsulation of cells between the layers during the deposition process; uniform cell distribution in turn increases cell survival and stimulates nitric oxide production, ultimately promoting angiogenesis (Liu & Gartner, 2012; Nemati et al., 2017). Bottom-up technologies provide gentle processing conditions, allowing for multiple polymer and cell types to be combined in a spatial manner. Chan et al. employed a stereolithography bottom-up approach with poly (ethylene glycol) diacrylate to encapsulate RGD peptides and NIH3T3 cells in a controlled distribution to monitor long-term cell viability, proliferation, and migration of cells (Chan, Zorlutuna, Jeong, Kong, & Bashir, 2010); this approach resulted in significant improvement of cell viability, proliferation, and migration over 14 days compared to the gels made with a top-down approach. Another study by Lee et al. used 3D printing to construct a hydrogel with a collagen base, two parallel gelatin channels on the surface of the collagen, and a human umbilical vein endothelial cell (HUVEC) and fibroblast loaded fibrin gel embedded on top of the collagen between the channels to create a vascular network connecting the channels; their results showed that flow medium at 10–100 μL/min through the channels enabled supporting cell viability for 2 weeks and promoted angiogenic sprouting from the printed channels on a scale of 400 μm over 7 days (Lee et al., 2014). While appealing, bottom-up approaches impose high material costs and extended production times and struggle to produce large, robust tissue scaffolds without compromising microfeatures (Nichol & Khademhosseini, 2009; Zhang, Sun, & Fang, 2004).

Post-gelation approaches consist of patterning a pre-formed hydrogel scaffold with biochemical cues and seeding cells thereupon, while initially arranged as a monolayer, cells migrate throughout the scaffold, ultimately forming a spatial distribution based on the biomolecular blueprint (Khetan & Burdick, 2011). Fabrication of hydrogels typically employ single or two-photon laser scanning photolithography, photoablation, or mask-based photolithography, which allow for simultaneous processing of samples and are relatively inexpensive (Khetan & Burdick, 2011). Following the incorporated biomolecular or mechanical cues, the cells migrate into the gel and contribute to forming the appropriate ECM for angiogenesis to occur (Nichol & Khademhosseini, 2009). To date, post-gelation tissue engineering has relied on recruiting blood capillaries and sprouting from the host into the scaffold, and vasculogenesis from endothelial precursor cells *in vitro* (Guillotin & Guillemot, 2011). Aizawa et al. fabricated agarose hydrogels functionalised with GRGDS and photochemically patterned with VEGF165 via multi-photon lasers to guide endothelial cells, which promoted successful guided migration of endothelial cells to form tubular structures for angiogenesis (Aizawa, Wylie, & Shoichet, 2010). Despite the advancements in surface patterning, post-gelation techniques have difficulty mimicking complex microstructures found on natural tissues and struggle to achieve the maturation and perfusion needed for 3D ETCs (Guillotin & Guillemot, 2011; Tiruvannamalai-Annamalai, Armant, & Matthew, 2014).

Finally, microfluidic systems are utilised to fine-tune the spatial distribution of biopolymers and the biomolecular patterning in pro-angiogenic scaffolds (Khetan & Burdick, 2011); they also enable concurrent loading of cells and their alignment along the patterns, which is particularly favourable to angiogenesis (Khetan & Burdick, 2011; Xiaolin Wang, Sun, & Pei, 2018). Microfluidic approaches can operate in the absence of UV light and introduce void space within the scaffold, which facilitates perfusion of nutrients and oxygen needed for cell viability. Additionally, this technology facilitates the integration of natural and synthetic biopolymers into complex architectures with fine-tuned mechanical and biochemical features (Minardi, Taraballi, Pandolfi, & Tasciotti, 2016). Zhen et al. developed an *in vitro* angiogenesis model by utilizing a microfluidic device with three main parallel channels perpendicularly connected by numerous smaller channels to precisely pattern 3D gels (Dai et al., 2011); endothelial cells were cultured and monitored in real time to evaluate the effects of pro-angiogenic factors on the cell proliferation, migration, and tube formation. Their results indicated that the proliferation rate was significantly promoted by the angiogenic factors and the cells migrated into the gel forming tubular structures. The main challenges in biopatterning using microfluidic systems are the fabrication of systems on physiological scales and the cost of customizing a microfluidic geometry for each application (Jafarkhani, Salehi, Aidun, & Shokrgozar, 2019; Tenje et al., 2020).

Upon developing 3D constructs, it is crucial to characterise cell adhesion, orientation, surface chemistry, gene expression, and proliferation in response to the morpho/topological, mechanical, and biomolecular features of the scaffolds (Ishii, Shin, Sueda, & Vacanti, 2005; Li & Folch, 2005; Raghunathan et al., 2013; Ulrich, De Juan Pardo, & Kumar, 2009; Yañez-Soto, Liliensiek, Murphy, & Nealey, 2013). Insufficient cellular adhesion can be addressed by artificially increasing functionalisation with bioactive molecules (Minardi et al., 2016), which may be insufficient in natural biopolymers and is missing in native synthetic polymers (Luo & Shoichet, 2004; Yan, Sun, & Ding, 2011).

The Cell 'Ingredients'

Vascular networks are formed by the differentiation, proliferation, and assembly of vascular cells, specifically endothelial cells and vascular smooth muscle cells (Tiruvannamalai Annamalai, Rioja, Putnam, & Stegemann, 2016). Typically, the incorporation of large amounts of cells in a scaffold is needed to form complex networks; therefore, cells are often attained from banked cell lines, cultured, and harvested prior to seeding in the scaffold (Sarker, Naghieh, Sharma, Ning, & Chen, 2019), although more recent efforts are aiming to utilise autologous induces pluripotent stem cells or endothelial cells derived from donors. The choice and source of the cell is a key role in guiding the formation of new blood vessels in ETCs. While many studies to date have utilised terminally differentiated cells (*e.g.*, HUVECs, BMVECs, etc.) (Howard, Buttery, Shakesheff, & Roberts, 2008), more recent work has investigated the co-culturing of stem (*e.g.*, mesenchymal), stromal, and progenitor (*e.g.*, endothelial) cells to enhance viability and density during vascular formation (Kook, Jeong, Lee, & Koh, 2017). In particular, the most successful scaffolds employed as implants utilise autologous cells (Nair et al., 2011); cell collection, however, is invasive and the risk of harvesting diseased cells has led researchers to include different types of stem cells for differentiation (Howard et al., 2008) (*e.g.*, mesenchymal or embryonic). Guided by the signal provided from the surrounding environment, stem cells differentiate into multiple different cell types allowing for the development of functional tissue ready for implantation (Evans, Gentleman, & Polak, 2006).

The two main cells typically used for angiogenesis are endothelial progenitor cells (EPC) and mesenchymal stem cells (MSCs) (Adair & Montani, 2010). EPCs are isolated form adult peripheral and umbilical cord blood and play a significant role in neovascularisation and ischemic tissue repair (Hristov, Erl, & Weber, 2003); MSCs isolated from adipose tissue and vessel walls are able to differentiate into a variety of cell types including chondrocytes, myocytes, adipocytes, and osteocytes (Brunt, Hall, Ward, & Melo, 2007). A variety of techniques can be used to isolate EPCs or MSCs, such as density gradient centrifugation, flow cytometry sorting, plastic adherence, and immunomagnetic selection (Wang et al., 2018). Terminally differentiated cells, such as HUVECs or BMVECs, are isolated from human umbilical cords, which are typically discarded after childbirth, thus avoiding contamination from other cells (Kocherova et al., 2019); these cells can also be co-cultured on scaffolds to enhance gene expression and vascularisation *in vivo* and *in vitro* (Sukmana & Vermette, 2010).

Substrate Materials

Biomaterials for tissue engineering include biopolymers (*i.e.*, oligo and polysaccharides, oligo and polypeptides, and proteins), along with their synthetic derivatives and mimetics, and synthetic polymers.

Polysaccharide-Based Substrates

Polysaccharides are abundant and affordable, biocompatible and biodegradable, feature high-hydrophilicity and high swelling ratio, and enable facile chemical functionalisation (Aminabhavi & Deshmukh, 2016; Guan, Liu, Zhao, Li, & Wang, 2020). The polysaccharides commonly used as ETCs are chitin and chitosan, cellulose, alginate, starch, dextran, agarose, and hyaluronic acid (Fan, Yan, Xu, & Ni, 2012; Hirano, Zhang, Chung, & Kim, 2000; Yu Luo et al., 2013; Robert, Robert, & Renard, 2010; Yu et al., 2010).

Chitosan is known for its bioadhesion and chemical versatility (Singh, Choi, Park, Akaike, & Cho, 2014). On the other hand, the phase stability and biochemical stability of chitosan hydrogels under physiological conditions is lower than that of other polysaccharides (Table 14.1) (Cheung, Ng, Wong, & Chan, 2015; Hadadi et al., 2018; Jiao, Huang, & Zhang, 2019). Current methods for producing chitosan hydrogels include ionic gelation, chemical crosslinking, and *in situ* gelation (Aminabhavi & Deshmukh, 2016). Chemical modification and *in situ* formation of covalent crosslinking have been widely utilised to stabilise and strengthen chitosan-based hydrogels (Jayakumar et al., 2010); a popular approach combines carboxymethyl-chitosan with dextran dialdehyde, which provides the rapid formation of a network of imine bonds without the need of additional reagents, allowing seamless *in vitro* encapsulation of cells and growth factors (Cheng et al., 2014; Jayakumar et al., 2010). Other biopolymers (collagen or gelatin, hyaluronic acid, etc. (Ahmadi, Oveisi, Samani, & Amoozgar, 2015) have been incorporated for tuning the phase behaviour and structural properties of chitosan-based hydrogels via physical or chemical crosslinking (Ahmadi et al., 2015). As a result, chitosan-based hydrogels are now widely used as scaffolds for tissue engineering *in vivo* and regeneration *in vivo* (Islam, Shahruzzaman, Biswas, Nurus Sakib, & Rashid, 2020). In a study by Deng et al., hydrogels containing collagen and chitosan at different ratios were fabricated via carbodiimide chemical crosslinking and loaded with HUVECs and human circulating progenitor cells (CPCs) to evaluate angiogenic activity (Deng et al., 2010). Upon addition of chitosan, the viability of HUVECs increased along with greater CPC differentiation and angiogenesis *in vitro*. This study also observed that *in vivo* vascular growth was enhanced and a great number of von Willebrand factor and C-X-C chemokine receptor type 4 ($CXCR4^+$) angiogenic cells were recruited; the hydrogels also showed more resistance to enzymatic degradation, denaturation, and suggests that the scaffold is promising for restoring vascularisation to damaged tissue.

TABLE 14.1

Biomaterials Commonly Used for ETCs and What Cell Functions They Enhance

		Cell Adhesion	Cell Proliferation	Cell Differentiation	Reference
Base Material	Chitosan	X	X	X	Rodríguez-Vázquez, Vega-Ruiz, Ramos-Zúñiga, Saldaña-Koppel, & Quiñones-Olvera (2015)
	Cellulose	X		X	Hickey & Pelling (2019)
	Alginate	X	X	X	Sun & Tan (2013)
	Dextran	X	X		Sun & Mao (2012)
	Hyaluronic acid	X		X	Zhu, Wang, Yang, & Luo, (2017)
	Collagen	X		X	Somaiah et al. (2015)
	Gelatin	X		X	Afewerki, Sheikhi, Kannan, Ahadian, & Khademhosseini (2018)
	GelMA	X	X	X	Lavrentieva et al. (2019)
	Silk	X			Jao, Mou, & Hu (2016)
	Polyacrylamides	X			Sanzari, Humphrey, Dinelli, Terracciano, & Prodromakis (2018)
	Polyvinyl alcohol	X	X		Teixeira, Amorim, & Felgueiras (2020)
	Polyethylene glycol	X			Zhu (2010)
	RGD peptide	X			Bellis (2011)
	REDV peptide	X	X		Bellis (2011)
	IKVAV peptide	X	X		Patel et al. (2019)

Another possible substrate are the cellulose-based hydrogels, which are commonly harvested from plant or bacterial sources. These scaffolds can be easily modified to provide excellent transport of oxygen and nutrients, loaded with biomolecules and cells, and injected as liquids that form gels *in situ* and support angiogenesis *in vivo* (Alfassi, Rein, Shpigelman, & Cohen, 2019; Hoffman, 2012). Cellulose-based hydrogels contain numerous hydroxyl groups amenable to physical and chemical modification or crosslinking to produce stable and functional 3D structures (Shen, Shamshina, Berton, Gurau, & Rogers, 2015). Myriad cellulose derivatives with different properties are available for building scaffolds for cell culture and tissue engineering, such as sodium carboxymethyl cellulose, cellulose acetate, and hydroxyethyl and hydroxypropyl cellulose (Nasatto et al., 2015). The hydroxyl and carboxyl groups in carboxymethyl cellulose increase the osmotic pressure and lead to swelling (Kabir et al., 2018); the methoxy groups of methyl cellulose provide an amphiphilic character and higher elastic modulus (Kabir et al., 2018; Nemazifard, Kavoosi, Marzban, & Ezedi, 2017); the acetyl groups of cellulose acetate decrease the intra- and inter-molecular hydrogen bonding, thus affecting the water solubility (Alfassi et al., 2019). Chemical crosslinking is widely employed to control swelling and prevent hydrolysis or enzymatic degradation of cellulose-based substrates (Kabir et al., 2018). Wang et al. loaded 3D bacterial cellulose gelatin scaffolds modified with heparin and VEGF to assess the pro-angiogenic effects on the proliferation and migration of pig iliac endothelium cells (Wang et al., 2018). Results indicated that the scaffolds were able to achieve a prolonged release of VEGF over 2 weeks with enhanced proliferation and migration, demonstrating that the promise for long-term implantation for angiogenesis *in vivo*.

Alginate, derived from brown seaweed and bacteria, is one of the most abundant biomaterials (Aminabhavi & Deshmukh, 2016). The abundance of carboxyl groups on its chains enables a broad swath of covalent (*e.g.*, carbodiimide, click chemistry) and ionic crosslinking strategies, whose choice is key to impart desired ECM-mimetic features to the resulting hydrogels (Hu, Wang, Xiao, Zhang, & Wang, 2019). Alginate gels formed by ionic crosslinking can be dissolved by releasing divalent ions that hold the gel together; chemically crosslinked alginate gels can be disassembled via hydrolysis of linking arms (*e.g.*, a disulphide or ester bonds). These gels show good size and shape retention *in vivo*, which makes them ideal when constructing complex 3D scaffolds (Chang et al., 2001). The incorporation of cells in alginate gels relies on the covalently linked molecules to promote cell adhesion to the gel (Lee & Mooney, 2012); proteins and peptides found in the ECM can also be conjugated to alginate to improve cell adhesion (*e.g.*, RGD, laminin), differentiation (*e.g.*, PAMAM, LRP5), and proliferation (*e.g.*, RGD, PAMAM) (Furth & Atala, 2013; Lee & Mooney, 2012; Schulz et al., 2018). Alginate can also be formed into protein-filled microcapsules that provide sustained release of signalling factors into a cell-seeded hydrogel; this has been demonstrated for angiogenic purposes using fibroblast growth factor (FGF) and VEGF, wherein the factor release was controlled over multiple weeks via the disassembly of the ionic network among alginate chains (Lee et al., 2004; Peters, Isenberg, Rowley, & Mooney, 1998); results indicated sustained release for 10 days *in vitro* suggesting a promising method for long-term supply of pro-angiogenic growth factors *in vivo*. Alginate has proven to be a viable biomaterial to use for regenerative medicine, but more efficient ways to prevent short-term degradation and enhance chemical and mechanical stability are still sought after.

A recent trend has emerged of combining different polysaccharides either by physical mixing or chemical ligation to engineer hybrids that feature multiple advantageous properties provided by the various ingredients. Waly et al. developed biodegradable chitosan–cellulose scaffolds with different cellulose ratio (up to 0.44 w/v%) and investigated their porosity, mechanical properties, biodegradation, and cytocompatibility (Waly, Abdel Hamid, Sharaf, Marei, & Mostafa, 2008), showing that higher cellulose contents provided smaller pores and higher mechanical strength; the scaffolds showed copious cell adhesion after 3 days of culture and high cell density after 8 days, concluding that the scaffolds promoted cell adherence along with cell proliferation, which are important factors in angiogenesis. Shi et al. developed collagen–chitosan hydrogels functionalised with angiogenin (ANG) and investigated the *in vivo* angiogenesis over 28 days (Shi, Han, Mao, Ma, & Gao, 2008); their results indicated that the ANG was released for a long period of time (14 days), enhancing angiogenesis; however, from day 14 through 28, the capillary density was less than the scaffolds without ANG, indicating the need to further adjust the scaffolds to achieve a longer-term release profile.

Protein- and Polypeptide-based Substrates

Proteins and polypeptides are an essential component of the ECM, to which they provide structure and functionality. While more expensive and less abundant than polysaccharides, proteins and peptides offer unique benefits in terms of chemical and biomolecular diversity. Specifically, the broad biochemical diversity described by the amino acid residues provides a rich toolbox to attain the desired physicochemical properties and offers a variety of reactive sites for chemical modification and crosslinking by either enzymatic or synthetic routes (Panahi & Baghban-Salehi, 2019). Furthermore, certain classes of natural polypeptides furnish responsiveness to external stimuli, such as temperature (*e.g.*, elastin-like polypeptides (ELPs), fibroin, and their mimetics), and ionic strength or pH (*e.g.*, elastin-like and polycationic peptides) (Table 14.1) (Augustine, Kalva, Kim, Zhang, & Kim, 2019). Also attractive is their ability to disassemble upon exposure to proteolytic enzymes, which makes them ideal bioresorbable materials for implantable applications (Augustine et al., 2019). On the other hand, protein-based substrates tolerate poor environmental conditions that cause denaturation and degradation (*e.g.*, sterilisation or storage) compared to polysaccharides (Jiang et al., 2019; Sun et al., 2015); other common limitations of protein-based substrates are their inability to undergo cycles of drying and rehydration-swelling or deformation (*e.g.*, compression or shear) during fabrication (Fisher, Baker, & Shoichet, 2017; Han et al., 2017; Hwang & Damodaran, 1996; Jiang et al., 2019; Kundu, Poole-Warren, Martens, & Kundu, 2012; Panahi & Baghban-Salehi, 2019; Sun et al., 2015; Xu et al., 2018; Zandi et al., 2021).

Protein-based hydrogels are commonly based on collagen, gelatin, silk, zein, keratin, fibronectin, and albumin. Collagen is a highly abundant protein in solid human tissues and forms 3D hydrogel structures that closely resemble the ECM (Jennings, 2020). Collagen is extracted from bones, skin, muscles, tendons, and ligaments, although synthetic peptide mimetics of collagens have been recently introduced (Karami et al., 2019; Yu, Li, & Kim, 2011). The synergy of cell instructive of mechanical properties and biochemical cues is a unique and particularly attractive feature of collagen. The molecular structure of collagen, in fact, comprises fibrils that form a covalently crosslinked, stable structure and features peptide motifs (*e.g.*, IKVAV, RGD) that promote cell adhesion (Gullberg et al., 1992; Smethurst et al., 2007). Collagen scaffolds are also particularly attractive owing to their porous structure, which facilitates cell migration, and diffusion of nutrients and biochemical cues (Abramo & Viola, 1992). Furthermore, collagen-based materials form ideal reservoirs for loading growth factors, antibiotics, and other therapeutic agents that accelerate angiogenesis and healing processes (Vo, Kasper, & Mikos, 2012). The main limitation of collagen-based substrates is their limited mechanical quality, specifically the lack of stability upon swelling (Lohrasbi, Mirzaei, Karimizade, Takallu, & Rezaei, 2020); furthermore, the low level of covalent crosslinking of collagen results in a collapse of the hydrogel structure upon variations in temperature, ionic strength, and pH, or trace amounts of proteases (Catoira, Fusaro, Di Francesco, Ramella, & Boccafoschi, 2019; Dinescu et al., 2019). Researchers have focused on improving the mechanical strength of collagen by controlling its network structure via enzymatic chain editing (cleavage or ligation) or chemical crosslinking (*e.g.*, using UV light) (Davidenko et al., 2015). Collagen contains a small fraction of lysine residues that provide the primary amino groups typically utilised for crosslinking and conjugation of biochemical cues (Gallop & Paz, 1975); in this context, glutaraldehyde, genipine, carbonyldiimidazole, and carbodiimide (EDC) have been adopted as cross-linking agents to increase hydrogel stability and mechanical strength (Panahi & Baghban-Salehi, 2019). These reactions form crosslinks between free amine (lysine) and carboxyl groups (glutamate and aspartate), allowing for tailoring the gel stiffness and limiting the natural swelling properties of collagen (Grover et al., 2012).

Gelatin, a collagen derivative, is a thermo-responsive material that is water soluble at physiological temperature and forms a hydrogel upon cooling (Chatterjee, Hui, & Kan, 2018). Being easy to crosslink, functionalise, and form, as well as biocompatible and biodegradable, gelatin hydrogels are ideal candidates for implantable scaffolds (Panahi & Baghban-Salehi, 2019). Most of the literature on gelatin modification has focused on improving the mechanical stability of hydrogels and reducing its native high sensitivity to proteolytic enzymes (Xing et al., 2014). On the other hand, a growing number of studies have introduced chemical modification methods aimed at installing biomolecular cues and/or receptors to enhance cell adhesion and function and stimulate angiogenic processes (Daniele, Adams, Naciri, North, & Ligler, 2014; Leucht, Volz, Rogal, Borchers, & Kluger, 2020; Norris, Delgado, & Kasko, 2019). The most popular chemical derivative of gelatin is gelatin methacryloyl (GelMA), which is prepared by reacting lysine residues with methacrylic anhydride. The resulting acrylate groups enable crosslinking of GelMA chains into a covalent hydrogel, whose stiffness is controlled by the substitution ratio and UV dose (Dong et al., 2019; Young, White, & Daniele, 2020). GelMA-based hydrogels provide excellent cell adhesion and proliferation along with tunable biodegradability (Su et al., 2018; Van Den Bulcke et al., 2000). Tunable stiffness is particularly desirable as it enables recapitulating the stem cell niche (Nikkhah et al., 2016). GelMA-based hydrogels have been extensively employed for building vascular networks (Chen et al., 2012; Lin, Chen, Moreno-Luna, Khademhosseini, & Melero-Martin, 2013; Nikkhah et al., 2012). One study by Nikkhah et al. used photolithography to create GelMA substrates with micropatterned features (50–150 μm in height) that significantly improved the proliferation, alignment, and formation of encapsulated endothelial cells and resulted in the assembly of endothelial cells expressing CD31 and VE-cadherin, which is a mark of forming vasculature (Nikkhah et al., 2012). The geometric features guided cell assembly into cord structures that were retained within the hydrogels for more than 2 weeks, indicated the ability to be used for organised vasculature. While gelatin-based hydrogels are among the most promising materials for vascular engineering, *in vivo* studies have only been short term (~ 4 weeks), and longer observation periods are needed to better evaluate cell toxicity, angiogenic response, and tissue integration, to elucidate the full translational potential of these substrates.

Silk is a protein-based biomaterial produced by a variety of silkworms with varying structure and properties and is of value for constructing tissue scaffolds since its resistance to protases is much higher than that of other natural biopolymers (Bandyopadhyay, Chowdhury, Dey, Moses, & Mandal, 2019). Silk comprises two main protein components, namely, fibroin and sericin: the former provides mechanical strength, while the latter promotes cell adhesion (Altman et al., 2003; Fu, Shao, & Fritz, 2009;

Kapoor & Kundu, 2016; Kim, Kim, Kang, & Kim, 2016; Kunz, Brancalhão, Ribeiro, & Natali, 2016; Legon'kova et al., 2016). The high crystallinity of silk retards the biodegradation process, and the absorption rate depends mostly on the implantation site, type of silk, and size of the implant (Kapoor & Kundu, 2016; Kearns, Macintosh, Crawford, & Hatton, 2008). Notably, silk extracted from specific sources (*e.g.*, *A. yamamai*and *A. pernyi* silk fibroins) contains RGD peptide sequences, which act as integrin-binding motifs and facilitate interaction with the ECM and cells (Kang et al., 2018; Minoura et al., 1995; Senta, Bergeron, Drevelle, Park, & Faucheux, 2011). In addition, silk features strong hydrogen bonding interactions between the crystallite regions and the α-helices of surrounding proteins, which provide high mechanical strength and make silk an ideal reinforcement material to create hybrid scaffolds (Bandyopadhyay et al., 2019). On the other hand, silk-based biomaterials pose limitations due to the lack of motifs providing signals directing cell development (Nguyen et al., 2019); this limitation has been addressed by incorporating signalling biomolecules, most commonly growth factors and their peptide mimetics (Vepari & Kaplan, 2007), that impart cell-directing function to silk-based hydrogels.

Finally, synthetic peptides represent special building blocks for tissue scaffold engineering, being biocompatible, biodegradable, bioactive, and low-toxicity protein mimetics (Du, Zhou, Shi, & Xu, 2015; Liu, Zhang, Zhu, Liu, & Chen, 2019). Peptides can assemble into hydrogels either physically, by forming a network of hydrogen bonds and electrostatic and hydrophobic interactions, or chemically, via site-selective crosslinking (Liu et al., 2019). Lastly, they can be manufactured in high volumes, affordably, and with no variability. This makes peptides ideal biomaterials for tissue engineering and wound healing (Barbosa & Martins, 2017). In this context, elastin-like polypeptides are of particular interest, as they feature a thermo-responsive phase behaviour, wherein the sol–gel transition temperature and the mechanical properties of the resulting hydrogel depend on the ELP's amino acid sequence (Urry, 1997). Engineered ELPs fused with cell-binding peptide domains or signalling proteins have been expressed and utilised as building blocks to construct different angiogenic substrates (*e.g.*, gels, films, or fibres) with tuned ELP: water ratio, temperature, and composition of the growth medium. Crosslinked ELP hydrogels have been constructed to present cell-binding motifs such as fibronectin-derived REDV, VEGF-derived QK, laminin-derived IKVAV, and integrin-binding RGD to act as ECM mimetics. Cai et al. developed ELP-based hydrogels encapsulating HUVECs and functionalised with the cell-binding RGD sequence and the VEGF mimetic QK peptide; the hydrogels maintained nearly 100% of cell viability along with significantly enhanced cell proliferation and 3D outgrowth (Cai, Dinh, & Heilshorn, 2014). Another study, by Santos et al., focused on cell- and factor-free hybrid hydrogels constructed with ELPs, polyethylene glycol (PEG), and the self-assembling IKVAV peptide and evaluated their ability to induce angiogenesis and innervation *in vivo* (dos Santos et al., 2019); notably, the hydrogel hosted a larger density of vessels 26 days post-implantation when the IKVAV peptide was present, indicating that integration of bioactive peptides in natural biomaterials provides a route towards long-term stability in pro-angiogenic scaffolds.

Synthetic-Based Biomaterials

Synthetic biomaterials include a wide range of polymers, ranging from biopolymer mimetics to fully synthetic products, that encompass a broad range of chemical, physical, and functional properties (Frassica & Grunlan, 2020). Achieving chemical, mechanical, and biomolecular mimicry of the ECM using synthetic polymers can be accomplished, thus overcoming many limitations associated to natural or protein-based polymers, although concerns remain regarding the long-term safety/biocompatibility and degradation/bioresorbability (Alaribe, Manoto, & Motaung, 2016; Bao, Le, Minh, Nguyen, & Minh, 2013). Synthetic polymer scaffolds typically include various polyacrylamides (*e.g.*, PAA, PNIPAm), poly(2-hydroxyethyl methacrylate) (pHEMA), polyvinyl alcohol and polyvinyl acetate (PVA and PVAc), and polyethylene glycol (Table 14.1) (Cross & Claesson-Welsh, 2001).

Polyacrylamides form soft, hydrophilic, non-degradable gels whose internal porosity and water content can be tuned by adjusting the number of crosslinking agents (*e.g.*, bisacrylamide) or comonomers (*e.g.*, acrylic acid) added during synthesis via radical polymerisation (Bromberg, Grosberg, Matsuo, Suzuki, & Tanaka, 1997; Caulfield, Qiao, & Solomon, 2002; Yang, Yin, & Suo, 2019). Owing to their high swelling and tunable stiffness, polyacrylamide matrices facilitate cell migration (Christensen, Breiting, Aasted, Jørgensen, & Kebuladze, 2003; Lee et al., 2007) while enabling mechanotactic effects under static, cyclic, or dynamic loads (Bai, Yang, Morelle, Yang, & Suo, 2018; Kolvin, Cohen, & Fineberg, 2018; Tanaka, Fukao, & Miyamoto, 2000). The integration of reactive comonomers into the chains provides routes for the physisorption, affinity, or chemical ligation of signalling factors for angiogenesis (Kulicke, Kniewske, & Klein, 1982; Ribeiro, Denisin, Wilson, & Pruitt, 2016). While not approved in the US, polyacrylamides have been approved in Europe and Asia for tissue engineering and regenerative medicine application in both humans and animals (Samavedi, Poindexter, Van Dyke, & Goldstein, 2014). Saunders et al. seeded HUVECs onto polyacrylamide hydrogels *in vitro* with varying stiffness (140–2500 Pa) functionalised with RGD and GFOGER, a peptide to control cell adhesion and gene expression of chondrocytes and differentiation of stem cells, and found that cells were able to form stable networks on soft gels when untreated with growth factors; this indicates the promise of polyacrylamide hydrogels with controlled stiffness towards enhancing blood vessel network formation (Saunders & Hammer, 2010).

pHEMA is a transparent, biocompatible, and amphiphilic polymer frequently used to construct substrates enabling the controlled release of proteins for cardiac tissue regeneration (Gloria, De Santis, Ambrosio, Causa, & Tanner, 2011; Hejčl et al., 2009; Madden et al., 2010; Meyvis, De Smedt, Stubbe, Hennink, & Demeester, 2001; Refojo & Yasuda, 1965). Like polyacrylamides, pHEMA is a non-degradable biomaterial, whose physical properties are tuned via copolymerisation or by varying the crosslinking density, enabling a broad range of long-term *in vivo* applications (Slaughter, Khurshid, Fisher, Khademhosseini, & Peppas, 2009); similarly, the bioresorbability of pHEMA-based scaffolds can be tuned by introducing enzymatically cleavable linkers (*e.g.*, oligosaccharides, polycaprolactone (PCL), peptides, etc.) (Atzet, Curtin, Trinh, Bryant, & Ratner, 2008; Meyvis et al., 2001). Amphiphilicity makes pHEMA ideal for biopatterning and sustained release of bioactive payloads (*e.g.*, drugs or proteins) to resident cells (Ayhan & Ayhan, 2018). Hu et al. fabricated pHEMA-based hydrogels with high aspect ratio surface microfeatures to evaluate the migration, orientation, and shape of human mesenchymal stem cells (hMSCs); they noticed that the cells grew into an interconnected layer and were able to divide for more than a month *in vitro*, showing the promise of pHEMA for long-term *in vivo* applications (Hu, You, & Aizenberg, 2016). Another study by Madden et al. focused on collagen-modified pHEMA-methacrylic acid hydrogels containing parallel channels seeded with human embryonic cell-derived cardiomyocytes supported by interconnected pores to enhance angiogenesis and reduce scaring (Madden et al., 2010); after 4 weeks of culture *in vivo*, the scaffolds with 30–40 μm pores showed significantly higher vascularisation with minimal fibrous encapsulation.

PVA and its derivatives (especially PVAc at different degrees of acetylation) are inexpensive, scalable, and FDA-approved polymers, widely employed in regenerative tissue engineering (Jyothibasu & Lee, 2018). PVA-based hydrogels are gas permeable, feature tunable hydrophilicity, and, depending on the degree of substitution, feature great mechanical properties that are ideal for tissue vascularisation (Nkhwa, Lauriaga, Kemal, & Deb, 2014). They also possess a rich chemical diversity that can be explored by tuning their degree of acetylation via conjugation to or crosslinking of the abundant hydroxyl groups (Bohl Masters, Leibovich, Belem, West, & Poole-Warren, 2002; Peppas, Hilt, Khademhosseini, & Langer, 2006; Schultz, Kyburz, & Anseth, 2015; Yon et al., 1994). As the degree of acetylation decreases from 100% (PVAc) to 0% (PVA), the hydrophilicity of the material increases, while concurrently cell adhesion and stiffness decrease in a gradual fashion, enabling accurate control of the bioadhesion and mechanotactic behaviour of these materials (Rynkowska, Fatyeyeva, Marais, Kujawa, & Kujawski, 2019). Adhesion of endothelial cells on soft (low acetylation) PVAc can be increased by hybridising the substrate with bioadhesive moieties (Vrana, Cahill, & McGuinness, 2010). *In situ* gelation of PVA is achieved using crosslinkable groups such as tyramines, methacrylates, hydrazides, and aldehydes (Alves, Young, Bozzetto, Poole-Warren, & Martens, 2012; Lim, Alves, Poole-Warren, & Martens, 2013; Nilasaroya, Poole-Warren, Whitelock, & Jo Martens, 2008). Functionalisation of PVA hydrogels, either surface or bulk, utilising biomolecules, cells, and antimicrobial agents enables substrates for tissue regeneration to be derived. A study by Zahid et al. produced chitosan-polyvinyl alcohol (CS-PVA) hydrogels loaded with 5% S-nitroso-N-acetyl-DL-penicillamine (SNAP) to accelerate angiogenesis by providing a continuous supply of NO (Zahid et al., 2019). 3T3 fibroblasts and HaCaT keratinocyte cells were seeded onto the hydrogel, and evaluated *in vitro* to assess cell viability, proliferation, migration, and angiogenic activity; the authors observed that incorporating SNAP into the hydrogels accelerated cell proliferation and angiogenesis, demonstrating the potential of additives like SNAP for promoting angiogenesis and accelerating wound healing. While numerous strategies are known to engineer PVA(c) into ECM mimics, balancing biodegradability, transport, and cell adhesion and proliferation remain challenging.

Lastly, PEG is a hydrophilic and biocompatible polymer typically employed in combination with other materials to build substrates for tissue culture (Lin & Anseth, 2009). Specifically, hyperbranched PEGs terminated with reactive moieties (*e.g.*, thiols, NHS ester, and acrylates) are often used as crosslinkers for mechanical and chemical reinforcement of natural biopolymers (*e.g.*, GelMA) (Zhu, 2010). Of particular interest is the ability of PEG-based hybrid ECM mimetic to enable homogeneous seeding of the cells, followed by UV-induced crosslinking (Fairbanks, Schwartz, Bowman, & Anseth, 2009; Gould & Anseth, 2016; Hu et al., 2019; Roberts, Nicodemus, Greenwald, & Bryant, 2011). The latter in particular is used to introduce both mechanical patterning – *e.g.*, via photolithography or 3D printing – and biomolecular patterning via UV-mediated chemical ligation (Liu Tsang & Bhatia, 2004), resulting in close mimetics of the ECM; notably, photo-polymerisation can be performed at physiological temperature and pH conditions, allowing for cell encapsulation in specific areas (Aurian-Blajeni & Ezra, 2016; Koh, Revzin, & Pishko, 2002). Oliviero et al. reported a 3D porous PEG diacrylate hydrogel crosslinked with heparin enabled controlled release of VEGF; *in vitro* and *in vivo* characterisation of the hydrogels indicated that the release of VEGF-induced HUVEC proliferation and the hydrogel maintained its bioactivity for 21 days, demonstrating significant angiogenic potential (Oliviero, Ventre, & Netti, 2012). This study demonstrates the potential of PEG-crosslinked biopolymer substrates as long-term scaffolds and their flexibility in enabling functionalisation with bioactive molecules.

Physical Properties

The physicochemical analysis of the biomaterials listed above is complemented by the investigation of morphological and topological features, whose role in governing physiologic phenomena has been widely recognised in the literature (Amani et al., 2019; Rahmati, Silva, Reseland, Heyward, & Haugen, 2020). In this context, most biopolymers manage to encompass a broad range of properties, enabled by a rich portfolio of chemical and mechanical processing methods (Table 14.2).

TABLE 14.2

Structural Properties Commonly Manipulated in ETCs and What Cell Functions They Enhance

		Cell Adhesion	Cell Proliferation	Cell Differentiation	Optimal Quantity	Reference
Structure	Porosity	X	X		~ 100 µm diameter	Murphy and O'Brien (2010)
	Stiffness	X	X	X	~ 15 kPa	Chen, Dong, Yang, and Lv (2015); Kaiser and Coulombe (2015)
	Morphology	X		X	~ 20 nm	Borghi, Podestà, Di Vece, Piazzoni, & Milani, (2018); Denchai, Tartarini, and Mele (2018)
	Topology	X		X	~ 200 µm	Kaiser and Coulombe (2015); Viswanathan et al. (2015)
	Surface roughness	X	X	X	—	Chang and Wang (2011)
	Degradability	X		X	~ months	Zhang, Dong, Bai, and Quan (2020)

Porosity

First, the porosity of biomaterials plays a critical role in guiding tissue formation and functionality (Cipitria et al., 2012). Porosity is classified in *(i)* open, having a clear entrance and exit, *(ii)* closed, having neither access nor end, *(iii)* blind, when dead ended, or *(iv)* through, when multiple pores are connected (Ahumada, Jacques, Calderon, & Martínez-Gómez, 2019). Homogeneous cell distribution in a scaffold requires a material with high porosity (>50%) and pore diameter (30–100 μm) (Mastrullo et al., 2020). High porosity is in fact needed for cell migration, and distribution of nutrients and oxygen, which are critical during the early stage of culture when the vascular system has not been yet established (Khademhosseini & Langer, 2007). Tuning the porosity and pore size also allow adjusting the stiffness of the material (Annabi et al., 2010) and can add 3D features that are conducive to cell survival, proliferation, migration, and differentiation, thereby resulting in an overall promotion of angiogenetic processes (Annabi et al., 2010; Gerecht et al., 2007; Lien, Ko, & Huang, 2009; Mandal & Kundu, 2009). Finally, the microarchitecture of hydrogel scaffolds for angiogenesis – and tissue engineering at large – can be adjusted by selecting appropriate processing techniques (Zhu, Luo, & Liu, 2020). Traditional methods include sol–gel transition, casting, freeze-drying, and electrospinning (Khademhosseini, Langer, Borenstein, & Vacanti, 2006; Wheeldon, Ahari, & Khademhosseini, 2010); more recently, micropatterning and micromolding techniques have been introduced, which provide advanced control of the microscopic features of the substrate and hold better promise to manufacture scaffolds with improved host–hydrogel integration (Du, Lo, Ali, & Khademhosseini, 2008; Khademhosseini, Eng, et al., 2006b; Ling et al., 2007). Chiu et al. fabricated PEG hydrogels with various pore sizes (25–150 μm) that were implanted onto mice to investigate cell invasion and vessel formation over a period of 3 weeks (Chiu et al., 2011); the hydrogels with smaller pores (25–50 μm) had limited cell invasion, whereas larger pore sizes (50–150 μm) supported the infiltration of a large number of cells, thus promoting the formation of mature vascularisation.

Elasticity

On par with porosity, elasticity (or stiffness) is a key property of the ECM's microenvironment that drives cells response (Allen, Cooke, & Alliston, 2012; Dinulovic, Furrer, & Handschin, 2017; Hiew, Simat, & Teoh, 2018; Lu Wang, Wang, Wu, Fan, & Li, 2020; Lu Wang et al., 2019; Xie et al., 2018). Upon formation of adhesion complexes between cell receptors and the ECM, the actin-myosin transfer directs the signalling within and between cells, allowing them to sense the stiffness of the surrounding environment (Discher, Janmey, & Wang, 2005). Substrates that mimic the elasticity of native tissues strongly influence the differentiation of seeded stem or progenitor cells, and the subsequent phenotypical fate of the resident cells (Allen et al., 2012; Discher et al., 2005; Engler, Sen, Sweeney, & Discher, 2006; Hiew et al., 2018). A study conducted by Wong et al. on the response of vascular progenitor cells (VPC) to polyacrylamide hydrogels of varying stiffness (10 kPa, 40 kPa, and >0.1 GPa) showed increased VPCs differentiation into endothelial cells on softer substrates due to the higher expression of genes associated with migration, vasculature development, and blood vessel formation (Wong et al., 2019). Another study by Stevenson et al. utilised self-assembling peptides to construct RGD-functionalised hydrogels with varying stiffness as substrates for human umbilical vein endothelial cells and human mesenchymal stem cells; the authors demonstrated that above a critical level of stiffness, substrates fail to promote the formation of a vascular network, despite the higher proliferation of hMSCs (Stevenson et al., 2013). Notably, the 3D profile of stiffness in the native ECMs frequently assumes a gradient-like distribution (Xia, Liu, & Yang, 2017). This has spurred a multitude of studies aimed at developing ECM mimetics with gradient-like distribution of stiffness employing 3D printing, UV photolithography, and infusion of composite microspheres or polymer materials to achieve complex patterns of elasticity (Bracaglia et al., 2017; Norris, Tseng, & Kasko, 2016; Singh et al., 2010).

Surface Morphology and Topology

Cell adhesion, migration, differentiation, and maturation are also impacted by the surface morphology and topology of the engineered scaffolds (Lee et al., 2020). Accordingly, different biological activities can be elicited on biomaterial surfaces by patterning features whose size ranges from nanometres to micrometres (Lim et al., 2007). Larger entities impact cell motility, while smaller ones impact the sensing of biomolecular cues and ultimately plays a role in directing proliferation, differentiation, and organisation (Bettinger, Zhang, Gerecht, Borenstein, & Langer, 2008; Kuen et al., 2004; Thomas, Collier, Sfeir, & Healy, 2002). The control of surface topology resides in the size and shape of the features, and the spacing between them, which can be varied in a homogeneous or gradient-like manner (Khang, 2015). In this context, micro-scale patterning has been the main focus in studying topology–biology correlations (Jana, Leung, Chang, & Zhang, 2014). Using grid-and-grating microstructures, Kim et al. achieved directional cell organisation and spatial density at the larger scale ($600 \times 600\ \mu m^2$) by promoting systematic migration and aggregation, which can be used to guide the formation of a vascular network (Kim et al., 2009). Jeon et al. showed that grid patterns of different ratios resulted in different cell motility, with rapid cell migration achieved when the grid size is smaller than a single cell, which is crucial for the initial steps of angiogenesis (Jeon, Hidai, Hwang, Healy, & Grigoropoulos, 2010). Surface nanoscale features have also been shown to affect cellular interactions. Giavaresi et al. patterned nanosized cylinders on polycaprolactone (PCL) by colloidal lithography and studied the response of fibroblast cells; the team demonstrated that patterned PCL was superior to control materials in achieving high vascular density, demonstrating promise for *in vitro* tissue regeneration (Giavaresi et al., 2006). Scaffolds that have both micro- and nanostructured patterns, known as 'hierarchical' substrates, have been investigated to elucidate the combined role of multi-scale surface roughness in angiogenesis (Lee et al., 2020). Santos et al. fabricated nano/micro-fibre-combined scaffolds seeded with endothelial cells and demonstrated that the 3D distribution of endothelial cells mirrored the underlying nano/micro-fibres arrangement providing structural stability for the cell migration and organisation of cells into blood vessel-like structures (Santos et al., 2008). While shown to be synergistic towards proliferation and differentiation, the specific roles of micro- and nanoscale features of hierarchical substrates in directing cell functions and interactions have not been fully elucidated (Kubo et al., 2009; Martins et al., 2009; Zhao et al., 2006).

Surface Chemistry

Cells sense, recognise, and respond to biomolecular cues displayed on the surrounding matrix (Urbanczyk, Layland, & Schenke-Layland, 2020). Accordingly, decorating the surface of scaffolds with various small-molecule and biomolecular moieties is key to achieve appropriate cell adhesion and function and direct the development of a vascular network and tissue integration (Grafahrend et al., 2011; Li, Chen, & Jiang, 2003; Ohya, Matsunami, & Ouchi, 2004; Thevenot, Hu, & Tang, 2008). Surface functionalisation enables the activation of cell surface receptors, thus triggering specific cell functions that stimulate angiogenesis. The predominant cell surface receptor for angiogenesis is VEGFR2, which is activated by VEGF molecules displayed on the surface of a scaffold, promoting endothelial migration, adhesion, and vessel formation (Cébe-Suarez, Zehnder-Fjällman, & Ballmer-Hofer, 2006). Besides VEGFR2, a large number of cell surface receptors exist (*e.g.*, $\alpha 1\beta 1$, $\alpha v\beta 3$) that recognise specific ECM motifs (*e.g.*, collagen, laminin, vitronectin) to guide cell adhesion and migration (Reinhart-King, 2008).

The integration of bioactive cues on the substrate relies on the availability of the groups that form its surface chemistry, such as hydroxyl, amine, and carboxyl. These can either simply adsorb or ligate pro-angiogenic proteins (*e.g.*, growth factors) (Kamble, Ahmed, Ramabhimaiaha, & Patil, 2013) or serve as handles for conjugating ligands (*e.g.*, peptide or aptamers) that capture them via affinity binding (Reverdatto, Burz, & Shekhtman, 2015). Yusuke et al. studied the formation of self-assembled monolayers of proteins mediating cell adhesion onto surfaces functionalised methyl, hydroxyl, carboxyl, and amino groups; the team observed that amino and carboxyl groups absorbed more fibronectin and vitronectin, improving the adhesion of HUVECs even in the presence of bovine serum albumin (BSA) (adhesion inhibitor) (Arima & Iwata, 2015). Another study done by Nillesen et al. prepared acellular collagen, collagen with heparin, collagen and heparin substrates with one growth factor, and collagen and heparin with two growth factors (rrFGF2 and rrVEGF) scaffolds via carbodiimide chemistry and implanted into 3-month-old Wistar rats to test the combination of growth factors displayed on the density of blood vessels formed and tissue response (Nillesen et al., 2007). The scaffold with two growth factors after 21 days of culture displayed the highest density and most mature blood vessels, suggesting that a large number of cells from the host were recruited by the scaffold to enhance angiogenesis at the site of injury.

It has also been observed that biomaterials whose surface is highly hydrophilic, neutrally charged, and poor of – or lacks altogether – hydrogen bonding groups prevent non-specific binding of biomolecules, undesired cell binding, or buildup of proteins and other biomolecules secreted by the resident cells, thus inhibiting undesired downstream effects (Ostuni, Chapman, Holmlin, Takayama, & Whitesides, 2001); this is particularly relevant in preventing the formation of layers of serum proteins (*e.g.*, immunoglobulins and albumin) that can inhibit or arrest the activity of growth factors loaded on implants to direct cell adhesion and activity. In a recent study, Rohman et al. developed PLGA scaffolds grafted with heparin-terminated PEG chains; as anticipated, PEG-ylation inhibited unspecific protein adsorption, while heparin enabled growth factor binding resulting in enhanced stromal cell adhesion and growth (Rohman, Baker, Southgate, & Cameron, 2009). Surface chemistry with specific motifs can also promote either the differentiation of resident cells (Thevenot et al., 2008) or the capture of host's cells that secrete factors necessary for the recruitment of cells (*e.g.*, circulating stem cells by fibroblast-secreted VEGF) that sustain angiogenesis and integration of the microvascular network with the host (Ko, Lee, Atala, & Yoo, 2013). Relying on the surface chemistry to recruit the host's stem and progenitor cells to complement the original cell payload in an implant has been shown to promote sustained angiogenic activity and substantially improve the efficiency of tissue integration (Su et al., 2020).

Despite the abundance of studies on the effect of surface chemistry *in vitro*, we noticed a dearth of analogous efforts focusing on its effects *in vivo*; in particular, the few studies available do not address the effects of surface chemistry of implants on preventing the formation of fibrotic tissues following injury or ischemic attack (Barbosa, Madureira, Barbosa, & Águas, 2006; Ratner & Bryant, 2004; Tang, Wu, & Timmons, 1998).

Degradability

Biodegradability and bioresorbability, while not strictly required, are frequently desired in pro-angiogenic implants (Sung, Meredith, Johnson, & Galis, 2004). Biodegradability suggests materials that can be broken down *in vivo* but does not imply the elimination of the material. On the other hand, bioresorbability refers to a material that can be broken down and eliminated from the body *in vivo* (Hutmacher, Hürzeler, 1996). The degradation rate of a wide variety of hydrogel and composite scaffolds has been studied to make implantable scaffolds more desirable for long-term tissue regeneration (Castaño et al., 2014; Chen et al., 2008; Ghanaati et al., 2010). *In vivo*, cells continuously restructure the ECM via enzymatic digestion of the pre-existing matrix, allowing cells to migrate therein and produce biopolymers that replenish the exhausted matrix, thereby ensuring tissue continuity and integrity. Accordingly, advanced ECM-mimetic materials comprise linkers prone to degradation by enzymes – whether secreted locally by the cells or injected *in situ* – in order for cells to migrate through the implant at rates comparable to those of the native ECM. Similarly, ideal pro-angiogenic implants are degraded at the same rate at which new vascularised tissue is formed. This is achieved by tailoring the formulation (*e.g.*, material chemistry, physicochemical modification, and water swelling) and fabrication (*e.g.*, porosity, specific surface, and shape factor) of the implant, balancing the degradation rate with the size of the scaffold (Bauer, Jackson, & Jiang, 2009; Lu Wang et al., 2020). Inadequate degradation kinetics can lead to local fractures and weakening in the scaffold, which conflicts with the cell functions and behaviour, and impacts negatively on the outcome of the vascularisation and regeneration processes (Pereira et al., 2016; Prado, Monteiro, Campos, Thim, & de Melo, 2020). It is therefore imperative to understand the degradation mechanism characteristic of the implant site and tailor scaffolds with appropriate degradation rate. In a recent study, Chwalek et al. incorporated matrix metalloproteinases with hydrolytic linkers to PEG hydrogels to mediate the degradation of the gel and the delivery of conjugated RGD and VEGF to understand the effects of morphogenesis and angiogenesis of endothelial cells (Chwalek et al., 2011). Their results indicated that the gels with accelerated degradation and protein delivery significantly enhance proliferation of cells *in vitro* and blood vessel density *in vivo*, demonstrating the importance of controlling the rate of degradation on endothelial cell functions.

Presentation and Delivery of Angiogenic Stimuli

The spatial presentation and delivery of chemical and biochemical cues (*e.g.*, inorganic ions, drug-like molecules, peptide/protein cytokines, chemokines, and growth factors (Nikolova & Chavali, 2019)) is key to direct cell adhesion, differentiation, growth, and proliferation towards the formation of microvascular networks (Table 14.3) (Samorezov & Alsberg, 2015; Wang et al., 2017). In this context, current methods include bolus delivery in media or patterning onto a substrate. The former, while relatively simple, may

TABLE 14.3

Functionalisation Moieties Commonly Incorporated in ETCs and What Cell Functions They Enhance

		Cell Adhesion	Cell Proliferation	Cell Differentiation	Concentration	Reference
Surface Functionalisation Group or Protein	Copper	X		X	~ 5 mM	Hu (1998); Rath et al. (2014)
	Magnesium	X			~ 4 mM	Adhikari et al. (2016); Shen et al. (2021)
	Silicate		X		~ 6 mM	Götz, Tobiasch, Witzleben, and Schulze (2019); Shkarina et al. (2018)
	VEGF			X	~ 10 ng/mL	Ge et al. (2018); Kim et al. (2016)
	FGF	X		X	~ 100 ng/mL	Khan et al. (2017); Xiong et al. (2017)
	HGF			X	~ 0.5 µg/mg	Brouwer et al. (2013); Forte et al. (2006)
	ANG			X	~ 1 µg/mL	Joo et al. (2011); Leuning et al. (2019)
	TNFα	X			~ 50 ng/mL	Mackay, Loetscher, Stueber, Gehr, and Lesslauer (1993); Mountziaris et al. (2013)
	TGF-β1	X		X	~ 10 ng/mL	Xu, Min, Dong, Li, and Zha (2003)
	Carboxyl	X	X		~ 6 wt%	Keselowsky, Collard, and García (2004); Novotna et al. (2013)
	Hydroxyl	X		X	—	Keselowsky et al. (2004)
	Amine		X	X	—	Thevenot et al. (2008)
	Methyl	X			—	Thevenot et al. (2008)

be inadequate to ensure sufficient concentrations or temporal sequencing of stimuli and may suffer from their rapid clearance or degradation; biopatterning, on the other hand, provides controlled and localised delivery, but its implementation is typically time and labour intensive (Chen & Mooney, 2003). Notably, a wide array of techniques is available for qualitative and quantitative analysis of the presentation and delivery of angiogenic stimuli, as listed below.

Inorganic Ions

Inorganic ions have been shown to promote angiogenesis and exhibit properties, like anti-inflammatory activity, that are critical to promote and support angiogenesis *in vitro* and *in vivo* (Pérez, Won, Knowles, & Kim, 2013). This has led to the integration of select ions into scaffolds as pro-angiogenic factors. Boron promotes the secretion of growth factors with pro-angiogenic function. Haro et al. reported that the controlled and localised release of boron from Matrigel promoted vascularisation *in vitro* and *in vivo* (Haro Durand et al., 2014). Similarly, Park et al. showed that the presence of boron activates the protein kinase signalling pathway, which increases cell proliferation at low concentrations and inhibits it at high concentration (Park, Li, Shcheynikov, Zeng, & Muallem, 2004). Magnesium has been integrated extensively in biomaterials for angiogenesis, as it plays a key role in ATP hydrolysis, trans-phosphorylation, and protein synthesis (Bernardini, Nasulewic, Mazur, & Maier, 2005; Wolf & Cittadini, 2003). Accordingly, endothelial cells are strongly influenced by the presence of magnesium (Maier, 2012). Bernardini et al. found that low concentrations of magnesium inhibit endothelial cell growth, whereas higher concentrations stimulate proliferation and migration, thus promoting angiogenesis (Bernardini et al., 2005). Furthermore, similarly to VEGF, magnesium induces nitric oxide production, which stimulates the permeation of nutrients to the forming microvasculature (Bernardini et al., 2005). Copper also coadjuvates angiogenesis by inducing the expression of pro-angiogenic proteins such as VEGF, FGF, and angiogenin and stimulates cell motility (Dingsheng, Zengbing, & Dong, 2016; Finney, Vogt, Fukai, & Glesne, 2009; Hu, 1998; Lee et al., 2020; Sen et al., 2002). Due to its inherent toxicity, the copper load in pro-angiogenic scaffolds has been the target of optimisation in numerous studies. Gerard et al. developed collagen scaffolds loaded with copper, VEGF, and FGF-2, which provided improved angiogenesis *in vitro* under optimal copper loading conditions (Gérard, Bordeleau, Barralet, & Doillon, 2010). Finally, silicates (SiO_4^{2-}) released by silica-based biomaterials promote cell functions and motility (chemotaxis) implicated in angiogenesis (Dashnyam et al., 2017; Gérard et al., 2010; Handel, Hammer, Nooeaid, Boccaccini, & Hoefer, 2013; Kong, Lin, Li, & Chang, 2014; Zhai et al., 2012). While relied upon, the fundamental mechanism of inorganic ions is not yet fully understood, making it challenging to assess its role in biomaterials for next-generation scaffolds (Dashnyam et al., 2017).

Cytokines and Chemokines

Cytokines and chemokines (CCs) support angiogenesis by stimulating cell migration and proliferation and modulating inflammation. The ECM contains in fact a myriad of CCs that intervene through the stages of cell activation, adhesion, migration, and crosstalk (Vaday & Lider, 2000). Furthermore, the implantation of scaffolds triggers inflammation and the expression of pro-inflammatory cytokines that activate M1 macrophages and the expression of anti-inflammatory cytokines that activate M2 macrophages (Liu et al., 2020; Saqib et al., 2018), which secrete the cytokines and chemokines to support local angiogenesis. Accordingly, integrating these signalling molecules in engineered biomaterials is a distinctive feature of new approaches to tissue regeneration (Vaday et al., 2001); specifically, interleukin 8 (IL-8), prostaglandin E1 and E2, and tissue necrotic factor α (TNF-α) are frequently employed in pro-angiogenic materials (Benelli, Lorusso, Albini, & Noonan, 2007). In a study by Li et al., collagen chitosan scaffolds (~185 μm) thick were fabricated with endothelial cells and macrophages cultured on the surface to promote cell migration and anti-inflammation (Li, Dai, Shen, & Gao, 2017). The macrophages, stimulated by varying concentrations of lipopolysaccharide (LPS), secreted VEGF and TNF-α, which promoted deep cell migration into the scaffold: gels without LPS were penetrated by cells only by 125 μm deep, whereas gels loaded with 300 ng/mL LPS featured appreciable titres of VEGF and TNF-α and cell migrate up to 220 μm deep. The migration and anti-inflammation encouraged by chemokines and cytokines is especially important for efficient tissue regeneration and vascularisation *in vivo*.

Bioactive Proteins and Protein-mimetic Peptides

Proteins and their peptide mimetics play an essential role in promoting and directing angiogenesis (Petrik et al., 2010). The key proteins employed in scaffolds for vascular tissues are VEGF, FGF, hepatocyte growth factor (HGF), angiopoietin (ANG), tissue necrosis factor-alpha (TNFα), and transforming growth factor-β1 (TGF-β1) (Baluk et al., 2009; Post, Laham, Sellke, & Simons, 2001). Their delivery and ECM-mimicking patterning have been extensively studied and demonstrated promising results for regenerative medicine (Joo et al., 2015), and so have growth factor-mimicking peptides and other cell adhesion peptides (RGD, KTL, PRG, RGI, RAD, etc.) (Lu et al., 2019). Each protein holds a unique role in the angiogenic process. The most prominent pro-angiogenic protein is vascular endothelial growth factor, which is secreted by fibroblasts, platelets, and keratinocytes, and promotes the formation and growth of new blood vessels (Ferrara, Gerber, & LeCouter, 2003); next, fibroblast growth factor regulates numerous cell activities, including cell proliferation, survival migration, and differentiation (Yun et al., 2010); hepatocyte growth factor, secreted by mesenchymal cells, acts as a cytokine and promotes cell motility and mitogenesis (Kajiya, Hirakawa, Ma, Drinnenberg, & Detmar, 2005); angiopoietin, which belongs to the vascular growth factor family, inhibits vascular inflammation and endothelial death and promotes the formation of new veins and blood vessels (Brindle, Saharinen, & Alitalo, 2006); finally, TGF-β1 is a peptide cytokine that controls cell growth, proliferation, differentiation, and apoptosis (Ferrari, Cook, Terushkin, Pintucci, & Mignatti, 2009). The controlled display and delivery – either matrix-bound or in solution – of angiogenic biomolecules is extensively employed when designing scaffolds for tissue regeneration, wound healing, and therapy (Lee et al., 2020). The sustained delivery of cocktails of signalling factors over the course of weeks-to-months has been hailed as the latest frontier in pro-angiogenic technology and has led researchers to focus on biomaterials that enable continuous release (Wang et al., 2017). In a recent study, Freeman et al. adsorbed VEGF, TGF-β1, and PDGF onto alginate-sulphate scaffolds enabling their continuous delivery *in vivo*; their results indicated that the scaffolds releasing all three proteins were far superior at inducing vascularisation than those loaded with only one factor over a period of 1–3 months (Freeman & Cohen, 2009), demonstrating the need for complex protein functionalisation of implants.

Protein Loading for In-Solution Release

Proteins for delivery in solution are typically integrated into the tissue scaffold via encapsulation in micro/nanoparticles from which they are released, or entrapment into the mesh of polymer chains through which they migrate to become available to the seeded cells or the surrounding tissue (Lee, Silva, & Mooney, 2011). Particulate vessels include liposomes, micelles, and micro/nanocapsules (Patra et al., 2018). A recent study by Xiong et al. reported the encapsulation of VEGF and ANG1 into polylactic acid microspheres and presented improvement in mesenchymal stem cell differentiation and proliferation to endothelial cells, ultimately promoting angiogenesis (Xiong et al., 2018). The techniques for forming protein-impregnated matrices include solvent casting, freeze-drying, phase separation, electrospraying, phase emulsion, or *in situ* polymerisation (Busby, Cameron, & Jahoda, 2001; Hsu et al., 1997; Hua, Kim, Lee, Son, & Lee, 2002; Mikos, Lyman, Freed, & Langer, 1994; Thomson, Yaszemski, Powers, & Mikos, 1995). The protein payload is either mixed with the substrate polymer during formation or is added subsequently by adsorption or injection; the latter being favoured for replenishing the released proteins and ensure sustained angiogenic activity (Lee et al., 2011).

Electrospraying is a single-step liquid atomisation-based technique that employs electric forces to create micro- or nanoparticles. This process has been widely used to form polymeric microparticles that contain different therapeutic molecules (*e.g.*, proteins and growth factors) (Jiamian Wang, Jansen, & Yang, 2019). Electrospraying provides accurate control over the characteristics of protein-impregnated particles (*i.e.*, size, shape, density, and coating with degradable layers) or surface-coating films (thickness and density) that enable tailoring the delivery of signalling factors (Jiamian Wang et al., 2019); on the other hand, the exposure to shear and electric field can cause denaturation or degradation of the biomolecular payload (Bock, Dargaville, Kirby, Hutmacher, & Woodruff, 2016; Nosoudi et al., 2020). More recently, the combination of electrospraying and electrospinning has been introduced for encapsulation and sustained delivery of growth factors (Nazarnezhad, Baino, Kim, Webster, & Kargozar, 2020). The combination of these techniques allows for easier infiltration of cells into the scaffolds, as demonstrated both *in vitro* and *in vivo* (Jinglei Wu & Hong, 2016). Devolder et al. developed electrosprayed PLGA microparticles that achieve sustained release of VEGF into a PLA microfiber matrix seeded with mouse endothelial cells; the team observed endothelial cell migration and alignment along the fibres resulting in a large number of blood vessels *in vitro* (Devolder, Bae, Lee, & Kong, 2011).

Emulsion utilises immiscible oil and aqueous polymer solutions to form oil-coated hydrogel droplets (0.1–1 μm) or fibres (10–100 mm) (Costa et al., 2019; Nazarnezhad et al., 2020); the droplets are crosslinked while in the oil phase and are later removed via washing, centrifugation, and filtration (Costa et al., 2019; Nazarnezhad et al., 2020). The emulsion process can be conducted either in a batch manner or in microfluidic devices, whereas the biomolecular payload is either mixed with the aqueous polymer prior to the emulsion or loaded via adsorption on the hydrogel nanoparticles/fibres (Nazarnezhad et al., 2020). Tian et al. developed a novel emulsion electrospinning-based method to fabricate poly (L-lactic acid-co-ε-captolactone) (PLCL) nanofibres that encapsulated VEGF with dextran (DEX) or BSA to evaluate cardiomyogenic differentiations and proliferation of MSCs *in vitro*; their results indicated that uniform nanofibres of PLCL-VEGF-DEX provided a continuous delivery and long-term bioactivity of VEGF enhancing the proliferation and differentiation of cells needed for angiogenesis (Tian, Prabhakaran, Ding, Kai, & Ramakrishna, 2013). Emulsion-based techniques, however, can cause issues of thermo-induced degradation of the biomolecular payload or cause cell toxicity due to electric discharge (Tian et al., 2013).

Matrix-Bound Proteins and Protein Mimetics

The immobilisation of biomolecular cues onto a cell culture scaffold significantly heightens the angiogenic activity of the resident cells (Ehrbar et al., 2004; Zisch, Schenk, Schense, Sakiyama-Elbert, & Hubbell, 2001). While on the substrate, the proteins are available for uptake by the cells seeded on the matrix, inducing a pattern of cell functions that mirrors the underlying biomolecular pattern (Lutolf & Hubbell, 2005). Three main approaches are known for immobilising biomolecules onto natural or synthetic biomaterials: physical adsorption, chemical ligation, or ligand-mediated affinity interaction (K. Lee et al., 2011). Protein display by physical adsorption has been pursued in many cell differentiation contexts to optimise the use of growth factors (Guex et al., 2014); by relying on large, bolus dosing of growth factors, however, this technique makes it challenging to predict the terminal differentiation efficiency. In general, incorporation on a solid phase by physical adsorption or loading is not adequate for growth factors because it does not ensure adequate orientation and presentation to the cells and results in 'burst' release, especially due to the high amounts of growth factors needed (Vo et al., 2012). Therefore, 'stable' immobilisation and presentation of growth factors have been introduced to explore and elucidate their cooperative roles and improve cell conditioning in biomanufacturing processes (Wang et al., 2017). Covalent conjugation relies on formation of covalent bonds either by activating the groups available on the substrate biopolymer (*e.g.*, using carbonyldiimidazole, cyanogen bromide, or epichlorohydrine) or click chemistry (*e.g.*, Michaels addition or Huisgen cycloaddition) (Meldal & Schoffelen, 2016; Mero, Clementi, Veronese, & Pasut, 2011; Rao, Fang, De Feyter, & Perepichka, 2017; Spicer, Pashuck, & Stevens, 2018). Chiu et al. immobilised VEGF and angiopoietin-1 onto collagen scaffolds loaded with endothelial cells, resulting in increased proliferation of cells compared to unmodified scaffolds and ultimately leading to improved tubule formation, which preludes to angiogenesis (Chiu & Radisic, 2010). This study demonstrated the potential of displaying growth factors in a more stable and localised method towards promoting the formation of vascular structures.

Peptides can be used in lieu of protein factors to direct cell adhesion and functions (C. C. Lin & Anseth, 2009). A very common peptide motif is fibrin-derived arginine-glycine-aspartic acid (RGD), which promotes cell adhesion on ECM mimetics by targeting the integrins displayed on the cell surface (Bellis, 2011). RGD peptides have become practically ubiquitous in tissue engineering scaffolds and have been used in combination with protein factors (*e.g.*, HGF and FGF) to enhance cell survival and migration (Hill, Boontheekul, & Mooney, 2006), for example, Wang et al. used a copper catalysed click reaction to functionalise decellularised ECM substrates with QK peptides and loaded them with HUVECs to evaluate the cell adhesion, proliferation, and tubule formation (Lin Wang et al., 2014); it was concluded that the decellularised ECM significantly promoted cellular adhesion and proliferation compared to matrigel, while the QK enhanced angiogenic responses. Notably, bioactive peptides and proteins exhibit synergism when used in combination, thus significantly enhancing cell function and the outcome of the angiogenic process (Bonnans, Chou, & Werb, 2014).

The latest frontier of ECM mimetics relies on ligand-mediated (non-covalent) immobilisation of growth factors (Guilak et al., 2009). This approach, inspired by the sequestration and storage of proteins by the binding sites distributed within the ECM, makes growth factors both appropriately oriented and available for uptake by the cells cultured on the substrate (Higuchi, Ling, Chang, Hsu, & Umezawa, 2013; Watt & Huck, 2013). Using biomolecular patterning technology, myriads of scenarios of combination, concentration, and presentation of growth factors on a single substrate can be achieved. Key to this end is the availability of ligands that selectively target proteins and cell receptors. In prior research, Menegatti et al. employed peptide ligands immobilised onto hydrogel substrates at optimal density and display for capturing proteins (*e.g.*, immunoglobulins and erythropoietin) (Menegatti, Hussain, Naik, Carbonell, & Rao, 2013) and developing cell mimetics for wound healing (Menegatti, Ward, et al., 2013). Assal et al. co-immobilised three different growth factors (FGF, EGF, and VEGF) onto a coiled-coil structure on a collagen scaffold to enhance vascular network formation; results showed stimulation in the growth and activity of HUVECs along with improved cell survival and growth factor release which explained the faster vascularisation of the wounded region over 7 days compared to scaffolds without growth factors (Assal, Mie, & Kobatake, 2013). More recently, our team has developed a method of affinity-mediated non-covalent display of growth factors on GelMA hydrogel substrates. This method consists of conjugating peptide ligands targeting growth factors onto the surface of the hydrogel. The peptides drive the affinity adsorption of growth factors with controlled orientation and density on the surface, ideal for cell binding and uptake. Recently, we discovered that complex geometrical VEGF patterns can be achieved by integrating this technique with photolithography or 3D printing on the micrometre scale (Figure 14.3).

The Next-Frontier of Matrix-Bound Protein: Gradients

In vivo, the ECM features overlapping gradients of signalling factors generated by resident cells, particularly in regions of directed angiogenesis. As a result, cells in close proximity are exposed to different biomolecular landscapes at the nano/micro-scale, which determine differences in cell differentiation and cell-to-cell crosstalk (Bonnans et al., 2014; Bowers, Banerjee, & Baudino, 2010; Gattazzo, Urciuolo, & Bonaldo, 2014; Guilak et al., 2009; Higuchi et al., 2013; Meredith, Fazeli, & Schwartz, 1993; Watt & Huck, 2013; Yue, 2014). ECM-bound signalling factors are released locally and interact with cell surface receptors. With this growth factor-mediated signalling, the ECM contributes to establishing gradients of soluble, diffusible growth factor morphogens that drive differentiation and developmental processes (Hussey, Keane, & Badylak, 2017; Losi et al., 2013; Mouw, Ou, & Weaver, 2014; Watt & Huck, 2013). Recapitulating this biomolecular tapestry into scalable devices is essential to achieve advanced and quantitative control of cell adhesion, proliferation, and differentiation. Accordingly, synthetic biomaterials aimed to control cell functions endeavour to mimic the mechanical and/or biomolecular patterns of the ECM and intercellular space (Huebsch et al., 2015; Schultz et al., 2015; Watt, 2016). Significant effort has been dedicated to understanding the effect of the spatiotemporal gradients of biochemical factors on the progression of cellular processes (Aman & Piotrowski, 2010; Augello & De Bari, 2010; Belair, Le, & Murphy, 2014; Burdick & Vunjak-Novakovic, 2009; Ferrara, 2010; Temple, 2001) and recapitulating these gradients on hydrogels to accurately modulate cell differentiation and functional outcome (Guex et al., 2014; Losi et al., 2013; Martino et al., 2015; Mohandas, Anisha, Chennazhi, & Jayakumar, 2015; Pompe, Salchert, Alberti, Zandstra, & Werner, 2010; Sachot, Castaño, Mateos-Timoneda, Engel, & Planell, 2013; Tallawi et al., 2015). Recent years have witnessed a rapid increase in technologies for generating gradients of signalling factors on hydrogels and study their effect on cellular processes (Mastrullo et al., 2020); these approaches include physical adsorption (Jindan Wu et al., 2012), printing (Wang, Lin, Jarman, Polacheck, & Baker, 2020), and microfluidics (Kim, Kasuya, Jeon, Chung, & Kamm, 2015). Liu et al. found that endothelial cells directionally migrated towards the more concentrated areas of a VEGF or fibronectin gradient; this migration further increased with both VEGF and fibronectin were present allowing for the control of angiogenesis distribution (Liu, Ratner, Sage, & Jiang, 2007). Another study done by Guarnieri et al. studied the speed of migration of mouse fibroblast on the gradients of RGD peptides with

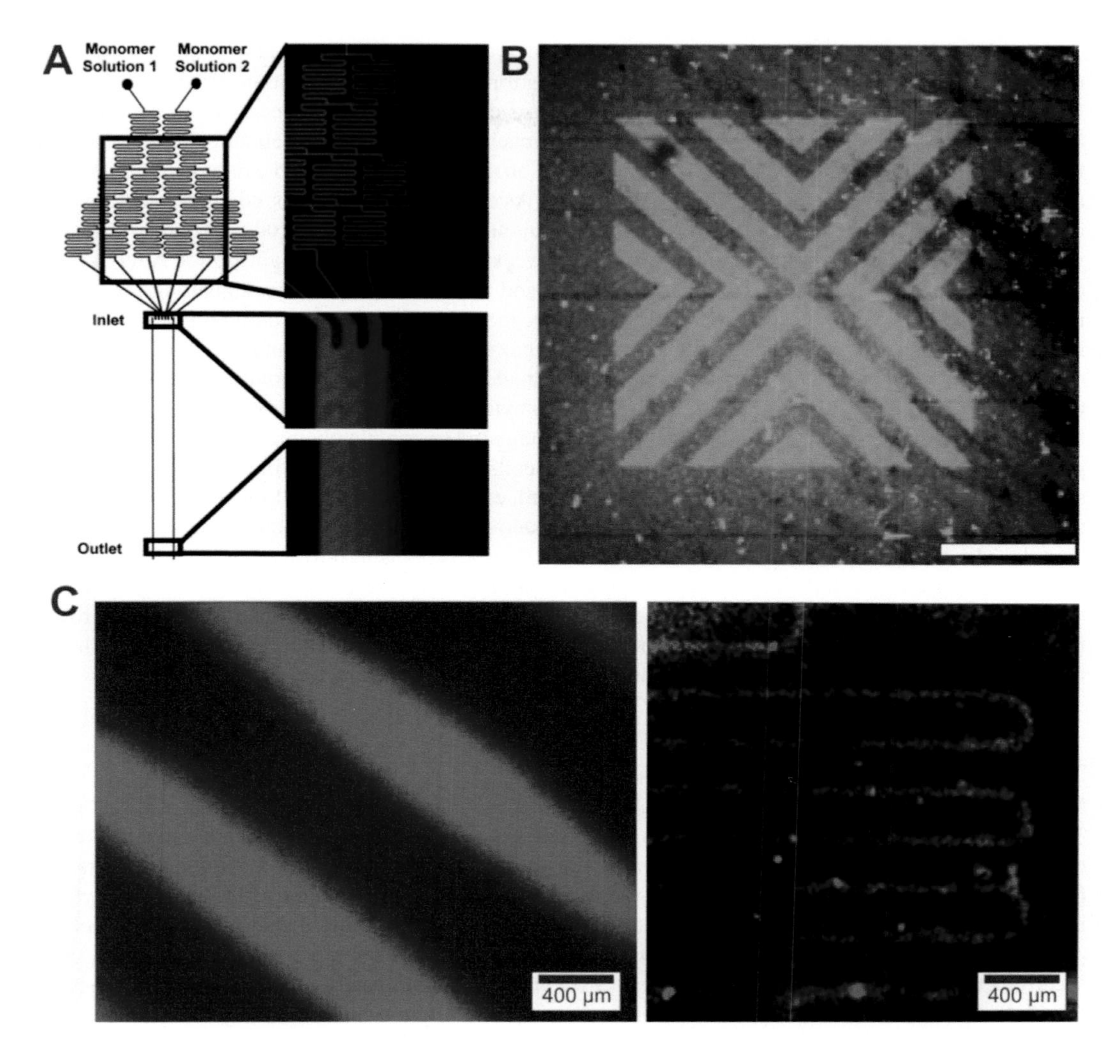

FIGURE 14.3 Experimental images demonstrating the use of microfluidic/photopolymerisation processes to create various patterns on scaffolds where (A) used two monomers and to create a gradient pattern, (B) used Protein A to enzymatically create a pattern, and (C) used cysteine terminated peptides create patterns.

different steepness on hydrogel surfaces and found that the cells tend to migrate towards high concentrations of RGD; they also discovered that the speed of migration increases as the RGD gradient steepness increased, which indicates prompt migration allowed for rapid angiogenesis (Guarnieri et al., 2010).

Controlled Dose and Release Rates

Advanced regenerative applications require the staggered release of biomolecular factors in response to sequences of environmental stimuli (Mastrullo et al., 2020). Complex temporal signalling mimics the natural vascular developmental processes. Innumerable control-release systems have been developed by integrating multiple stimuli-responsive components – whether natural or synthetic – in the substrate biomaterial (Esaki, Marui, Tabata, & Komeda, 2007). The encapsulation of multiple proteins via different physical mechanisms (*e.g.*, electrostatic vs. hydrophobic) has been widely studied to control the sequence and rate of release of signalling factors. Richardson et al. developed a 3D PLG scaffold that allows local release of VEGF and PDGF to induce rapid vessel formation and maturation *in vivo*, while the release of either VEGF or PDGF alone was insufficient, the dual delivery over 4 weeks after implantation led to much higher blood vessel density and maturity allowing for rapid tissue regeneration (Richardson, Peters, Ennett, & Mooney, 2001). The release of growth factors can also be tuned by controlling the rate(s) of hydrogel degradation (Fournier, Passirani, Montero-Menei, & Benoit, 2003). Yamamoto et al. absorbed FGF and TGF-β1 to gelatin hydrogels, implanted them on mice, and monitored the release of growth factors during the progressive degradation of gels *in vivo*. As expected, ionically bound growth factors were only released only when the gel was degraded (Yamamoto, Ikada, & Tabata, 2001); the growth factors were retained in the mice for longer periods of time (2 weeks) as the degradation got slower. Alternatively, external triggers can be adopted to control the dose and release rate of various signalling proteins in response to variations in the surrounding environment; these include pH, temperature, electric or magnetic fields, presence of specific ions, biomolecules, or enzymes (Yoshida & Okano, 2010). Garbern et al. developed pH- and temperature-responsive hydrogels loaded with VEGF and monitored the release over time at different values of pH (Garbern, Hoffman, & Stayton, 2010). When hydrogels were incubated at a pH of 5–6, the release occurs slowly over 28 days compared to completely dissolving after 4 days at a pH of 7.4. Finally, the release rates can be elicited by cell-directed mechanic stimulation; as cells apply a traction force, the entrapped biomolecules are released, enabling enhanced cell functions for angiogenesis (Mastrullo et al., 2020). A recent study by Stejskalová et al. anchored aptamer-peptide constructs complexed with growth factors to various base materials (*i.e.*, collagen, polyacrylamide, and glass), on which microvascular endothelial cells and human dermal fibroblasts were seeded to regulate growth factor release and cell proliferation (Mastrullo et al., 2020). The aptamer-peptide constructs trap the growth factors and release them when exposed to mechanical stimuli; the traction induced by peptide binding to the cell surface triggers the aptamer to unfold, thus releasing the growth factors at a controlled rate.

Analytical Techniques

Myriad analytical techniques are now available for supporting the development of biomaterial scaffolds. Techniques for probing the chemistry in bulk include gel permeation or gas chromatography, Fourier transform infrared and Raman spectroscopy, nuclear magnetic resonance (Pradhan, Rajamani, Agrawal, Dash, & Samal, 2017), whereas the analysis of surface chemistry employs time of flight secondary ionisation mass spectrometry, X-ray photoelectron spectroscopy, matrix-assisted laser desorption electrospray ionisation time of flight and infrared matrix-assisted laser desorption electrospray ionisation mass spectrometry, and energy-dispersive X-ray spectroscopy (Kingshott, Andersson, McArthur, & Griesser, 2011). Bulk morphological properties are probed via dynamic light scattering, X-ray and neutron scattering, X-ray diffraction (Bose & Bandyopadhyay, 2013), and scanning confocal fluorescence microscopy, whereas surface topology is typically probed high-resolution of scanning and transmission electron microscopy (SEM and TEM) and laser scanning microscopy (Merrett, Cornelius, McClung, Unsworth, & Sheardown, 2002). Currently, cell response is evaluated via immunohistochemical staining, SEM imaging, and gene expression assays to evaluate cell viability, proliferation, differentiation, and angiogenic activity (Łopacińska et al., 2012). The collective work suggests promise for 3D engineered hydrogels with controlled heterogeneity and controlled cell behaviour.

Future Outlook and Perspective

To date a large body of work has been conducted to increase the efficiency and rapidity of tissue regeneration by focusing on promoting angiogenesis and vasculogenesis using ETCs *in vitro* and *in vivo*. The incorporation of various biomaterials, cell types, proteins, and peptides onto ETCs has significantly aided in these accomplishments; much, however, is yet to be optimised. As 'ECM-inspired' bio-patterned substrates and their fabrication become better understood, the transition towards true 'ECM-mimetics' for next generation tissue engineering and regenerative medicine become feasible. The long-term goal of 3D biomaterials, instrumental to achieve complete biomimicry, is to trigger the full ensemble of endocrine, paracrine, autocrine, juxtracrine, and intracrine signalling of the ECM (Babensee, McIntire, & Mikos,

2000; Barrientos, Stojadinovic, Golinko, Brem, & Tomic-Canic, 2008). Recent efforts have been focused on biomolecular patterning ETCs in order to mimic matrix-bound ECM signals. In particular, a novel and promising approach is to decorate ETCs via affinity-mediated patterning of growth factors and subsequent seeding of endothelial and/or stem cells atop. This method holds true promise to present effectively signalling factors to cells, thus promoting their growth and differentiation in a spatially directed manner and promoting rapid and optimal angiogenesis; additionally, this method may allow continuous replenishing of the scaffold with signalling proteins both *in vivo* and *in vitro*, enabling long-term effects. Together with matrix-bound signalling, cells (*e.g.*, fibroblasts) incorporated in the ETC can secrete other angiogenic factors, contributing a continuous solution-phase patterning of factors that further sustains long-term pro-angiogenic differentiation and arrangement of cells. Ultimately, these physical and biochemical developments must coincide with the extensive integration and analyses of materials as well as cellular characterisation data. With the material and data technologies unremittingly improving, these goals are at hand and can be implemented in a number of tissue engineering and regenerative medicine applications.

REFERENCES

Abramo, A.C., & Viola, J.C. 1992. *British Journal of Plastic Surgery*. *45*(2): 117–122. DOI: 10.1016/0007-1226(92)90170-3.

Adair, T.H., & Montani, J.P. 2010. *Angiogenesis*. San Rafael (CA): Morgan & Claypool Life Sciences. Available from: https://www.ncbi.nlm.nih.gov/books/NBK53242/

Adhikari, U., Rijal, N.P., Khanal, S., & Pai, D., et al. 2016. *Bioactive Materials*. 23; *1*(2): 132–139. DOI: 10.1016/j.bioactmat.2016.11.003.

Afewerki, S., Sheikhi, A., Kannan, S., & Ahadian, S., et al. 2018. *Bioengineering & Translational Medicine*. 28; *4*(1): 96–115. DOI: 10.1002/btm2.10124.

Ahmadi, F., Oveisi, Z., Samani, M., & Amoozgar, Z. 2015. *Research in Pharmaceutical Sciences*. *10*(1): 1–16.

Ahumada, M., Jacques, E., Calderon, C., & Martínez-Gómez, F, et al. 2019a. In *Handbook of Ecomaterials*. DOI: 10.1007/978-3-319-68255-6_162.

Ahumada, M., Jacques, E., Calderon, C., & Martínez-Gómez, F. 2019b. Porosity in biomaterials: A key factor in the development of applied materials in biomedicine. In *Handbook of Ecomaterials*. DOI: 10.1007/978-3-319-68255-6_162.

Aisenbrey, E. A., & Murphy, W. L. 2020. *Nature Reviews Materials*. *5*(7): 539–551. DOI: 10.1038/s41578-020-0199-8.

Aizawa, Y., Wylie, R., & Shoichet, M. 2010. *Advanced Materials*. *22*(43): 4831–4835. DOI: 10.1002/adma.201001855.

Alfassi, G., Rein, D. M., Shpigelman, A., & Cohen, Y. 2019. *Polymers (Basel)*. *11*(11): 1734. DOI: 10.3390/polym11111734.

Allen, J. L., Cooke, M. E., & Alliston, T. 2012. *Molecular Biology of the Cell*. *23*(18): 3731–3742. DOI: 10.1091/mbc.E12-03-0172.

Altman, G. H., Diaz, F., & Jakuba, C., et al. 2003. *Biomaterials*. *24*(3): 401–416. DOI: 10.1016/s0142-9612(02)00353-8.

Alves, M. H., Young, C. J., Bozzetto, K., Poole-Warren, L. A., & Martens, P. J. 2012. *Biomedical Materials*. *7*(2): 024106. DOI: 10.1088/1748-6041/7/2/024106.

Aman, A., & Piotrowski, T. 2010. *Developmental Biology*. *341*(1): 20–33 DOI: 10.1016/j.ydbio.2009.11.014.

Amani, H., Arzaghi, H., Bayandori, M., Shiralizadeh Dezfuli, A., Pazoki-Toroudi, H., Shafiee, A., & Moradi, L. 2019. *Advanced Materials Interfaces*. *6*: 1900572. DOI: 10.1002/admi.201900572.

Aminabhavi, T.M., & Deshmukh, A.S. 2016. *Polymeric Hydrogels as Smart Biomaterials*. Springer: Berlin/Heidelberg, Germany, ISBN 978-3-319-25320-6.

Annabi N, Nichol JW, & Zhong X, et al. 2010. *Tissue Engineering Part B: Reviews*. *16*(4): 371–383. DOI: 10.1089/ten.TEB.2009.0639.

Arima, Y., & Iwata, H. 2015. *Acta Biomaterialia*. *26*: 72–81. DOI: 10.1016/j.actbio.2015.08.033.

Assal, Y, Mie, M., & Kobatake, E. 2013. *Biomaterials*. *34*(13): 3315–3323. DOI: 10.1016/j.biomaterials.2013.01.067.

Atala, A. 2009. *Current Opinion in Biotechnology*. *20*(5): 575–592. DOI: 10.1016/j.copbio.2009.10.003.

Atzet, S., Curtin, S., Trinh, P., Bryant, S., & Ratner, B. 2008. *Biomacromolecules*. *9*(12): 3370–3377. DOI: 10.1021/bm800686h.

Augello, A., & De Bari, C. 2010. *Human Gene Therapy*. *21*(10): 1226–1238. DOI: 10.1089/hum.2010.173.

Augustine, R., Kalva, N., Kim, H.A., Zhang, Y., & Kim, I. 2019. *Molecules*. *24*(16): 2961. Published 2019 Aug 15. DOI: 10.3390/molecules24162961.

Aurian-Blajeni, Dan Ezra. 2016. *Open Wet Ware*. https://openwetware.org/mediawiki/index.php?title=Poly(ethylene_glycol)-based_Biomaterials_in_Tissue_Engineering,_by_D._Ezra_Aurian-Blajeni&oldid=939519 [accessed 15 August 2021]

Ayhan, Hakan & Ayhan, Fatma. 2018. *Turkish Journal of Biochemistry*. *43*(3): 228–239. DOI: 10.1515/tjb-2017-0250.

Babensee, J.E., McIntire, L.V., & Mikos, A.G. 2000. *Pharmaceutical Research*. *17*(5): 497–504. DOI: 10.1023/a:1007502828372.

Bai, R., Yang, J., Morelle, X. P., Yang, C., et al. 2018*ACS Macro Letters*. *7*(3): 312–317. DOI: 10.1021/acsmacrolett.8b00045.

Baluk, P., Yao, L.C., Feng, J., & Romano, T., et al. 2009. *Journal of Clinical Investigation*. DOI: 10.1172/JCI37626.

Bandyopadhyay, A., Chowdhury, S.K., Dey, S., Moses, J.C., et al. 2019. DOI: 10.1007/s41745-019-00114-y.

Bao, Ha, Le, T., Minh, T., Nguyen, D., & Minh, D. 2013. In *Regenerative Medicine and Tissue Engineering*. DOI: 10.5772/55668.

Barbosa, M. A., & Martins, M. C. L. *Peptides and Proteins as Biomaterials for Tissue Regeneration and Repair*. 2017. DOI: 10.1016/C2015-0-01811-1.

Barbosa, J. N., Madureira, P., Barbosa, M. A., & Águas, AP. 2006. *Journal of Biomedical Materials Research - Part A*. DOI: 10.1002/jbm.a.30602.

Barnabas, O., Wang, H., & Gao, X.M. 2013. DOI: 10.3969/j.issn.1671-5411.2013.04.008.

Barrientos, S., Stojadinovic, O., Golinko, M.S., & Brem, H., et al. 2008. DOI: 10.1111/j.1524-475X.2008.00410.x.

Bauer, A.L., Jackson, T.L., & Jiang, Y. 2009. *PLoS Computational Biology*. DOI: 10.1371/journal.pcbi.1000445.

Belair, D.G., Le, N.N., & Murphy, W.L. 2014. *Chemical Communications*. DOI: 10.1039/c4cc04317k.

Bellis, S.L. 2011. *Biomaterials*. DOI: 10.1016/j.biomaterials.2011.02.029.

Benelli, R., Lorusso, G., Albini, A., & Noonan, D. 2007. *Current Pharmaceutical Design*. DOI: 10.2174/138161206777947461.

Bernardini, D., Nasulewic, A., Mazur, A., & Maier, J.A.M. 2005. *Frontiers in Bioscience : A Journal and Virtual Library*. DOI: 10.2741/1610.

Berthiaume, F., Maguire, T.J., & Yarmush, M.L. 2011. DOI: 10.1146/annurev-chembioeng-061010-114257.

Bettinger, C. J., Zhang, Z., Gerecht, S., & Borenstein, J.T., et al. 2008. *Advanced Materials*. DOI: 10.1002/adma.200702487.

Bishop, E. S., Mostafa, S., Pakvasa, M., Luu, H. H., et al. 2017. DOI: 10.1016/j.gendis.2017.10.002.

Bock, N., Dargaville, T.R., Kirby, G.T.S., & Hutmacher, D.W., et al. 2016. *Tissue Engineering - Part C: Methods*. DOI: 10.1089/ten.tec.2015.0222.

Bohl Masters, K.S., Leibovich, S.J., Belem, P., West, J.L., et al. 2002. *Wound Repair and Regeneration*. DOI: 10.1046/j.1524-475X.2002.10503.x.

Bonnans, C., Chou, J., & Werb, Z. 2014. DOI: 10.1038/nrm3904.

Borghi, F., Podestà, A., Di Vece, M., & Piazzoni, C., et al. 2018. In *Encyclopedia of Interfacial Chemistry: Surface Science and Electrochemistry*. DOI: 10.1016/B978-0-12-409547-2.12935-X.

Bose, S., & Bandyopadhyay, A. 2013. In *Characterization of Biomaterials*. DOI: 10.1016/B978-0-12-415800-9.00001-2.

Bowers, S.L.K., Banerjee, I., & Baudino, T.A. 2010. DOI: 10.1016/j.yjmcc.2009.08.024.

Bracaglia, L.G., Smith, B.T., Watson, E., Arumugasaamy, N., et al. 2017. DOI: 10.1016/j.actbio.2017.03.030.

Brindle, N.P.J., Saharinen, P., & Alitalo, K. 2006. DOI: 10.1161/01.RES.0000218275.54089.12.

Bromberg, L., Grosberg, A.Y., Matsuo, E.S., Suzuki, Y., et al. 1997. *Journal of Chemical Physics*. DOI: 10.1063/1.474084.

Brouwer, K.M., Wijnen, R.M., Reijnen, D., Hafmans, T.G., et al. 2013. *Organogenesis*. DOI: 10.4161/org.25587.

Brunt, K.R., Hall, S.R.R., Ward, C.A. & Melo, L.G. 2007. *Methods in Molecular Medicine*. DOI: 10.1007/978-1-59745-571-8_12.

Van Den Bulcke, A.I., Bogdanov, B., De Rooze, N., Schacht, E.H., et al. 2000. *Biomacromolecules*. DOI: 10.1021/bm990017d.

Burdick, J.A., & Vunjak-Novakovic, G. 2009. DOI: 10.1089/ten.tea.2008.0131.

Busby, W., Cameron, N.R. & Jahoda, C.A.B. 2001. Biomacromolecules. DOI: 10.1021/bm0000889.

Cai, L., Dinh, C.B. & Heilshorn, S.C. 2014. *Biomaterials Science*. DOI: 10.1039/c3bm60293a.

Canver, A.C., Ngo, O., Urbano, R.L. & Clyne, A.M. 2016. *Journal of Biomechanics*. DOI: 10.1016/j.jbiomech.2015.12.037.

Carson, A.E., & Barker, T.H. 2009. DOI: 10.2217/rme.09.30.

Castaño, O., Sachot, N., Xuriguera, E., Engel, E., et al. 2014. *ACS Applied Materials and Interfaces*. DOI: 10.1021/am500885v.

Catoira, M.C., Fusaro, L., Di Francesco, D., Ramella, M., et al. 2019. *Journal of Materials Science: Materials in Medicine*. DOI: 10.1007/s10856-019-6318-7.

Caulfield, M.J., Qiao, G.G., & Solomon, D.H. 2002. *Chemical Reviews*. DOI: 10.1021/cr010439p.

Cébe-Suarez, S., Zehnder-Fjällman, A., & Ballmer-Hofer, K. 2006. DOI: 10.1007/s00018-005-5426-3.

Chan, B.P., & Leong, K.W. 2008. In *European Spine Journal*. DOI: 10.1007/s00586-008-0745-3.

Chan, V., Zorlutuna, P., Jeong, J.H., Kong, H., et al. 2010. *Lab on a Chip*. DOI: 10.1039/c004285d.

Chang, H.-I., & Wang, Y. 2011. In *Regenerative Medicine and Tissue Engineering - Cells and Biomaterials*. DOI: 10.5772/21983.

Chang, S.C.N., Rowley, J.A., Tobias, G., Genes, N.G., et al. 2001. *Journal of Biomedical Materials Research*. DOI: 10.1002/1097-4636(20010615)55:4<503::AID-JBM1043>3.0.CO;2-S.

Chatterjee, S., Hui, P.C.L., & Kan, C. Wai. 2018. DOI: 10.3390/polym10050480.

Chen, F.M., & Liu, X. 2016. DOI: 10.1016/j.progpolymsci.2015.02.004.

Chen, R.R., & Mooney, D.J. 2003. DOI: 10.1023/A:1025034925152.

Chen, G., Dong, C., Yang, L., & Lv, Y. 2015. *ACS Applied Materials and Interfaces*. DOI: 10.1021/acsami.5b02662.

Chen, Y.C., Lin, R.Z., Qi, H., Yang, Y., et al. 2012. *Advanced Functional Materials*. DOI: 10.1002/adfm.201101662.

Chen, Y.W., Shi, G.Q., Ding, Y.L., Yu, X.X., et al. 2008. *Journal of Materials Science: Materials in Medicine*. DOI: 10.1007/s10856-007-3350-9.

Cheng, Y., Nada, A.A., Valmikinathan, C.M., Lee, P., et al. 2014. *Journal of Applied Polymer Science*. DOI: 10.1002/app.39934.

Cheung, R.C.F., Ng, T.B., Wong, J.H., & Chan, W.Y. 2015. DOI: 10.3390/md13085156.

Chiu, L.L.Y., & Radisic, M. 2010. *Biomaterials*. DOI: 10.1016/j.biomaterials.2009.09.039.

Chiu, Y.C., Cheng, M.H., Engel, H., Kao, S.W., et al. 2011. *Biomaterials*. DOI: 10.1016/j.biomaterials.2011.04.066.

Christensen, L.H., Breiting, V.B., Aasted, A., Jørgensen, A., et al. 2003. *Plastic and Reconstructive Surgery*. DOI: 10.1097/01.PRS.0000056873.87165.5A.

Chwalek, K., Levental, K.R., Tsurkan, M. V., Zieris, A., et al. 2011. *Biomaterials*. DOI: 10.1016/j.biomaterials.2011.08.078.

Cipitria, A., Lange, C., Schell, H., Wagermaier, W., et al. 2012. *Journal of Bone and Mineral Research*. DOI: 10.1002/jbmr.1589.

Costa, C., Medronho, B., Filipe, A., Mira, I., et al. 2019. DOI: 10.3390/polym11101570.

Cross, M.J., & Claesson-Welsh, L. 2001. *Trends in pharmacological sciences*. *22*(4): 201–207. DOI: 10.1016/S0165-6147(00)01676-X.

Cushing, M.C., & Anseth, K.S. 2007. DOI: 10.1126/science.1140171.

Dai, X.Z., Cai, S.X., Ye, Q.F., Jiang, J.H., et al. 2011. *Chinese Science Bulletin*. DOI: 10.1007/s11434-011-4717-3.

Daniele, M.A., Adams, A.A., Naciri, J., North, S.H., et al. 2014. *Biomaterials*. DOI: 10.1016/j.biomaterials.2013.11.009.

Dashnyam, K., Jin, G.Z., Kim, J.H., Perez, R., et al. 2017. *Biomaterials*. DOI: 10.1016/j.biomaterials.2016.11.053.

Davidenko, N., Schuster, C.F., Bax, D. V., Raynal, N., et al. 2015. *Acta Biomaterialia*. DOI: 10.1016/j.actbio.2015.07.034.

Denchai, A., Tartarini, D., & Mele, E. 2018. DOI: 10.3389/fbioe.2018.00155.

Deng, C., Zhang, P., Vulesevic, B., Kuraitis, D., et al. 2010. *Tissue Engineering - Part A*. DOI: 10.1089/ten.tea.2009.0504.

Devolder, R.J., Bae, H., Lee, J. & Kong, H. 2011. *Advanced Materials*. DOI: 10.1002/adma.201100823.

Dhandayuthapani, B., Yoshida, Y., Maekawa, T., & Kumar, D.S. 2011. DOI: 10.1155/2011/290602.

Dinescu, S., Albu Kaya, M., Chitoiu, L., Ignat, S., et al. 2019. DOI: 10.1007/978-3-319-77830-3_54.

Dingsheng, L., Zengbing, L., & Dong, H. 2016. *Iranian Journal of Basic Medical Sciences*. DOI: 10.22038/ijbms.2016.7815.

Dinulovic, I., Furrer, R., & Handschin, C. 2017. In *Advances in Experimental Medicine and Biology*. DOI: 10.1007/978-3-319-69194-7_8.

Discher, D.E., Janmey, P., & Wang, Y.L. 2005. DOI: 10.1126/science.1116995.

Dong, Z., Yuan, Q., Huang, K., Xu, W., et al. 2019. DOI: 10.1039/c9ra02695a.

Du, X., Zhou, J., Shi, J., & Xu, B. 2015. DOI: 10.1021/acs.chemrev.5b00299.

Du, Y., Lo, E., Ali, S., & Khademhosseini, A. 2008. *Proceedings of the National Academy of Sciences of the United States of America*. DOI: 10.1073/pnas.0801866105.

Eberli, D., Filho, L.F., Atala, A., & Yoo, J.J. 2009. *Methods*. DOI: 10.1016/j.ymeth.2008.10.014.

Edelman, E.R., Mathiowitz, E., Langer, R., & Klagsbrun, M. 1991. *Biomaterials*. DOI: 10.1016/0142-9612(91)90107-L.

Ehrbar, M., Djonov, V.G., Schnell, C., Tschanz, S.A., et al. 2004. *Circulation Research*. DOI: 10.1161/01.RES.0000126411.29641.08.

Engler, A.J., Sen, S., Sweeney, H.L., & Discher, D.E. 2006. *Cell*. DOI: 10.1016/j.cell.2006.06.044.

Esaki, J., Marui, A., Tabata, Y., & Komeda, M. 2007. DOI: 10.1517/17425247.4.6.635.

Evans, N.D., Gentleman, E., & Polak, J.M. 2006. *Materials Today*. DOI: 10.1016/S1369-7021(06)71740-0.

Fairbanks, B.D., Schwartz, M.P., Bowman, C.N., & Anseth, K.S. 2009. *Biomaterials*. DOI: 10.1016/j.biomaterials.2009.08.055.

Fan, W., Yan, W., Xu, Z., & Ni, H. 2012. *Colloids and Surfaces B: Biointerfaces*. *90*: 21–27. DOI: 10.1016/j.colsurfb.2011.09.042.

Ferrara, N. 2010. *Molecular Biology of the Cell*. DOI: 10.1091/mbc.E09-07-0590.

Ferrara, N., Gerber, H.P. & LeCouter, J. 2003. DOI: 10.1038/nm0603-669.

Ferrari, G., Cook, B.D., Terushkin, V., Pintucci, G., et al. 2009. *Journal of Cellular Physiology*. DOI: 10.1002/jcp.21706.

Finney, L., Vogt, S., Fukai, T., & Glesne, D. 2009. DOI: 10.1111/j.1440-1681.2008.04969.x.

Fisher, S.A., Baker, A.E.G., & Shoichet, M.S. 2017. DOI: 10.1021/jacs.7b00513.

Forte, G., Minieri, M., Cossa, P., Antenucci, D., et al. 2006. *Stem Cells*. DOI: 10.1634/stemcells.2004-0176.

Fournier, E., Passirani, C., Montero-Menei, C.N., & Benoit, J.P. 2003. DOI: 10.1016/S0142-9612(03)00161-3.

Frassica, M.T., & Grunlan, M.A. 2020. DOI: 10.1021/acsbiomaterials.0c00753.

Freeman, I., & Cohen, S. 2009. *Biomaterials*. DOI: 10.1016/j.biomaterials.2008.12.057.

Fu, C., Shao, Z., & Fritz, V. 2009. DOI: 10.1039/b911049f.

Furth, M.E., & Atala, A. 2013. In *Principles of Tissue Engineering: Fourth Edition*. DOI: 10.1016/B978-0-12-398358-9.00006-9.

Gallop, P.M., & Paz, M.A. 1975. DOI: 10.1152/physrev.1975.55.3.418.

Garbern, J.C., Hoffman, A.S., & Stayton, P.S. 2010. *Biomacromolecules*. DOI: 10.1021/bm100318z.

Gattazzo, F., Urciuolo, A., & Bonaldo, P. 2014. DOI: 10.1016/j.bbagen.2014.01.010.

Ge, Q., Zhang, H., Hou, J., Wan, L., et al. 2018. *Molecular Medicine Reports*. DOI: 10.3892/mmr.2017.8059.

Geckil, H., Xu, F., Zhang, X., Moon, S., et al. 2010. DOI: 10.2217/nnm.10.12.

Gérard, C., Bordeleau, L.J., Barralet, J., & Doillon, C.J. 2010. *Biomaterials*. DOI: 10.1016/j.biomaterials.2009.10.009.

Gerecht, S., Townsend, S.A., Pressler, H., Zhu, H., et al. 2007. *Biomaterials*. DOI: 10.1016/j.biomaterials.2007.07.039.

Ghanaati, S.M., Thimm, B.W., Unger, R.E., Orth, C., et al. 2010. *Biomedical Materials*. DOI: 10.1088/1748-6041/5/2/025004.

Ghasemi-Mobarakeh, L. 2015. *World Journal of Stem Cells*. DOI: 10.4252/wjsc.v7.i4.728.

Giavaresi, G., Tschon, M., Daly, J.H., Liggat, J.J., et al. 2006. *Journal of Biomaterials Science, Polymer Edition*. DOI: 10.1163/156856206778937226.

Gloria, A., De Santis, R., Ambrosio, L., Causa, F., et al. 2011. *Journal of Biomaterials Applications*. DOI: 10.1177/0885328209360933.

Goli-Malekabadi, Z., Tafazzoli-Shadpour, M., Tamayol, A., & Seyedjafari, E. 2017. *BioImpacts*. DOI: 10.15171/bi.2017.06.

Götz, W., Tobiasch, E., Witzleben, S., & Schulze, M. 2019. DOI: 10.3390/pharmaceutics11030117.

Gould, S.T., & Anseth, K.S. 2016. *Journal of Tissue Engineering and Regenerative Medicine*. DOI: 10.1002/term.1836.

Grafahrend, D., Heffels, K.H., Beer, M. V., Gasteier, P., et al. 2011. *Nature Materials*. DOI: 10.1038/nmat2904.

Grimes, D.R., Kelly, C., Bloch, K., & Partridge, M. 2014. *Journal of the Royal Society Interface*. DOI: 10.1098/rsif.2013.1124.

Grover, C.N., Gwynne, J.H., Pugh, N., Hamaia, S., et al. 2012. *Acta Biomaterialia*. DOI: 10.1016/j.actbio.2012.05.006.

Guan, N., Liu, Z., Zhao, Y., Li, Q., et al. 2020. DOI: 10.1080/10717544.2020.1831104.

Guarnieri, D., De Capua, A., Ventre, M., Borzacchiello, A., et al. 2010. *Acta Biomaterialia*. DOI: 10.1016/j.actbio.2009.12.050.

Guex, A.G., Hegemann, D., Giraud, M.N., Tevaearai, H.T., et al. 2014. *Colloids and Surfaces B: Biointerfaces*. DOI: 10.1016/j.colsurfb.2014.10.016.

Guilak, F., Cohen, D.M., Estes, B.T., Gimble, J.M., et al. 2009. DOI: 10.1016/j.stem.2009.06.016.

Guillotin, B., & Guillemot, F. 2011. DOI: 10.1016/j.tibtech.2010.12.008.

Gullberg, D., Gehlsen, K.R., Turner, D.C., Ahlen, K., et al. 1992. *EMBO Journal*. DOI: 10.1002/j.1460-2075.1992.tb05479.x.

Guo, B., & Ma, P.X. 2014. *Science China Chemistry*. DOI: 10.1007/s11426-014-5086-y.

Hadadi, A., Whittaker, J.W., Verrill, D.E., Hu, X., et al. 2018. *Biomacromolecules*. DOI: 10.1021/acs.biomac.8b00903.

Han, L., Xu, J., Lu, X., Gan, D., et al. 2017. *Journal of Materials Chemistry B*. DOI: 10.1039/c6tb02348g.

Handel, M., Hammer, T.R., Nooeaid, P., Boccaccini, A.R., et al. 2013. *Tissue Engineering - Part A*. DOI: 10.1089/ten.tea.2012.0707.

Harada, S.I., Nagy, J.A., Sullivan, K.A., Thomas, K.A., et al. 1994. *Journal of Clinical Investigation*. DOI: 10.1172/JCI117258.

Haro Durand, L.A., Góngora, A., Porto López, J.M., Boccaccini, A.R., et al. 2014. *Journal of Materials Chemistry B*. DOI: 10.1039/c4tb01043d.

Hasan, A. 2016. *Tissue Engineering for Artificial Organs: Regenerative Medicine, Smart Diagnostics and Personalized Medicine*. DOI: 10.1002/9783527689934.

Hejčl, A., Lesný, P., Přádný, M., Šedý, J., et al. 2009. *Journal of Materials Science: Materials in Medicine*. DOI: 10.1007/s10856-009-3714-4.

Hellmund, K.S., & Koksch, B. 2019. DOI: 10.3389/fchem.2019.00172.

Hench, L.L. 1980. *Science*. DOI: 10.1126/science.6246576.

Hench, L.L., & Polak, J.M. 2002. DOI: 10.1126/science.1067404.

Hench, L.L., & Thompson, I. 2010. DOI: 10.1098/rsif.2010.0151.focus.

Hickey, R.J., & Pelling, A.E. 2019. DOI: 10.3389/fbioe.2019.00045.

Hiew, V.V., Simat, S.F.B., & Teoh, P.L. 2018. DOI: 10.1007/s12015-017-9764-y.

Higuchi, A., Ling, Q.D., Chang, Y., Hsu, S.T., et al. 2013. DOI: 10.1021/cr300426x.

Hill, E., Boontheekul, T., & Mooney, D.J. 2006. *Proceedings of the National Academy of Sciences of the United States of America*. DOI: 10.1073/pnas.0506004103.

van Hinsbergh, V.W.M. 2016. In *Vascularization for Tissue Engineering and Regenerative Medicine*. DOI: 10.1007/978-3-319-21056-8_1-1.

Hirano, S., Zhang, M., Chung, B.G., & Kim, S.K. 2000. *Carbohydrate Polymers*. DOI: 10.1016/S0144-8617(99)00081-8.

Hoffman, A.S. 2012. DOI: 10.1016/j.addr.2012.09.010.

Howard, D., Buttery, L.D., Shakesheff, K.M., & Roberts, S.J. 2008. DOI: 10.1111/j.1469-7580.2008.00878.x.

Hristov, M., Erl, W., & Weber, P.C. 2003. DOI: 10.1016/S1050-1738(03)00077-X.

Hsu, Y.Y., Gresser, J.D., Trantolo, D.J., Lyons, C.M., et al. 1997. *Journal of Biomedical Materials Research*. DOI: 10.1002/(SICI)1097-4636(199704)35:1<107::AID-JBM11>3.0.CO;2-G.

Hu, G.F. 1998. *Journal of Cellular Biochemistry*. DOI: 10.1002/(SICI)1097-4644(19980601)69:3<326::AID-JCB10>3.0.CO;2-A.

Hu, W., Wang, Z., Xiao, Y., Zhang, S., et al. 2019. DOI: 10.1039/c8bm01246f.

Hu, Y., You, J.O., & Aizenberg, J. 2016. DOI: 10.1021/acsami.5b12268.

Hua, F.J., Kim, G.E., Lee, J.D., Son, Y.K., et al. 2002. *Journal of Biomedical Materials Research*. DOI: 10.1002/jbm.10121.
Huang, N.F., Okogbaa, J., Lee, J.C., Jha, A., et al. 2013. *Biomaterials*. DOI: 10.1016/j.biomaterials.2013.02.036.
Huebsch, N., Lippens, E., Lee, K., Mehta, M., et al. 2015. *Nature Materials*. DOI: 10.1038/nmat4407.
Hussey, G.S., Keane, T.J. & Badylak, S.F. 2017. DOI: 10.1038/nrgastro.2017.76.
Hutmacher D, Hürzeler MB, & Schliephake, H. 1996. *The International Journal of Oral Maxillofac Implants*. *11*(5):667–678.
Hwang, D.C., & Damodaran, S. 1996. *Journal of Agricultural and Food Chemistry*. DOI: 10.1021/jf9503826.
Ishii, O., Shin, M., Sueda, T., & Vacanti, J.P. 2005. *Journal of Thoracic and Cardiovascular Surgery*. DOI: 10.1016/j.jtcvs.2005.05.048.
Islam, M.M., Shahruzzaman, M., Biswas, S., Nurus Sakib, M., et al. 2020. DOI: 10.1016/j.bioactmat.2020.01.012.
Jafarkhani, M., Salehi, Z., Aidun, A., & Shokrgozar, M.A. 2019. *Iranian Biomedical Journal*. DOI: 10.29252/IBJ.23.1.9.
Jana, S., Leung, M., Chang, J., & Zhang, M. 2014. *Biofabrication*. DOI: 10.1088/1758-5082/6/3/035012.
Jao, D., Mou, X., & Hu, X. 2016. *Journal of Functional Biomaterials*. DOI: 10.3390/jfb7030022.
Jayakumar, R., Prabaharan, M., Nair, S. V., Tokura, S., et al. 2010. DOI: 10.1016/j.pmatsci.2010.03.001.
Jennings, K.-A. 2020. *Collagen- What Is It and What Is It Good For?*
Jeon, H., Hidai, H., Hwang, D.J., Healy, K.E., et al. 2010. *Biomaterials*. DOI: 10.1016/j.biomaterials.2010.01.103.
Jiang, L.B., Su, D.H., Ding, S.L., Zhang, Q.C., et al. 2019. *Advanced Functional Materials*. DOI: 10.1002/adfm.201901314.
Jiao, J., Huang, J., & Zhang, Z. 2019. DOI: 10.1002/app.47235.
Joo, H.J., Kim, H., Park, S.W., Cho, H.J., et al. 2011. *Blood*. DOI: 10.1182/blood-2010-12-323907.
Joo, S., Yeon Kim, J., Lee, E., Hong, N., et al. 2015. *Scientific Reports*. DOI: 10.1038/srep13043.
Jyothibasu, J.P., & Lee, R.H. 2018. *Polymers*. DOI: 10.3390/polym10111247.
Kabir, S.M.F., Sikdar, P.P., Haque, B., Bhuiyan, M.A.R., et al. 2018. *Progress in Biomaterials*. DOI: 10.1007/s40204-018-0095-0.
Kaiser, N.J., & Coulombe, K.L.K. 2015. DOI: 10.1088/1748-6041/10/3/034003.
Kajiya, K., Hirakawa, S., Ma, B., Drinnenberg, I., et al. 2005. *EMBO Journal*. DOI: 10.1038/sj.emboj.7600763.
Kamble, S., Ahmed, M.Z., Ramabhimaiaha, S. & Patil, P. 2013. *Biomedical and Pharmacology Journal*. DOI: 10.13005/bpj/420.
Kang, Z., Wang, Y., Xu, J., Song, G., et al. 2018. Polymers. DOI: 10.3390/polym10111193.
Kapoor, S., & Kundu, S.C. 2016. DOI: 10.1016/j.actbio.2015.11.034.
Karami, A., Tebyanian, H., Sayyad Soufdoost, R., Motavallian, E., et al. 2019. *World Journal of Plastic Surgery*. DOI: 10.29252/wjps.8.3.352.
Kearns, V., Macintosh, A.C., Crawford, A., & Hatton, P. V. 2008. [Book].
Keselowsky, B.G., Collard, D.M., & García, A.J. 2004. *Biomaterials*. DOI: 10.1016/j.biomaterials.2004.01.062.
Khademhosseini, A., & Langer, R. 2007. DOI: 10.1016/j.biomaterials.2007.07.021.
Khademhosseini, A., Langer, R., Borenstein, J., & Vacanti, J.P. 2006a. DOI: 10.1073/pnas.0507681102.
Khademhosseini, A., Eng, G., Yeh, J., Fukuda, J., et al. 2006b. *Journal of Biomedical Materials Research - Part A*. DOI: 10.1002/jbm.a.30821.
Khan, S., Villalobos, M.A., Choron, R.L., Chang, S., et al. 2017. *Journal of Vascular Surgery*. DOI: 10.1016/j.jvs.2016.04.034.
Khang, G. 2015. *Biosurface and Biotribology*. DOI: 10.1016/j.bsbt.2015.08.004.
Khetan, S., & Burdick, J.A. 2011. DOI: 10.1039/c0sm00852d.
Kim, B.S., & Mooney, D.J. 1998. DOI: 10.1016/S0167-7799(98)01191-3.
Kim, C., Kasuya, J., Jeon, J., Chung, S., et al. 2015. *Lab on a Chip*. DOI: 10.1039/c4lc00866a.
Kim, D.H., Seo, C.H., Han, K., Kwon, K.W., et al. 2009. *Advanced Functional Materials*. DOI: 10.1002/adfm.200801174.
Kim, J.H., Kim, T.H., Kang, M.S., & Kim, H.W. 2016. *BioMed Research International*. DOI: 10.1155/2016/9676934.
Kingshott, P., Andersson, G., McArthur, S.L., & Griesser, H.J. 2011. DOI: 10.1016/j.cbpa.2011.07.012.
Klimek, K., & Ginalska, G. 2020. DOI: 10.3390/POLYM12040844.
Ko, I.K., Lee, S.J., Atala, A., & Yoo, J.J. 2013. DOI: 10.1038/emm.2013.118.

Kocherova, I., Bryja, A., Mozdziak, P., Angelova Volponi, A., et al. 2019. *Journal of Clinical Medicine*. DOI: 10.3390/jcm8101602.
Koh, W.G., Revzin, A., & Pishko, M. V. 2002. *Langmuir*. DOI: 10.1021/la0115740.
Kolte, D., McClung, J.A., & Aronow, W.S. 2016. *Translational Research in Coronary Artery Disease: Pathophysiology to Treatment*. DOI: 10.1016/B978-0-12-802385-3.00006-1.
Kolvin, I., Cohen, G., & Fineberg, J. 2018. *Nature Materials*. DOI: 10.1038/NMAT5008.
Kong, N., Lin, K., Li, H., & Chang, J. 2014. *Journal of Materials Chemistry B*. DOI: 10.1039/c3tb21529f.
Kook, Y.M., Jeong, Y., Lee, K., & Koh, W.G. 2017. *Journal of Tissue Engineering*. DOI: 10.1177/2041731417724640.
Krafts, K.P. 2010. DOI: 10.4161/org.6.4.12555.
Kubo, K., Tsukimura, N., Iwasa, F., Ueno, T., et al. 2009. *Biomaterials*. DOI: 10.1016/j.biomaterials.2009.06.021.
Kuen, Y.L., Alsberg, E., Hsiong, S., Comisar, W., et al. 2004. *Nano Letters*. DOI: 10.1021/nl0493592.
Kulicke, W.M., Kniewske, R., & Klein, J. 1982. DOI: 10.1016/0079-6700(82)90004-1.
Kundu, J., Poole-Warren, L.A., Martens, P., & Kundu, S.C. 2012. *Acta Biomaterialia*. DOI: 10.1016/j.actbio.2012.01.004.
Kunz, R.I., Brancalhão, R.M.C., Ribeiro, L.D.F.C., & Natali, M.R.M. 2016. DOI: 10.1155/2016/8175701.
Lavrentieva, A., Kirsch, M., Birnstein, L., Pepelanova, I., et al. 2019. *Bioengineering*. DOI: 10.3390/bioengineering6030076.
Lee, K.Y., & Mooney, D.J. 2012. DOI: 10.1016/j.progpolymsci.2011.06.003.
Lee, J.H., Parthiban, P., Jin, G.Z., Knowles, J.C., et al. 2020. DOI: 10.1016/j.pmatsci.2020.100732.
Lee, K., Silva, E.A., & Mooney, D.J. 2011. DOI: 10.1098/rsif.2010.0223.
Lee, K.W., Yoon, J.J., Lee, J.H., Kim, S.Y., et al. 2004. In *Transplantation Proceedings*. DOI: 10.1016/j.transproceed.2004.08.078.
Lee, P.P., Cerchiari, A., & Desai, T.A. 2014a. *Nano Letters*. DOI: 10.1021/nl501523v.
Lee, S.W., Son, Y.I., Kim, C.H., Lee, J.Y., et al. 2007. *Laryngoscope*. DOI: 10.1097/MLG.0b013e3180caa1b1.
Lee, V.K., Lanzi, A.M., Ngo, H., Yoo, S.S., et al. 2014b. *Cellular and Molecular Bioengineering*. DOI: 10.1007/s12195-014-0340-0.
Legon'kova, O.A., Savchenkova, I.P., Belova, M.S., Korotaeva, A.I., et al. 2016. *Polymer Science - Series D*. DOI: 10.1134/S199542121602012X.
Leucht, A., Volz, A.C., Rogal, J., Borchers, K., et al. 2020. *Scientific Reports*. DOI: 10.1038/s41598-020-62166-w.
Leuning, D.G., Witjas, F.M.R., Maanaoui, M., de Graaf, A.M.A., et al. 2019. *American Journal of Transplantation*. DOI: 10.1111/ajt.15200.
Li, N., & Folch, A. 2005. *Experimental Cell Research*. DOI: 10.1016/j.yexcr.2005.10.007.
Li, L., Chen, S., & Jiang, S. 2003. *Langmuir*. DOI: 10.1021/la0262957.
Li, X., Dai, Y., Shen, T., & Gao, C. 2017. *Regenerative Biomaterials*. DOI: 10.1093/rb/rbx005.
Lien, S.M., Ko, L.Y., & Huang, T.J. 2009. *Acta Biomaterialia*. DOI: 10.1016/j.actbio.2008.09.020.
Lim, J.Y., Dreiss, A.D., Zhou, Z., Hansen, J.C., et al. 2007. *Biomaterials*. DOI: 10.1016/j.biomaterials.2006.12.020.
Lim, K.S., Alves, M.H., Poole-Warren, L.A., & Martens, P.J. 2013. *Biomaterials*. DOI: 10.1016/j.biomaterials.2013.06.005.
Lin, C.C., & Anseth, K.S. 2009. DOI: 10.1007/s11095-008-9801-2.
Lin, H.I., Kuo, Y.M., Hu, C.C., Lee, M.H., et al. 2018. *Scientific Reports*. DOI: 10.1038/s41598-018-32462-7.
Lin, R.Z., Chen, Y.C., Moreno-Luna, R., Khademhosseini, A., et al. 2013. *Biomaterials*. DOI: 10.1016/j.biomaterials.2013.05.060.
Ling, Y., Rubin, J., Deng, Y., Huang, C., et al. 2007. *Lab on a Chip*. DOI: 10.1039/b615486g.
Liu, J.S., & Gartner, Z.J. 2012. DOI: 10.1016/j.tcb.2012.09.004.
Liu, C., Zhang, Q., Zhu, S., Liu, H., et al. 2019. DOI: 10.1039/c9ra05934b.
Liu, L., Ratner, B.D., Sage, E.H., & Jiang, S. 2007. *Langmuir*. DOI: 10.1021/la701435x.
Liu, S.Q., Tian, Q., Hedrick, J.L., Po Hui, J.H., et al. 2010. *Biomaterials*. DOI: 10.1016/j.biomaterials.2010.06.001.
Liu, W., Zhang, G., Wu, J., Zhang, Y., et al. 2020. DOI: 10.1186/s12951-019-0570-3.
Liu Tsang, V., & Bhatia, S.N. 2004. DOI: 10.1016/j.addr.2004.05.001.
Lohrasbi, S., Mirzaei, E., Karimizade, A., Takallu, S., et al. 2020. *Cellulose*. DOI: 10.1007/s10570-019-02841-y.

Łopacińska, J.M., Grădinaru, C., Wierzbicki, R., Købler, C., et al. 2012. *Nanoscale*. DOI: 10.1039/c2nr11455k.
Losi, P., Briganti, E., Errico, C., Lisella, A., et al. 2013. *Acta Biomaterialia*. DOI: 10.1016/j.actbio.2013.04.019.
Lovett, M., Lee, K., Edwards, A., & Kaplan, D.L. 2009. DOI: 10.1089/ten.teb.2009.0085.
Lu, J., Yan, X., Sun, X., Shen, X., et al. 2019. *Nanoscale*. DOI: 10.1039/c9nr04521j.
Luo, Y., & Shoichet, M.S. 2004. *Nature Materials*. DOI: 10.1038/nmat1092.
Luo, Y., Wang, S., Shen, M., Qi, R., et al. 2013. *Carbohydrate Polymers*. DOI: 10.1016/j.carbpol.2012.08.069.
Lutolf, M.P., & Hubbell, J.A. 2005. DOI: 10.1038/nbt1055.
Ma, Z., Kotaki, M., Inai, R., & Ramakrishna, S. 2005. *Tissue Engineering*. DOI: 10.1089/ten.2005.11.101.
Mackay, F., Loetscher, H., Stueber, D., Gehr, G., et al. 1993. *Journal of Experimental Medicine*. DOI: 10.1084/jem.177.5.1277.
Madden, L.R., Mortisen, D.J., Sussman, E.M., Dupras, S.K., et al. 2010. *Proceedings of the National Academy of Sciences of the United States of America*. DOI: 10.1073/pnas.1006442107.
Maier, J.A.M. 2012. DOI: 10.1042/CS20110506.
Manabe, R.I., Tsutsui, K., Yamada, T., Kimura, M., et al. 2008. *Proceedings of the National Academy of Sciences of the United States of America*. DOI: 10.1073/pnas.0803640105.
Mandal, B.B., & Kundu, S.C. 2009. *Biomaterials*. DOI: 10.1016/j.biomaterials.2009.02.006.
von der Mark, K., & Park, J. 2013. DOI: 10.1016/j.pmatsci.2012.09.002.
Martin, T.A., Caliari, S.R., Williford, P.D., Harley, B.A., et al. 2011. *Biomaterials*. DOI: 10.1016/j.biomaterials.2011.02.018.
Martino, M.M., Brkic, S., Bovo, E., Burger, M., et al. 2015. DOI: 10.3389/fbioe.2015.00045.
Martins, A., Chung, S., Pedro, A.J., Sousa, R.A., et al. 2009. *Journal of Tissue Engineering and Regenerative Medicine*. DOI: 10.1002/term.132.
Mastrullo, V., Cathery, W., Velliou, E., Madeddu, P., et al. 2020. DOI: 10.3389/fbioe.2020.00188.
Meldal, M. & Schoffelen, S. 2016. *F1000Research*. DOI: 10.12688/f1000research.9002.1.
Menegatti, S., Hussain, M., Naik, A.D., Carbonell, R.G., et al. 2013a. *Biotechnology and Bioengineering*. DOI: 10.1002/bit.24760.
Menegatti, S., Ward, K.L., Naik, A.D., Kish, W.S., et al. 2013b. *Analytical Chemistry*. DOI: 10.1021/ac401954k.
Meredith, J.E., Fazeli, B., & Schwartz, M.A. 1993. *Molecular Biology of the Cell*. DOI: 10.1091/mbc.4.9.953.
Mero, A., Clementi, C., Veronese, F.M., & Pasut, G. 2011. In *Methods in Molecular Biology*. DOI: 10.1007/978-1-61779-151-2_8.
Merrett, K., Cornelius, R.M., McClung, W.G., Unsworth, L.D., et al. 2002. DOI: 10.1163/156856202320269111.
Meyvis, T., De Smedt, S., Stubbe, B., Hennink, W., et al. 2001. *Pharmaceutical Research*. DOI: 10.1023/A:1013038716373.
Mhanna, R., Öztürk, E., Vallmajo-Martin, Q., Millan, C., et al. 2014. *Tissue Engineering - Part A*. DOI: 10.1089/ten.tea.2013.0519.
Mikos, A.G., Lyman, M.D., Freed, L.E., & Langer, R. 1994. *Biomaterials*. DOI: 10.1016/0142-9612(94)90197-X.
Minardi, S., Taraballi, F., Pandolfi, L., & Tasciotti, E. 2016. DOI: 10.3389/fbioe.2016.00045.
Minoura, N., Aiba, S.I., Higuchi, M., Gotoh, Y., et al. 1995. *Biochemical and Biophysical Research Communications*. DOI: 10.1006/bbrc.1995.1368.
Mitchell, A.C., Briquez, P.S., Hubbell, J.A. & Cochran, J.R. 2016. DOI: 10.1016/j.actbio.2015.11.007.
Mohandas, A., Anisha, B.S., Chennazhi, K.P., & Jayakumar, R. 2015. *Colloids and Surfaces B: Biointerfaces*. DOI: 10.1016/j.colsurfb.2015.01.024.
Montrucchio, G., Lupia, E., Battaglia, E., Passerini, G., et al. 1994. *Journal of Experimental Medicine*. DOI: 10.1084/jem.180.1.377.
Mountziaris, P.M., Dennis Lehman, E., Mountziaris, I., Sing, D.C., et al. 2013. *Journal of Biomaterials Science, Polymer Edition*. DOI: 10.1080/09205063.2013.803455.
Mouw, J.K., Ou, G., & Weaver, V.M. 2014. DOI: 10.1038/nrm3902.
Murphy, C.M., & O'Brien, F.J. 2010. DOI: 10.4161/cam.4.3.11747.
Murugan, R., Molnar, P., Rao, K.P. & Hickman, J.J. 2009. *International Journal of Biomedical Engineering and Technology*. DOI: 10.1504/IJBET.2009.022911.
Nair, A., Shen, J., Lotfi, P., Ko, C.Y., et al. 2011. *Acta Biomaterialia*. DOI: 10.1016/j.actbio.2011.06.050.
Nakayama, K.H., Hong, G., Lee, J.C., Patel, J., et al. 2015. *ACS Nano*. DOI: 10.1021/acsnano.5b00545.
Nakayama, K.H., Surya, V.N., Gole, M., Walker, T.W., et al. 2016. *Nano Letters*. DOI: 10.1021/acs.nanolett.5b04028.

Narayan, R.J. 2010. DOI: 10.1098/rsta.2010.0001.

Nasatto, P.L., Pignon, F., Silveira, J.L.M., Duarte, M.E.R., et al. 2015. DOI: 10.3390/polym7050777.

Nazarnezhad, S., Baino, F., Kim, H.W., Webster, T.J., et al. 2020. DOI: 10.3390/nano10081609.

Nelson, P.R., Yamamura, S., Kent, K.C., Callow, A.D., et al. 1996. *Journal of Vascular Surgery*. DOI: 10.1016/S0741-5214(96)70141-6.

Nemati, S., Rezabakhsh, A., Khoshfetrat, A.B., Nourazarian, A., et al. 2017. *Biotechnology and Bioengineering*. DOI: 10.1002/bit.26395.

Nemazifard, M., Kavoosi, G., Marzban, Z., & Ezedi, N. 2017. *International Journal of Food Properties*. DOI: 10.1080/10942912.2016.1219369.

Nguyen, E.H., & Murphy, W.L. 2018. DOI: 10.1016/j.biomaterials.2018.07.041.

Nguyen, T.P., Nguyen, Q.V., Nguyen, V.H., Le, T.H., et al. 2019. DOI: 10.3390/polym11121933.

Nichol, J.W., & Khademhosseini, A. 2009. *Soft Matter*. DOI: 10.1039/b814285h.

Nikkhah, M., Eshak, N., Zorlutuna, P., Annabi, N., et al. 2012. *Biomaterials*. DOI: 10.1016/j.biomaterials.2012.08.068.

Nikkhah, M., Akbari, M., Paul, A., Memic, A., et al. 2016. *Biomaterials from Nature for Advanced Devices and Therapies*. DOI: 10.1002/9781119126218.ch3.

Nikolova, M.P., & Chavali, M.S. 2019. *Bioactive Materials*. DOI: 10.1016/j.bioactmat.2019.10.005.

Nilasaroya, A., Poole-Warren, L.A., Whitelock, J.M., & Jo Martens, P. 2008. *Biomaterials*. DOI: 10.1016/j.biomaterials.2008.08.011.

Nillesen, S.T.M., Geutjes, P.J., Wismans, R., Schalkwijk, J., et al. 2007. *Biomaterials*. DOI: 10.1016/j.biomaterials.2006.10.029.

Nkhwa, S., Lauriaga, K.F., Kemal, E., & Deb, S. 2014. *Conference Papers in Science*. DOI: 10.1155/2014/403472.

Norris, S.C.P., Tseng, P., & Kasko, A.M. 2016. *ACS Biomaterials Science and Engineering*. DOI: 10.1021/acsbiomaterials.6b00237.

Norris, S.C.P., Delgado, S.M., & Kasko, A.M. 2019. *Polymer Chemistry*. DOI: 10.1039/c9py00308h.

Nosoudi, N., Jacob, A.O., Stultz, S., Jordan, M., et al. 2020. *Bioengineering*. DOI: 10.3390/bioengineering7010021.

Novotna, K., Havelka, P., Sopuch, T., Kolarova, K., et al. 2013. *Cellulose*. DOI: 10.1007/s10570-013-0006-4.

Ohya, Y., Matsunami, H., & Ouchi, T. 2004. *Journal of Biomaterials Science, Polymer Edition*. DOI: 10.1163/156856204322752264.

Oliviero, O., Ventre, M., & Netti, P.A. 2012. *Acta Biomaterialia*. DOI: 10.1016/j.actbio.2012.05.019.

Ostuni, E., Chapman, R.G., Holmlin, R.E., Takayama, S., et al. 2001. DOI: 10.1021/la010384m.

Panahi, R., & Baghban-Salehi, M. 2019. DOI: 10.1007/978-3-319-77830-3_52.

Pandey, A.K., Singhi, E.K., Arroyo, J.P., Ikizler, T.A., et al. 2018. DOI: 10.1161/HYPERTENSIONAHA.117.10271.

Park, J.B. & Bronzino, J.D. 2002. *Biomaterials: Principles and Applications*.

Park, M., Li, Q., Shcheynikov, N., Zeng, W., et al. 2004. *Molecular Cell*. DOI: 10.1016/j.molcel.2004.09.030.

Patel, R., Santhosh, M., Dash, J.K., Karpoormath, R., et al. 2019. DOI: 10.1002/pat.4442.

Patra, J.K., Das, G., Fraceto, L.F., Campos, E.V.R., et al. 2018. *Journal of Nanobiotechnology*. DOI: 10.1186/s12951-018-0392-8.

Peng, L., Barczak, A.J., Barbeau, R.A., Xiao, Y., et al. 2010. *Nano Letters*. DOI: 10.1021/nl903043z.

Peppas, N.A., Hilt, J.Z., Khademhosseini, A., & Langer, R. 2006. DOI: 10.1002/adma.200501612.

Pereira, G.K.R., Venturini, A.B., Silvestri, T., Dapieve, K.S., et al. 2016. DOI: 10.1016/j.jmbbm.2015.10.017.

Pérez, R.A., Won, J.E., Knowles, J.C., & Kim, H.W. 2013. DOI: 10.1016/j.addr.2012.03.009.

Peters, M.C., Isenberg, B.C., Rowley, J.A. & Mooney, D.J. 1998. *Journal of Biomaterials Science, Polymer Edition*. DOI: 10.1163/156856298X00389.

Petrik, J., Campbell, N.E., Kellenberger, L., Greenaway, J., et al. 2010. DOI: 10.1155/2010/586905.

Pina, S., Ribeiro, V.P., Marques, C.F., Maia, F.R., et al. 2019. DOI: 10.3390/ma12111824.

Pompe, T., Salchert, K., Alberti, K., Zandstra, P., et al. 2010. *Nature Protocols*. DOI: 10.1038/nprot.2010.70.

Post, M.J., Laham, R., Sellke, F.W., & Simons, M. 2001. DOI: 10.1016/S0008-6363(00)00216-9.

Pradhan, S., Rajamani, S., Agrawal, G., Dash, M., et al. 2017. *Characterization of Polymeric Biomaterials*. DOI: 10.1016/B978-0-08-100737-2.00007-8.

Prado, P.H.C.O., Monteiro, J.B., Campos, T.M.B., Thim, G.P., et al. 2020. *Journal of the Mechanical Behavior of Biomedical Materials*. DOI: 10.1016/j.jmbbm.2019.103482.

Qiong Liu, S., Tian, Q., Wang, L., Hedrick, J.L., et al. 2010. *Macromolecular Rapid Communications*. DOI: 10.1002/marc.200900818.

Raghunathan, V., McKee, C., Cheung, W., Naik, R., et al. 2013. *Tissue Engineering - Part A*. DOI: 10.1089/ten.tea.2012.0584.

Rahmati, M., Silva, E.A., Reseland, J.E., Heyward, Catherine A., et al. 2020. DOI: 10.1039/d0cs00103a.

Rana, D., Ramasamy, K., Leena, M., Pasricha, R., et al. 2017. *Biology and Engineering of Stem Cell Niches*. DOI: 10.1016/B978-0-12-802734-9.00021-4.

Rao, M.R., Fang, Y., De Feyter, S., & Perepichka, D.F. 2017. *Journal of the American Chemical Society*. DOI: 10.1021/jacs.6b12005.

Rath, S.N., Brandl, A., Hiller, D., Hoppe, A., et al. 2014. *PLoS One*. DOI: 10.1371/journal.pone.0113319.

Ratner, B.D., & Bryant, S.J. 2004. DOI: 10.1146/annurev.bioeng.6.040803.140027.

Refojo, M.F., & Yasuda, H. 1965. *Journal of Applied Polymer Science*. DOI: 10.1002/app.1965.070090707.

Reinhart-King, C.A. 2008. DOI: 10.1016/S0076-6879(08)02003-X.

Reverdatto, S., Burz, D., & Shekhtman, A. 2015. *Current Topics in Medicinal Chemistry*. DOI: 10.2174/1568026615666150413153143.

Ribeiro, A.J.S., Denisin, A.K., Wilson, R.E., & Pruitt, B.L. 2016. DOI: 10.1016/j.ymeth.2015.08.005.

Richardson, T.P., Peters, M.C., Ennett, A.B., & Mooney, D.J. 2001. *Nature Biotechnology*. DOI: 10.1038/nbt1101-1029.

Rivas, M., Del Valle, L.J., Alemán, C., & Puiggalí, J. 2019. DOI: 10.3390/gels5010014.

Robert, L., Robert, A.-M., & Renard, G. 2010. *Pathologie Biologie*. DOI: 10.1016/j.patbio.2009.09.010.

Roberts, J.J., Nicodemus, G.D., Greenwald, E.C., & Bryant, S.J. 2011. *Clinical Orthopaedics and Related Research*. DOI: 10.1007/s11999-011-1823-0.

Rodríguez-Vázquez, M., Vega-Ruiz, B., Ramos-Zúñiga, R., Saldaña-Koppel, D.A., et al. 2015. DOI: 10.1155/2015/821279.

Rohman, G., Baker, S.C., Southgate, J., & Cameron, N.R. 2009. *Journal of Materials Chemistry*. DOI: 10.1039/b911625g.

Rynkowska, E., Fatyeyeva, K., Marais, S., Kujawa, J., et al. 2019. *Polymers*. DOI: 10.3390/polym11111799.

Sachot, N., Castaño, O., Mateos-Timoneda, M.A., Engel, E., et al. 2013. *Journal of the Royal Society Interface*. DOI: 10.1098/rsif.2013.0684.

Salinas, C.N. & Anseth, K.S. 2008. *Journal of Tissue Engineering and Regenerative Medicine*. DOI: 10.1002/term.95.

Samavedi, S., Poindexter, L.K., Van Dyke, M., & Goldstein, A.S. 2014. *Regenerative Medicine Applications in Organ Transplantation*. DOI: 10.1016/B978-0-12-398523-1.00007-0.

Samorezov, J.E., & Alsberg, E. 2015. DOI: 10.1016/j.addr.2014.11.018.

Santos, M.I., Tuzlakoglu, K., Fuchs, S., Gomes, M.E., et al. 2008. *Biomaterials*. DOI: 10.1016/j.biomaterials.2008.07.033.

dos Santos, B.P., Garbay, B., Fenelon, M., Rosselin, M., et al. 2019. *Acta Biomaterialia*. DOI: 10.1016/j.actbio.2019.08.028.

Sanzari, I., Humphrey, E.J., Dinelli, F., Terracciano, C.M., et al. 2018. *Scientific Reports*. DOI: 10.1038/s41598-018-30360-6.

Saqib, U., Sarkar, S., Suk, K., Mohammad, O., et al. 2018. DOI: 10.18632/oncotarget.24788.

Sarker, M.D., Naghieh, S., Sharma, N.K., Ning, L., et al. 2019. DOI: 10.1155/2019/9156921.

Saunders, R.L., & Hammer, D.A. 2010. *Cellular and Molecular Bioengineering*. DOI: 10.1007/s12195-010-0112-4.

Sawkins, M.J., Saldin, L.T., Badylak, S.F., & White, L.J. 2018. DOI: 10.1007/978-3-319-77023-9_2.

Schultz, K.M., Kyburz, K.A., & Anseth, K.S. 2015. *Proceedings of the National Academy of Sciences of the United States of America*. DOI: 10.1073/pnas.1511304112.

Schulz, A., Katsen-Globa, A., Huber, E.J., Mueller, S.C., et al. 2018. *Journal of Materials Science: Materials in Medicine*. DOI: 10.1007/s10856-018-6113-x.

Sekiya, S., Shimizu, T., Yamato, M., Kikuchi, A., et al. 2006. *Biochemical and Biophysical Research Communications*. DOI: 10.1016/j.bbrc.2005.12.217.

Semino, C.E., Kamm, R.D., & Lauffenburger, D.A. 2006. *Experimental Cell Research*. DOI: 10.1016/j.yexcr.2005.10.029.

Sen, C.K., Khanna, S., Venojarvi, M., Trikha, P., et al. 2002. *American Journal of Physiology - Heart and Circulatory Physiology*. DOI: 10.1152/ajpheart.01015.2001.

Senta, H., Bergeron, E., Drevelle, O., Park, H., et al. 2011. *Canadian Journal of Chemical Engineering*. DOI: 10.1002/cjce.20453.

Shen, J., Chen, B., Zhai, X., Qiao, W., et al. 2021. *Bioactive Materials*. DOI: 10.1016/j.bioactmat.2020.08.025.

Shen, X., Shamshina, J.L., Berton, P., Gurau, G., et al. 2015. DOI: 10.1039/c5gc02396c.

Shi, J., & Wei, P.K. 2016. *Oncology Letters*. DOI: 10.3892/ol.2015.4035.

Shi, H., Han, C., Mao, Z., Ma, L., et al. 2008. *Tissue Engineering - Part A*. DOI: 10.1089/ten.tea.2007.0007.

Shimizu, T., Sekine, H., Yang, J., Isoi, Y., et al. 2006. *The FASEB Journal*. DOI: 10.1096/fj.05-4715fje.

Shkarina, S., Shkarin, R., Weinhardt, V., Melnik, E., et al. 2018. *Scientific Reports*. DOI: 10.1038/s41598-018-27097-7.

Singh, B., Choi, Y.J., Park, I.K., Akaike, T., et al. 2014. DOI: 10.1166/jnn.2014.9079.

Singh, M., Dormer, N., Salash, J.R., Christian, J.M., et al. 2010. *Journal of Biomedical Materials Research - Part A*. DOI: 10.1002/jbm.a.32765.

Slaughter, B. V., Khurshid, S.S., Fisher, O.Z., Khademhosseini, A., et al. 2009. DOI: 10.1002/adma.200802106.

Smethurst, P.A., Onley, D.J., Jarvis, G.E., O'Connor, M.N., et al. 2007. *Journal of Biological Chemistry*. DOI: 10.1074/jbc.M606479200.

Smith, L.R., Cho, S., & Discher, D.E. 2018. DOI: 10.1152/physiol.00026.2017.

Smithmyer, M.E., Sawicki, L.A., & Kloxin, A.M. 2014. DOI: 10.1039/c3bm60319a.

Somaiah, C., Kumar, A., Mawrie, D., Sharma, A., et al. 2015. *PLoS One*. DOI: 10.1371/journal.pone.0145068.

Song, W., Chiu, A., Wang, L.H., Schwartz, R.E., et al. 2019. *Nature Communications*. DOI: 10.1038/s41467-019-12373-5.

Spicer, C.D., Pashuck, E.T., & Stevens, M.M. 2018. DOI: 10.1021/acs.chemrev.8b00253.

Stevenson, M.D., Piristine, H., Hogrebe, N.J., Nocera, T.M., et al. 2013. *Acta Biomaterialia*. DOI: 10.1016/j.actbio.2013.04.002.

Su, T., Huang, K., Daniele, M.A., Hensley, M.T., et al. 2018. *ACS Applied Materials and Interfaces*. *10*(39):33088–33096. DOI: 10.1021/acsami.8b13571.

Su, T., Huang, K., Mathews, K.G., Scharf, V.F., et al. 2020. *ACS Biomaterials Science and Engineering*. DOI: 10.1021/acsbiomaterials.0c00942.

Sukmana, I., & Vermette, P. 2010. *Biomaterials*. DOI: 10.1016/j.biomaterials.2010.02.076.

Sun, G., & Mao, J.J. 2012. DOI: 10.2217/nnm.12.149.

Sun, J., & Tan, H. 2013. DOI: 10.3390/ma6041285.

Sun, X., Zhao, X., Zhao, L., Li, Q., et al. 2015. *Journal of Materials Chemistry B*. DOI: 10.1039/c5tb00645g.

Sung, H.J., Meredith, C., Johnson, C. & Galis, Z.S. 2004. *Biomaterials*. DOI: 10.1016/j.biomaterials.2004.01.066.

Tallawi, M., Rosellini, E., Barbani, N., Grazia Cascone, M., et al. 2015. DOI: 10.1098/rsif.2015.0254.

Tanaka, Y., Fukao, K., & Miyamoto, Y. 2000. *European Physical Journal E*. DOI: 10.1007/s101890070010.

Tang, L., Wu, Y., & Timmons, R.B. 1998. *Journal of Biomedical Materials Research*. DOI: 10.1002/(SICI)1097-4636(199810)42:1<156::AID-JBM19>3.0.CO;2-J.

Teixeira, M.A., Amorim, M.T.P., & Felgueiras, H.P. 2020. DOI: 10.3390/polym12010007.

Temple, S. 2001. DOI: 10.1038/35102174.

Tenje, M., Cantoni, F., Porras Hernández, A.M., Searle, S.S., et al. 2020. *Organs-on-a-Chip*. DOI: 10.1016/j.ooc.2020.100003.

Thevenot, P., Hu, W., & Tang, L. 2008. *Current Topics in Medicinal Chemistry*. DOI: 10.2174/156802608783790901.

Thomas, C.H., Collier, J.H., Sfeir, C.S., & Healy, K.E. 2002. *Proceedings of the National Academy of Sciences of the United States of America*. DOI: 10.1073/pnas.032668799.

Thomlinson, R.H., & Gray, L.H. 1955. *British Journal of Cancer*. DOI: 10.1038/bjc.1955.55.

Thomson, R.C., Yaszemski, M.J., Powers, J.M. & Mikos, A.G. 1995. *Journal of Biomaterials Science, Polymer Edition*. DOI: 10.1163/156856295X00805.

Tian, L., Prabhakaran, M.P., Ding, X., Kai, D., et al. 2013. *Journal of Materials Science: Materials in Medicine*. DOI: 10.1007/s10856-013-5003-5.

Tiruvannamalai-Annamalai, R., Armant, D.R., & Matthew, H.W.T. 2014. *PLoS One*. DOI: 10.1371/journal.pone.0084287.

Tiruvannamalai Annamalai, R., Rioja, A.Y., Putnam, A.J., & Stegemann, J.P. 2016. *ACS Biomaterials Science and Engineering*. DOI: 10.1021/acsbiomaterials.6b00274.

Ulrich, T.A., De Juan Pardo, E.M., & Kumar, S. 2009. *Cancer Research*. DOI: 10.1158/0008-5472.CAN-08-4859.

Urbanczyk, M., Layland, S.L., & Schenke-Layland, K. 2020. DOI: 10.1016/j.matbio.2019.11.005.

Urry, D.W. 1997. DOI: 10.1021/jp972167t.

Vacanti, J.P., & Vacanti, C.A. 2000. *Tissue Engineering Intelligence Unit.*

Vaday, G.G., & Lider, O. 2000. DOI: 10.1002/jlb.67.2.149.

Vaday, G.G., Franitza, S., Schor, H., Hecht, I., et al. 2001. *Journal of Leukocyte Biology*. DOI: 10.1189/jlb.69.6.885.

Vepari, C., & Kaplan, D.L. 2007. DOI: 10.1016/j.progpolymsci.2007.05.013.

Viola, J., Lal, B., & Grad, O. 2003. *The emergence of tissue engineering as a research field.*

Viswanathan, P., Ondeck, M.G., Chirasatitsin, S., Ngamkham, K., et al. 2015. *Biomaterials*. DOI: 10.1016/j.biomaterials.2015.01.034.

Vo, T.N., Kasper, F.K., & Mikos, A.G. 2012. DOI: 10.1016/j.addr.2012.01.016.

Vrana, N.E., Cahill, P.A., & McGuinness, G.B. 2010. *Journal of Biomedical Materials Research - Part A*. DOI: 10.1002/jbm.a.32790.

Waly, G.H., Abdel Hamid, I.S., Sharaf, M.A., Marei, M.K., et al. 2008. *Proceedings of the ASME 2nd Multifunctional Nanocomposites and Nanomaterials Conference, MN2008*. DOI: 10.1115/MN2008-47068.

Wang, J., & Borenstein, J.T. 2017. *Biology and Engineering of Stem Cell Niches.*

Wang, B., Lv, X., Chen, S., Li, Z., et al. 2018a. *Carbohydrate Polymers*. DOI: 10.1016/j.carbpol.2017.11.055.

Wang, J., Jansen, J.A., & Yang, F. 2019a. DOI: 10.3389/fchem.2019.00258.

Wang, L., Zhao, M., Li, S., Erasquin, U.J., et al. 2014. *ACS Applied Materials and Interfaces*. DOI: 10.1021/am501309d.

Wang, L., Wu, S., Cao, G., Fan, Y., et al. 2019b. *Journal of Materials Chemistry B*. DOI: 10.1039/c9tb01539f.

Wang, L., Wang, C., Wu, S., Fan, Y., et al. 2020a. DOI: 10.1039/d0bm00269k.

Wang, W.Y., Lin, D., Jarman, E.H., Polacheck, W.J., et al. 2020b. *Lab on a Chip*. DOI: 10.1039/c9lc01170f.

Wang, X., Sun, Q., & Pei, J. 2018b. DOI: 10.3390/mi9100493.

Wang, X., Zhao, Z., Zhang, H., Hou, J., et al. 2018c. *Experimental and Therapeutic Medicine*. DOI: 10.3892/etm.2018.6844.

Wang, Z., Wang, Z., Lu, W.W., Zhen, W., et al. 2017. DOI: 10.1038/am.2017.171.

Watt, F.M. 2016. DOI: 10.1016/j.devcel.2016.08.010.

Watt, F.M., & Huck, W.T.S. 2013. *Nature Reviews Molecular Cell Biology*. DOI: 10.1038/nrm3620.

Wheeldon, I., Ahari, A.F., & Khademhosseini, A. 2010. *JALA - Journal of the Association for Laboratory Automation*. DOI: 10.1016/j.jala.2010.05.003.

Williams, J., Gilchrist, M., Strain, D., Fraser, D., et al. 2020. DOI: 10.1111/micc.12613.

Wolf, F.I., & Cittadini, A. 2003. DOI: 10.1016/S0098-2997(02)00087-0.

Wong, L., Kumar, A., Gabela-Zuniga, B., Chua, J., et al. 2019. *Acta Biomaterialia*. DOI: 10.1016/j.actbio.2019.07.030.

Wu, J., & Hong, Y. 2016. DOI: 10.1016/j.bioactmat.2016.07.001.

Wu, J., Mao, Z., Tan, H., Han, L., et al. 2012. DOI: 10.1098/rsfs.2011.0124.

Xia, T., Liu, W., & Yang, L. 2017. DOI: 10.1002/jbm.a.36034.

Xie, J., Zhang, D., Zhou, C., Yuan, Q., et al. 2018. *Acta Biomaterialia*. DOI: 10.1016/j.actbio.2018.08.018.

Xing, Q., Yates, K., Vogt, C., Qian, Z., et al. 2014. *Scientific Reports*. DOI: 10.1038/srep04706.

Xiong, B.J., Tan, Q.W., Chen, Y.J., Zhang, Y., et al. 2018. *Aesthetic Plastic Surgery*. DOI: 10.1007/s00266-017-1062-1.

Xiong, S., Zhang, X., Lu, P., Wu, Y., et al. 2017. *Scientific Reports*. DOI: 10.1038/s41598-017-04149-y.

Xu, R., Ma, S., Lin, P., Yu, B., et al. 2018. *ACS Applied Materials and Interfaces*. DOI: 10.1021/acsami.7b04290.

Xu, Z., Min, X.S., Dong, Z.M., Li, Y.W., et al. 2003. *Cell Research*. DOI: 10.1038/sj.cr.7290179.

Yamamoto, M., Ikada, Y., & Tabata, Y. 2001. *Journal of Biomaterials Science, Polymer Edition*. DOI: 10.1163/156856201744461.

Yan, C., Sun, J., & Ding, J. 2011. *Biomaterials*. DOI: 10.1016/j.biomaterials.2011.01.078.

Yañez-Soto, B., Liliensiek, S.J., Murphy, C.J., & Nealey, P.F. 2013. *Journal of Biomedical Materials Research - Part A*. DOI: 10.1002/jbm.a.34412.

Yang, C., Yin, T., & Suo, Z. 2019. *Journal of the Mechanics and Physics of Solids*. DOI: 10.1016/j.jmps.2019.06.018.

Yon, S.H.H., Cha, W.I., Ikada, Y., Kita, M., et al. 1994. *Journal of Biomaterials Science, Polymer Edition*. DOI: 10.1163/156856294X00103.

Yoshida, R., & Okano, T. 2010. *Biomedical Applications of Hydrogels Handbook*. DOI: 10.1007/978-1-4419-5919-5_2.
Young, A.T., White, O.C., & Daniele, M.A. 2020. *Macromolecular Bioscience*. DOI: 10.1002/mabi.202000183.
Yu, J., Du, K.T., Fang, Q., Gu, Y., et al. 2010. *Biomaterials*. DOI: 10.1016/j.biomaterials.2010.05.078.
Yu, S.M., Li, Y., & Kim, D. 2011. DOI: 10.1039/c1sm05329a.
Yue, B. 2014. DOI: 10.1097/IJG.0000000000000108.
Yun, Y.R., Won, J.E., Jeon, E., Lee, S., et al. 2010. DOI: 10.4061/2010/218142.
Zahid, A.A., Ahmed, R., Raza Ur Rehman, S., Augustine, R., et al. 2019. *International Journal of Biological Macromolecules*. DOI: 10.1016/j.ijbiomac.2019.06.136.
Zandi, N., Sani, E.S., Mostafavi, E., Ibrahim, D.M., et al. 2021. *Biomaterials*. DOI: 10.1016/j.biomaterials.2020.120476.
Zant, E., & Grijpma, D.W. 2016. *Biomacromolecules*. DOI: 10.1021/acs.biomac.5b01721.
Zhai, W., Lu, H., Chen, L., Lin, X., et al. 2012. *Acta Biomaterialia*. DOI: 10.1016/j.actbio.2011.09.008.
Zhang, C., Dong, P., Bai, Y., & Quan, D. 2020. *European Polymer Journal*. DOI: 10.1016/j.eurpolymj.2020.109647.
Zhang, X., Sun, C., & Fang, N. 2004. *Journal of Nanoparticle Research*. DOI: 10.1023/B:NANO.0000023232.03654.40.
Zhao, D., Xue, C., Li, Q., Liu, M., et al. 2018. *Journal of Cellular Physiology*. DOI: 10.1002/jcp.26189.
Zhao, G., Zinger, O., Schwartz, Z., Wieland, M., et al. 2006. *Clinical Oral Implants Research*. DOI: 10.1111/j.1600-0501.2005.01195.x.
Zhu, J. 2010. DOI: 10.1016/j.biomaterials.2010.02.044.
Zhu, J., & Marchant, R.E. 2011. DOI: 10.1586/erd.11.27.
Zhu, L., Luo, D., & Liu, Y. 2020. DOI: 10.1038/s41368-020-0073-y.
Zhu, Z., Wang, Y.-M., Yang, J., & Luo, X.-S. 2017. *Plastic and Aesthetic Research*. DOI: 10.20517/2347-9264.2017.71.
Zisch, A.H., Schenk, U., Schense, J.C., Sakiyama-Elbert, S.E., et al. 2001. *Journal of Controlled Release*. DOI: 10.1016/S0168-3659(01)00266-8.

15

Nanopharmaceuticals in Alveolar Bone and Periodontal Regeneration

Mark A. Reynolds
University of Maryland School of Dentistry, Baltimore, USA

Zeqing Zhao
University of Maryland School of Dentistry, Baltimore, USA
Capital Medical University, Beijing, China

Michael D. Weir
University of Maryland School of Dentistry, Baltimore, USA

Tao Ma
University of Maryland School of Dentistry, Baltimore, USA

Jin Liu
University of Maryland School of Dentistry, Baltimore, USA
Key Laboratory of Shanxi Province for Craniofacial Precision Medicine Research, College of Stomatology, Xi'an Jiaotong University, China

Hockin H. K. Xu
University of Maryland School of Dentistry, Baltimore, USA
Member, Marlene and Stewart Greenebaum Cancer Center, University of Maryland School of Medicine, Baltimore, USA
Center for Stem Cell Biology & Regenerative Medicine, University of Maryland School of Medicine, Baltimore, USA

Abraham Schneider
University of Maryland School of Dentistry, Baltimore, USA
Member, Marlene and Stewart Greenebaum Cancer Center, University of Maryland School of Medicine, Baltimore, USA

CONTENTS

DOI: 10.1201/9781003153504-15

Introduction

The need for predictable and robust therapeutic strategies to achieve alveolar bone, periodontal, and craniofacial regeneration have continued to increase with advancing life expectancy. This chapter focuses on frontiers in tissue engineering and regenerative medicine, including novel developments in nanotechnology and nanostructured biomaterials for the delivery of pharmaceuticals and stem cells in alveolar bone and periodontal regeneration.

The periodontium is comprised of alveolar bone, cementum, periodontal ligament (PDL), and gingiva (Bottino et al. 2012; Sowmya et al. 2013). Cementum and alveolar bone are mineralised tissues. PDL is a fibrous tissue that attaches the root cementum of a tooth to the host alveolar bone (Liu et al. 2019). Periodontal disease is initiated by pathogenic bacteria, which triggers an inflammatory response. Inflammation of the gingiva without clinical evidence of breakdown of the periodontium is considered reversible and characteristic of gingivitis. Periodontitis, however, involves an irreversible breakdown of the connective tissue attachment to the root of the tooth and alveolar bone resorption, attributable primarily to the immune and inflammatory response to bacterial pathogens. Progressive periodontal destruction results in tooth mobility (loose teeth) and tooth loss. In nearly 50% of adults, the host response to oral bacteria leads to periodontitis, with progressive destruction of tooth-supporting apparatus. Severe periodontitis is relatively prevalent, affecting as many as 8–15% of the entire global population (Frencken et al. 2017). Moreover, alveolar bone loss and periodontal defects due to congenital birth defects, traumatic injury, tumours, and other infectious conditions may lead to the need for alveolar bone reconstruction, periodontal regeneration, or both. Indeed, alveolar bone defects have been associated with a decrease in the health and quality of life for millions of people (Bottino et al. 2012).

Therefore, extensive research has been carried out on alveolar bone and periodontal tissue engineering to regenerate the hierarchical and complex anatomy of the periodontium, including new alveolar bone, PDL, and cementum (Bosshardt and Sculean 2010). Currently, however, the development of therapeutic strategies to achieve predictable and robust alveolar bone and periodontal regeneration has remained elusive (Polimeni et al. 2010). Conventional debridement of periodontal defects with open flap debridement results in only limited, if any, new alveolar bone, cementum, and PDL formation (Illueca et al. 2006; Rosling et al. 1976).

Stem cells and tissue engineering approaches are highly attractive for alveolar bone and periodontal regeneration (Shimauchi et al. 2013). Biomaterials, active agents (e.g., growth factors), and stem cells represent important strategies in regenerative medicine. Nanotechnology, such as nanostructured biomaterials, have shown promise in providing excellent capabilities of being biocompatible, biodegradable and osteoinductive, in order to enhance stem and mesenchymal cell attachment, proliferation, and differentiation (Bottino et al. 2011; Liu et al. 2018). In addition, bioactive factors, other proteins, and pharmaceuticals can be added into biomaterials (Meirelles et al. 2010). For example, platelet-rich growth factor (PDGF), bone morphogenetic proteins (BMPs), metformin, human platelet lysate (HPL), and enamel matrix derivatives have demonstrated potential to promote periodontal wound healing and regeneration (Lyngstadaas et al. 2009).

This chapter focuses on emerging research investigating novel nanotechnology-based scaffolds and biomaterial carriers for cell and drug delivery, including nanoparticles, nanogels, nanofibres, and nanofillers, to potentiate alveolar bone and periodontal regeneration. We also present recent evidence highlighting the development of nanoapatite scaffolds that mimic minerals in natural bone for the targeted delivery of stem cells in combination with bioactive agents, such as metformin, to stimulate alveolar bone and periodontal regeneration. Furthermore, this chapter includes compelling results supporting the use of bioactive resins containing calcium phosphate and calcium fluoride nanoparticles for the promotion of tissue regeneration with stem cells harvested in the periodontal ligament and dental pulp.

Harvesting Human Periodontal Ligament Stem Cells (hPDLSCs) from Extracted Teeth for Osteogenic, Fibrogenic, and Cementogenic Differentiation

The root surfaces of healthy human teeth following extraction provide a significant source of hPDLSCs. The PDL tissues contain various cell types including fibroblasts, cementoblasts, osteoblasts, endothelial cells, and epithelial cells (Lekic et al. 2010). Regeneration of the periodontal attachment apparatus, including new bone, cementum, and periodontal ligament, involves highly coordinated osteogenesis, cementogenesis, and fibrogenesis because a new fibrous attachment is dependent on formation of new bone and cementum (Zeichner-David 2006).

The multipotent capacity of hPDLSCs has been demonstrated when being cultured under special induction media conditions, resulting in cellular differentiation towards adipogenic, osteogenic, chondrogenic, and myofibroblastic phenotypes (Seo et al. 2004; Zhu and Liang 2015). Interestingly, hPDLSCs showed cementogenicability when induced with vitamin C (Gauthier et al. 2017). In contrast, osteogenic stimulus containing dexamethasone and β-glycerophosphate induced hPDLSCs to osteogenic, but not cementogenic, differentiation (Gauthier et al. 2017). Furthermore, in preclinical studies using animal models, hPDLSCs transplanted into various periodontal defects led to the regeneration of the complex cementum-PDL-bone-like periodontal structures (Seo et al. 2004; Tsumanuma et al. 2016; Xie and Liu 2012; Zhang et al. 2016).

In addition, a recent study showed for the first time that hPDLSCs harvested from the same teeth differentiated to osteogenic, fibrogenic, and cementogenic types of cells (Liu et al. 2020). These promising data demonstrate the potential of hPDLSCs to be induced in a periodontal defect to regenerate all three supporting tissues of the periodontium: alveolar bone, PDL, and cementum. The following were main findings of that study (Liu et al. 2020). (1) hPDLSCs were derived from healthy human premolars (Figures 15.1A and B). (2) hPDLSCs were cultured in osteogenic medium, had high expression levels of bone-related genes: runt-related transcription factor 2 (RUNX2), alkaline phosphatase (ALP), osteopontin (OPN), and collagen1 (COL1). In addition, the hPDLSCs synthesised bone minerals (Figures 15.1C and D). (3) hPDLSCs cultured in a connective tissue growth factor medium showed elevated PDL fibrogenic gene expressions of COL1, COL3, fibroblast-specific protein-1 (FSP-1), periodontal ligament-associated protein-1 (PLAP-1), and elastin. Samples were treated using fluorescent antibodies against F-actin and fibronectin and produced PDL fibres (Figures 15.1E and F). (4) hPDLSCs cultured in cementogenic medium yielded mineral cementum and elevated levels of cementum types of genes, including cementum attachment protein (CAP), CEMP1, and bone sialoprotein (Figures 15.1G and H) (Liu et al. 2020).

These studies underscore the potential of using hPDLSCs for periodontal regeneration. Healthy teeth are extracted for various reasons, including partially erupted or impacted third molars (aka. wisdom teeth) and orthodontic crowding. Therefore, there is an ongoing availability of extracted teeth from which hPDLSCs can be obtained, without performing an additional secondary medical procedure, such as in the case of harvesting bone marrow stem cells. The use of hPDLSCs also avoids the controversy associated with the use of embryonic and foetal stem cells. Hence, exciting research is underway to investigate the use of hPDLSCs for alveolar bone and periodontal tissue regeneration. Further studies are needed to investigate the formation of the cementum-PDL-bone structure in periodontal defects in animal models.

Calcium Phosphate Cement Scaffold and Interactions with Stem Cells for Bone Tissue Engineering

As a water-soluble compound, calcium phosphate cement (CPC) powder can be mixed with an aqueous solution to form an injectable paste for use in dental, periodontal, and maxillofacial applications. Through minimally invasive surgical procedures, the CPC paste could be injected, molded, and sculpted into the desired shapes for aesthetics, and then harden *in situ*. Indeed, the mechanical, physical, and biological properties of CPC scaffolds can be enhanced through the incorporation of absorbable fibres (Zhou et al. 2011), chitosan (Weir and Xu 2010), mannitolporogen (Tang et al. 2012), gas-foaming porogen (Chen

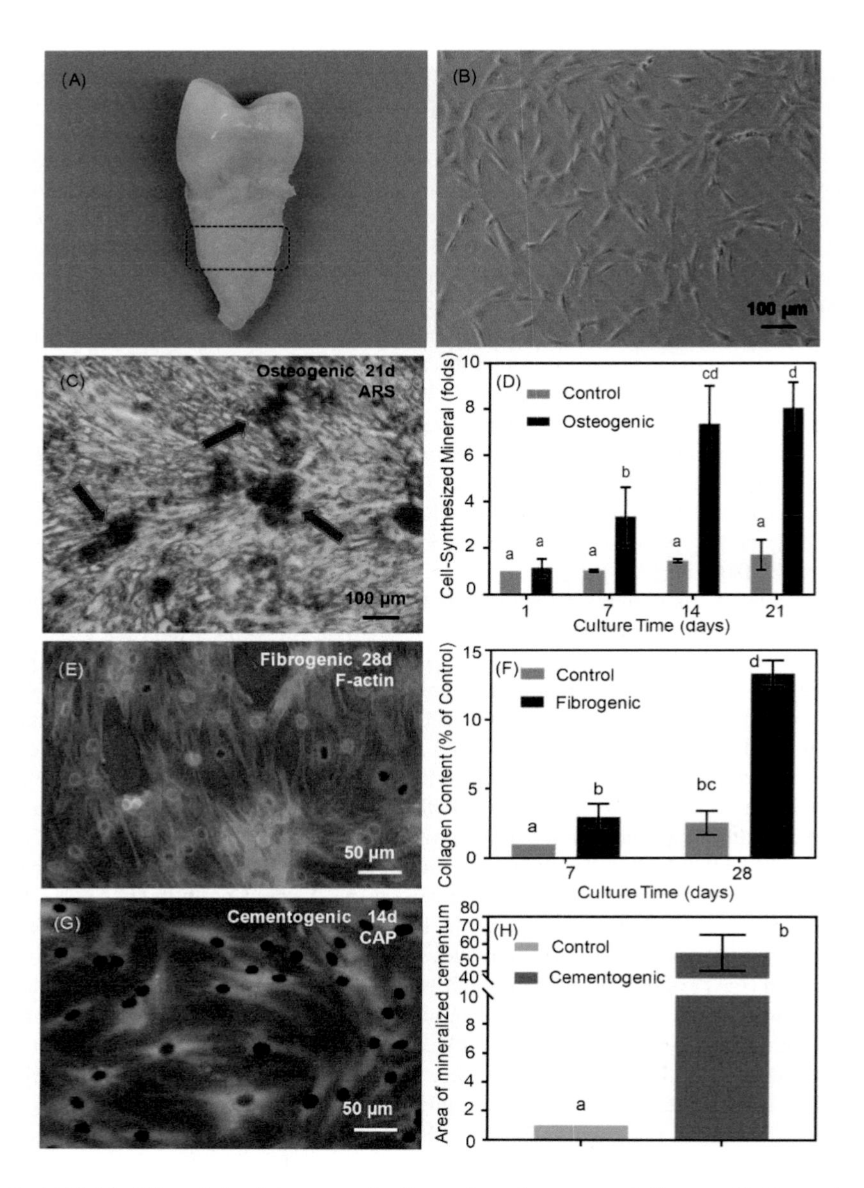

FIGURE 15.1 hPDLSCs differentiation into osteogenic, fibrogenic, and cementogenic lineages. (A) Harvesting hPDLSCs from extracted human premolar. PDL was harvested from the middle third of the root (red rectangle). (B) Colonies were digested to obtain cells which formed spindle shapes at 10 days. (C) hPDLSCs in osteogenic medium produced greater amounts of mineral nodules than in control medium (red staining via ARS). (D) Mineral production by cells. (E) Fibrogenic group had more red fluorescence for F-actin than control. (F) Collagen amount produced by hPDLSCs in control and fibrogenic medium. (G) Higher expression of CAP in hPDLSCs in cementoblastic medium than control. (H) At 21 days, mineral cementum area in the cementogenic group was 50 times that in the control group. (Adapted from (Liu et al. 2020) with permission)

et al. 2012), hydrogel microbeads (Weir and Xu 2010), and biofunctionalisation agents (Thein-Han et al. 2012, Thein-Han et al. 2013). Recent evidence shows that CPC with nanoapatite minerals demonstrated excellent properties for cell delivery, cell attachment, and differentiation, as well as the delivery of growth factors and other bioactive agents (Zhang et al. 2014).

Human umbilical cord mesenchymal stem cells (hUCMSCs) were found to readily attach to a CPC-chitosan-fibre scaffold, with cells anchoring to the fibres and the nanosized CPC-containing apatite minerals (Figure 15.2A). The incorporation of fibres into CPC greatly increased the gene expression levels of ALP (Figure 15.2B), compared to those without fibres (Xu et al. 2010). In addition, osteoblasts, human bone marrow MSCs (hBMSCs), human embryonic stem cell-derived MSCs (hESC-MSCs), and human induced pluripotent stem cell-derived MSCs (hiPSC-MSCs) all showed excellent performance when seeded on and attaching to the bone mineral-mimicking nanoapatite CPC scaffolds (Wang et al. 2014).

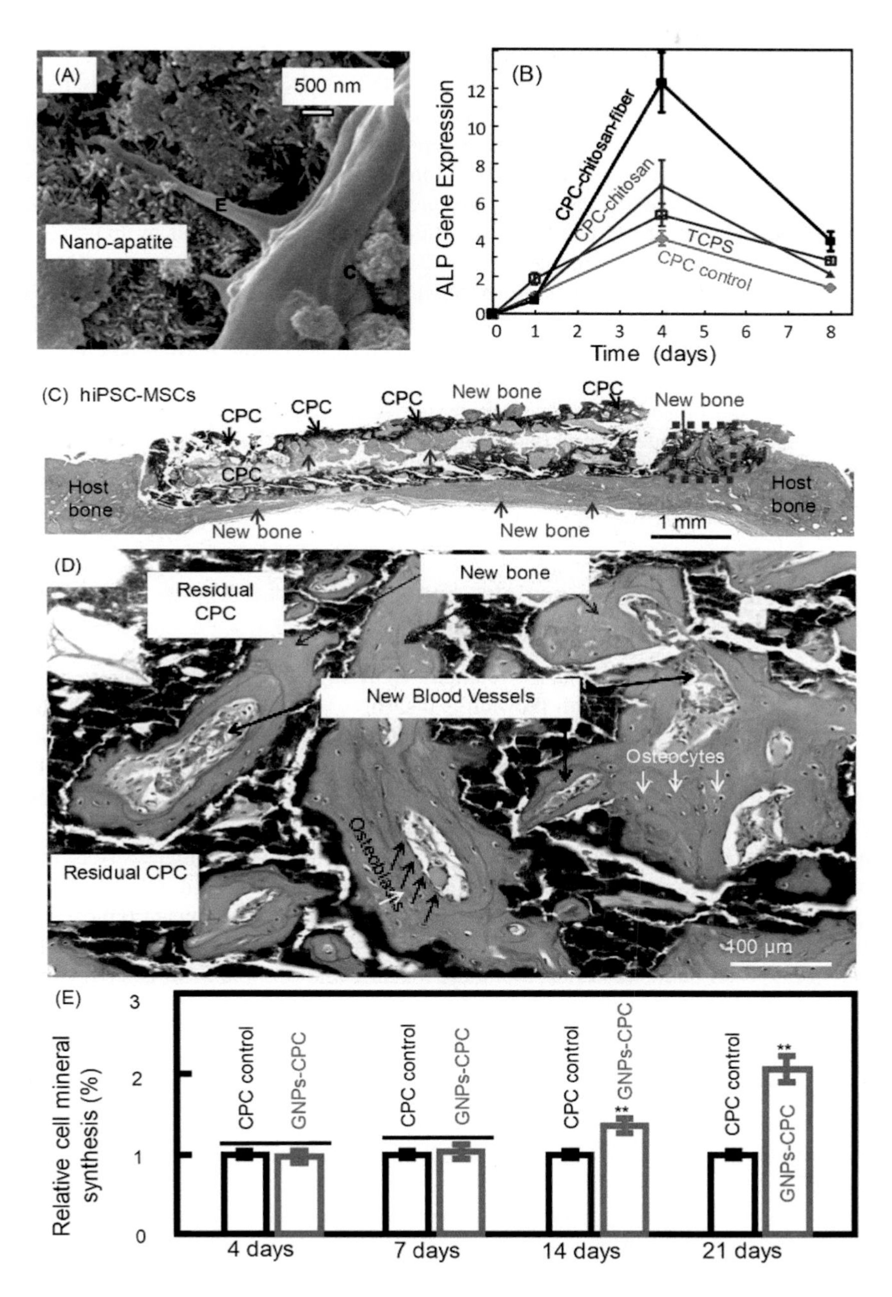

FIGURE 15.2 Nanoapatite CPC for cell delivery. (A) hUCMSC extensions attaching to nanoapatite CPC. (B) Osteogenic differentiation of hUCMSCs on CPC showed high ALP expression. (C) H&E staining of hiPSC-MSC-CPC implanted in critical-sized cranial defects in nude rats. (D) High magnification of an area in (C). (E) Bone mineral synthesis by cells. (A and B were adapted from (Xu et al. 2010), C and D from (Wang et al. 2014), and E from (Xia et al. 2017), all with permissions.)

A recent study (Wang et al. 2014) compared the *in vivo* bone regeneration efficacy of hUCMSCs, hBMSCs, and hiPSC-MSC delivered via a biofunctionalised CPC-chitosan-arginine-glycine-aspartic acid (RGD) scaffold in a critical-sized cranial bone defect model in the nude rat. When delivered using macroporous CPC scaffold, hiPSC-MSCs had excellent osteogenic ability when compared to hUCMSCs and the gold-standard hBMSCs (Figure 15.2C). The new bone percentage in the entire defect at 12 weeks for hiPSC-MSC-CPC constructs reached 30.4%, which was 2.8-fold greater than the 11.0% new bone for CPC control having no cell delivery. The new bone exhibited an organised morphology with osteocytes and new blood vessels as well as having osteoblasts lining the newly formed bone (Figure 15.2D).

Moreover, gold nanoparticles (GNPs) were added in CPC (GNP-CPC scaffold) (Xia et al. 2017). The incorporation of GNPs improved the attachment and viability of human dental pulp stem cells (hDPSCs) on the scaffold. This included a greater cell attachment (2-fold increase in cell spreading) and proliferation, and greater osteogenic differentiation (2-3-fold increase) (Figure 15.2E). GNPs endowed the CPC with a unique nanostructure, thereby increasing the surface characteristics for cell attachment. Moreover, the GNPs were diffused out of the GNP-CPC scaffold and were internalised by hDPSCs, which was demonstrated to enhance cellular function (Xia et al. 2017).

Therefore, nanoapatite CPC constructs are promising biomimetic scaffolds to deliver a wide range of stem cells, to direct cell behaviour and cell fate via incorporation of bioactive agents, and to enhance osteogenic differentiation for alveolar bone and periodontal regeneration. Further preclinical studies are needed to examine CPC performance in alveolar bone and periodontal tissue repair in animal models delivering stem cells and bioactive agents. Current studies are underway to deliver stem cells and antibacterial agents via CPC to achieve these two needed functions: (1) reduce bacterial biofilms and control infections and (2) simultaneously promote stem cell viability and regenerative capability.

Dental Resins Containing Nanoparticles and Antibacterial Agents to Inhibit Oral pathogens/Infection and Strengthen Tooth Roots

An increasingly greater number of adults are retaining teeth throughout the life span, in part, due to the successful implementation of oral health preventive measures including improved oral hygiene and the availability of fluoridated drinking water. Nevertheless, the severity of periodontitis often increases with age, in part, because of periodontal pocket formation and compositional shifts bacterial biofilms. With tooth root exposure and increased incidence of root caries, dental restorations referred to as Class V, are often indicated in these patients. However, Class V restorations with subgingival margins can prevent good oral hygiene, thus facilitating the formation of periodontitis types of pathogens (Hellyer et al. 1990). This in turn may lead to periodontal attachment loss and progression of periodontitis.

Currently, commercial resin-based filling materials for Class V restorations are typically not antibacterial. Quite the contrary, commercial dental resins were shown to grow greater amounts of biofilms and plaque than other types of dental materials including metallic and ceramic materials (Beyth et al. 2007). Using such a resin for Class V restorations could aggravate the initiation and progression of periodontitis. Furthermore, recurrent caries at the interface between the tooth structure and the restoration is still a major cause for dental restoration failures (Deligeorgi et al. 2001).

To address these problems, filler particles were added to bonding agents to increase the load-bearing ability of the adhesive resin, minimise the shrinkage due to polymerisation, and render a stronger resin-tooth interfacial bond (Belli et al. 2014; Hikita et al. 2007). Moreover, to make resins that have remineralisation ability, calcium phosphate fillers were added to the resins. The release of calcium (Ca) and phosphate (P) ions favoured hydroxyapatite (HA) precipitation and caused dental lesion remineralisation (Dunn 2007). Nanoparticles of amorphous calcium phosphate (NACP) were made using a spray-drying facility and loaded into dental resins (Xu et al. 2006). Notably, the NACP composite provided the releases of Ca and P ions that matched conventional CaP-resin composites; however, the NACP composite exhibited much greater load-bearing properties (Xu et al. 2010).

Furthermore, antibacterial dimethylaminododecyl methacrylate was synthesised and combined with nanoparticles of silver (NAg) in a bonding agent that also had NACP (Zhang et al. 2013). Because of the fine size of NACP and NAg, these nanoparticles easily infiltrated into the tubules in dentine and produced resin tags (Figure 15.3A). The latter property could eliminate residual bacteria in the dentinal tubules and remineralise carious lesions (Figure 15.3B) (Cheng et al. 2013; Zhang et al. 2013; Zhang et al. 2017). It was shown that after immersion in water for half a year, the material maintained its potent antibacterial activity, with no loss in biofilm-inhibition power, than that at 1 day (Zhang et al. 2013). Moreover, adding NACP, methacryloyloxyethylphosphorylcholine (MPC) and DMAHDM into the bonding agent did not compromise the dentin bond strength. Live-dead staining images of periodontal biofilms composed of *Streptococcus gordonii, Actinomyces naeslundii, Porphyromonas gingivalis*, and *Fusobacterium*

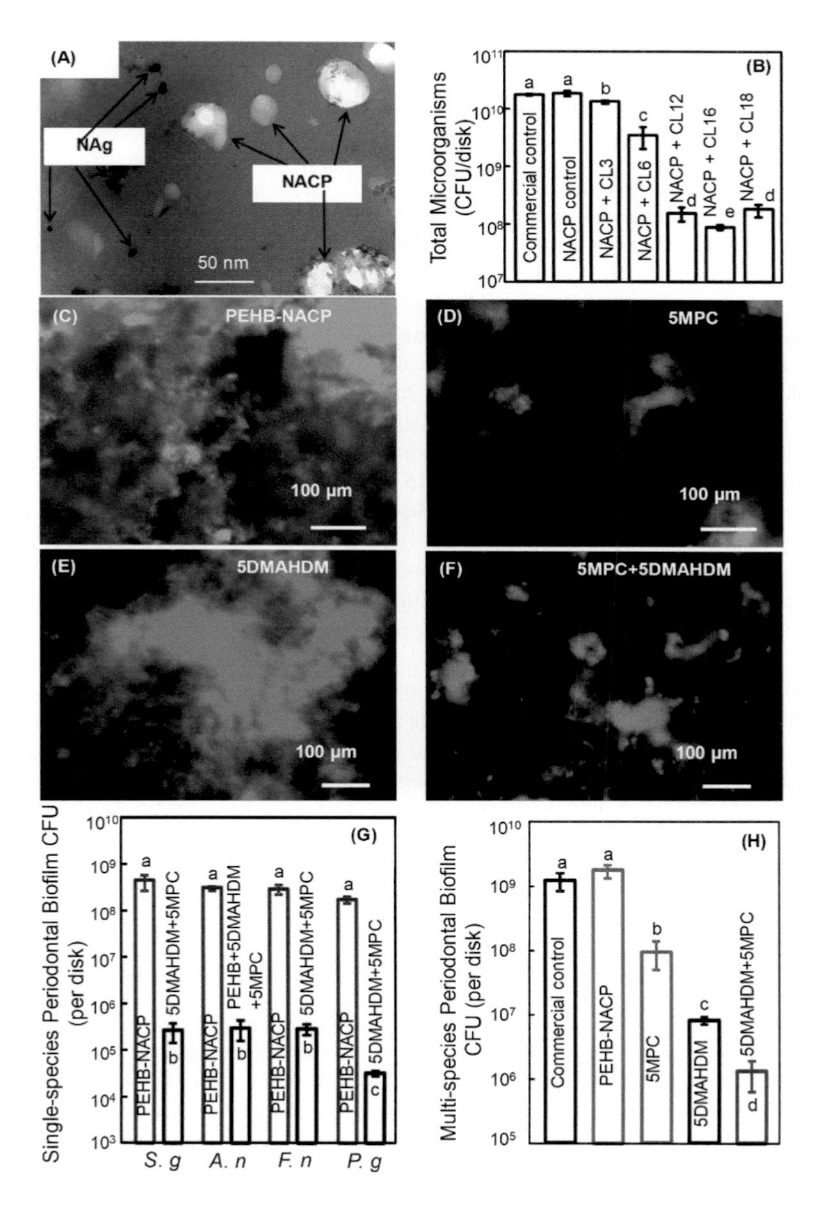

FIGURE 15.3 Dental resins to remineralise lesions and inhibit periodontal biofilms. (A) TEM indicating NAg and NACP in resin tag in human dentin. (B) Dental plaque microcosm 2-day biofilms on composites. (C–F) Live/dead images of 3-day periodontal biofilms on four adhesive resins. (I) CFU of periodontal biofilms. (J) CFU of four single-species biofilms on resins. (A and B were adapted from (Zhang et al. 2013) and C to H from (Wang et al. 2017), all with permissions.)

nucleatum are shown in Figures 15.3C–F. Indeed, biofilm colony-forming units (CFU) were reduced by 3–4 log via DMAHDM and MPC, and the metabolic activities and polysaccharide production of the biofilms showed great reductions as well (Figures 15.3G and H) (Wang et al. 2017).

These results show that novel nanocomposites inhibited oral biofilms harbouring periodontal pathogens, while NACP remineralised lesions and strengthen tooth structures. Hence, these nanocomposites exhibit the capacity to reduce disease-inducing bacterial pathogens (oral acid-producing biofilms and periodontal pathogens) and promote hard-tissue repair (remineralising tooth structures via NACP) via the same nanostructured dental resin; however, more investigations are warranted to determine these properties *in vivo*.

Resin Containing Calcium Fluoride Nanoparticles (Nano-CaF_2) to Promote Periodontal Regeneration

In addition to NACP, Nag, and DMAHDM, recent studies also have focused on the incorporation of novel nano-CaF_2 into dental resins (Weir et al. 2012, Xu et al. 2008, Xu et al. 2010). Nano-CaF_2 were developed using a spray-drying machine, which yielded particles with a diameter of approximately 50 nm (Figure 15.4A) (Dai et al. 2021). The nano-CaF_2 composite released fluoride (F) and Ca ions for the formation of fluoroapatite and suppression of caries (Sun and Chow 2008). In addition, the nano-CaF_2 composite was 'smart' with respect to ion release, as it released few ions at a neutral pH, but could raise the levels of F and Ca ion release substantially at a caries-inducing low pH, when such ions are especially beneficial to deter the caries progression (Xu et al. 2010).

A nanocomposite containing nano-$CaF_2$was recently formulated to deliver hPDLSCs. For the first time, this nano-CaF_2 composite was shown to support hPDLSC viability and proliferation as well asosteogenic and cementogenic differentiation (Figures 15.4B–G) (Liu et al. 2020). The following are the key findings of this study. (1) The nano-CaF_2 composite had mechanical properties that exceeded those of a commercially marketed (control) composite. (2) High levels of Ca and F ion release from nano-CaF_2 composite were achieved. (3) The nano-CaF_2 composite substantially increased the hPDLSC expressions of osteogenic and cementogenic genes, induced a high level of ALP activity, and promoted bone mineralisation viahPDLSCs that was nearly 100% greater than that without nano-CaF_2 (Figures 15.4B–G) (Liu et al. 2020).

Therefore, the potential application of novel nano-CaF_2 composite resin in Class V restorations adjacent to the periodontal tissues possess two promising therapeutic functions: (1) repair of the damaged surface and release F and Ca ions to remineralise dental lesions and prevent recurrent caries and (2) increase hPDLSC function, promote the osteogenic and cementogenic differentiation, and enhance the periodontal repair and regeneration. *In vivo* studies are needed to validate these favourable properties.

The following section examines the potential role of other types of nanocomposite scaffolds, fibres, and membranes for delivering hPDLSCs.

Nanostructured Scaffolds with hPDLSCs for Periodontal Regeneration

In addition to nanoapatite and resin materials, previous studies also developed other types of novel nanocomposite scaffolds, fibres, and membranes, which created three-dimensional (3D) environments to guide cell adhesion, proliferation, and differentiation to regenerate the periodontal tissues (Liu et al. 2018). Among them, nanocomposite scaffolds based on electrospun nanofibres have received research activity because of their capability to emulate natural extracellular matrix to influence cell viability, adhesion, and organisation (Lu et al. 2018). Important properties of nanofibrous scaffolds include a greater ratio of surface area/volume as well as the promotion of protein absorption, cell reaction, triggering of special osteogenic gene expressions, and cell–cell signals (Jia et al. 2020). Electrospun nanofibres observed by scanning electron microscopy (SEM) indicated the presence of appropriate microstructures suitable for cellular interactions (Figure 15.5A). In addition, *in vivo* experiments showed a relatively homogeneous

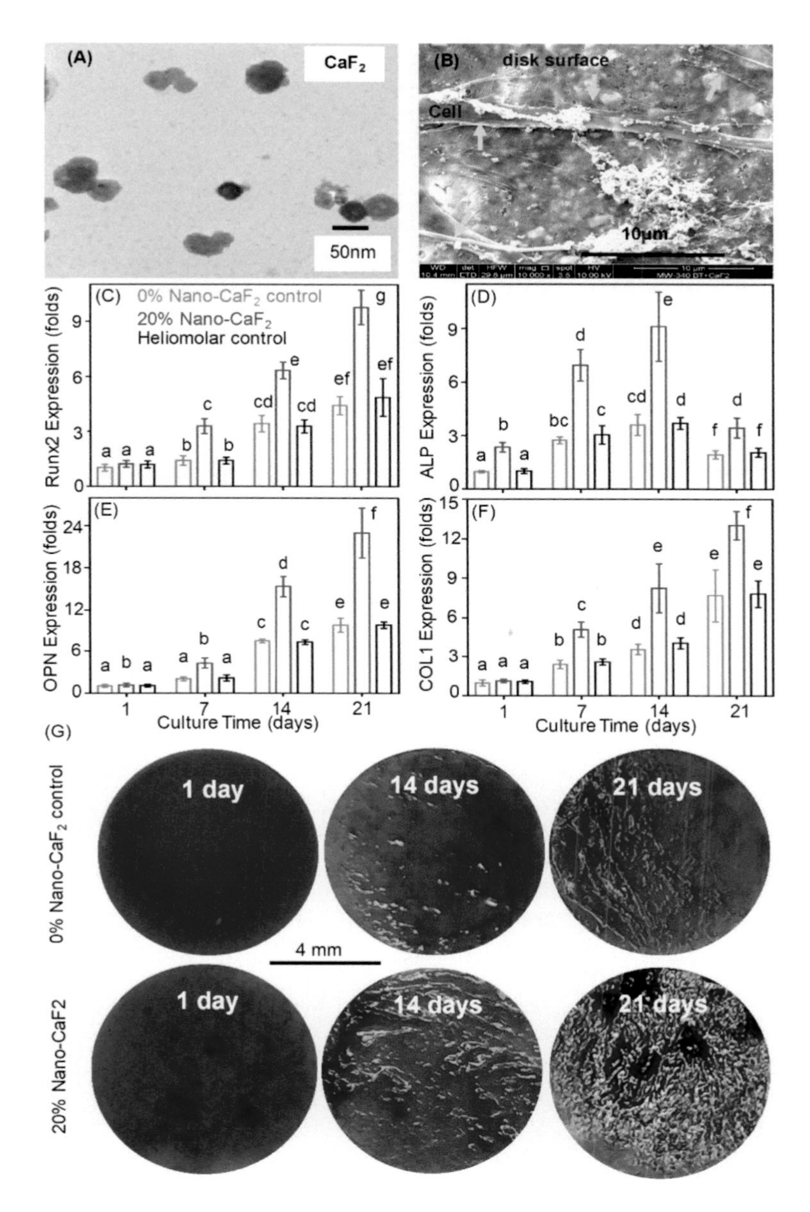

FIGURE 15.4 Nano-CaF_2 for periodontal regeneration. (A) TEM image of nano-CaF_2. (B) SEM of hPDLSCs attaching to the nano-CaF_2 composite. (C–F) Expression of osteogenic genes of hPDLSCs on composite. (G) ARS images of hPDLSC-synthesised bone mineral nodules (stained red). (A was adapted from (Dai et al. 2021) and B to G from (Liu et al. 2020), all with permissions.)

bone and cartilage structure with almost no visible residual fibres and membranes which were resorbed, while new tissues were regenerated (Figure 15.5B).

Another study reported a three-dimensional scaffold with multiple layers (Jiang, Li, Zhang, Huang, Jing and Wu 2015). Biodegradable poly (ε-caprolactone)-poly (ethylene glycol) copolymer electrospun nanofibrous mats were embedded into porous chitosan (CHI) to yield topographic signals to induce periodontal tissue formation in an organised manner (Figure 15.5C) (Jiang et al. 2015). The scaffolds were designed to contain cross-sectioned sections and put against the exposed root surfaces. Then, a bovine-derived scaffold with porosity (Bio-Oss) was implanted in the alveolar bone defect (Jiang et al.

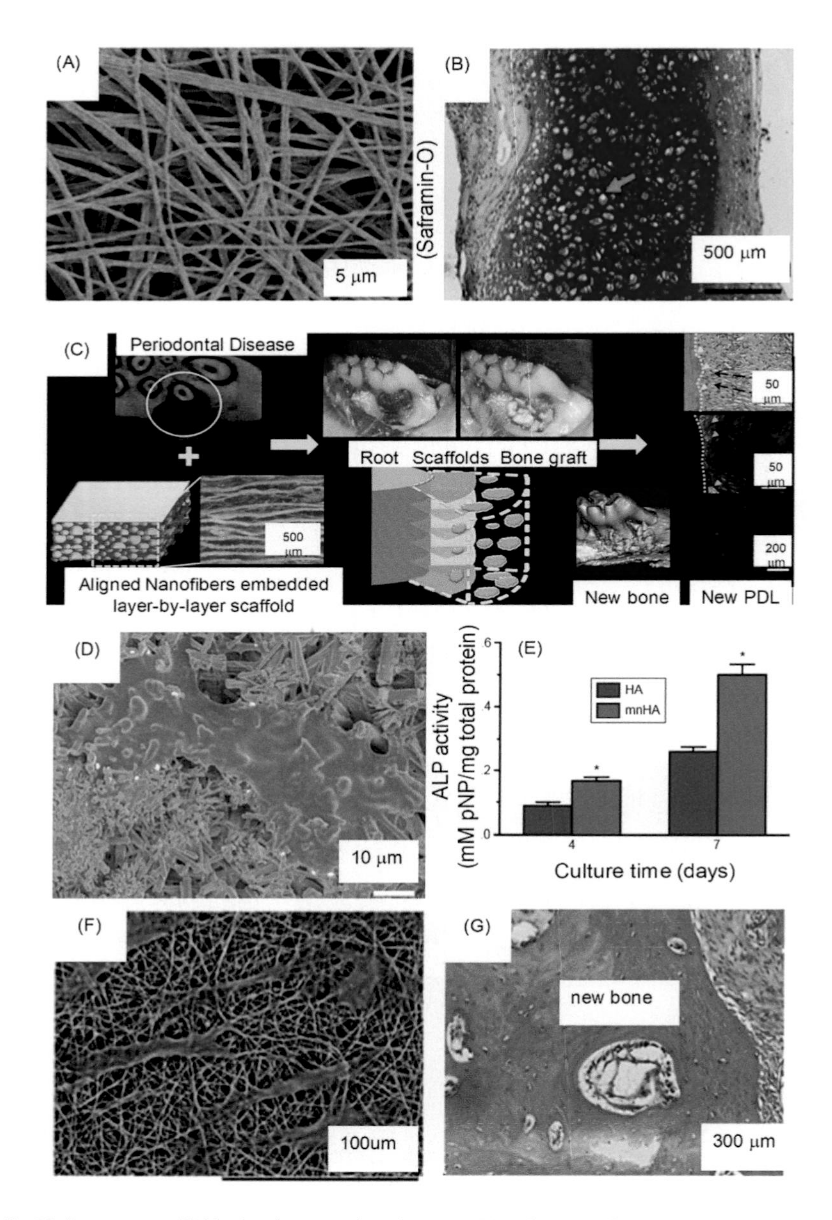

FIGURE 15.5 Various nanoscaffolds for tissue engineering. (A) SEM images of membranes. (B) *In vivo* neocartilage evaluation at 8 weeks (safranin-O staining): homogeneous cartilage-specific extracellular matrix distribution and residual electrospun membranes. (C) Process of nanofibre-embedding scaffold for ordered PDL generation. (D) SEM of hPDLSCs on mnHA after 6 h of seeding. (E) ALP activity of hPDLSCs on mnHA for 4 and 7 days. (F) SEM of HPDLCs at 1 day. (G) Mesiodistal H&E staining of a class II furcation defect at 60 d using Col/BG/CS membrane. (A and B were adapted from (Jia et al. 2020), C from (Jiang et al. 2015), D and E from (Mao et al. 2015), and F and G from (Zhou et al. 2017), all with permissions.)

2015). *In vitro*, compared with random group and porous control, the scaffold with nanofibre embedding yielded the oriented and elongated cells with greater cell infiltration. This was accompanied by greater cell viability and higher expressions of periodontal ligament genes. In an *in vivo* rat model, the scaffold with nanofibre embedding had greater organisation of PDL tissue formation that was perpendicular to the root surface while also producing a greater amount of mature collagen fibres when compared to a random group and porous control (Jiang et al. 2015). Moreover, greater collagen I/collagen III ratio, greater expression level of periostin, and greater amounts of tooth-supporting mineralised tissue were achieved in the new periodontium of the aligned scaffold group (Zhuang et al. 2019). The addition of the aligned nanofibres into porous CHI was shown to increase the viability and differentiation of BMSCs, producing greater amounts of an organised PDL. This system has great potential to be useful in tissue engineering of a wide range of human tissues with specialised orientations (Zhuang et al. 2019).

In addition, bioactive agents, including bioactive ceramics and functionalised polymeric materials, could be mixed with nanofibres to induce the physical-chemical-biological properties and increase the tissue engineering potential (Masoudi Rad et al. 2017; Yang et al. 2015). In addition, bioactive factors, other proteins, and pharmaceuticals could also be incorporated to properly induce cell behaviour and control the inflammatory microenvironment at a local site, which could enhance periodontal tissue regeneration and performance (Furtos et al. 2017; Tang et al. 2016).

Furthermore, HA having a micro-nano-hybrid surface (mnHA) was developed using a hydrothermal reaction, where α-tricalcium phosphate particles served as precursors in an aqueous solution (Mao et al. 2015). The influence of mnHA on the proliferation and cementogenic and osteogenic differentiation of PDLSC were investigated (Mao et al. 2015). The scaffold contained nanorods and microrods. The nanorod diameters ranged from 80 to 120 nm. The microrod diameters ranged from 1 to 4 μm. PDLSCs on mnHA exhibited a fibroblast-like appearance (Figure 15.5D). The mnHA bioceramics promoted cell proliferation, increased ALP activity (Figure 15.5E), and upregulated the expression of osteogenic and cementogenic markers. The studies showed that the mnHA scaffold had a good potential as a graft material for bone and cementum repairs in periodontal regeneration.

In another study, a biomimetic electrospun fish collagen/bioactiveglass/chitosan (Col/BG/CS) composite nanofibre membrane was developed for periodontal regeneration (Zhou et al. 2017). The morphology of the membrane was examined using SEM, with a fibre diameter of 159 ± 59 nm. hPDLCs were able to attach firmly and spread well on the membrane (Figure 15.5F). This nanomembrane exhibited a biomimetic structure with a high hydrophilicity and a relatively high tensile strength. The nanomembrane promoted cell growth with high expressions of osteogenic genes for hPDLSCs, including RUNX2 and OPN. This nanocomposite membrane was found to promote PDL and bone formation in a canine class II (Glickman's) furcation defect model (Zhou et al. 2017). Compared to the control group, the composite membrane formed greater amounts of new bone with less inflammation (Figure 15.5G). New bone formation percentage by Col/BG/CS group was 69%, higher than the control group's 45%, at 60 days after treatment. The novel nanofibre membrane exhibited more macroporosity and greater surface area to induce cell–cell and cell–matrix interactions, thus promoting osteogenesis and periodontal regeneration.

Novel Delivery of Metformin in Nanoapatite Scaffolds for Stem Cell-Based Bone and Periodontal Regeneration

Several reports have recently proposed the potential application of the pharmaceutical agent metformin (1,1-dimethylbiguanide hydrochloride) in bone and periodontal regeneration. Metformin, a member of the biguanide family, was approved by the United States Food and Drug Administration (FDA) in 1995 to be prescribed for the treatment of type 2 diabetes (Tanja and Langlass 1995). Currently, metformin is the most widely used oral anti-diabetic medication worldwide, in part, due to its reliable clinically proven effect on reducing hyperglycaemia, low cost, and an impressive safety record after long-term use (Stein et al. 2013). Early preclinical evidence demonstrated that metformin is capable of inducing osteogenesis and inhibiting adipogenesis through the osteoblastic differentiation of BMSCs and pre-osteoblastic cell lines (Cortizo et al. 2006; Gao et al. 2008; Jang et al. 2011; Kanazawa et al. 2008; Wang et al. 2016).

The cellular and molecular mechanisms underlying the osteogenic effects of metformin have centred on the activation of the AMP-activated protein kinase (AMPK) pathway, a conserved signalling pathway that modulates multiple metabolic pathways and acts as master sensor of cellular energetics (Chen et al. 2017; Jeyabalan et al. 2012; Ma et al. 2018; Molinuevo et al. 2010; Wang et al. 2016; Zhou et al. 2001). In fact, the osteogenic commitment of MSCs has been shown to be dependent on AMPK activation since the expression levels of osteogenic differentiation markers, such as ALP, OCN, and RUNX2, are significantly downregulated in calvarial osteoblasts isolated from $AMPK^{-/-}$ mice (Kanazawa et al. 2018; Kim et al. 2012). These findings underscore the key role played by the AMPK pathway in bone development and imply that metformin, as a potent AMPK activating drug, could be safely repurposed as a new pharmacological enhancer of MSC-based bone regeneration. Currently, metformin is less costly than other regenerative biologics, such as BMPs, making it potentially more cost-effective. Collectively, these features suggest that metformin is a promising, safe, and relatively inexpensive pharmaceutical to potentiate the efficacy of bone tissue engineering.

Compelling evidence demonstrates that metformin in doses ranging from 10 to 100 μM has the capacity to induce the differentiation of human MSCs derived from tissue sources other than the bone marrow, including hiPSC-MSCs, hUCMSCs, hDPSCs, and hPDLSCs (Al Jofi et al. 2018, Houshmand et al. 2018, Jia et al. 2020, Qin et al. 2018, Wang et al. 2018, Wang et al. 2019, Yang et al. 2019, Zhao et al. 2019). As a highly hydrophilic cationic drug, the cellular uptake of metformin is known to be facilitated by a group of polyspecific cell membrane organic cation transporters (OCTs: OCT-1, OCT-2, and OCT-3) belonging to the solute carrier 22A gene family. In addition to metformin, OCTs mediate transport of structurally diverse, small hydrophilic organic cationic endogenous compounds, toxins, and several drugs (Nies et al. 2011; Shu 2011). Noteworthy, the expression of functional OCT-1 in hepatocytes is critical to mediate the therapeutic anti-diabetic action of metformin, as reported by Shu et al., who found that AMPK activation and the glucose-lowering effect of metformin were completely abolished in OCT-1-deficient mice (Shu et al. 2007). Therefore, a significant translationally relevant factor that could impact the uptake of metformin by MSCs, and ultimately osteogenesis, is the expression and function of OCTs. Recent evidence demonstrates that functional OCTs are present in a variety of MSCs harvested from humans (Al Jofi et al. 2018, Qin et al. 2018, Wang et al. 2018). Proper OCT function appears to be a biological prerequisite for the osteogenic action of metformin in MSCs, given that inhibition of OCT by pharmacological or genetic approaches markedly prevented AMPK activation, mineralised nodule formation, and RUNX2 upregulation (Al Jofi et al. 2018).

Relative to the promising application of metformin in alveolar bone and periodontal tissue engineering, a factor that could compromise the local therapeutic action of metformin within an osseous or periodontal defect is rapid drug dilution (Kajbaf et al. 2016). Thus, developing novel strategies to incorporate metformin into biocompatible carriers could lead to a more sustained and significant increase in its local release efficacy. This concept has been recently advanced by formulating CPC scaffolds containing metformin seeded with different MSCs, including hiPSC-MSCs, hDPSCs, and hPDLSCs (Qin et al. 2018, Wang et al. 2018, Zhao et al. 2019). As mentioned previously, CPC contains nanoapatite which mimics the mineral composition present in natural bone (Figure 15.6A) (Jin et al. 2014). Other beneficial attributes include low cost, excellent biocompatibility, MSC adhesion to nanoapatite, optimal mechanical strength, osteoconductivity, and biodegradability (Jin et al. 2014; Xu et al. 2017).

Recent work by Wang et al. demonstrated that treatment of hiPSC-MSCs with metformin supported cell viability, significantly stimulated ALP gene expression and activity, upregulated RUNX2 and osterix (OSX) gene and protein expression levels, and also enhanced mineralised nodule formation (Wang et al. 2018). This study further explored the role of the AMPK pathway in metformin-induced RUNX2 expression. Given that serine/threonine liver kinase B1 (LKB1) is responsible for phosphorylating and activating AMPK in response to energetic stress, the authors transiently overexpressed a LKB1 wild-type (LKB1-WT) or a catalytically inactive LKB1 kinase dead (LKB1-KD) plasmid construct and exposed hiPSC-MSCs to metformin or vehicle control (Shaw et al. 2004). While metformin triggered AMPK activation and increased RUNX2 expression in LKB1-WT cells, these responses were markedly downregulated in LKB1-KD cells. These findings were verified through immunofluorescence staining, where LKB1-WT cells treated with metformin showed a strong RUNX2 nuclear expression, which was undetected in LKB1-KD cells. These data highlight the critical link between the AMPK signalling pathway and the osteogenic action of metformin in OCT-expressing MSCs. To further investigate the use of metformin and stem cells in bone engineering, iPSC-MSCs seeded on CPC scaffolds were treated with metformin for 3 weeks. Remarkably, a 3-fold increase in mineralised nodule formation was observed in metformin-exposed iPSC-MSCs when compared to cells seeded on control CPC scaffolds. Collectively, this work provided the first evidence supporting the potential use of metformin-loaded CPC formulations combined with MSCs as a tissue engineering platform for alveolar bone and periodontal regeneration (Wang et al. 2018).

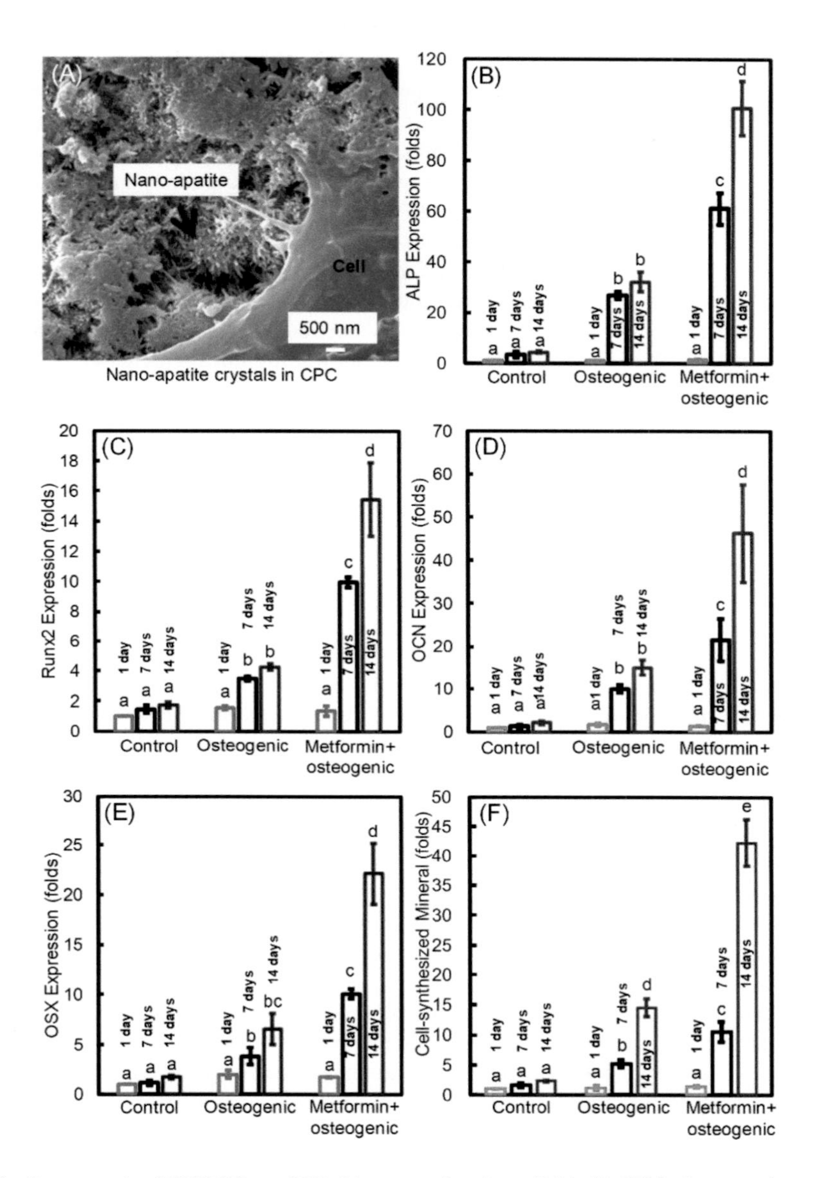

FIGURE 15.6 Osteogenesis of hPDLSCs on CPC-chitosan-metformin scaffold. (A) SEM of nanoapatite crystals in CPC. (B–E) qRT-PCR assay of osteogenic differentiation of hPDLSCs at 1, 7, and 14 days, (B) ALP, (C) Runx2, (D) OCN, and (E) OSX gene expressions. (F) Mineral synthesis by cells. (A was adapted from (Wang et al. 2014), B–F from (Zhao et al. 2019), all with permissions.)

The human dental pulp from healthy extracted teeth offers a readily accessible source of post-natal, autologous MSCs. hDPSCs have inherent proliferative potential and plasticity to differentiate into several cell types, including osteoblasts. Because of these features, hDPSCs have received special attention that justifies their use as a unique, tissue-specific stem cell niche for bone tissue engineering (Anitua et al. 2018; Gronthos et al. 2000; Gronthos et al. 2002). Recent results on the osteogenic responses of hDPSCs seeded on a novel metformin-releasing CPC scaffold reinforced with the natural biopolymer chitosan have been reported (Qin et al. 2018). In the study by Qin et al., CPC scaffolds were initially prepared with three different concentrations of metformin (10, 30, or 50 µg). Based on the initial burst release of [^{14}C]-labelled metformin and its continuous release overtime, the 50-µg concentration was shown to be the most effective formulation. The release profile of metformin reached a sustained rate of ~1,750 ng mL after 7–21 days, or about 11 μM, a dose consistent with the plasma concentration of metformin found in diabetics within 2–4 h after an oral dose of approximately 1000 mg (Bailey and Turner 1996; Jang et al. 2011). Furthermore, these findings provided the initial evidence of continuous metformin release from the CPC-chitosan scaffold over several weeks, offering potential for the sustained stimulation of osteogenic differentiation of hDPSCs during early wound healing. It is likely that the sustained release of metformin from these formulations was related to the distribution of metformin into chitosan, whereas the initial burst release was possibly associated with the binding of metformin to the CPC surface. When hDPSCs were seeded on metformin-CPC-chitosan scaffolds for a maximum of 7 days, cells proliferated and remained viable in a similar manner as cells cultured on CPC-chitosan scaffolds not containing metformin. Interestingly, in comparison to control scaffolds fabricated only with CPC-chitosan, hDPSCs cultured on metformin-CPC-chitosan scaffolds for 14 days showed significant increases in ALP activity and the expression of osteogenic/odontogenic gene markers including OCN, RUNX2, dentin matrix protein-1 (DMP-1), and dentin sialophosphoprotein. In addition, mineralised nodule formation was also markedly enhanced in hDPSCs seeded on metformin-CPC-chitosan scaffolds for 21 days (Qin et al. 2018).

The clinical application of metformin in combination with hPDLSCs for periodontal bone tissue regeneration has been recently addressed by several groups (Jia et al. 2020, Kuang et al. 2020, Yang et al. 2019, Zhang et al. 2019, Zhao et al. 2019). However, only two research groups have so far focused on the potential application of hPDLSCs with metformin in combination with nanobiomaterials (Yang et al. 2019, Zhao et al. 2019). Yang et al. reported that metformin added to hPDLSCs seeded on polydopamine-templated hydroxyapatite substantially increased AMPK activation, reduced reactive oxygen species production and apoptosis, and enhanced cell viability and osteogenic differentiation (Yang et al. 2019). Zhao et al. have also obtained promising results by combining metformin-CPC-chitosan scaffolds with hPDLSCs. In their study, they tested three different conditions: control group (CPC-chitosan scaffold cultured in growth medium); osteogenic group (CPC-chitosan scaffold cultured in osteogenic medium); and metformin+osteogenic group (metformin-CPC-chitosan scaffold cultured in osteogenic medium). The following were the main findings of the study: (1) metformin loaded in CPC-chitosan scaffold greatly increased ALP activity and the expression levels of ALP, OCN, RUNX2, and OSX. As shown in Figures 15.6B–E, the peak values of osteogenic gene expression in the metformin+osteogenic group were 3- to 4-fold higher than the osteogenic group without metformin; (2) At 7 and 14 days, the amount of bone mineral synthesised by hPDLSCs in the metformin+osteogenic group was 2-fold and 3-fold higher, respectively, than observed in the osteogenic group without metformin (Figure 15.6F) (Zhao et al. 2019).

These studies strongly demonstrate the innovative application of bone mineral mimicking formulations as nanobiomaterials capable of sustaining the local release of metformin. Metformin is an effective AMPK pathway activating agent that, when incorporated into MSC-based tissue engineering platforms, may greatly enhance the regeneration of mineralised tissues affected by dental, craniofacial, and orthopaedic conditions. Further pre-clinical testing of this experimental approach appears warranted.

Nanoapatite Scaffold to Deliver Human Platelet Lysate and Stem Cells

Bone regeneration is a complicated process involving a range of growth factors and cytokines (Man et al. 2012). Because the dynamic response of cells to external biologic agents is dependent on many factors, such as dosage, it is critically important to design therapeutic strategies that optimise cellular responses and regenerative outcomes (Santo et al. 2012). Thus, optimal regenerative outcomes appear highly dependent on the controlled and coordinated delivery of proteins, such as growth factors and cytokines, to targeted tissue defects (Santo et al. 2012). Platelets play a key role in wound healing, serving as a repository for proteins critical for tissue repair and regeneration (Santo et al. 2012). The disruption of platelet membranes results in the release HPL, which contains a variety of growth factors, including vascular endothelial growth factor (VEGF), transforming growth factor (TGF), fibroblast growth factor (FGF), insulin-like growth factor (IGF), PDGF, and platelet-derived angiogenesis factor (Albanese et al. 2013, Fréchette et al. 2005, van den Dolder et al. 2006). The growth factors and growth factor concentrations found in the HPL are considered by some to reflect an optimal composition and dosing for promoting tissue regeneration (Babo et al. 2016a). In clinical practice, concentrations of platelets (e.g., platelet-rich plasma) have been widely applied to promote wound healing and regeneration. Therefore, HPL is highly promising for tissue engineering.

Several studies (Babo et al. 2016a, 2016b) incorporated HPL into hyaluronic acid particles, which were blended into CPC to form a composite scaffold. However, hyaluronic acid particles lowered the load-bearing ability of CPC scaffold (Babo et al. 2016b). Another study coated HPL on hydroxyapatite/β-tricalcium phosphate scaffolds through immersion of the scaffolds inside a HPL solution (Leotot et al. 2013). The disadvantage of this method was that it only produced a HPL coating on the surface layer, with insufficient concentration and inadequate duration for release.

Recently, HPL was added into a chitosan liquid which was used to form a CPC-chitosan-HPL scaffold, where HPL was contained throughout the scaffold (Zhao et al. 2019). This strategy maintained the mechanical properties of CPC scaffold while providing a sustained release of HPL. hPDLSCs were seeded on the scaffold, and the effects of HPL on hPDLSC viability, osteogenic differentiation, and bone mineral synthesis were investigated (Zhao et al. 2019). Five groups were tested: control group (CPC-chitosan+0% HPL); 2.66% HPL group (CPC-chitosan+2.66% HPL); 5.31% HPL group (CPC-chitosan+5.31% HPL); 10.63% HPL group (CPC-chitosan+10.63% HPL); and 21.25% HPL group (CPC-chitosan+21.25% HPL) (Zhao et al. 2019).

The following were the main findings from this study (Zhao et al. 2019). (1) As shown in Figure 15.7A, the results of live/dead staining show that the live cell densities of the HPL groups were greater when compared to that of the control group at 7 and 14 days. The live cell density increased with HPL dosage up to 10.63% HPL. (2) In Figure 15.7B, cell proliferation of HPL groups was faster than the control group. The CPC-chitosan+10.63% HPL construct exhibited the greatest cell proliferating rate, increasing by 4-folds from 1 to 14 days, when compared to a 2-fold increase for the control group. (3) Figures 15.7C and D, respectively, show how the 14-day ALP activity and bone mineral synthesis change with the HPL mass fraction. The results demonstrate that, compared to the group without HPL, the addition of 10.65% HPL into the CPC scaffold resulted in a 2- to 3-fold increase in the ALP activity and bone mineral production by the hPDLSCs.

These results were likely due to the following mechanisms. First, HPL contains a wide spectrum of growth factors that can promote the MSC proliferating rate (Altaie et al. 2016). These include VEGF, TGF, PDGF, FGF, and IGF (Altaie et al. 2016). However, high concentrations of HPL (21.25% HPL) appeared to inhibit cell proliferation. The latter suggests a differential dose response to one or more proteins in HPL (Chen et al. 2012, Lee et al. 2011). Negative feedback is a mechanism that can block activation of the incoming signalling pathway to prevent inappropriate cellular response to excessive extracellular signals (such as growth factors) (Perrimon and McMahon 1999). Second, regarding the factors in HPL, TGF, IGF, FGF, and PDGF have been shown to enhance the osteogenesis of MSCs (Xia et al. 2011). TGF induces osteogenesis by activating receptor-regulated Smads and initiating mitogen-activated protein kinase signalling cascade (Iwasaki et al. 2018). TGF can also increase the Runx2 expression (Iwasaki et al. 2018). Moreover, IGF can upregulate type I collagen transcription and stabilise β-catenin, which is essential for osteoblastogenesis (Giustina et al. 2008). FGF can activate Runx2 and upregulate the anabolic function of osteoblasts (Majidinia et al. 2018). Furthermore, PDGF can promote the osteogenic differentiation by activating the BMP-Smad1/5/8- Runx2/Osterix pathway (Majidinia et al. 2018).

Therefore, HPL delivered via a CPC-chitosan scaffold induced a strong and tailored effect on the proliferation and osteogenic differentiation of hPDLSCs. The cellular response was dependent on the concentration of HPL. Indeed, the HPL concentration in CPC-chitosan scaffolds was directly proportional to the osteogenesis of hPDLSCs. The CPC-chitosan-HPL-hPDLSC construct may provide an important therapeutic strategy for orthopaedic and maxillofacial regeneration. By combining HPL with CPC-chitosan scaffold, HPL could be delivered to bone defects for local release and maximal efficacy. This is a realizable strategy for the clinical application of HPL. Further pre-clinical testing is necessary to determine the beneficial dosing amounts of HPL in the CPC-chitosan scaffold and to determine the potential of the construct to support alveolar bone and periodontal regeneration.

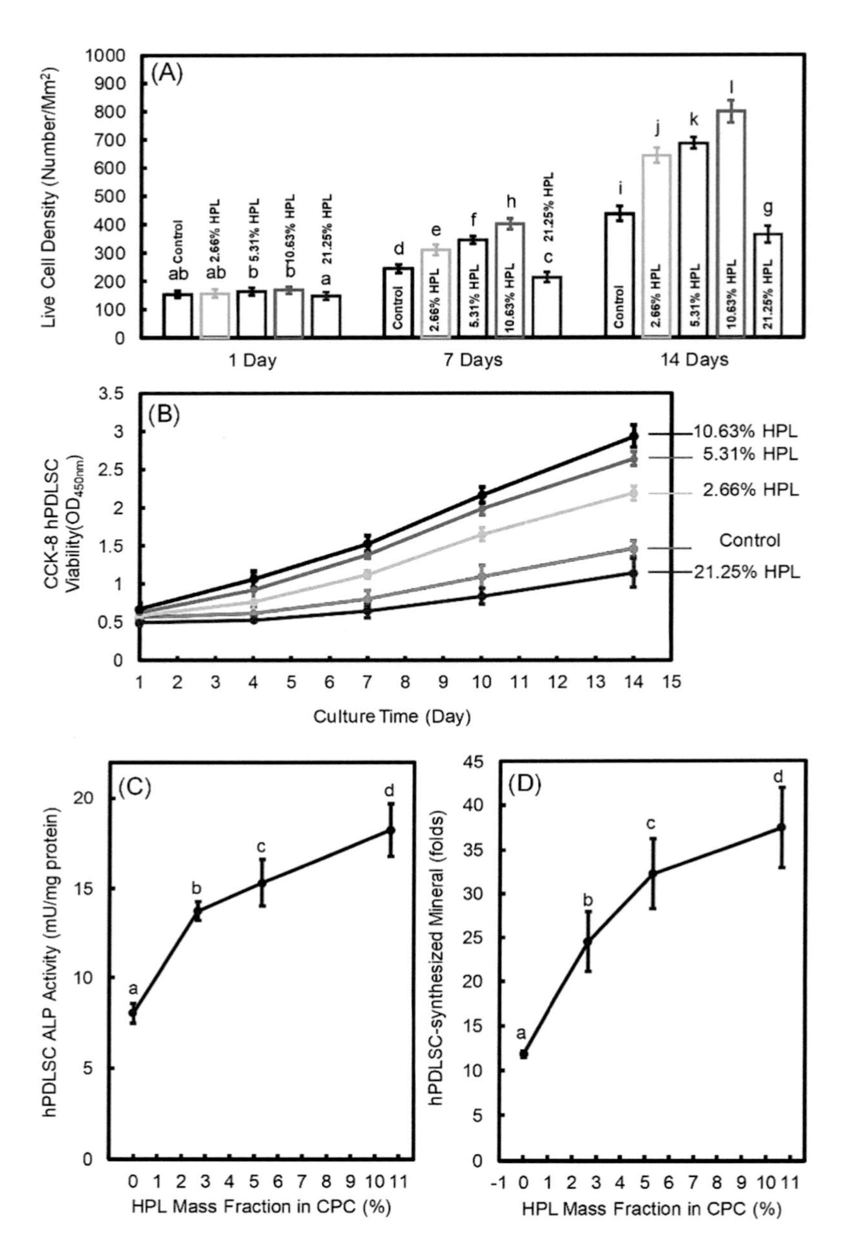

FIGURE 15.7 Proliferation and osteogenesis of hPDLSCs on CPC-chitosan-HPL scaffolds. (A) Live cell density increased with time. (B) hPDLSC viability. (C) The influence of HPL weight concentration in CPC-chitosan scaffold on hPDLSC ALP activity at 2 weeks. (D) The influence of HPL weight concentration in CPC-chitosan scaffold on the production of bone mineral matrix byhPDLSCs at 2 weeks. (Adapted from (Zhao et al. 2019), with permission)

Conclusions

Dental stem cells including hPDLSCs and hDPSCs were isolated from extracted human teeth. hPDLSCs were differentiated into osteoblasts, fibroblasts, and cementoblasts for tissue engineering the periodontium to generate the cementum-PDL-bone structure. In addition, stem cells showed exciting potential for maxillofacial bone tissue engineering applications. In particular, the novel hPDLSC-CPC-metformin construct displayed excellent properties for bone tissue engineering, and the bone mineral-mimicking nanoapatite CPC scaffold supported the stem cell performance. Furthermore, bioactive agents were delivered via the scaffolds. Metformin delivered using CPC greatly increased the osteogenic differentiation and mineral production of hPDLSCs. The metformin+osteogenic construct had 3–4 times of increase in expression levels of osteogenic genes, ALP activity, and mineral production. Indeed, CPC-metformin-hPDLSC construct showed a high promise for bone engineering in dental, craniofacial, and orthopaedic repairs. In addition, CPC-chitosan scaffold demonstrated a great potential for HPL delivery, and HPL in CPC promoted hPDLSCs for bone regeneration. Moreover, nano-CaF_2 composite demonstrated excellent potential for tooth restorations to release calcium and fluoride ions to promote hPDLSCs for osteogenic and cementogenic differentiations. Therefore, novel developments in stem cell technology, bioactive and nanostructured scaffolds, and nanopharmaceuticals are highly promising for periodontal regeneration and therapies in the oral and maxillofacial regions.

Acknowledgement

We thank Drs. Ping Wang and Wei Qin for discussions and experimental contributions, and NIH for funding (R21DE029611-01A1and R01DE023578).

REFERENCES

AlJofi FE, Ma T, Guo D, Schneider MP, Shu Y, Xu HHK and Schneider A. 2018. *Cytotherapy 20* (5): 650–659.
Albanese A, Licata ME, Polizzi B and Campisi G. 2013. *Immun Ageing 10* (1): 23.
Altaie A, Owston H and Jones E. 2016. *World Journal of Stem Cells 8* (2): 47–55.
Anitua E, Troya M and Zalduendo M. 2018. *Cytotherapy 20* (4): 479–498.
Babo PS, Carvalho PP, Santo VE, Faria S, Gomes ME and Reis RL. 2016a. *Journal of Biomaterials Applications 31* (5): 637–649.
Babo PS, Santo VE, Gomes ME and Reis RL. 2016b. *Macromolecular Bioscience 16* (11): 1662–1677.
Bailey CJ and Turner RC. 1996. *The New England Journal of Medicine 334* (9): 574–579.
Belli R, Kreppel S, Petschelt A, Hornberger H, Boccaccini AR and Lohbauer U. 2014. *Journal of the Mechanical Behavior of Biomedical Materials 37*: 100–108.
Beyth N, Domb AJ and Weiss EI. 2007. *Journal of Dentistry 35* (3): 201–206.
Bosshardt DD and Sculean A. 2010. *Periodontology 51* (1): 208–219.
Bottino MC, Thomas V and Janowski GM. 2011. *Acta Biomaterialia 7* (1): 216–224.
Bottino MC, Thomas V, Schmidt G, Vohra YK, Chu TMG, Kowolik MJ and Janowski GM. 2012. *Dental Materials Official Publication of the Academy of Dental Materials 28* (7).
Chen SC, Brooks R, Houskeeper J, Bremner SK, Dunlop J, Viollet B, Logan PJ, Salt IP, Ahmed SF and Yarwood SJ. 2017. *Molecular and Cellular Endocrinology 440*: 57–68.
Chen B, Sun HH, Wang HG, Kong H, Chen FM and Yu Q. 2012a. *Biomaterials 33* (20): 5023–5035.
Chen W, Zhou H, Tang M, Weir MD, Bao C and Xu HHK. 2012b. *Tissue Engineering Part A 18* (7–8): 816.
Cheng L, Zhcmg K, Weir MD, Liu H, Zhou X and Xu HHK. 2013. *Dental Materials Official Publication of the Academy of Dental Materials 29* (4): 462–472.
Cortizo AM, Sedlinsky C, McCarthy AD, Blanco A and Schurman L. 2006. *European Journal of Pharmacology 536* (1–2): 38–46.
Dai Q, Weir MD, Ruan J, Liu J, Gao J, Lynch CD, Oates TW, Li Y, Chang X and Xu HHK. 2021. *Dental Materials 37* (6): 1009–1019.
Deligeorgi V, Mjör IA and Wilson NHH. 2001. *Primary Dental Care Journal of the Faculty of General Dental Practitioners 8* (1): 5–11.
Dunn WJ. 2007. *American Journal of Orthodontics and Dentofacial Orthopedics 131* (2): 243–247.
Fréchette J, Martineau I and Gagnon G. 2005. *Journal of Dental Research 84*: 434–439.
Frencken JE, Sharma P, Stenhouse L, Green D, Laverty D and Dietrich T. 2017. *Journal of Clinical Periodontology 44*: S94–S105.
Furtos G, Rivero G, Rapuntean S and Abraham GA. 2017. *Journal of Biomedical Materials Research Part B Applied Biomaterials 105* (5): 966–976.
Gao Y, Xue J, Li X, Jia Y and Hu J. 2008. *The Journal of Pharmacy and Pharmacology 60* (12): 1695–1700.

Gauthier P, Yu Z, Tran QT, Bhatti FUR, Zhu X and Huang TJ. 2017. *Cell & Tissue Research 368* (1): 79–92.

Giustina A, Mazziotti G and Canalis E. 2008. *Endocrine Reviews 29* (5): 535–559.

Gronthos S, Brahim J, Li W, Fisher LW, Cherman N, Boyde A, DenBesten P, Robey PG and Shi S. 2002. *Journal of Dental Research 81* (8): 531–535.

Gronthos S, Mankani M, Brahim J, Robey PG and Shi S. 2000. *Proceedings of the National Academy of Sciences of the United States of America 97* (25): 13625–13630.

Hellyer PH, Beighton D, Heath MR and Lynch EJ. 1990. *British Dental Journal 169* (7): 201–206.

Hikita K, Meerbeek BV, Munck JD, Ikeda T, Landuyt KV, Maida T, Lambrechts P and Peumans M. 2007. *Dental Materials 23* (1): 71–80.

Houshmand B, Tabibzadeh Z, Motamedian SR and Kouhestani F. 2018. *Archives of Oral Biology 95*: 44–50.

Illueca FMA, Vera PB, Cabanilles PDG, Fernandez VF and Loscos FJG. 2006. *Medicina Oral Patología Oral Y Cirugía Bucaledinglesa 11* (4): E382–E392.

Iwasaki Y, Yamato H and Fukagawa M. 2018. *International Journal of Molecular Sciences 19* (8): 2352.

Jang WG, Kim EJ, Bae IH, Lee KN, Kim YD, Kim DK, Kim SH, Lee CH, Franceschi RT, Choi HS and Koh JT. 2011. *Bone 48* (4): 885–893.

Jeyabalan J, Shah M, Viollet B and Chenu C. 2012. *The Journal of Endocrinology 212* (3): 277–290.

Jia L, Xiong Y, Zhang W, Ma X and Xu X. 2020a. *Experimental Cell Research 386* (2): 111717.

Jia Z, Li H, Cao R, Xiao K and Chen R. 2020b. *American Journal of Translational Research 12* (7): 3754–3766.

Jiang W, Li L, Zhang D, Huang S, Jing Z, Wu Y, etal. 2015. *Acta Biomaterialia 25*: 240–252.

Jin Q, Cheng J, Liu Y, Wu J, Wang X, Wei S, Zhou X, Qin Z, Jia J and Zhen X. 2014. *Brain, Behavior, and Immunity 40*: 131–142.

Kajbaf F, Bennis Y, Hurtel-Lemaire AS, Andrejak M and Lalau JD. 2016. *Diabetic Medicine 33*(1): 105–110.

Kanazawa I, Takeno A, Tanaka KI, Notsu M and Sugimoto T. 2018. *Endocrinology 159* (2): 597–608.

Kanazawa I, Yamaguchi T, Yano S, Yamauchi M and Sugimoto T. 2008. *Biochemical and Biophysical Research Communications 375* (3): 414–419.

Kim EK, Lim S, Park JM, Seo JK, Kim JH, Kim KT, Ryu SH and Suh PG. 2012. *Journal of Cellular Physiology 227* (4): 1680–1687.

Kuang Y, Hu B, Feng G, Xiang M, Deng Y, Tan M, Li J and Song J. 2020. *Biogerontology 21* (1): 13–27.

Lee UL, Jeon SH, Park JY and Choung PH. 2011. *Regenerative Medicine 6* (1): 67–79.

Lekic P, Rojas J, Birek C, Tenenbaum H and McCulloch CAG. 2010. *Journal of Periodontal Research 36*(2).

Leotot J, Coquelin L, Bodivit G, Bierling P, Hernigou P, Rouard H and Chevallier N. 2013. *Acta Biomaterialia 9* (5): 6630–6640.

Liu G, Ding Z, Yuan Q, Xie H and Gu Z. 2018. *Frontiers in Chemistry 6*: 439.

Liu J, Dai Q, Weir MD, Schneider A, Zhang C, Hack GD, Oates TW, Zhang K, Li A and Xu HHK. 2020. *Materials 13* (21): 4951.

Liu J, Ruan J, Weir MD, Ren K and Xu HHK. 2019. *Cells 8* (6): 537.

Liu J, Zhao Z, Ruan J, Weir MD, Ma T, Ren K, Schneider A, Oates TW, Li A and Zhao L. 2020. *Journal of Dentistry 92*: 103259.

Lu Z, Qu Z, Wang C, Wei Z, Xu L and Gui Y. 2018. *Frontiers in Chemistry 6*: 540.

Lyngstadaas SP, Wohlfahrt JC, Brookes SJ, Paine ML and Snead ML. 2009. *Orthodontics & Craniofacial Research 12*: 243–253.

Ma J, Zhang ZL, Hu XT, Wang XT and Chen AM. 2018. *European Review for Medical and Pharmacological Sciences 22* (22): 7962–7968.

Majidinia M, Sadeghpour A and Yousefi B. 2018. *Journal of Cellular Physiology 233* (4): 2937–2948.

Man Y, Wang P, Guo Y, Xiang L, Yang Y, Qu Y, Gong P and Deng L. 2012. *Biomaterials 33* (34): 8802–8811.

Mao LX, Liu J, Zhao J, Xia L, Jiang L, Wang X, Lin K, Chang J and Fang B. 2015. *International Journal of Nanomedicine 10*: 7031–7044.

Masoudi Rad M, Nouri Khorasani S, Ghasemi-Mobarakeh L, Prabhakaran MP, Foroughi MR, Kharaziha M, Saadatkish N and Ramakrishna S. 2017. *Materials Science & Engineering C 80*: 75–87.

Meirelles L, Arvidsson A, Andersson M, Kjellin P, Albrektsson T and Wennerberg A. 2010. *Journal of Biomedical Materials Research Part A 87A* (2): 299–307.

Molinuevo MS, Schurman L, McCarthy AD, Cortizo AM, Tolosa MJ, Gangoiti MV, Arnol V and Sedlinsky C. 2010 *Journal of Bone and Mineral Research 25* (2): 211–221.

Nies AT, Koepsell H, Damme K and Schwab M. 2011. *Handbook of Experimental Pharmacology* (201): 105–167.

Perrimon N and McMahon AP. 1999. *Cell 97* (1): 13–16.

Polimeni G, Xiropaidis AV and Wikesjö UME. 2010. *Periodontology 41* (1).

Qin W, Chen JY, Guo J, Ma T, Weir MD, Guo D, Shu Y, Lin ZM, Schneider A and Xu HHK. 2018a. *Stem Cells International 2018*: 7173481.

Qin W, Gao X, Ma T, Weir MD, Zou J, Song B, Lin Z, Schneider A and Xu HHK. 2018b. *Journal of Endodontics 44* (4): 576–584.

Rosling B, Nyman S, Lindhe J and Jern B. 1976. *Journal of Clinical Periodontology 3* (4): 233–250.

Santo VE, Duarte AR, Popa EG, Gomes ME, Mano JF and Reis RL. 2012. *Journal of Controlled Release 162* (1): 19–27.

Seo B-M, Miura M, Gronthos S, Bartold PM, Batouli S, Brahim J, Young M, Robey PG, Wang CY and Shi S. 2004. *The Lancet 364* (9429): 149–155.

Shaw RJ, Kosmatka M, Bardeesy N, Hurley RL, Witters LA, DePinho RA and Cantley LC. 2004. *Proceedings of the National Academy of Sciences of the United States of America 101* (10): 3329–3335.

Shimauchi H, Nemoto E, Ishihata H and Shimomura M. 2013. *Japanese Dental Science Review 49* (4): 118–130.

Shu Y. 2011. *Zhong Nan Da Xue Xue Bao. Yi Xue Ban = Journal of Central South University. Medical Sciences 36* (10): 913–926.

Shu Y, Sheardown SA, Brown C, Owen RP, Zhang S, Castro RA, Ianculescu AG, Yue L, Lo JC, Burchard EG, Brett CM and Giacomini KM. 2007. *Journal of Clinical Investigation 117* (5): 1422–1431.

Sowmya S, Bumgardener JD, Chennazhi KP, Nair SV and Jayakumar R. 2013. *Progress in Polymer Science 38* (10–11): 1748–1772.

Stein SA, Lamos EM and Davis SN. 2013. *Expert Opinion on Drug Safety 12* (2): 153–175.

Sun L and Chow LC. 2008. *Dental Materials 24* (1): 111–116.

Liu J, Dai Q, Weir MD, Schneider A, Zhang C, Hack GD, Oates TW, Zhang K, Li A and Xu HHK. 2020. *Materials 13* (21): 4951.

Tang M, Weir MD and Xu HHK. 2012. *Journal of Tissue Engineering and Regenerative Medicine 6*: 214–224.

Tang Y, Chen L, Zhao K, Wu Z, Wang Y and Tan Q. 2016. *Composites Science & Technology 125* (March 23): 100–107.

Tanja JJ and Langlass TM. 1995. *The Diabetes Educator 21* (6): 509–510, 13–14.

Thein-Han WW, Liu J and Xu HHK. 2012. *Dental Materials 28* (10): 1059–1070.

Thein Han W, Liu J, Tang M, Chen W, Cheng L and Xu HHK. 2013. *Bone Research 4*: 371–384.

Tsumanuma Y, Iwata T, Kinoshita A, Washio K, Yoshida T, Yamada A, Takagi R, Yamato M, Okano T and Izumi Y. 2016. *Bioresearch Open Access 5*(1): 22–36.

van denDolder J, Mooren R, Vloon AP, Stoelinga PJ and Jansen JA. 2006. *Journal of Tissue Engineering 12* (11): 3067–3073.

Wang L, Li C, Weir MD, Zhang K, Zhou Y, Xu HHK and Reynolds MA. 2017. *RSC Advances 7* (46): 29004–29014.

Wang P, Ma T, Guo D, Hu K, Shu Y, Xu HHK and Schneider A. 2018. *Journal of Tissue Engineering and Regenerative Medicine 12* (2): 437–446.

Wang P, Zhao L, Liu J, Weir MD, Zhou X and Xu HHK. 2014. *Bone Research 2*: 14017.

Wang S, Xia Y, Ma T, Weir MD, Ren K, Reynolds MA, Shu Y, Cheng L, Schneider A and Xu HHK. 2019. *Drug Delivery and Translational Research 9* (1): 85–96.

Wang YG, Qu XH, Yang Y, Han XG, Wang L, Qiao H, Fan QM, Tang TT and Dai KR. 2016. *Cell Signal 28* (9): 1270–1282.

Weir MD, Moreau JL, Leuine ED, Strassler HE, Chow LC and Xu HHK. 2012. *Dental Materials 28* (6): 642–652.

Weir MD and Xu HHK. 2010a. *Acta Biomaterialia 6* (10): 4118–4126.

Weir MD and Xu HHK. 2010b. *Journal of Biomedical Materials Research Part A 94A* (1): 223–233.

Xia W, Li H, Wang Z, Xu R, Fu Y, Zhang X, Ye X, Huang Y, Xiang AP and Yu W. 2011. *Cell Biology International 35* (6): 639–643.

Xia Y, Chen H, Zhang F, Bao C, Weir MD, Reynolds MA, Ma J, Gu N and Xu HHK. 2017. *Nanomedicine Nanotechnology Biology & Medicine* S1549963417301612.

Xie H and Liu H. 2012. *Journal of Periodontology 83* (6): 805.

Xu HHK, Moreau JL, Sun L and Chow LC. 2008. *Biomaterials 29* (32): 4261–4267.

Xu HHK, Sun L, Weir MD, Antonucci JM, Takagi S, Chow LC and Peltz M. 2006. *Journal of Dental Research 85* (8): 722–727.

Xu HHK, Wang P, Wang L, Bao C, Chen Q, Weir MD, Chow LC, Zhao L, Zhou X and Reynolds MA. 2017. *Bone Research 5*: 17056.

Xu HHK, Moreau JL, Sun L and Chow LC. 2010a. *Journal of Dental Research 89* (7): 739–745.

Xu HHK, Weir MD, Sun L, Moreau JL, Takagi S, Chow LC and Antonucci JM. 2010b. *Journal of Dental Research 89*(1): 19–28.

Xu HHK, Zhao L and Weir MD. 2010c. *Journal of Dental Research 89* (12): 1482–1488.

Yang K, Zhang J, Ma X, Ma Y, Kan C, Ma H, Li Y, Yuan Y and Liu C. 2015. *Materials Science & Engineering C 56* (November): 37–47.

Yang Z, Gao X, Zhou M, Kuang Y, Xiang M, Li J and Song J. 2019. *European Journal of Oral Sciences 127* (3): 210–221.

Zeichner-David M. 2006. *Periodontol 2000 41* (1): 196–217.

Zhang H, Liu S, Zhu B, Xu Q, Ding Y and Jin Y. 2016. *Stem Cell Research & Therapy 7* (1): 168.

Zhang J, Liu W, Schnitzler V, Tancret F and Bouler JM. 2014. *Acta Biomaterialia 10* (3): 1035–1049.

Zhang K, Cheng L, Wu EJ, Weir MD, Bai Y and Xu HHK. 2013. *Journal of Dentistry 41* (6): 504–513.

Zhang K, Zhang N, Weir MD, Reynolds MA, Bai Y and Xu HHK. 2017. *Dental Clinics of North America 61* (4): 669–687.

Zhang R, Liang Q, Kang W and Ge S. 2019. *Cell Biology International 44* (1): 70–79.

Zhao Z, Liu J, Schneider A, Gao X, Ren K, Weir MD, Zhang N, Zhang K, Zhang L, Bai Y and Xu HHK. 2019a. *Journal of Dentistry 91*: 103220.

Zhao Z, Liu J, Weir MD, Zhang N, Zhang L, Xie X, Zhang C, Zhang K, Bai Y and Xu HHK. 2019b. *RSC Advances 9* (70): 41161–41172.

Zhou G, Myers R, Li Y, Chen Y, Shen X, Fenyk-Melody J, Wu M, Ventre J, Doebber T, Fujii N, Musi N, Hirshman MF, Goodyear LJ and Moller DE. 2001. *Journal of Clinical Investigation 108* (8): 1167–1174.

Zhou H, Weir MD and Xu HHK. 2011. *Tissue Engineering Part A 17* (21–22): 2603–2613.

Zhou T, Liu X, Sui B, Liu C, Mo X and Sun J. 2017. *Biomedical Materials 12* (5): 05504.

Zhu W and Liang M. 2015. *Stem Cells International 2015*: 972313.

Zhuang Y, Lin K and Yu H. 2019. *Frontiers in Chemistry 7*: 495.

16

Nanopharmaceuticals in Cardiovascular Medicine

Ramandeep Singh, Anupam Mittal, Maryada Sharma, Ajay Bahl, and Madhu Khullar
Postgraduate Institute for Medical Education and Research, Chandigarh, India

CONTENTS

Cardiovascular diseases (CVDs) constitute disorders of the circulatory system involving the blood vessels and heart. CVDs include coronary artery disease, myocardial infarction, cardiomyopathy, rheumatic valve disease, and more. CVDs are the major cause of mortality around the globe, especially Indian sub-continent. The unhealthy lifestyle, obesity, poor living conditions, and malnutrition exacerbate the conditions and augment the cases of CVDs in developing countries. The number of deaths caused by cardiovascular diseases annually is higher than any other disease representing a major percentage of the global deaths. Out of these, 85% are caused by heart attack and stroke. The major limitation in the treatment of CVDs is the delivery of drugs to the heart and their availability in the heart for a long time.

With the advancement of technology, nanoscience became an integral part of the health system. Because of their small size, nanoparticles have potential to be used as therapeutic modality for drug delivery for CVDs.

What Are Nanoparticles

Over the last few decades, there has been a lot of progress in the nanoparticle technology field and has been applied to various disciplines. The researchers have been able to generate nanoparticles of sizes ranging from 1 to 100 nm using different biomolecules like proteins, carbohydrates, lipids, DNA, RNA, and inorganic particles like gold, iron, silica, and carbon. They exhibit high dissolution rates owing to the high ratios of surface to volume. Nanoparticle superstructures due to their exceptionally small sizes can penetrate easily through physiological and biological barriers, which are generally impermeable to large-sized biomolecules. Nanoparticles can be assembled into different shapes, i.e., nanocapsules or nanospheres, depending on the method of preparation and the drug targets. In recent years, their physiochemical properties have been studied extensively to understand the interactions of their shape, size, and surface chemistry with the targeted cells and tissues. For example, it was found that large-sized nanoparticles are required to target tumours and improve their retention inside the body. The nanoparticles can be assembled to form superstructures comprising of several layers of biomolecules such as DNA, RNA, and proteins with an inorganic core of metal ions like gold or iron. These layers owe to the surface chemistry, sizes, and shapes of the nanoparticle superstructures. Zeta potential is used as a common parameter to assess the charge on the surface of nanopharmaceutical drug delivery systems (NDDSs), which relies

DOI: 10.1201/9781003153504-16

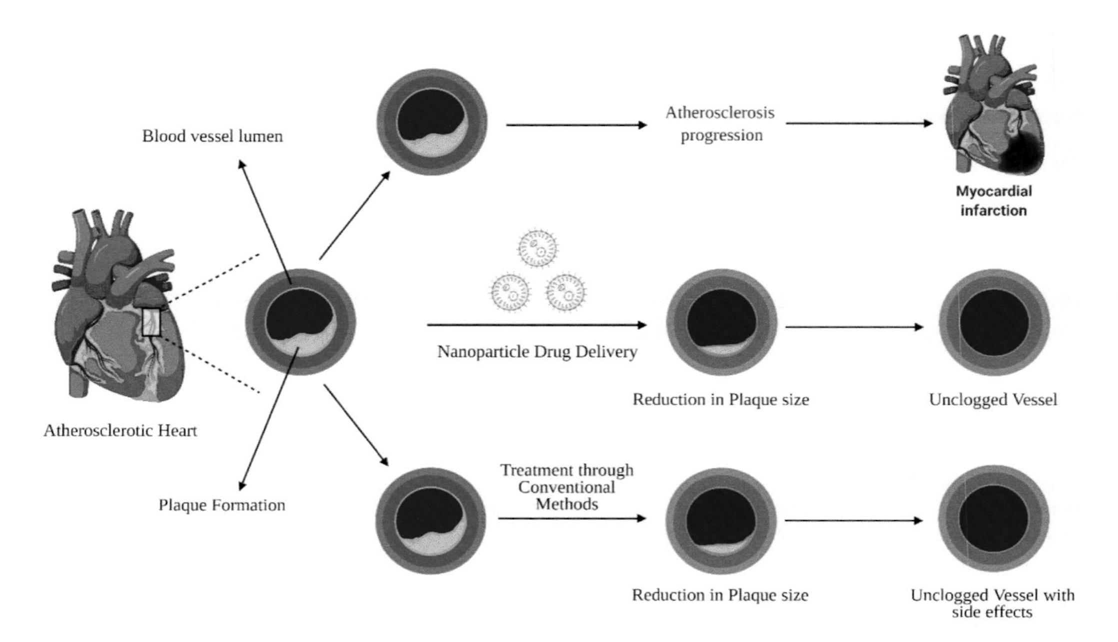

FIGURE 16.1 Pictorial illustration of a comparison between the nanoparticle drug delivery system and the conventional methods used for the treatment of atherosclerosis heart disease. Untreated condition can lead to myocardial infarction. Conventional drugs lower plaque volumes but may show side effects. NDDSs have shown supremacy over conventional drugs in terms of overcoming their side effects.

on the constitution of the multilayered structure. NDDS which possess a Zeta potential of more than (±) 30mV are known to be stable in the suspension and can also help determine whether the drug has been absorbed on the surface or encapsulated within the structure. To improve the site specificity and target of pharmacologically active compounds, nanoparticle superstructures can be coated with polymer polyethylene glycol (PEG), which helps to mitigate nonspecific interactions between nanoparticles and nontargeted cells or tissues. These polymers also help NDDs to escape the immune system of the body. These layers also help in the retention of drugs inside the body and prolong the clearance. Depending on the structures of the nanoparticle drug delivery system, drug is either entrapped inside or attached to the surface. These structural complexities confer the properties like the slow and prolonged release of the pharmacologically active compound in the targeted tissue, site specificity, and relatively higher doses of the drug. These nanoparticles can be administered through various routes, i.e., nasal, oral, and parenteral. All the mentioned qualities make nanoparticle drug delivery systems a potential candidate for the treatment of cardiovascular diseases (Kim et al., 2019). A comparison between the two therapeutic strategies has been illustrated in the form of a diagram (Figure 16.1).

Types of Nanoparticles

A repertoire of various NDDSs has been developed in the past two decades exhibiting different shapes, sizes, surface properties, multilayered interaction systems, solubilities, roughness, drug targeting abilities, etc., which has been summarised below.

a. **Liposomes**: These biomolecules were first introduced to the world about five decades back. Since then, they have been extensively studied and used in various fields including medicinal usage. These are composed of a phospholipid layer with an aqueous core. They can be easily metabolised by the body because of the bilipid layer composition and thus have low retention rate inside the body. This can be prevented by coating the surface with PEGs, which confers stealth like properties, increases the retention time, and cannot be cleared from the body by phagocytic cells. The surface exhibits hydrophobic properties; therefore, hydrophobic drugs can be incorporated inside the lipid layer and aqueous hydrophilic core can be used for encapsulation of drugs, which are hydrophilic in nature (Bulbake et al., 2017). These are extensively used as a carrier for chemotherapeutic drugs, anticancer therapy, and antibodies. Liposomes can be versatile in their shape and size and based on this have been classified into three major categories, i.e., smallest vesicles ranging between 20 and 40 μm, larger vesicles, and multilayered vesicles comprising of several lipid layers, which are differentiated by aqueous layers between them. The lipid layers can be easily manipulated conferring to various properties like robustness, retention rate, and specificity (Wilczewska et al. 2012) (Figure 16.2).

Besides being formed with the naturally present bilipid layer, they might not be able to penetrate deep into the skin when administered through the transdermal route. This can be enhanced by the incorporation of some anionic surfactants and ethanol which can increase the fluidity of the bilipid layer, thus enhancing the penetration of the liposomes. Earlier for gene transfer, DNA or gene of interest was assembled with liposomes but was prone to degradation by endocytosis mediated transfer. Thus, cationic liposomes were developed. Electrostatic interactions between positive cationic liposomes and DNA, which are negatively charged, stabilize the complex and improve the rate of transfection as well. These positively charged cationic liposomes also protected the DNA or gene of interest against degradation (Marcato et al., 2008).

In case of cardiovascular diseases, there is an elevation of various biomarkers specific to each disorder. These cytokines such as TNF-alpha and interleukin-8 are known to elevate the inflammation and necrosis. Designing NDDS targeting the inhibition of these cytokines can reduce the damage caused by the cardiovascular disease. Similarly, liprostin developed by Endovasc Limited is an encapsulated liposome used in the treatment of restenosis (recurrence of narrowing of blood vessel leading to restricted blood flow). It delivers prostaglandin E-1 which is antiinflammatory, antithrombotic, and used as a vasodilator. It has been approved conditionally by FDA and is in the third phase of clinical trials. This product can provide

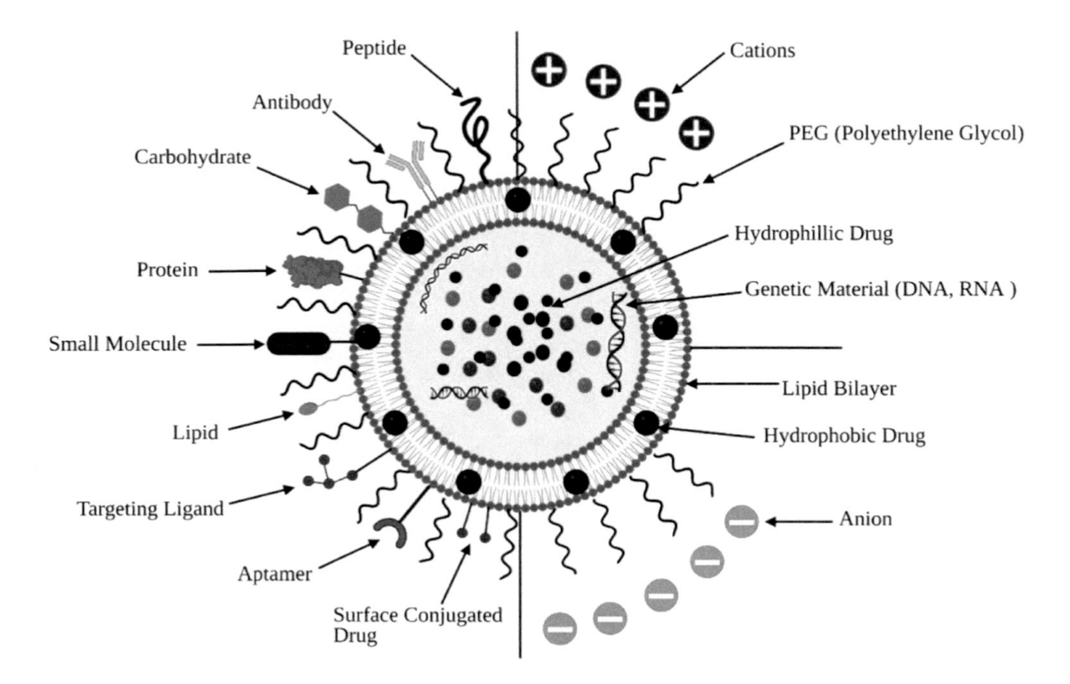

FIGURE 16.2 Representation of liposome with various biomolecules encapsulated within the lipid bilayer and adsorbed on the surface according to their hydrophilic and hydrophobic nature.

great relief to the patients with restricted blood flow along with the angioplasty procedure (Huang et al., 2020). NDDSs have been used for the treatment of thrombosis, which is a disease obstructing the blood flow in the blood vessels eventually leading to cardiac arrest or stroke. Conventional medicines used for the treatment of thrombosis, lack site specificity, and have short lives, therefore, fail to target thrombus. Cyclic RGD functionalized liposomes were assembled incorporating urokinase, which is cheap, effective, and has been widely used in clinics. This delivery system has also reduced the dosage by 75% as identical thrombolytic results can be achieved by free urokinase (Zhang et al., 2018).

b. **Micelles**: Micelles nanoparticles are amphiphilic colloidal molecules comprising of both hydrophilic and hydrophobic groups. They can be synthesized using amphiphilic polymers or lipids with a hydrophobic core and can range from 5 to 100 nm in size. They have been widely used for delivery of low-molecular-weight hydrophobic pharmacologically active compounds, proteins, and DNA (Figure 16.3). Their small sizes have been advantageous in delivering anticancer drugs as they can penetrate deep inside the tumours and prolong the retention time of the drugs. Critical micelle concentration is required for the formation of micelle nanoparticles. Lesser micelle concentrations can increase the number of undesired interactions and weaken the micelle assembly. Micelle concentrations are mitigated when injected directly into the blood stream and may get opsonized by the proteins present in the blood. This can dissociate micelle structures releasing the encapsulated pharmacologically active compound in the bloodstream and diluting the drug. Dilution of the drug may have the same results as that of the free drug that may have been administered through different routes. Therefore, site specificity may not be achieved (Lu et al., 2018).

Micelle structures possess more structural stability as compared to liposomes and other low-molecular-weight drug delivery systems owing to their low excipient to drug ratios. This confers the advantage of lesser cytotoxicity and cost effectiveness. The synthesis of micelle structures can be manipulated to great extents and is easily controllable.

Micelles synthesised with polyethylene glycol/phosphatidyl ethanolamine coating of size ranging from 7 to 20 nm and a Zeta potential of −18 mV were used as therapeutic agents for rabbits induced myocardial

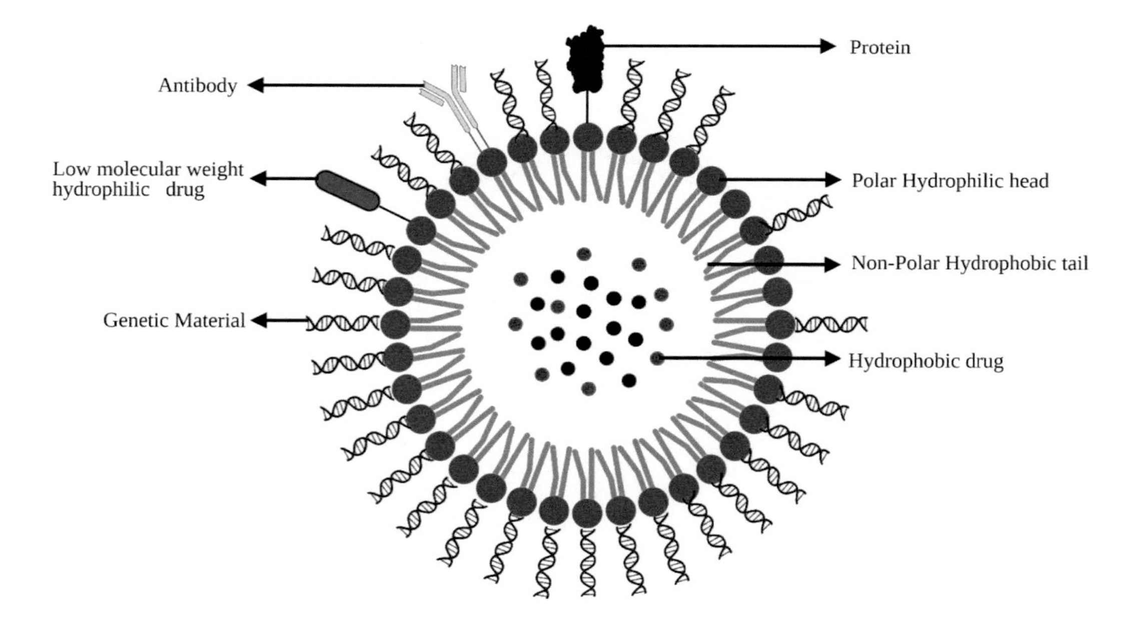

FIGURE 16.3 Illustration of a micellar structure with polar hydrophilic head and hydrophobic tail used for drug and genetic material delivery.

infarction. Experiments revealed that the radiolabelled micelle nanoparticles amassed at the infarcted regions of heart eight times more efficiently than the normal regions of the heart with increased half-life. These results suggest that micelle nanoparticles can be used for drug delivery efficiently in deep tissues as well (Lukyanov et al., 2004).

c. **Dendrimers**: These unique macromolecules offer to aid in different areas due to a range in their shapes, size, and functionality. They are three-dimensional, highly branched structures with repeating units. They possess a central core and the repeating internal units form dendrimer arms that are attached to the core through covalent bonds. They can accommodate large quantities of drugs due to high number of functional groups on the surface such as RGD peptide (arginylglycylaspartic acid), folic acid, PEG, antimicrobial agents based on silver salts, antibodies, etc., which offer a high number of ligand attachment (Sandoval-Yañez et al., 2020). They are highly stable, provide better solubility to pharmacologically active compounds, and are great alternatives for the nonviral transfer of DNA/RNA (Kaneda et al., 2006). These exclusive properties make dendrimers an effective candidate for drug delivery systems. They are more suited for the oral and transdermal route of drug administration (Figure 16.4).

Researchers have been trying to explore the potential use of nitric oxide (NO) in the treatment of cardiovascular diseases as it is an active vasodilator. It is synthesized in the body by nitric oxide synthases by using L-arginine. It serves a critical role in the cardiovascular system as one of the major NO synthase isoform is actively present in the vascular endothelium (Oliveira-Paula et al., 2017). Studies have shown that poly propylene imine dendrimers are able to encapsulate NO and release it in a sustained manner. These dendrimers can be efficiently used in the treatment of atherosclerosis and maintenance of the homeostasis in endothelial vasculature (Lu et al., 2011).

d. **Metallic Nanoparticles**: The class of metallic nanoparticles consists of silver, iron, gold, zinc, silica, copper, and other metals. These multilayered structures can be molded into different shapes, i.e., nanowires, nanorods, nanospheres, nanocapsules, etc. Gold nanoparticles are being used long in a range of biomedical techniques because of their unique physical, chemical, and optical properties. They have been used for controlled biological delivery of DNA assemblies, photothermal treatment of arthritis and various types of tumours, photodynamic therapy, probes for imaging and biosensors, Raman spectroscopy, etc. (Jiang et al., 2007). They have also been used for endocytosis mediated delivery of CRISPR Cas-9 protein for homology-directed DNA

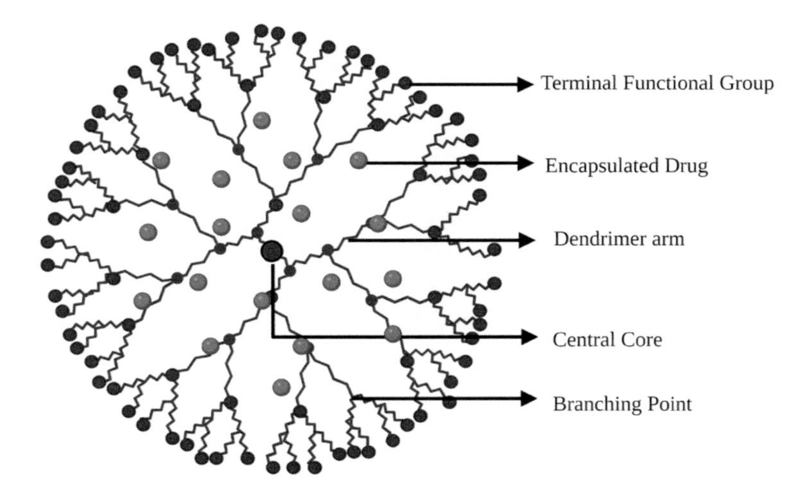

FIGURE 16.4 Structure of a dendrimer with dendrimer arms attached to the central core through covalent bonds. The drugs are embedded inside the 3D structure with functional groups at the periphery.

repair. The gold nanoparticles used for this application comprised of donor DNA, Cas9, guide RNA, and PAsp(DET) coating, which were released into the cytoplasm (Lee et al., 2017).

Similarly, silver nanoparticles can used for targeted drug delivery, but their major applications have been as antimicrobial and antitumour agents. Studies have shown that metal nanoparticles can be reduced extracellularly to obtain stable metal nanoparticles (Figure 16.5). These stable silver nanoparticles help generateg sterile environments and due to their antimicrobial effects can be used in hospitals to reduce and prevent infections (Durán et al., 2007). The iron nanoparticles are being used as a diagnostic agent for different ailments including tumours, atherosclerosis, arthritis, etc., because of their paramagnetism. These stable nanoparticles exhibit low cytotoxicity and increased biocompatibility.

e. **Carbon Nanotubes**: Carbon nanoparticles or cylindrical fullerenes emerged as a great candidate for use in biomedical applications after their discovery in the 1980s. They can be synthesized using fullerenes or graphite. They are classified based on number of walls, i.e., carbon nanotubes with single wall (SWCNTs) and carbon nanotubes with multiple walls (MWCNTs) (Saad et al., 2012). They have a large surface area with good thermal and electrical conductivity. Their biocompatibility is flexible and can be changed with chemical alterations on the surface. These properties render them to enter cells with ease through diffusion. They are known to generate an immune response in murine via activation of a compliment system through carbon nanoparticles incorporating peptides in it.

The drugs and pharmacologically active compounds can be incorporated into carbon nanocarriers in three ways, i.e., encapsulation inside carbon nanotubes, adsorption on the surface, and attaching the drugs to functionalized carbon nanocarriers. The selectivity of carbon nanotubes can be improved by attaching homing agents to them like folic acid. The ends of carbon nanotubes can be sealed with polypyrrole films or fullerenes to avert their unwanted and untimely release (Figure 16.6).

They have been used in cancer therapy for boron atoms delivery. Boron atoms conjugated with carbon nanoparticles accumulate around tumours multiple times as compared to the free boron administered through blood.

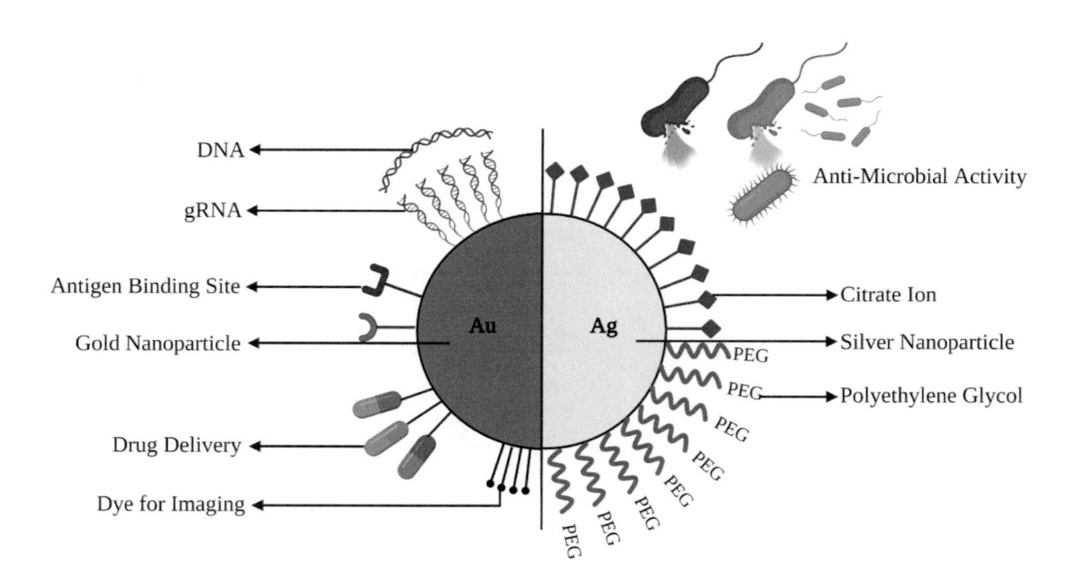

FIGURE 16.5 Pictorial representation of gold and silver nanoparticle with various surface ligands. Gold nanoparticles have been used for CRISPER-guided DNA transfer, drug delivery, and Imaging. Silver nanoparticles are known for their antibacterial properties.

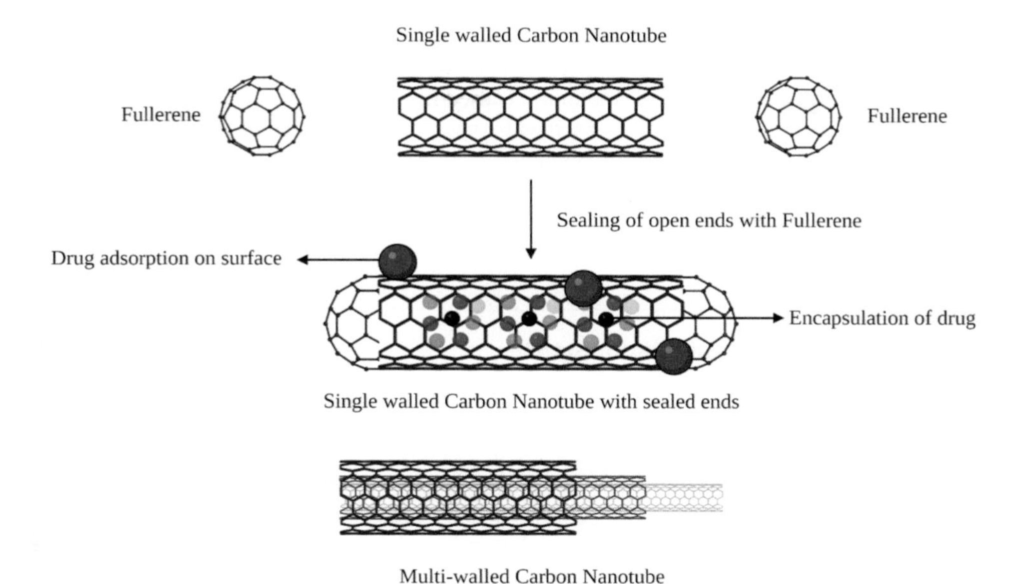

FIGURE 16.6 Schematic representation of various types of carbon nanotubes. The open ends on either side of the nanotube are sealed with fullerenes or polypyrrole to avoid untimely release of the drugs.

Nanoparticles in Diagnosis and Imaging

Nanotechnology applied to the biomedical field has left researchers with many opportunities to develop diagnostic, imaging techniques and biosensors for a variety of diseases.

Nanoparticle systems have exhibited their potential use as biomarkers in various clinical tests. Researchers have shown that a quantum dots assembly of zinc sulphide and cadmium selenide conjugated with monoclonal antibody on a glass substrate coated with indium tin oxide can detect abnormal levels of procalcitonin protein, which is generally overexpressed in case of urinary tract infections (Ghrera et al., 2019). A recent study found that higher levels of C-reactive protein are related to chances of myocardial infarction. It also acts as a biomarker for several diseases that show inflammation as one of the symptoms. Antibodies against C-reactive protein conjugated on the surface of gold nanoparticles have shown promising results as a detection method for the inflammation in different diseases (António et al., 2018).

Troponins are a group of proteins found in the cardiac muscles and skeletal muscles that help to control and regulate the contraction of the muscles. There are three variants of the protein troponin, i.e., troponin I, troponin C, and troponin T. In case of cardiovascular diseases where cardiomyocytes are damaged, troponin C and troponin I are expressed in higher quantities in blood. The levels of both troponin C and troponin I are undetectable in normal conditions; therefore, they serve as potential biomarkers for cardiovascular diseases. Efforts are being made into developing these two biomolecules into biomarkers (Park et al., 2020).

These biomarkers-based systems are being developed on electrochemical sensors, which are relatively low cost and simple than other systems used for the diagnosis of various diseases.

The imaging approach of nanoparticle systems involves modifications of the surface with ligands that are specific for some biomarkers. They are also incorporated with functional groups or targets that are easily detectable like strong paramagnetic compounds, radioactive materials, fluorescent dyes, etc. For example, a research group used cyclic RGD peptide dendrimer conjugated with PEG and Cu for imaging of the hindlimb ischemic model of murine. They demonstrated that the nanoparticle system accumulated in the aorta, where plaque formation had taken place (Wang et al., 2016).

Inflammation is a common representation in many CVDs, which is detectable through techniques like cardiovascular magnetic resonance (CMR). This noninvasive method observes the changes in the configuration of the myocardial tissue. Gadolinium is used in CMR, which aids to improve the clarity of the pictures taken of the myocardium. It is retained by necrotic tissues, whereas healthy tissues clean it. It improves the visibility of the plaques, inflammation, tumours, necrotic tissues, etc. (Park et al., 2020).

Besides this, gold nanoparticles due to their stability and inertness can be used for photoacoustic imaging which uses an excitation source of light and the signals generated by the target are detected by the ultrasonography (Chandarana et al., 2018; Miller et al., 2017).

Cardiovascular Diseases and Nanoparticles

With the advancements in drug/pharmaceutical formulations and surgical procedures in cardiovascular therapies, cardiovascular health could be boosted worldwide. Along with therapeutic development, rudimentary preventive measures need to be included in day to day lives for instance improved lifestyle and diet (Fuster et al., 2014). The contemporary strategies used in the therapeutics of cardiovascular diseases are focused on restoring the oxygenated blood flow and avoiding the recurrence of episodes of cardiovascular diseases. These include attenuating high cholesterol levels and reducing plaque volumes by using statin therapy, which causes a significant reduction in plaque volumes and dense fibrous calcium volumes but is ineffective in mitigating necrotic tissues (Banach et al., 2015; Scolaro et al., 2018). Antiplatelets such as aspirin and aggrenox and anticoagulants such as warfarin and rivaroxaban may exhibit common adverse effects in some serious cases like intracranial haemorrhage and gastrointestinal bleeding. Though both groups of medications prevent the formation of blood clots, their intrinsic pathways of inhibiting the biomolecules involved in the genesis of blood clots are different. These medications present a discrepancy between the dosage offering maximum prevention and causing adverse effects. They have been known to show resistance in some patients and have varied effects on a large population. There is always a risk factor associated with the multiple drug usage which cannot be ignored. These medications used for the prevention of cardiovascular diseases may have adverse effects ranging from serious muscle damage, memory problems, elevation of liver enzymes to elevated complications in diabetic patients, and skin-related allergies. Other treatments for cardiovascular diseases involve invasive surgeries that require highly trained and skilled physicians (Kapil et al., 2017). All these drawbacks along with the low water solubility, low biological efficacies, and nontargeting of the conventional medications and therapies reinforce the requirement of advancements in the cardiovascular medicines and drug delivery systems. This presents a need for improvement in the NDDS.

The following table describes some of the studies done in past few years to evaluate the efficacy of the NDDSs in targeted delivery of the drug/pharmaceutical formulations and Imaging of the cardiovascular diseases (Table 16.1).

Limitations of NDDSS

Despite the innumerable advancements in the nanotechnology field, there are certain drawbacks associated with the active use of nanotechnology and the nanoparticle drug delivery systems. Though many of these nanoparticles are comprised of natural biodegradable and biocompatible compounds, some of them are not biodegradable which raises a concern about their prolonged usage. The extended usage of such micromolecular nanoparticles may lead to their accumulation inside the body, which can be toxic in higher quantities. A detailed study on the relation between the size of the nanoparticles and their toxicity is required. The size of the nanoparticle may influence the cytotoxicity to great extent. A decrease in the size of nanoparticle may destabilize the assembly and increase the surface area. This may lead to an increase in the oxidation and degradation of DNA by nanoparticles. Over the past few years, advanced liposomes have been developed such as cationic liposomes possessing a positive charge on the surface. However, high positive charges may lead clumping of platelets together and hemolysis causing severe toxicity (Agarwal et al., 2018). There are various nanoparticle superstructures that may comprise

TABLE 16.1

Studies of Nanoparticles Used for CVD Treatment and Imaging

Sr. No.	Types of Nanoparticle Drug Delivery Systems	Technique Used	Drug Used for Treatment of CVDs	Method Used	Result/Outcome of the Study	Reference
1	Biocompatible superparamagnetic iron-hydroxyapatite nanoparticles (FeHAs)	Targeted delivery of drug	Ibuprofen	Evaluation of biological stress in isolated cardiomyocytes and rat model of electromagnetic exposure	No change observed in electrophysiological properties *in vitro* as well as *in vivo* proposingFeHAs, a potential candidate for delivering drug in failing heart	Marrella, A. et al. (2018)
2	Gold nanoparticles	Targeted drug delivery and imaging	Vascular endothelial growth factor	Fluorescent gold nanoparticles injected intravenously for imaging and treatment of ischemia	1.7-fold increment in blood perfusion by VEGF conjugated gold nanoparticle delivery as compared to control mice	Kim et al. (2011)
3	Poly (anhydride-ester) nanoparticles based on ferulic acid linked with diglycolic acid (PFAG)	Targeted delivery of PFAG	PFAG	Inhibition of macrophagal uptake of oxidised low-density lipoprotein (oxLDL)	Downregulation in expression of macrophage scavenger receptors and lowered reactive oxygen species (ROS) production	Chmielowski, R.A. et al. (2017)
4	Cyclic arginine-glycine-aspartic (cRGD) peptide-functionalised liposomes	Targeted delivery of urokinase	Urokinase	Thrombolysis potential of urokinase was estimated *in vitro* as well as *in vivo*	A good thrombolysis effect was witnessed both *in vitro* and *in vivo* studies, showed reduction in dose by 75% for achieving the same results as free urokinase in blood	Zhang et al. (2018)
5	Berberine-loaded liposomes	Targeted delivery of berberine	Berberine	Antiinflammatory and antioxidative, along with cardioprotective properties of berberine were evaluated both *in vitro* and *in vivo*	IL-6 secretion was inhibited *in vitro* and cardiac ejection fraction was preserved significantly on 28th day after myocardial infarction by 64% as compared to control	Allijn et al. (2017)

(*Continued*)

TABLE 16.1 (Continued)

Sr. No.	Types of Nanoparticle Drug Delivery Systems	Technique Used	Drug Used for Treatment of CVDs	Method Used	Result/Outcome of the Study	Reference
6	Adenosine (Ade) prodrug-based, atrial natriuretic peptide (ANP) conjugated lipid nanocarriers (ANP Ade/LNCs)	Site directed delivery of cardioprotective drugs	Adenosine prodrug	Drug release in the presence of plasma was tested *in vitro*. Intravenous injection of nanoparticles was administered to evaluate tissue distribution, pharmacokinetics, and inhibition effects of drug on infarct size in rats	The efficiency of (ANP Ade/LNCs) was better in all aspects than nonmodified Ade/LNCs	Yu et al. (2018)
7	PEG-coated dendrimer (Gn-PEG-Gn) constituting polyglutamic acid	Delivery of nattokinase	Nattokinase	Thrombus dissolving activity of nattokinase was assessed *in vitro*	All dendrimer formulations exhibited optimal enzyme activity for thrombolysis *invitro*	Zhang et al. (2017)
8	Gold nanoparticles	Delivery of mesenchymal stem cells for therapeutics	Mesenchymal stem cells (MSCs) pretreated with 5-aza	Evaluation of cardiogenic differentiation of MSCs for regeneration of myocardial infarcted tissues	Mesenchymal stem cells pretreated with 5-aza may induce enhanced cardiac differentiation in therapeutics of infarcted heart	Ravichandran et al. (2014)
9	Porous silicon-based nanoparticles conjugated with atrial natriuretic peptide (ANP)	Imaging and autoradiography of ischemic heart	Atrial natriuretic peptide (ANP)	SPECT/CT (single-photon emission computed tomography) imaging followed by autoradiography of ischemic heart *in vivo*	Enhanced accumulation of ANP-conjugated silicon nanoparticles in endothelial layer of the left ventricle of ischemic heart	Ferreira et al. (2017)
10	Liposomal nanoparticle drug delivery	Site-directed drug delivery and therapeutics	Prednisolone	Intravenous administration of liposomal nanoparticle-conjugated prednisolone for the treatment of atherosclerosis in human	Successful delivery of nanoparticle to the atherosclerotic plaques in patients; however, no antiinflammatory effect was witnessed	Van der Valk et al. (2015)
11	Collagen adhesin (CNA35)-conjugated gold nanoparticles	Computed tomographic (CT) imaging	Collagen homing peptide, collagen adhesin (CNA35) incorporating gold nanoparticles	Left anterior descending coronary artery was reperfused to create a myocardial scar. Ungated CT imaging was done after 30 days	Prolonged and uniform opacification was observed for 6 h. Myocardial scar formation and accumulation of nanoparticles was confirmed by histological staining	Danila et al. (2013)

Sr. No.	Types of Nanoparticle Drug Delivery Systems	Technique Used	Drug Used for Treatment of CVDs	Method Used	Result/Outcome of the Study	Reference
12	Glycol Chitosan (GC)-conjugated gold nanoparticles	Microcomputed tomographic imaging	Gold nanoparticles incorporated with glycol chitosan (GC-AuNPs)	Intravenous injection of the nanoparticle system in experimental models of thrombosis	Nanoparticle accumulation of gold nanoparticles-aided visualisation of thrombi presence in the carotid arteries of mouse	Kim et al. (2013)
13	High density-lipoprotein-conjugated gold nanoparticle (Au-HDL)	Spectral computed tomography	High-density lipoprotein-conjugated gold nanoparticles	Intravenous injection of gold nanoparticles at a concentration of 500 mg per Kg of mice body weight. After 24 h a contrast material formulation based on Iodine was injected for contrast and imaged	Detection of Au-HDL in aorta of Apolipoprotein E knockout mice and visualisation of bone and vasculature due to contrast agent based on iodine and tissues possessing high concentrations of calcium	Cormode et al. (2010)
14	Porphyrin nanoparticle conjugated with Folate (Porphysome)	Noninvasive active tracking of macrophages in multiple modes by radioligand and fluorescent imaging	Porphyrin nanoparticle incorporated with folate	Cardiac fluorescence signal was checked post folate-porphysome injection after nine days in myocardial infarction-induced mice	Porphysomes aid in noninvasive imaging of the macrophages located in the heart and investigating their active behaviour	Ni et al. (2016)

of nonbiodegradable repeating units for instance carbon nanotubes, which incorporate fullerenes. They are known to transport nucleic acids and other pharmacologically active compounds. However, they are nonbiodegradable and may have adverse effects on their accumulation inside the body. There is limited knowledge available about the potential side effects of the accumulation of nanoparticles in various organs of the body like liver and kidney.

These nanoparticle systems offer prolonged retention in the system through various coatings on the outer surface and escape the immune system of the body. As we discussed that these nanoparticles offer diagnosis of diseases, they may remain in the body for longer periods. This may lead to difficulty in the acquisition of the image and background noise in the form of nonspecific signals.

A lot of experimental results suggest that nanoparticles may show promising results in the treatment of cardiovascular diseases like atherosclerosis, myocardial infarction, etc. Till date, there is no evidence of a reduction in the necrotic tissues or cells after the episode of cardiovascular disease. There is still a requirement for systems that may help in the vascularization of the dead tissues and regain functional complexities. The systems must not alter their physical and chemical characteristics but rather should be focused more on endothelial infiltration.

The nanoparticle drug delivery systems are multilayered complex superstructures and most of their physiochemical properties depend on the monomeric units they are comprised of, and the functional groups incorporated in them. It is challenging to achieve the complexity of the system for commercial development. Nearly 250 nanoparticle-based products commercialised or in the clinical trials have elementary compositions, but their complex delivery systems make it challenging for increment in their commercial production and maintaining the quality of the products. With time, more and more complex biomolecules and therapeutic agents are being developed and their delivery through the nanoparticle drug delivery systems demand continuous efforts to make it safer for prolonged use (Ragelle et al., 2017).

Conclusion and Future Perspectives

The advancements in the field of NDDSs are focused on improving the quality and health standards of life. However, there is a need for continuous efforts for development in treatment through NDDSs. As compared to the conventional therapeutics and diagnosis of various CVDs, NDDSs have shown their supremacy. The treatment of cardiovascular diseases through gene delivery is a promising sector. NDDSs offer virus-free gene delivery systems with great efficiency. They have been known to deliver DNA with protein assemblies for DNA repair and may improve the vascularity of necrotic tissues through gene delivery in future. They may also provide real-time tracking of the tissue repair with stem cell therapy. To enhance the bioavailability or retention time of the stem cells in the body, NDDSs provide the cells with an extracellular matrix specific to the cardiovascular microenvironment (Sun et al., 2020).

The NDDSs have shown tremendous possibilities in therapeutics and imaging of cardiovascular diseases. Studies have shown that they can enhance the targeting capacity and efficacy of traditional drugs. As far as concerned to the cost and dosage of the conventional medicines for cardiovascular diseases, these advanced drug delivery systems have emerged as a better alternative. They possess the capability of attracting a considerable amount of expenditure used for ongoing treatments and imaging procedures.

The investment in the research, development, and commercial production of drug delivery systems has boosted in past two decades and the products have the potential to go through the meticulous clinical trials. The advanced drug delivery systems may flood the markets in the coming few years around the globe. However, the need for scrupulous studies of the effects of NDDSs *in vivo* still remains, and this might be a major reason for reluctance in the acceptance of these advanced delivery systems by patients.

REFERENCES

Agarwal, V., Bajpai, M. and Sharma, A., 2018. *Recent Patents on Drug Delivery & Formulation*, *12*(1), pp. 40–52.

Allijn, I.E., Czarny, B.M., Wang, X., Chong, S.Y., Weiler, M., DaSilva, A.E., Metselaar, J.M., Lam, C.S.P., Pastorin, G., deKleijn, D.P. and Storm, G., 2017. *Journal of Controlled Release*, *247*, pp. 127–133.

António, M., Nogueira, J., Vitorino, R. and Daniel-da-Silva, A.L., 2018. *Nanomaterials*, *8*(4), p. 200.

Banach, M., Serban, C., Sahebkar, A., Mikhailidis, D.P., Ursoniu, S., Ray, K.K., Rysz, J., Toth, P.P., Muntner, P., Mosteoru, S. and Garcia-Garcia, H.M., 2015. *BMC Medicine*, *13*(1), pp. 1–21.

Bulbake, U., Doppalapudi, S., Kommineni, N. and Khan, W., 2017. *Pharmaceutics*, *9*(2), p. 12.

Chandarana, M., Curtis, A. and Hoskins, C., 2018. *Applied Nanoscience*, *8*(7), pp. 1607–1619.

Chmielowski, R.A., Abdelhamid, D.S., Faig, J.J., Petersen, L.K., Gardner, C.R., Uhrich, K.E., Joseph, L.B. and Moghe, P.V., 2017. *Acta Biomaterialia*, *57*, pp. 85–94.

Cormode, D.P., Roessl, E., Thran, A., Skajaa, T., Gordon, R.E., Schlomka, J.P., Fuster, V., Fisher, E.A., Mulder, W.J., Proksa, R. and Fayad, Z.A., 2010. *Radiology*, *256*(3), pp. 774–782.

Danila, D., Johnson, E. and Kee, P., 2013. *Nanomedicine: Nanotechnology, Biology and Medicine*, *9*(7), pp. 1067–1076.

Durán, N., Marcato, P.D., DeSouza, G.I., Alves, O.L. and Esposito, E., 2007. *Journal of Biomedical Nanotechnology*, *3*(2), pp. 203–208.

Ferreira, M.P., Ranjan, S., Kinnunen, S., Correia, A., Talman, V., Mäkilä, E., Barrios-Lopez, B., Kemell, M., Balasubramanian, V., Salonen, J. and Hirvonen, J., 2017. *Small*, *13*(33), p. 1701276.

Fuster, V., 2014. *Top 10 Nature Reviews Cardiology*, *11*(11), p. 671.

Ghrera, A.S., 2019. *Analytica Chimica Acta*, *1056*, pp. 26–33.

Huang, J., Wang, D., Huang, L.H. and Huang, H., 2020. *International Journal of Molecular Sciences*, *21*(3), p. 739.

Jiang, W., Kim, B.Y., Rutka, J.T. and Chan, W.C., 2007. *Expert Opinion on Drug Delivery*, *4*(6), pp. 621–633.

Kaneda, Y. and Tabata, Y., 2006. *Cancer Science*, *97*(5), pp. 348–354.

Kapil, N., Datta, Y.H., Alakbarova, N., Bershad, E., Selim, M., Liebeskind, D.S., Bachour, O., Rao, G.H. and Divani, A.A., 2017. *Clinical and Applied Thrombosis/Hemostasis*, *23*(4), pp. 301–318.

Kim, D.E., Kim, J.Y., Sun, I.C., Schellingerhout, D., Lee, S.K., Ahn, C.H., Kwon, I.C. and Kim, K., 2013. *Annals of Neurology*, *73*(5), pp. 617–625.

Kim, J., Cao, L., Shvartsman, D., Silva, E.A. and Mooney, D.J., 2011. *Nano Letters*, *11*(2), pp. 694–700.

Kim, K.S., Song, C.G. and Kang, P.M., 2019. *Antioxidants & Redox Signaling*, *30*(5), pp. 733–746.

Lee, K., Conboy, M., Park, H.M., Jiang, F., Kim, H.J., Dewitt, M.A., Mackley, V.A., Chang, K., Rao, A., Skinner, C. and Shobha, T., 2017. *Nature Biomedical Engineering*, *1*(11), pp. 889–901.

Lu, Y., Sun, B., Li, C. and Schoenfisch, M.H., 2011. *Chemistry of Materials*, *23*(18), pp. 4227–4233.

Lu, Y., Zhang, E., Yang, J. and Cao, Z., 2018. *Nano Research*, *11*(10), pp. 4985–4998.

Lukyanov, A.N., Hartner, W.C. and Torchilin, V.P., 2004. *Journal of Controlled Release*, *94*(1), pp. 187–193.

Marcato, P.D. and Durán, N., 2008. *Journal of Nanoscience and Nanotechnology*, *8*(5), pp. 2216–2229.

Marrella, A., Iafisco, M., Adamiano, A., Rossi, S., Aiello, M., Barandalla-Sobrados, M., Carullo, P., Miragoli, M., Tampieri, A., Scaglione, S. and Catalucci, D., 2018. *Journal of the Royal Society Interface*, *15*(144), p. 20180236.

Miller, M.R., Raftis, J.B., Langrish, J.P., McLean, S.G., Samutrtai, P., Connell, S.P., Wilson, S., Vesey, A.T., Fokkens, P.H., Boere, A.J.F. and Krystek, P., 2017. *ACS Nano*, *11*(5), pp. 4542–4552.

Ni, N.C., Jin, C.S., Cui, L., Shao, Z., Wu, J., Li, S.H., Weisel, R.D., Zheng, G. and Li, R.K., 2016. *Molecular Imaging and Biology*, *18*(4), pp. 557–568.

Oliveira-Paula, G.H., Lacchini, R. and Tanus-Santos, J.E., 2017. *Nitric Oxide*, *63*, pp. 39–51.

Park, J.H., Dehaini, D., Zhou, J., Holay, M., Fang, R.H. and Zhang, L., 2020. *Nanoscale Horizons*, *5*(1), pp. 25–42.

Ragelle, H., Danhier, F., Préat, V., Langer, R. and Anderson, D.G., 2017. *Expert Opinion on Drug Delivery*, *14*(7), pp. 851–864.

Ravichandran, R., Sridhar, R., Venugopal, J.R., Sundarrajan, S., Mukherjee, S. and Ramakrishna, S., 2014. *Macromolecular Bioscience*, *14*(4), pp. 515–525.

Saad, M.Z.H., Jahan, R. and Bagul, U., 2012. *Asian Journal of Biomedical and Pharmaceutical Sciences*, *2*(14), p. 11.

Sandoval-Yañez, C. and Castro Rodriguez, C., 2020. *Materials*, *13*(3), p. 570.

Scolaro, B., Soo Jin Kim, H. and deCastro, I.A., 2018. *Critical Reviews in Food Science and Nutrition*, *58*(6), pp. 958–971.

Sun, Y., Lu, Y., Yin, L. and Liu, Z., 2020. *Frontiers in Bioengineering and Biotechnology*, *8*, p. 947.

van derValk, F.M., vanWijk, D.F., Lobatto, M.E., Verberne, H.J., Storm, G., Willems, M.C., Legemate, D.A., Nederveen, A.J., Calcagno, C., Mani, V. and Ramachandran, S., 2015. *Nanomedicine: Nanotechnology, Biology and Medicine*, *11*(5), pp. 1039–1046.

Wang, Y.J., Larsson, M., Huang, W.T., Chiou, S.H., Nicholls, S.J., Chao, J.I. and Liu, D.M., 2016. *Progress in Polymer Science*, *57*, pp. 153–178.

Wilczewska, A.Z., Niemirowicz, K., Markiewicz, K.H. and Car, H., 2012. *Pharmacological Reports*, *64*(5), pp. 1020–1037.

Yu, J., Li, W. and Yu, D., 2018. *Drug Design, Development and Therapy*, *12*, p. 1697.

Zhang, N., Li, C., Zhou, D., Ding, C., Jin, Y., Tian, Q., Meng, X., Pu, K. and Zhu, Y., 2018. *Acta Biomaterialia*, *70*, pp. 227–236.

Zhang, S.F., Gao, C., Lü, S., He, J., Liu, M., Wu, C., Liu, Y., Zhang, X. and Liu, Z., 2017. *Colloid Surface B Biointerface*, *159*, pp. 284–292.

17

Nanoparticles for Cardiovascular Medicine: Trends in Myocardial Infarction Therapy

Yifan Tai and Adam C. Midgley
Nankai University, Tianjin, China

CONTENTS

Introduction: Cardiovascular Disease in Brief

Conditions affecting the cardiovascular system, such as coronary heart disease, congestive heart failure, myocardial infarction (MI) (heart attack, heart ischemia), inflammatory heart disease, myocardial fibrosis, diseases of the vasculature itself (e.g., atherosclerosis, thrombosis, calcification), and associated ischemic diseases, are grouped under the collective umbrella term of cardiovascular diseases (CVDs). CVDs hold claim to being the leading cause of mortality and morbidity the world over (Sacco et al. 2016), with a World Health Organisation (WHO) estimate that 31% of global deaths are caused by CVDs (WHO 2017). In 2017, Europe had an estimated 45% of deaths attributable to CVDs (Timmis et al. 2018); and in China, 40% of death were recorded as CVD related, and this is predicted to increase with the emergence of diabetes, obesity, and the aging population (Liu et al. 2019a). The socioeconomic burden of CVD healthcare and treatments is substantial (Mensah and Brown 2007). CVDs are the costliest disease group in the USA and the American Heart Association estimated that health care expenditure and lost productivity resulting from CVDs exceeded $350 billion in 2014–2015 (Benjamin et al. 2019).

DOI: 10.1201/9781003153504-17

Therefore, despite promising outcomes of therapeutic options and pharmacological advancements, the discovery of novel and efficient treatment modalities for CVDs is welcomed as a critical requirement. The field of nanomedicine is a rapidly developing research topic, and research teams around the world are investigating the use of nanomedicines to fill the gap left by pharmaceutical limitations, aiming to revolutionise CVD healthcare.

An Overview of Nanoparticles and Prospects in Cardiovascular Medicine

Nanoparticles are nanoscale carrier materials aimed at improving the targeted delivery of payloads, enhancing pharmacokinetics, and increasing the efficacy of diagnostics and therapeutics (Pelaz et al. 2017). The term, 'nanoparticle' is generally applied to particulates that fall within the nanoscale range (1–100 nm). However, the current consensus is that the term is also acceptable for particulates that fall within the sub-micrometre range (>100 nm, <1 μm) (Boverhof et al. 2015). Advancements in nanoparticle technology have sought to overcome challenges facing drug delivery, such as biodistribution, elimination, adverse reactions, systemic degradation and byproduct generation, drug instability, and bioactivity half-life. Essential benefits of nanoparticle-mediated drug delivery include the improvement of solubility of hydrophobic drugs and thus the capacity to overcome physiological barriers, increase permeability, offer controlled release, and enhance the circulation lifetime (Kalepu and Nekkanti 2015; Din et al. 2017). Limitation of nonspecific drug uptake in off-target tissues, which would otherwise lead to toxicity, while facilitating the enhancement of drug bioactivity and bioavailability, is a key feature in the design of nanoparticles. Moreover, the important advantages of enhanced permeation and retention effect improve circulation and accumulation of the loaded drug or biologic agent at target sites; higher drug concentrations at the target site minimise the overall dose, frequency, and curbs adverse drug reactions. Thus, nanoparticle formulations have received extensive attention in recent years as attractive approaches for specific design in improving CVD diagnosis and treatment.

Nanoparticles can be designed to convey multifunctionality through modification of nanoparticle materials themselves (e.g., inclusion of surface modification with targeting moieties, tuned size, and charge). This emergent goal of nanoparticle multimodality or multifunctionality aims to realise codelivery of multiple bioactive factors as adjunct combinatory therapies and to take advantages of the intrinsic properties and bioactivity conveyed by the nanoparticle materials themselves. Multifunctionality can enable bioimaging (Lu et al. 2019), improve cell growth/cytocompatibility (Midgley et al. 2020b), and serve to reduce immunogenicity (Fam et al. 2020), whilst improving drug delivery to target sites. Nonetheless, obstacles posed by technical and regulatory bottlenecks have hindered the advancement towards clinical use of multifunctional nanoparticle therapies for CVDs, in part due to the increased complexity of multifunctional nanoparticle formulations. According to ClinicalTrials.gov database (search performed 25 November 2020; terms '*cardiovascular diseases*' and '*nanoparticle*'), there are less than 20 clinical trials exploring the use of nanoparticles for CVDs. Despite the great interest and investment in developing treatments for CVDs, current CVD nanoparticle trials largely remain limited to simplistic drug carriers, monitoring, and detection methods.

The use of nanoparticles to deliver cardiovascular medicines and achieve sustained release of biologics has emerged as a promising utility to rescue compromised heart and vascular function. Depending on the physicochemical properties of the biologic, the nanoparticle carrier, predicted effects of the treatment and the intended target, the pursuit of different administration routes can help achieve optimal nanoparticle therapy for CVDs. Intracardiac and intravenous injections, inhalation, and functionalised heart patches are typically the administration routes opted for use in preclinical investigations. Figure 17.1 illustrates the popularised choices of nanoparticle administration routes, each provides intrinsic advantages and disadvantages (Chenthamara et al. 2019), but they have opened options for targeted treatment of several pathologies associated with CVDs, such as dysregulated vascularisation, progressive fibrosis, inflammatory conditions, and ischemic heart disease while simultaneously improving cardiac functionality.

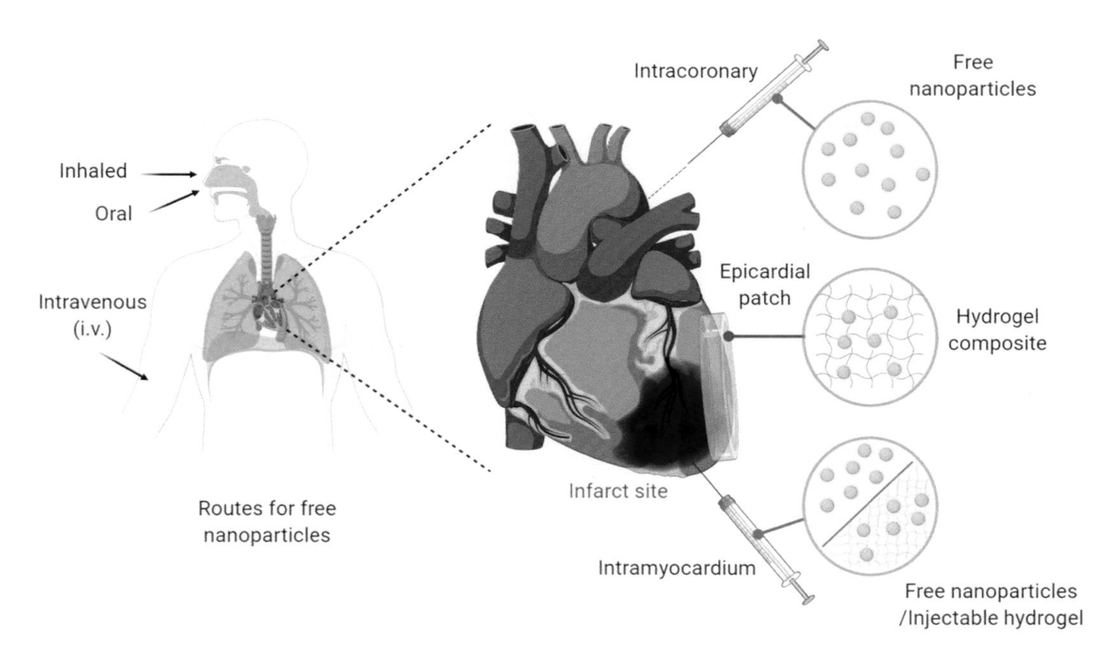

FIGURE 17.1 Routes of nanoparticle administration for MI therapy. Multiple routes for nanoparticle administration for MI therapy have been implemented, including noninvasive routes: inhalable nanoparticles; orally administered nanoparticles; and intravenous (i.v.) injection of free nanoparticles. More popularised routes used in recent preclinical investigations were intravenous or intracoronary injection of free nanoparticles; intramyocardial injection (near infarct border) of free nanoparticles or nanoparticles within an injectable hydrogel suspension; and nanoparticles entrapped within adhesive hydrogel composites for application as a cardiac patch and adhered to the LV epicardium.

Trends in Nanoparticle Use for Myocardial Infarction Therapy

Ischemic heart diseases are typically a result of coronary artery occlusion and are a major contributor to CVDs (Lloyd-Jones et al. 2010). An ischemic event within the heart is otherwise known as myocardial infarction. The onset of MI is followed by a series of events including myocardial hypertrophy, dilatation, mitochondrial dysfunction, and the generation of a large amount of reactive oxygen species (ROS). These ROS drive downstream cardiomyocyte apoptosis and the excessive deposition of extracellular matrix by cardiac fibroblasts, cumulating in the development of myocardial fibrosis and a dysfunctional myocardium (Abbate et al. 2006; Talman and Ruskoaho 2016).

Currently, the use of surgical interventions (stenting, coronary angioplasty) and pharmacological treatments (statins, antiplatelet therapy, β-blockers) for post-MI therapy is aimed at mitigation of symptoms without tissue repair. Alternative therapeutic avenues for MI have received substantial attention from the research community. Options such as stem cell (Terashvili and Bosnjak 2019), gene (Oggu et al. 2017), and cytokine therapies (Reboucas et al. 2016) have shown promising outcomes but still face numerous challenges that stifle the progress of these therapeutic technologies towards clinical application (Egwim et al. 2017). Nanoparticles offer advantages that have been shown to overcome some of the challenges facing sustained released and targeted delivery of therapies to the region of MI (Oduk et al. 2018). As of 25 November 2020, there are only three registered clinical trials exploring the use of nanoparticles in MI, focusing on improved imaging modalities, diagnostics, and trafficking of inflammatory cells post-MI (ClinicalTrials.gov search terms *'myocardial infarction'* and *'nanoparticle'*). However, the journey towards the clinical translational of nanoparticles for MI therapy has begun, as evidenced by numerous preclinical investigations discussed hereafter, each seeking to exploit the advantageous properties of nanoparticles and to establish innovative solutions to address the challenges of current MI treatments.

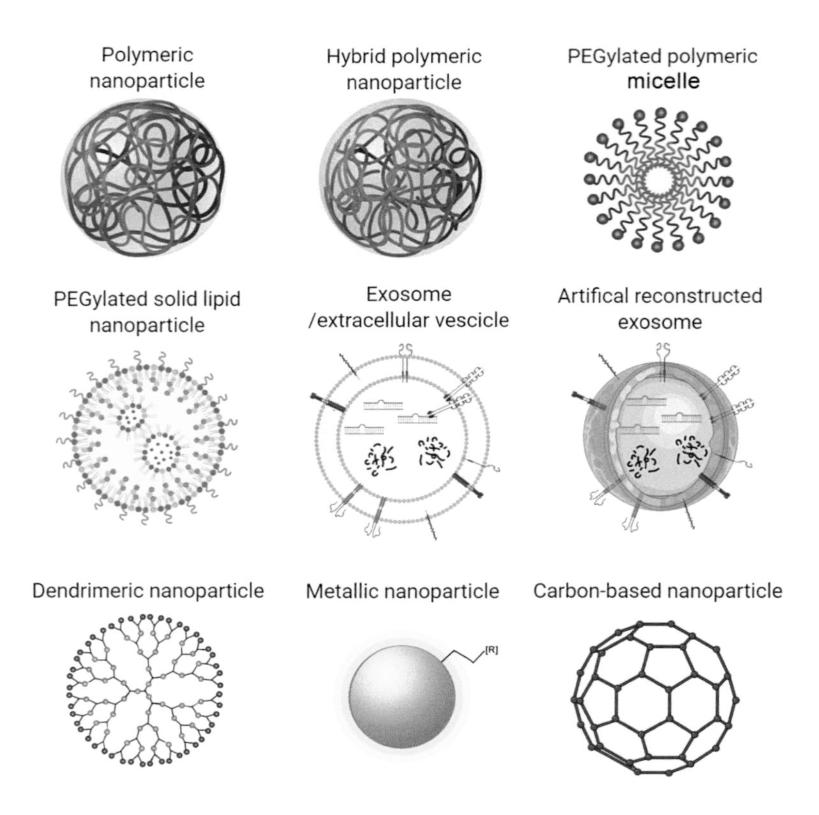

FIGURE 17.2 Recently utilised nanoparticle types used in preclinical investigations for MI therapy. Basic representative images of various nanoparticle types that have been used in recent preclinical investigations for MI therapy. Typically, payloads such as drug molecules, bioactive peptides, or nucleic acids are encapsulated within the core of polymeric, solid lipid nanoparticles, dendrimeric, and carbon-based nanocages. Exosomes and artificial reconstructed exosomes contain cell-derived or enriched contents within the nanoparticle cavity and cell membrane coatings are reflective of the parental cell. Metallic nanoparticles convey not only their own multimodality and functionalities but can also be conjugated onto payload molecules via functionalisation of chemical side chains. Modifications on sites such as PEGylations or dendrimer branches enable prolonged circulation and additional sites engraftment of targeting moieties.

In the following sections, we provide summaries of the recent preclinical applications of different types of nanoparticles that have been utilised for MI therapeutics. Multiple types of nanoparticles have recently been employed in preclinical MI therapeutic research; many of these investigations have sought to design polymeric and lipid-based nanoparticles for the improved bioavailability, stability, and safety of already existing cardiovascular medicines. On the other hand, emergent studies have employed the use of exosomes, dendrimer nanocarriers, metallic-based, and carbon-based nanoparticles to achieve multimodality of nanoparticle therapeutics. Figure 17.2 provides an overview of recent and frequently used nanoparticles for preclinical MI investigations.

Polymeric Nanoparticles

Polymeric nanoparticles can be synthesised from a singular type of polymer material or from various kinds of polymers in combination. Typically, the use of polymers for nanoparticle generation results in spherical, porous nanostructures. The relatively simplistic synthesis process of polymeric nanoparticles offers the added advantage of facile loading and controlled release of drugs or biologics, which is dependent on the hydrophobicity and charge of the polymer-payload combination. Depending on preparation method, polymeric nanoparticles can be either nanospheres or nanocapsules (Rawat et al. 2006).

Nanospheres consist of uniformly dispersed payload within the polymer matrix, whereas in nanospheres, the payload is encapsulated within in a cavity surrounded by a membrane-like polymer. In addition, the payload can also be adsorbed onto the polymeric nanoparticle surface by taking advantage of electrostatic interactions or chemical conjugation with the polymers. The choice of biocompatible, biodegradable, nonimmunogenic, and nontoxic polymers gives rise to many opportunities for the use of polymeric nanoparticles in biomedical applications. Table 17.1 lists the recent preclinical investigations that have utilised polymeric nanoparticles as carrier systems for controlled delivery and release of drugs, biologics, peptides, cytokines, and nucleic acids to the infarcted heart.

Synthetic Polymeric Nanoparticles

Synthetic copolymers such as poly(lactic-co-glycolic acid) (PLGA), poly(lactic acid) (PLA) and poly(glycolic acid) are FDA-approved polymers, making them popular choices for use as nanocarrier systems for drug delivery in CVDs (Oduk et al. 2018; Pascual-Gil et al. 2017). Polymeric nanoparticle-mediated delivery of typical cardiovascular medicines has received considerable attention. Given the status of PLGA as an FDA-approved copolymer, multiple studies have pursued the use of PLGA nanoparticles

TABLE 17.1

Recent Preclinical Investigation Trends Using Polymeric Nanoparticles for MI Therapy

Polymeric Nanoparticle	Payload	Model	Author. (Reference)
Poly(lactic-coglycolic acid) (PLGA)	Statin (Pitavastatin)	Mouse	(Mao et al. 2017)
		Porcine	(Ichimura et al. 2016)
	Statin (Simvastatin) to support ADSC therapy	Mouse	(Yokoyama et al. 2019)
	TAK-242 (TLR4 inhibitor)	Mouse	(Fujiwara et al. 2019)
	Pioglitazone (PPARγ agonist)	MousePorcine	(Tokutome et al. 2019)
	Irbesartan (PPARγ agonist)	Mouse	(Nakano et al. 2016)
	VEGF	Mouse	(Oduk et al. 2018)
	FGF1, CHIR99021 (Wnt1 agonist/GSK-3β antagonist)	Mouse Porcine	(Fan et al. 2020b)
	Wogonin	Rat	(Bei et al. 2020)
	Resveratol	Rat	(Sun et al. 2020b)
	Mitochondrial division inhibitor 1 (Mdivi1)	Mouse	(Ishikita et al. 2016)
Poly(lactic acid) (PLA)	Curcumin, nisin	Guinea pig	(Nabofa et al. 2018)
Polyvinyl copolyoxalate (PVAX)	—	Mouse	(Bae et al. 2016)
	Neuropeptide-$Y_{3\text{-}36}$	Mouse	(Mahmood et al. 2020)
PLGA/poly(ethylene glycol) (PEGylated-PLGA)	Liraglutide (GLP-1 receptor agonist)	Rat	(Qi et al. 2017)
	Melatonin-supported codelivery of adipose-derived stem cells (ADSCs)	Rat	(Ma et al. 2018)
PEGylated-RGD-PLA	MicroRNA(miR)-133	Rat	(Sun et al. 2020a)
Gelatin (Gel)	6-Bromoindirubin-3-oxime (BIO), insulin-like growth factor 1 (IGF-1)	Rat	(Fang et al. 2015)
Methacrylic anhydride-Gel (GelMA)	Polypyrrole	Rat	(He et al. 2018)
Heparin (Hep)	VEGF-A, VEGF-C	Mouse	(Qiao et al. 2020)
Poly(glycidyl methacrylate) ethanolamine (PGEA) shell/ Unlockable Hep core	miR-499, VEGF plasmid DNA (pDNA)	Mouse	(Nie et al. 2018)
Hyaluronan sulphate	miR-21, Ca^{2+}	Mouse	(Bejerano et al. 2018)

as carriers for various payloads in MI therapeutic applications. PLGA nanoparticles offer nonimmunogenicity, higher drug loading efficiency, and sustained release profiles. Statins are commonly prescribed cardiovascular medicines prescribed to lower triglycerides and detrimental cholesterol levels but have been shown to bear antiinflammatory effects (Nagaoka et al. 2015). Therefore, multiple research groups have sought to incorporate statin into PLGA nanoparticles to improve targeted delivery and release to inflammatory cells. The inflammatory process during MI is essential for cardiac repair, but chronic inflammation promotes inappropriate left ventricular (LV) hypertrophy and remodelling (Frangogiannis 2012). Monocytes are the first inflammatory cells mobilised to the infarcted myocardium (Swirski et al. 2009), followed by bone marrow-derived monocyte accumulation (Robbins et al. 2012). To address this, Mao et al. developed PLGA nanoparticles loaded with pitavastatin for MI therapy (Mao et al. 2017), inspired by their previous studies that showed the therapeutic inhibition of the MCP1-CCR2 pathway by nanoparticles loaded with pitavastatin in ischemia-reperfusion (I/R) injury models (Nagaoka et al. 2015). PLGA nanoparticles were shown to be selectively taken-up by inflammatory cells (mainly monocytes) after intravenous administration (Katsuki et al. 2014). Mao and colleagues suggested that bone marrow-derived monocyte/macrophage recruitment was the target of pitavastatin-loaded PLGA nanoparticles, which resulted in attenuated pathology of ligated left anterior descending coronary artery (LAD) models of MI in mice (Mao et al. 2017). Ichimura et al. utilised the bioabsorbable properties of PLGA nanoparticles to delivery pitavastatin to obtain preclinical evidence of therapeutic effects on myocardial I/R injury in conscious and anesthetised pig models (Ichimura et al. 2016). The researchers found that an additional action of pitavastatin was the inhibition of the PI3K-Akt pathway and subsequent inflammatory mechanisms. Furthermore, statins are suggested to have pleiotropic effects that can augment stem cell therapy while suppressing pathological events (Yang et al. 2008). Yokoyama et al. sought to utilise PLGA nanoparticles loaded with simvastatin, to support adipose-derived stem cells (ADSCs) in mouse models of MI (Yokoyama et al. 2019). Simvastatin-conjugated PLGA nanoparticles significantly promoted ADSC migration and promoted the expression of beneficial growth factors. ADSCs were loaded with the nanoparticles *in vitro* and then intravenously administered. The researchers demonstrated that nanoparticle-loaded ADSCs improved cardiac function following MI and induced endogenous cardiac regeneration via activation of epicardial cells.

Fujiwara et al. also targeted the inflammatory response in MI injury using PLGA nanoparticles (Fujiwara et al. 2019). The researchers loaded PLGA with TAK-242, a toll-like receptor 4 (TLR4) inhibitor. TAK-242-loaded PLGA nanoparticles attenuated MI through targeting monocytes/macrophages present within the spleen, blood, and heart. The nanoparticles inhibited Ly-6C^{high} monocytes recruitment to infarcts but did not affect infarct size in TLR4/CCR2-deficient mice, indicating a TLR4-specific mechanism and monocyte/macrophage-mediated inflammation were primary therapeutic targets of TAK-242-PLGA nanoparticles in regulating MI in mice. Tokutome et al. focused on peroxisome proliferator-activated receptor-γ (PPARγ) (Tokutome et al. 2019). Binding of pioglitazone to PPARγ induces receptor heterodimerisation with retinoid X receptor (RXR). The PPARγ/RXR heterodimer complex binds PPAR response elements to regulate gene expression associated with adipogenesis, lipid/glucose metabolism (Ahmadian et al. 2013; Libby and Plutzky 2007), chemokine expression, and macrophage infiltration (Ricote and Glass 2007). In monocytes/macrophages, PPARγ activation drives pro-healing M2 polarisation of macrophages (Odegaard et al. 2007). Pioglitazone administered orally before I/R reduced infarct size in mouse and rat MI models (Ye et al. 2008); however, intraperitoneal injection of pioglitazone at the time of reperfusion resulted in lost effect (Honda et al. 2008). Cardioprotective agents are typically administered during reperfusion. Tokutome et al. determined that it was essential to use a carrier to facilitate pioglitazone delivery to effector cells in I/R injury (Tokutome et al. 2019). In lieu of reports that PLGA nanoparticles accumulate in I/R myocardium and were taken up by circulating macrophages (Nagaoka et al. 2015; Gustafson et al. 2015), researchers developed PLGA nanoparticles as carriers for pioglitazone and showed that they exerted an antiinflammatory effect, which in turn limited the infarct size in I/R mouse and porcine models. The rationale to target PPARγ was supported by the work of Nakano et al., which employed irbesartan as a therapeutic agent for MI (Nakano et al. 2016). Irbesartan blocks angiotensin II type 1 receptor and conveys a partial agonistic effect on PPARγ, and an irbesartan derivative was shown to reduce myocardial inflammation by reducing proinflammatory cytokine expression by cardiomyocytes after MI and conveyed an antihypertensive effect (Ugdutt et al. 2010). Thus, Nakano et al. developed

PLGA nanoparticle-mediated delivery of irbesartan, which accumulated at the MI site when administered intravenously at the time of reperfusion, where the enhanced permeability and retention effect improved vascular permeability (Acharya and Sahoo 2011) and subsequently allowed nanoparticles to be rapidly taken up by mononuclear and phagocytic cells that had infiltrated the MI site. The mechanism of action was shown to be through irbesartan antagonising the additional recruitment of Ly6C^{high}CCR2^{+} inflammatory monocytes into the MI site.

Growth factors such as VEGF and FGF1 have established roles in promoting angiogenesis, which following MI injury, is a critical step in realizing cardiac repair and recover of functional ability. Oduk et al. described the use of PLGA nanoparticles for the loading and delivery of VEGF for restoration of post-MI vascularisation (Oduk et al. 2018). VEGF release was continuous for over 30 days following injection into the peri-infarct region of heart tissue in mouse models of MI. This sustained release improved cardiac function over the course of four weeks posttreatment, largely due to the gradual biodegradation of PLGA, controlled release of VEGF and subsequent enhanced angiogenesis in the infarct site. Fan et al. combined cytokine therapy with co-delivery of a Wnt1 agonist/GSK-3β antagonist, CHIR99021 (Fan et al. 2020b). Following encapsulation of both therapeutics into PLGA nanoparticles, the researchers found that slow release over 4 weeks had a synergistic effect on the enhancement of cardiomyocyte growth and proliferation *in vitro*. Intramyocardial injection of the nanoparticles protected the myocardium against MI injury and reduced the infarct size in both mouse and pig models of post-MI left ventricular remodelling. In addition, cardiac contractile function was maintained, cardiomyocyte apoptosis was reduced, and an increased rate of angiogenesis was observed.

Traditional medicines and natural products are widely used in cardiovascular medicine due to intrinsic antioxidant and antiinflammatory properties. Hence, it was inevitable that these bioactive compounds would be paired with nanoparticles to improve their bioactivity and targeted effects. Bei et al. incorporated wogonin, a flavonoid compound, into PLGA nanoparticles and demonstrated a protective effect against isoproterenol induced MI in rats (Bei et al. 2020). Pre-treatment with wogonin-PLGA nanoparticles significantly reduced infarct size and serum cardiac markers, as well as lipid peroxidation and inflammatory markers. Sun et al. aimed to evaluate the effects of pretreatment with PLGA nanoparticles loaded with resveratrol in isoproterenol-induced MI in rats (Sun et al. 2020b). The researchers noted that high doses of resveratrol offered a cardioprotective effects, whereas nanoparticles loaded with varying concentrations could reciprocate the effects. Treatment prevented cardiac troponin T expression, lactate dehydrogenase, and aspartate aminotransferase activities in cardiomyocytes. Oxidative stress parameters were improved, proinflammatory markers were attenuated, and neutrophil infiltration was reduced. Nabofa and colleagues used formulated curcumin and nisin-based poly(lactic acid) nanoparticles to ascertain the cardioprotective effects of these dietary products on isoproterenol-induced MI in guinea pigs (Nabofa et al. 2018). Pretreatments prevented atrial fibrillation and decreased the expression of MI marker cardiac troponin I (CTnI) and kidney injury molecule-1 (KIM-1). Taken together, exploring avenues of nanoparticle technology could help significantly improve the efficacy of traditional and natural cardiovascular medicines.

Mitochondria-initiated cell death is suggested to play a key role in MI pathogenesis (Vander Heide and Steenbergen 2013; Konstantinidis et al. 2012). Ishikita et al. formulated PLGA nanoparticles containing mitochondrial division inhibitor 1 (Mdivi1), to target mitochondria and regulate responses to ROS following MI (Ishikita et al. 2016). The researchers showed that PLGA nanoparticles were taken-up and present in the cytosol and mitochondria of cardiomyocytes under I/R conditions, which ameliorated cell death. Mdivi1-PLGA nanoparticle treatment through coronary artery administration during reperfusion attenuated I/R injury and infarct size. Other researchers have sought to develop nanoparticles using polymers with intrinsic activity beneficial to the consequences of MI pathogenesis. Copolyoxalate containing vanillyl alcohol (PVAX) is an antiinflammatory polymer with antioxidant activity due to sensitivity to specific ROS and oxidative stress byproducts (Bae et al. 2016; Eshun et al. 2017; Mahmood et al. 2020; Takagawa et al. 2007). Bae et al. developed chemically engineered PVAX containing hydrogen peroxide (H_2O_2)-responsive peroxalate ester linkages (Bae et al. 2016). The prodrug nanoparticles elicited rapid activation by H_2O_2 at the MI site. The researchers observed that the H_2O_2-scavenging peroxalate ester bonds in combination with antioxidant and antiinflammatory effects of PVAX polymer possessed antiapoptotic activity and prevented I/R injury-induced MI in mouse models. Mahmood et al. utilised

PVAX as a nanoparticle carrier of neuropeptide Y fragment ($NPY_{3\text{-}36}$), derived from a highly conserved and abundant heart neuropeptide involved in the pathological remodelling of the heart through sympathetic overactivity during MI (Mahmood et al. 2020). Selective stimulation of angiogenesis by $NPY_{3\text{-}36}$ in combination with ROS-scavenging PVAX nanocarriers overcame the short half-life of administered $NPY_{3\text{-}36}$, significantly decreased infarction size and mortality, and improved heart function, which was determined to be an outcome of upregulated pro-angiogenic cytokine release and enhanced angiogenesis in the infarct site in mouse MI models.

Poly(ethyl glycol) (PEG)-modified Polymeric Nanoparticles

Polymeric nanoparticles enable facile modification to enhance specific functions or bioactivities. A common modification used is the addition of varying lengths of PEG, otherwise known as PEGylation. PEGylation of nanoparticles (PEGylated nanoparticles) offer reduced particle aggregation, lowered immunogenicity, and reduced recognition and phagocytosis by immune cells (Gref et al. 2000; Wattendorf and Merkle 2008). Thus, PEGylated-nanoparticles persist for longer within systemic circulation, which is of benefit to heart-targeted nanomedicines. Qi et al. fabricated PEGylated-PLGA nanoparticles formed micelle-like nanoparticles to improve the delivery of liraglutide, a clinically approved drug for type 2 diabetes (Qi et al. 2017). Liraglutide has poor bioavailability with undesired systemic side effects (Marre et al. 2009) but has been suggested to be beneficial for MI (Richards et al. 2014; Inoue et al. 2015). Therefore, the researchers established spatiotemporal delivery of liraglutide-loaded PEGylated-PLGA nanoparticles, locally injected into the infarcted myocardium to treat MI. Intramyocardial delivery in LAD ligated rats reduced undesired systemic side effects and maximised therapeutic effects on MI, including relieved adverse cardiac remodelling, induced angiogenesis, and inhibited cardiac cell apoptosis.

Ma and colleagues employed PEGylated-PLGA nanoparticles to realise the controlled and sustained release of melatonin to support the codelivery of ADSCs (Ma et al. 2018). Melatonin pretreatment of stem cells was previously shown to increase survivability (Khan et al. 2016) and resistance to oxidative stress through the direct detoxification of ROS and stimulation of antioxidant enzymes (Zhou et al. 2015; Reiter et al. 2016). Thus, the researchers used melatonin encapsulated within nanoparticles to promote stem cell survival in the MI environment of rat hearts. The PEGylated PLGA/melatonin nanoparticles protected ADSCs against apoptosis and necrosis and prolonged ADSC persistence following injection into rat infarcted heart tissue.

Sun et al. fabricated PEGylated and RGD-functionalised PLA nanoparticles for microRNA (miR) gene therapy (Sun et al. 2020a). Important regulators of the myocardial tissue microenvironment, miRs have broad roles in regulating angiogenesis, apoptosis, and fibrosis following MI. However, miRs are sensitive to degradation and often require carriers to serve as effective therapeutics. The researchers used the nanoparticles to delivery miR-133, an inhibitor of apoptotic genes in myocardial cells (Chen et al. 2017; Yu et al. 2019), thus reducing the extent of MI. In addition to PEG, the researchers used the arginine-glycine-aspartic acid tripeptide (RGD) sequence to target nanoparticles to activated platelets at the thrombus of the MI lesion site (Wang et al. 2018; Duro-Castano et al. 2017). The combinatory modification strategy resulted in higher distribution of nanoparticles in the MI heart following intravenous administration, and the subsequent alleviation of injury, reduced myocardial cell apoptosis, reduced inflammation, and limited oxidative stress.

Natural Polymeric Nanoparticles

Gelatin (Gel) is a natural, versatile, polypeptide biopolymer; it is an appealing carrier material due to its low production cost, abundant supply, biodegradability, biocompatibility, and ease of modification through its many active groups (Elzoghby 2013; Azarmi et al. 2006). Gel is obtained from the hydrolysis of collagen and has shown strong potential as a drug carrier for controlled release (Bajpai and Choubey 2006). The different mechanical properties (thermal range, swelling) of Gel rely on its amphoteric interactions and degree of cross-linking density (Saxena et al. 2005). Fang et al. loaded gel nanoparticles with mollusc-derived 6-bromoindirubin-3-oxime (BIO) (Fang et al. 2015). BIO was shown to inhibit glycogen synthase kinase-3, induce cardiomyocyte and endothelial cell dedifferentiation, and lead to the promotion of mature cardiomyocyte proliferation (Tseng et al. 2006; Leri et al. 2014). In addition, Gel nanoparticles loaded with insulin-like growth factor 1 (IGF-1), a cytokine involved in cardiomyocyte growth and survival (Torella et al. 2004; Shafiq et al. 2018), were also synthesised. Codelivery by daily intramyocardial injection of BIO and IGF-1 to the MI area maintained nondetrimental levels of therapeutic agents and increased resident cardiac cells in rat MI models. BIO also promoted dedifferentiation and proliferation of terminally differentiated cardiomyocytes, as evidenced through cytoskeletal rearrangement and inhibited cardiac troponin-T expression (Tseng et al. 2006). Local delivery and sustained expression of IGF-1 increased angiogenesis and rescued cardiac function. Codelivery of BIO and IGF-1 significantly improved revascularisation, heart functional recovery, and supported the proliferation of resident cardiomyocytes (Fang et al. 2015). He et al. took advantage of the facile adaptability of gel by loading polypyrrole, a conductive and electrically stable polymer, into methyl acrylic anhydride-modified gel (GelMA) nanoparticles (He et al. 2018). This modification enabled the facile dopamine cross-linking of nanoparticles into a mussel-inspired GelMA/polycaprolactone hydrogel delivery scaffolds or heart patches. GelMA nanoparticles shaped and neutralised toxicity of polypyrrole during oxidative polymerisation, thus conveying preferable biocompatibility at high concentrations. The high conductivity of heart patches promoted cardiomyocyte function, enhanced revascularisation, and reduced inflammatory infiltration into the MI area – both in the presence and absence of exogenous cardiomyocytes.

Heparin (Hep) is a glycosaminoglycan commonly used as an anticoagulant and thrombolytic agent (Heusch and Gersh 2017). For this reason, intravenous Hep is often used post-MI to prevent blood clots and reduce early reemergence/extension of MI. VEGF has central roles in angiogenesis and vasculogenesis (Robinson 2001). VEGF-A promotes angiogenesis through regulation of endothelial cell proliferation (Lanahan et al. 2010) and has been widely used ischemic disease treatment, but has a short half-life, and thus requires high and frequent doses to achieve the desired effect (Oduk et al. 2018; Lin et al. 2012; Formiga et al. 2010). This results in issues, evident from several clinical trials that ended when VEGF-A cytokine therapy caused elevated permeability and tissue oedema (Henry et al. 2003; Kusumanto et al. 2006). VEGF-C was shown to help alleviate oedema by stimulating lymphangiogenesis and helped to reduce heart injury (Henri et al. 2016; Karkkainen et al. 2004; Cui 2010). Negatively charged Hep targeted interacting peptides, improving their stability, function, and specific targeting to tissue sites (Garcia-Dorado et al. 2012). Thus, Qiao et al. employed a dual-phase treatment regime, using nanoparticles synthesised from heparin and coassembled with VEGF-C or VEGF-A, for intravenous MI therapy in mouse models (Qiao et al. 2020). Hep nanoparticles loaded with VEGF-C were administered during the acute phase of MI (30 minutes to 1-h post-MI) to promote lymphangiogenesis and reduce oedema, whilst VEGF-A-loaded Hep nanoparticles were administered 5 days post-MI to promote angiogenesis and repair. The staggered delivery achieved better therapeutic results and served to greatly reduce requisite protein doses compared to free protein delivery. Nie et al. sought to take advantage of Hep bioactivity in combination with gene therapy to promote cardiac repair (Nie et al. 2018). Overexpression of cardiomyocyte-protective miRs, such as miR-499 (Wang et al. 2011), inhibited apoptosis or promoted proliferation of cardiomyocytes. Delivery vehicles are vital for successful miR and plasmid (pDNA)-based gene therapies. Polycations condense negatively charged nucleic acids through electro-interactions, thus protecting them from degradation and improving cell uptake (Han et al. 2014; Wang et al. 2017b; Peng et al. 2016). Hydroxyl-rich ethanolamine-modified poly(glycidyl methacrylate) (PGEA) can serve as versatile gene carriers. PGEA has low cytotoxicity and high transfection efficiency, and once inside cells, their redox-responsive disulphide bonds are reduced in the presence of glutathione (GSH) (Zhou et al. 2016; Zou et al. 2017; Li et al. 2016; Shao et al. 2017b). Nie and colleagues developed a gene delivery vehicle with the core-shell nanocomplex. An unlockable Hep nanoparticle core was surrounded by a cationic PGEA shell (Nie et al. 2018). Upon breakdown of the redox-responsive Hep core within cells, the unlocked Hep displaced the nucleic acids from outer PGEA shell and accelerated their release. The researchers explored the use of these nanoparticle systems for chronologically staged miR-499 and VEGF pDNA gene therapy. Cardiomyocyte apoptosis was inhibited by miR-499 and a decrease in the infarct size during the initial stage of MI was observed, while the later delivery of VEGF pDNA promoted angiogenesis at the ischemic border. Heart tissue expression of GSH increases with the accumulation of ROS during MI progression. Thus, the staged gene therapy strategy using unlockable Hep cores produced impressive results in the rescue of cardiac functions whilst limiting myocardial fibrosis and hypertrophy (Nie et al. 2018).

Other natural polymeric nanoparticles, such as hyaluronan sulphate nanoparticles have been successfully employed as nanocarriers for gene therapy. Bejerano et al. used hyaluronan sulphate nanoparticles to codeliver miR-21 and Ca^{2+} to infarcted myocardium for targeted macrophage uptake and regulation (Bejerano et al. 2018). Intravenous injection of the nanoparticles promoted a polarisation switch in macrophages, from M1 to M2 phenotypes, thus promoting angiogenesis and limiting fibrosis and cardiac cell apoptosis. However, mechanisms of macrophage and hyaluronan sulphate nanoparticle interactions for targeted regulation of macrophages are not fully understood.

Lipid-Based Nanoparticles

Lipid-based nanoparticles are promising candidates for MI therapy. Lipid nanoparticles are composed of single or bilayers of lipid molecules with an inner water cavity and can be used to incorporate lipophilic or hydrophilic drug molecules, imaging agents, peptides, or nucleic acids (Saludas et al. 2018; Cheraghi et al. 2017). In recent applications for MI therapy, solid lipid nanoparticles have received increasing attention as surrogates to colloidal delivery systems, combining advantages of polymeric nanoparticles with the less stable liposomes, while evading the disadvantages of acute and chronic toxicity. Furthermore, the use of stem or cardiac cell-derived exosome therapy and the development of artificial exosome-like nanoparticles from cell membrane-coated polymeric nanoparticles are beginning to emerge as powerful candidates for use as MI nanomedicines. Table 17.2 lists recent preclinical investigations that have utilised lipid-based nanoparticles as carrier systems for MI therapy, including PEGylated solid lipid nanoparticles, exosomes or extracellular vesicles, and artificial reconstructed exosome-like nanoparticles.

Solid Lipid Nanoparticles

Solid lipid nanoparticles combine the advantages of colloidal liposome or nano-emulsion systems with polymeric nanoparticles. Generally, solid lipid nanoparticles have superior biocompatibility and biodegradability; can be synthesised without the use of organic solvents that may otherwise damage payloads; have high physical stability, thereby allowing for ease of sterilisation and storage; can control drug release and targeting; can encapsulate both lipophilic and hydrophilic drugs; and they can be manufactured in large scale. Solid lipid nanoparticles are widely used to improve hydrophobic drug delivery to target cells, either through passive mechanisms dependent on the tissue microenvironment, through active mechanisms promoted by the use of surface modification of solid lipid nanoparticles, or via codelivery mechanisms. A popularised modification of solid lipid nanoparticles is PEGylation to facilitate improved circulation time and reduced immune recognition, often resulting in improved bioactivity of loaded drugs in vivo and improved MI therapy (Zhang et al. 2016; Guo et al. 2019). In addition, PEGylation provides an additional modification site (also known as a PEG linker) to further functionalise the nanoparticles.

Shao et al. synthesised matrix metalloproteinase (MMP)-sensitive peptide-modified and PEGylated-solid lipid nanoparticles (Shao et al. 2017a). The nanoparticles were used to load Schisandrin B, a type of naturally derived dibenzocyclooctadiene used to treat hepatitis and ischemia. The study aimed to achieve drug targeting, and improvements in the solubility and bioavailability of Schisandrin B, for the treatment of MI in partial coronary artery ligation rat models. Matrix metalloproteinase (MMP)-2 and -9 are proteolytic enzymes present in the myocardium and have importance in myocardial remodelling and restructuring post-MI (Phatharajaree et al. 2007). Activated MMP-2 and MMP-9 were shown to be present within the extracellular space in infarct areas (Nguyen et al. 2015). By modifying solid-lipid nanoparticles with PEG and MMP-sensitive peptide, CPLGLAGG, circulation time, and drug release at infarct site were achieved, respectively. Reductions in infarct size were observed. Similarly, Dong et al. (2017) and Qiu et al. (2017) aimed to improve the bioavailability and stability of loaded drug components by modifying RGD-PEG onto solid-lipid nanoparticles, to treat MI in rat models. In tissue ischemia, it was reported that during angiogenesis, endothelial cells exhibited a high cell surface expression of integrin $\alpha_v\beta_3$ (Lee et al. 2016b). Cyclic arginyl-glycyl-aspartic acid (RGD) peptide enabled binding and interactions with integrin $\alpha_v\beta_3$ (Shan et al. 2015), promotion of cell anchorage (Rosso et al. 2005), and influence the tissue microenvironment to enhance angiogenesis (Davis et al. 2005). Thus, PEG modifications increased

TABLE 17.2

Recent Preclinical Investigation Trends Using Lipid-Based Nanoparticles for MI Therapy

Lipid Nanoparticle	Payload	Model	Reference
PEGylated solid lipid	Schisandrin B (dibenzocyclooctadiene)	Rat	(Shao et al. 2017a)
	Puerarin (flavonoid)	Rat	(Dong et al. 2017)
	Baicalin (flavonoid)	Rat	(Zhang et al. 2016)
	Tanshinone, puerarin-prodrug	Rat	(Guo et al. 2019)
	Salvianolic acid B (benzofuran), panax notoginsenoside (ginsenoside)	Rat	(Qiu et al. 2017)
	Adenosine prodrug	Rat	(Yu et al. 2018)
	miR-199a-3p mimic	Rat	(Yang et al. 2019)
	siRNA targeting *Sdf1* or *Mcp1*	$ApoE^{-/-}$ Mouse	(Krohn-Grimberghe et al. 2020)
Extracellular vesicles/exosomes	Mesenchymal stromal cell (MSC) factors, miR-182	Mouse	(Zhao et al. 2019)
	Human embryonic kidney (HEK) cell factors, miR-21	Mouse	(Song et al. 2019)
	MSC factors, magnetic iron oxide nanoparticles	Rat	(Lee et al. 2020)
	Atorvastatin pretreated MSC factors, supported MSC codelivery	Rat	(Huang et al. 2019)
	Induced pluripotent stem cell (iPSC)-derived cardiovascular progenitor cell factors	Mouse	(Wu et al. 2020a)
	iPSC-derived cardiomyocyte (iCM) factors	Mouse	(Santoso et al. 2020)
	iPSC-derived mixed cardiac cell factors	Porcine	(Gao et al. 2020)
	Cardiosphere-derived cell factors, cardiac homing peptide (CHP)	Rat	(Vandergriff et al. 2018)
Cell membrane-coated PLGA	MSC 3D-spheroid membrane and secretome	Mouse	(Zhang et al. 2020)
Cell membrane-coated mesoporous silica (MS)	MSC membrane, miR-21	Mouse	(Yao et al. 2020)

circulation time, whereas RGD modification enabled the controlled release and cardiac accumulation of drug molecule payloads at the ischemic MI site and enhanced angiogenesis, which in turn reduced the size of the MI region. Yu et al. utilised solid lipid nanoparticles to achieve a prodrug strategy to the treatment of MI (Yu et al. 2018). The researchers conjugated adenosine, a ubiquitous nucleoside implicated in physiological heart functions during IR injury (Kazemzadeh-Narbat et al. 2015; Ernens et al. 2015; Chin et al. 2017), to oleic acid lipids, which resulted in a lipid prodrug-based nanosystems. The strategy reduced toxicity before oleic acid separation from adenosine, increased adenosine stability *in vivo*, and provided prolonged adenosine release. Furthermore, the researchers attached cardioprotective atrial natriuretic peptide (ANP), which enables specific targeting to endocardial expressed natriuretic peptide receptors during MI (Ferreira et al. 2017; Gaudin et al. 2014). ANP-DSPE-PEG-oleic acid adenosine was self-assembled before intravenous injection into rats with myocardial ischemia. The results showed that ANP enhanced the efficiency and efficacy of adenosine in MI treatment.

Yang et al. loaded miR-199a-3p into PEGylated solid-lipid nanoparticles (Yang et al. 2019), following reports that miR-199a-3p stimulated rodent cardiomyocyte (via HOMER1 and CLIC5) and endothelial cell (via caveolin-2) proliferation (Lesizza et al. 2017; Shatseva et al. 2011). In addition, miR-199a-3p was also shown to trigger p-Akt-mediated cardioprotection during ischemic cardiomyopathy (Park et al. 2016) and suppressed fibrosis pathways in lung and kidney tissues (Savary et al. 2019). Intramyocardial injection of a hydrogel containing the miR-loaded and PEGylated solid-lipid nanoparticles allowed for robust *in vivo* and localised miR delivery for post-MI treatment. In addition to protection of miRs against exogenous breakdown, nanoparticles facilitated the high uptake and retention of miRs in stem cell-derived cardiomyocytes and endothelial cells under hypoxic conditions. Injectable hydrogel-nanoparticle composites significantly improved cardiac function in post-MI I/R rat models, restored contractility, reduced myocardial fibrosis, and promoted angiogenesis at the infarct border zone. Krohn-Grimberghe and colleagues (Krohn-Grimberghe et al. 2020) sought to treat MI through targeting barrier cells responsible for releasing stem cells, leukocytes, and progenitor cells from bone marrow (Morrison and Scadden 2014; Mendelson and Frenette 2014; Itkin et al. 2016). Systemically injected PEGylated lipid nanoparticles encapsulating small interfering RNA (siRNA) targeted bone-marrow endothelial cells within the haematopoietic stem-cell niche to silence stromal-derived factor 1 (Sdf1) or monocyte chemotactic protein 1 (Mcp1). Sdf1 enhanced cell release, whereas Mcp1 inhibited cell release from bone marrow. Silencing of Mcp1 reduced the presence of leukocytes in ischemic hearts, which resulted in improved cardiac repair and attenuated heart failure. Superior avidity and uptake efficiency in bone marrow endothelial cells was enhanced by optimising PEG molecular weight, lipid chain, and PEG surface density.

Extracellular Vesicles and Exosomes

Mesenchymal stromal cells (MSCs) have been suggested to confers functional and structural benefits in the regeneration of ischemic heart tissues, through the modification of almost every mechanism essential to cardiac repair (Golpanian et al. 2016). The current consensus is that MSCs exert their beneficial effect on the myocardium through paracrine effect, as opposed to their differentiation and replenishment of myocardial cells (Kishore and Khan 2016). One route of MSC communication to other cells is through the production of exosomes. Exosomes are a subset of extracellular vesicles, are typically 30–150 nm in diameter (Vestad et al. 2017), are abundant in the secretome of many types of cells and stem cells (Jung et al. 2017) and have membrane compositions that are reflective of their cell of origin. Exosomes functionally alter recipient cells through delivery of their contents, which contain small RNAs and proteins (Sluijter et al. 2014). Recent studies have shown that MSC-derived exosomes could rescue tissue injury and regulate inflammation in multiple disease models (Aliotta et al. 2016; Zhang et al. 2018; Xiao et al. 2018). In a very recent study, Zhao et al. improved the immunomodulatory effect of MSC exosomes on macrophages in LAD ligation I/R models of MI, through forced over expression of the TLR4/NF-κB pathway inhibiting miR-182, in parental MSCs (Zhao et al. 2019). The result was exosomes with an elevated miR-182 payload, which were taken up my macrophages in the MI tissue microenvironment when injected in the peri-infarct myocardial region. Song et al. used a similar approach and overexpressed miR-21 (Song et al. 2019), which is a reported inhibitor of PDCD4/AP-1-mediated apoptosis in cardiomyocytes (Dong et al. 2009; Xiao et al. 2016), and has been shown to promote angiogenesis through the PTEN/Akt/VEGF axis in endothelial cells (Wang et al. 2017a). The miR-21-enriched exosomes were produced by genetically modified HEK293T cells and delivered by intramyocardial injection into mouse MI models to promote cardiac functional recovery. The exosomes were shown to protect miR-21 against RNase activity and improved the efficient delivery of miR-21 to recipient cells, which decreased PDCD4 expression, reduced apoptosis, and improved cardiac function. This study highlighted that cells other than MSCs could be utilised to produce enriched exosomes for use as therapeutic vehicles.

Lee et al. opted to develop exosome-mimetic nanovesicles derived from MSCs that had been preloaded with iron oxide nanoparticles (Lee et al. 2020). This allowed for nanovesicles to have extended retention within the infarcted hearts of mice through application of a magnetic field. Interestingly, treatment of MSCs with iron oxide nanoparticles increased the levels of therapeutic molecules produced by MSCs and, thus, also increased the content within the resultant nanovesicles. The nanovesicles induced a shift from an inflammation to a reparative state and were antiapoptotic, antifibrotic, proangiogenic, and cardioprotective. Moreover, the researchers overcame the problem of low exosome yield limitations accounted by previous studies (Katsuda et al. 2013; Dai et al. 2008), by producing exosome-mimics with higher proportions of therapeutic factors, through a hypoxia-like mechanism of hypoxia-inducible factor 1α activation in MSCs (Krock et al. 2011; Minet et al. 1999). Huang et al. found that intramyocardial injection of exosomes from statin pretreated MSCs, followed by the sequential transplantation of MSCs further improved cardiac function, reduced infarct size, and increased angiogenesis, compared to exosome therapy or MSC therapy alone (Huang et al. 2019). The researchers showed that intramyocardial injection of primed exosomes 30 min after MI, combined with MSC transplantation 3 days after, achieved the highest improvement in heart function and reduced scar size. It was speculated that the improvement was likely due to an improved microenvironment, primed by the exosomes and more conducive to MSC transplantation. Reduced inflammation by exosome therapy provided better MSC survival, recruitment, and retention.

In addition to MSCs, pluripotent embryonic stem cells (ESCs) and induced pluripotent stem cells (iPSCs) are attractive sources for stem cell-mediated therapies. Further to this, iPSCs and other pluripotent stem cells (PSCs) can be forced to differentiate into cardiovascular progenitors and other cardiac cells with relative ease and reproducibility. The efficacy of PSC-derived cardiac cell-secreted exosomes in the repair of MI has started to come to light. Wu et al. reported the cardioprotective effects of exosomes secreted by ESC-derived cardiac progenitor cells (Wu et al. 2020a). Intramyocardial injection of exosomes/extracellular vesicles into acutely infracted mouse hearts significantly improved cardiac function, reduced fibrosis, and improved vascularisation and cardiomyocyte survival at infarct border zones. The researchers revealed that a high abundance of the long noncoding RNA, MALAT1, was present and upregulated in the infarcted myocardium and cardiomyocytes treated with exosomes. MALAT1 expression improved cell viability, while its knockdown inhibited exosome-promoted proangiogenic activities, through targeting miR-497. Santoso et al. harvested exosomes from iPSCs that had differentiated into contractile cardiomyocytes (Santoso et al. 2020). The iPSC-derived cardiomyocyte-secreted exosomes preserved cardiac function, myocyte viability, and exerted antiapoptotic effects in mouse MI models. Furthermore, exosomes upregulated autophagy activity to maintain cardiac homeostasis. The researchers implicated autophagic flux impairment as a key characteristic of ischemia, and that it could be restored by exosome treatment. Gao et al. compared intramyocardial injection of a mixture of iPSC-derived cardiomyocyte, endothelial cell, and smooth muscle cell exosomes (2:1:1 ratio) into the pig models of MI (Gao et al. 2020). Measurements of myocardial function and bioenergetics were improved, whilst myocardial hypertrophy and fibrosis were ameliorated in MI regions at 1-month postexosome injection, suggesting an attractive method of harvesting high yields of exosomes to provide acellular therapeutic options for MI. The simultaneous harvest of exosomes from multiple cardiac cell types has been revealed to be a promising source of abundant yields of therapeutic exosomes that exert cardioprotective effects. Cardiospheres are self-assembled multicellular constructs that form after cellular outgrowth from cardiac explants on a nonadhesive substrate. Vandergriff et al. harvested exosomes from cardiospheres and conjugated the exosomes to cardiac homing peptide (peptide sequence CSTSMLKAC) (Vandergriff et al. 2018). The researchers demonstrated increased retention of the exosomes within the MI heart in rat I/R injury models, which was accompanied by decreased infarct size and fibrosis, increased cellular proliferation, and enhanced angiogenesis.

Artificial Extracellular Vesicles and Exosomes

Recently, biomaterial-based therapies have shown promising routes to overcome the obstacles facing stem cell-based therapy (Murphy et al. 2014; Midgley et al. 2020a; Trappmann et al. 2017). Through the packaging of factors secreted by *in vitro* cultures of MSCs into PLGA particles, Luo and colleagues prepared materials that mimicked stem cell function (Luo et al. 2017). These synthetic MSC-like particles had therapeutic effects comparable to MSC therapy in various diseases (Tang et al. 2018; Liang et al. 2018). Building on this, Zhang et al. harvested paracrine release factors from 3D spheroid cultures of MSCs (Zhang et al. 2020), which more adequately represented paracrine components released from *in vivo* engraftment of stem cells into tissues (Laschke and Menger 2017). The researchers artificially reconstructed stem cell exosome-like nanoparticles with biomimetic functionality, by encapsulating paracrine factors from 3D MSC spheroids within PLGA nanoparticles coated with a mixture of reassembled platelet and red blood cell membranes. The artificial exosome-like nanoparticles exhibited elevated retention in MI mouse hearts, after systemic tail vein injection, attributable to increased circulation time and vascular damage recognition from the red blood cell and platelet membrane coatings, respectively. In mice with acute MI, cardiac functional recovery was achieved, scar size was significantly reduced, and evidence of proangiogenic activity was demonstrated. The membrane coating of the nanoparticles enhanced delivery of the paracrine factors to infarcted hearts. The strategy offered a new alternative to cell-based therapeutic options for the regeneration of ischemic tissue, by addressing issues that hinder the clinical advancement of stem cell therapies.

Yao et al. used a similar method of coating nanoparticles with cell membranes to improve therapeutic efficacy (Yao et al. 2020). The researchers used self-assembled stem cell membrane-camouflaged and exosome-mimicking nanocomplexes to replicate exosome functionality and to achieve the efficient delivery of miR-21 for MI repair. The nanocomplex was constructed by coating miR-21-loaded mesoporous silica nanoparticles with reassembled MSC membranes, which enabled efficient miR entrapment and protection from degradation. The nanocomplexes evaded the immunologic system and targeted ischemic cardiomyocytes to enhance their proliferation. Intravenous administration of exosome-mimicking nanocomplexes preserved the viable myocardium and augmented cardiac function. The researchers noted that an interesting unknown arose from this investigation, how the nanocomplex escaped from vasculature proximal to the MI region. They speculated that inflammation near the infarct site, together with mesoporous silica cores, may have induced oxidative stress in nearby endothelial cells, which then led to increased vessel permeability (Tee et al. 2019; Setyawati et al. 2016) and escape of nanocomplexes from the vasculature into the MI site.

Other Types of Nanoparticles

Various other types of nanoparticles have been used as strategies in the treatment of MI. Table 17.3 lists the recent preclinical investigations that have utilised various nanoparticles for treatment of the infarcted heart. In the following section, we highlight the recently used nanoparticles that include dendrimeric nanoparticles, metallic nanoparticles, and carbon-based nanoparticles.

Dendrimers

Nanoparticles assembled from repetitively branched polymers, with symmetrical 'tree-like' branches originating from a core node and spherical 3D morphology, are termed dendrimers. The numerous branches of dendrimers can facilitate a high degree of functionalisation. The multitude of active sites available for modification makes dendrimers ideal nanocarriers that can be modified to possess one or more targeting ligand attachments on each branch's terminal site.

Previously, it was shown that miR-1 accelerated cardiomyocyte apoptosis via inhibition of PKCε, Bcl-2, and other associated antiapoptotic proteins (Pan et al. 2012). Xue et al. sought to overcome complications

TABLE 17.3

Recent Preclinical Investigation Trends Using Other Types of Nanoparticles for MI Therapy

Nanoparticle	Payload	Model	Reference
Dendrimer			
PEGylated-dendrigraft poly-L-lysine (DGL)	AMO-1 (miRNA-1 inhibitor)	Mouse	(Xue et al. 2018)
Linear dendronised polymer (Denpol)	AID-peptide	Mouse	(Viola et al. 2020)
Metallic			
Gold (AuNPs)	—	Rat	(Dong et al. 2020)
	2-methoxy-isobutyl-isonitrile (MIBI)	Rat	(Tartuce et al. 2020)
PEGylated-AuNPs	—	Mouse	(Tian et al. 2018)
PEGylated silica-coated iron oxide (Fe_3O_4)	Antibody capture of circulating exosomes and binding to injured cardiomyocytes	Rabbit Rat	(Liu et al. 2020a)
Carbon-based			
Graphene oxide	—	Rat	(Zhou et al. 2018)
Fullerenol	Supported ADSC delivery	Rat	(Hao et al. 2017)

associated with administration of free antisense molecules (Yang et al. 2007; Miyagi et al. 2010; Zhang et al. 2013), using anti-miR-1 oligonucleotides (AMO-1) nanocarriers to inhibit miR-1 and promote cell survival in MI (Xue et al. 2018). The researchers developed an approach for early myocardium-targeting capacity, via recognition of the angiotensin II type 1 receptors (AT1R) that are overexpressed during early stages of MI (Molavi et al. 2006; Dvir et al. 2011). A peptide domain with high affinity for AT1R, located within angiotensin II type 1 (AT1) was identified as a ligand for early targeting to MI (Hong et al. 2014; Liu et al. 2014). Dendrigraft poly-L-lysine (DGL) dendrimers were used as gene therapy nanovectors due to their highly branched and monodispersed structure. The high density of amines on their surface facilitates strong nucleic acid binding capacity (Mann et al. 2008; Li et al. 2013). AT1 peptide conjugation enabled the targeted delivery to the ischemic myocardium via specific binding to AT1R during the initial stage of MI. The combined properties of the nanoparticles exhibited improved targeting ability and attenuated cardiomyocyte apoptosis, significantly reducing the infarct size in MI mice after a single injection. Viola et al. demonstrated the use of optimised linear dendronised polymers (denpols) as materials for nanoparticle-mediated delivery of bioactive peptides to target α-interaction domain of the cardiac L-type calcium channel (AID), which subsequently regulated calcium channel activity and treated animal models of MI (Viola et al. 2020). The denpol-AID complexes were used to specifically target the MI heart in mice and *ex vivo* MI hearts from guinea pig models. Within 2 h of denpol-AID administration, maximal cardiac uptake and comprehensive distribution throughout the myocardium were achieved, which resulted in reduced myocardial damage (Clemons et al. 2013).

Metallic Nanoparticles

Metallic nanoparticles are nanosized clusters of metals, often prepared to be modifiable with functional chemical groups to allow binding of ligands or drugs. Metallic nanoparticles allow for a broad range of application in CVD therapeutics, including modalities that other nanoparticles lack, such as catalyst enzyme-like activities, imaging by MRI and magnetisation for guided delivery. Another advantage of metallic-based materials is their innate conductivity of electrical charge. Major challenges in repair of MI are how to increase electrical behaviours of implanted biomaterials and how to maximise electrical interactions between cells and scaffolds. Dong et al. utilised the distribution of electrically and biologically active gold nanoparticles throughout tissue regeneration-conducive ECM/silk fibroin scaffolds (Dong et al. 2020). The supply of gold nanoparticles within ECM matrices can provide corrugation of cardiac cells and the effective organisation of electrical coupling with cells (Somasuntharam et al. 2016; Nair et al. 2017). The researchers used different ratios of ECM, silk fibroin, and gold nanoparticles to investigate changes in morphology, mechanical properties, and cell compatibility with MSCs. The optimised composite scaffolds containing gold nanoparticles, or cardiac patches, supported MSC survivability and promoted myocardium regeneration in MI cryo-injury rat models. The researchers concluded that their developed heart patches were conducive to supporting cardiomyocyte cell behaviour, essential to myocardial function.

Tartuce et al. sought to increase myocyte viability after MI by targeted drug delivery by using 2-methoxy-isobutyl-isonitrile (MIBI)-conjugated gold nanoparticles (Tartuce et al. 2020). MIBI was captured by myocytes and inflammatory marker expression was decreased in rat hearts after MI. The therapeutic mechanism of action was determined to be decreased glutathione oxidation in treated rat heart tissue post-MI. The gold nanoparticles possessed antioxidant and antiinflammatory properties, as evidenced by suppressed detoxification from GSH system interaction and reduced inflammatory cytokine (IL-6, TNF-α) mRNA expression, respectively. The gold nanoparticles easily crossed cell membranes and reached the mitochondria, and these actions were enhanced in cardiomyocytes through conjugation of nanoparticles to cationic lipophilic compound MIBI. Tian et al. showed that PEGylated gold nanoparticles dampened the inflammatory response and subsequent fibrosis in MI mouse models, following systemic injection via the tail vein (Tian et al. 2018). Cardiac function was improved, infarct size was reduced, profibrotic collagen deposition was attenuated, and inflammation was inhibited.

Recently, Liu et al. exploited the magnetic guidance properties of ferro-ferric oxide (Fe_3O_4) nanoparticles and used them as 'vesicle shuttles' for the capture of circulating exosomes and subsequent release at the MI site (Liu et al. 2020a). The Fe_3O_4 cores were enclosed within silica shells and coated with stimuli-cleavable PEG chains conjugated to two types of antibody (one for endogenous exosome capture and one for targeting recipient injured cells). The researchers showed that the vesicle shuttles effectively collected, transported, and released circulating exosomes at the designated MI region in rat models. The released exosomes enhanced cardiac function, improved cardiomyocyte survival, and promoted angiogenesis. The investigation is one of the leading pioneers in the concept of manipulating endogenously produced exosomes *in vivo* and redirecting their biodistribution for MI therapy.

Carbon-Based Nanoparticles

Carbon-based nanoparticles and nanotubes have received acclaim for their beneficial results in a myriad of environmental and biomedical applications. Recent emergent studies have sought to apply carbon-based nanoparticles to MI therapy, whilst overcoming safety concerns often associated with carbon nanoparticles. Zhou et al. combined graphene oxide (GO) nanoparticles with oligo (PEG) fumarate (OPF) to create conductive hydrogels (Zhou et al. 2018). The injectable OPF/GO hydrogels improved cardiac electrical propagation in MI rat hearts. At 1-month post-MI there was enhanced Ca^{2+} signal conduction within cardiomyocytes located at the MI site, which resulted in improved cardiac function. Furthermore, composite hydrogels provided mechanical support and electrical conductivity to cardiomyocytes in healthy and infarcted myocardium, which led to the activation of canonical Wnt signalling. GO has good electrical conductivity, mechanical properties, and could enhance electric signal propagation, which improved cardiac repair. Hao et al. investigated fullerenol/alginate hydrogels loaded with ADSCs to modulate the MI tissue microenvironment for improved efficacy of stem cell therapy (Hao et al. 2017). Ionotropic alginate hydrogels have been widely used as vehicles for stem cell delivery, owing to its structural resemblance to the ECM (Wu et al. 2020b). However, alginate's high degree of hydrophilicity is associated with poor cellular adhesion and support for proliferation. Furthermore, alginate has low antioxidant activity, which results in failure to protect cells from damage under ROS microenvironments, such as MI (Alijani-Ghazyani et al. 2020). Fullerenol is an antioxidant nanoparticle and derivative of fullerene. Unlike other carbon-based nanomaterials, fullerenol's high degree of water solubility is beneficial for the preparation of homogeneous hydrogels, lower cytotoxicity, and antioxidant activity (He and Chen 2020; Majid et al. 2020). The surface of fullerenol has electron-deficient positions, which can easily adhere to ROS. Electron transfer from ROS on adjacent electron-deficient position to the fullerenol cage may induce the destruction of ROS in a similar manner to the ROS quenching activity of superoxide dismutase. Thus, fullerenol and fullerene derivatives have been explored for their effective suppression activity of ROS and oxidant stress damage in cells (An et al. 2020; Mancuso et al. 2020). Fullerenol nanoparticles readily penetrate cell membranes and further modulate expression of signalling proteins related to stem cell growth, survival, and cardiomyogenesis (He and Chen 2020; Liu et al. 2020b). Therefore, researchers hypothesised that fullerenol/alginate hydrogels would suppress oxidative stress and improve cardiac repair after MI. The fullerenol/alginate hydrogels prevented ROS-induced damage and inhibited the activation of adverse cell-signalling cascades after MI. The fullerenol/alginate hydrogels inhibited JNK signalling pathways but activated antiapoptotic and prosurvival signalling pathways, such as ERK and p38, in codelivered ADSCs. In addition, fullerenol improved the cardiomyogenic differentiation capacity of ADSCs and promoted gap junction formation between cells. The fullerenol/alginate hydrogels combined with ADSCs reduced infarct size, increased wall thickness, and improved cardiac functions.

Future Perspectives

The series of biochemical and cellular events that follow MI incidence are relatively well understood, but there remain some unknowns in the in-depth cellular mechanisms during the onset and consequences of MI tissue pathology. Until these mechanisms are fully understood, nanotechnologists and biomaterial scientists can select key events in MI to target for intervention, including but not limited to: immunoregulation; mitochondrial regulation; cardiomyocyte apoptosis mitigation; dampening ROS accumulation and effects; angiogenesis at the infarct site; prevention or reversal of myocardial fibrosis; and restoration of cardiac functionality. Nanoparticles and nanotechnology can facilitate the production of a series of nanoproducts loaded with small-molecule drugs, growth factors, cell products, etc., to enhance the delivery efficiency of drugs, through physical and chemical modification of targeting, and to inhibit pathogenesis in MI with increased efficacy.

When selecting appropriate nanomaterials, disease-driven approaches rather than formulation-driven approaches may be the preferred option for MI therapy. In addition, choice of drug administration route provides nanoparticle systems for MI with augmented support for better benefit-to-risk ratios, for example, invasive intramyocardial injection or cardiac patch application in combination with I/R intervention *versus* noninvasive oral, inhaled, and intravenous routes with nanoparticles modified to traverse tissue barriers and to target the myocardium but at the risk of interacting with off-target tissues and organs (i.e., liver, kidneys, spleen, brain, etc.). Additional routes of nanoparticle administration to the heart have emerged that may inform future nanoparticle-based MI treatment. Therapeutic targeting of the heart with biocompatible and inhalable nanoparticles displayed lung-to-heart transmission, as bioinspired from heart tissue accumulation of ultrafine combustion-derived particulates from air pollution was described. These nanoparticles crossed biological barriers and arrived at the myocardium in animal models of diabetic cardiomyopathy, highlighting the potential for more efficient cardiac-specific nanoparticle delivery (Miragoli et al. 2018).

An intended purpose of nanoparticle drug carriers is to prevent adverse side effects. The incorporation of ligand- or receptor-targeting approaches add additional layers of complexity to nanoparticle therapeutic products, which improves therapeutic efficacy but may in turn creates more obstacles regarding authorisation by global drug administrations and slowing the progress towards clinical application. The shared aim to reduce off-target toxicity whilst increasing therapeutic selectivity heart tissue will likely give rise to more complex nanoparticle systems exhibiting functionalisation with cell-specific targeting moieties (peptides, aptamers, etc.) to allow for selective tuning of nanoparticle biodistribution towards the MI site. Establishment of a targeted cellular point-of-interaction will lead to more efficient nanoparticles that exploit specific cell internalisation mechanisms and thus enhance the release of payload within cardiomyocytes or other cardiac cells. Stimuli-responsive nanoparticles, such as magnetic or MI tissue microenvironment sensitive nanoparticles, may enhance the controlled guidance of nanoparticles to the MI site and facilitate further enactment of localised payload release. Currently, there have been no clinically approved nanoparticle-drug therapeutics available for the above purposes. Functionalisation of nanoparticles with targeting moieties may lead to concerns more commonly raised over resultant undesirable *in vivo* physicochemical changes affecting the nanoparticle itself. The application prospects of nanoparticles in MI patients are still too early to assess, and there is a requirement for substantial effort to realise clinical breakthrough. Synergy between biocompatibility, biodegradability, and efficient administration strategies may represent an attractive approach for pharmacological treatment of MI.

Exosome-based therapy, although receiving extensive interest as a viable alternative to cell-based therapies, has clinical applications that are still challenged and limited by several issues. First, the large amounts of exosome production time are required, and the poor purification and isolation efficiency of current techniques has failed to produce high quantities of exosomes. Second, exosomes contain large amounts of bioactive factors, some of which could result in undesirable or counter-productive side effects in cardiac tissues. Third, heterogeneous components in exosomes may give rise to potential risks for tumour formation and immunogenicity, based on nature of the parental cells. Finally, the exact therapeutic effect of exosomes is unclear in intervention during disease due to their complex structure and composition. Accordingly, investigations into the use of stem cell-derived exosomes are taking precedence due to their demonstrated functional and therapeutic activity. However, several research groups have shown that exosomes from various cardiac source and even circulating exosomes convey therapeutic effects. Therefore, it remains critical that exploited functional roles, precise components of exosomes, and methods to enrich their production are fully understood before therapeutic delivery in patients can be realised.

Utilisation of various smart materials combined with therapeutic technologies may give rise to advanced therapeutic strategies with strengthened applicability for MI therapy, over the use of single-phase materials only. For example, human iPSC-derived cardiomyocytes combined with injectable erythropoietin-loaded nanohydrogels facilitated cell survival and increased post-MI tissue remodelling (Chow et al. 2017). Injectable hydrogels can also be used as scaffolds for the promotion of endogenous repair. Nguyen et al. used MMP-responsive hydrogels that exhibited the capacity for retention at the infarct site upon enzymatically triggered biotransformation (Nguyen et al. 2016), thus indicating suitability for sustained delivery of therapeutic biologics. The use of a biocompatible epicardial patch placed on the border zone of the infarct achieved sustainable delivery of cells, bioactive macromolecules, and drugs for cardiac therapy (Whyte et al. 2018). Feasibly, these materials can also be used for improved efficiency and controlled release of therapeutic nanoparticles *in situ*, thereby increasing nanoparticle retention within the MI site.

Representation for new and appealing opportunities in the development of MI therapies can be found in the modulation of cardiac metabolism and gene expression, by pharmacological intervention and manipulation of miR-mediated epigenetic regulatory networks. In this direction, a possible switch for the control of cardiomyocyte proliferation has implicated the Hippo/YAP pathway (Torrini et al. 2019). The Hippo/YAP pathway has been strongly associated with the initiation and progression of cardiomyopathies (Clippinger et al. 2019). The specific manipulation of Hippo/YAP in the heart offers a novel option for mitigating cardiomyocyte cell death and treating MI. Research developing nanoparticle modulation of cardiomyocyte Hippo/YAP activity remains in its infancy, but a breakthrough may herald revolutionary MI therapies.

Gene editing technologies hold potential for restoring cardiac function. The CRISPR/Cas9 system can be exploited to introduce dsDNA cleavage and sequence-specific loci (Jinek et al. 2012). Nanoparticles are being explored as nanocarriers for mediating CRISPR/Cas9 components and to guide genetic reprogramming (Liu et al. 2019b). However, nanocarrier specificity and associated technical challenges of efficiency and safety are yet to be characterised. Moreover, the packaging of the multiple vital components (donor DNA, guide RNA, and the Cas9 ribonucleoprotein) requires complex nanoparticle-based systems, which are posed with the subsequent issues of difficulty in scalability and escalating expense.

Alternative material strategies for nanoparticle generation, although not recently employed for MI therapy, may serve as suitable approaches to diversify MI therapy. DNA nanostructures can be precisely assembled to exhibit chemical and morphological characteristics and can offer considerable prospects to delineate information concerning the nanoparticle–biological interface (Lee et al. 2016a). Structural DNA nanotechnology is considered a promising stratagem for drug delivery (Ke et al. 2018). Artificially constructed nucleic acid nanodevices have been engineered to release payloads upon sensing certain tissue microenvironments (Fan et al. 2020a). Moreover, several studies proposed the use of various DNA nanostructure strategies to encapsulate biomolecular drugs for their specific release within cardiac tissues (Zhang et al. 2019).

An application of nanoparticles not discussed here is MI diagnostics. Engineered of multifunctional nanoparticles that target specific tissues or diseases whilst conveying bioimaging capabilities can enable their use in both therapy and diagnosis (often termed theranostics). One example of the nanoparticles for CVD diagnosis is their use for early detection of atherosclerosis. Atherosclerosis is often a precursor to MI and can be identified through the detection of inflammatory markers associated with the early stage of the disease. Noninvasive diagnostics are key requirements to successfully determine the onset of atherosclerosis and establish treatment regimens to prevent progression to MI. This chapter defers to an in-depth review on the potential of nanoparticles as diagnostic tools for atherosclerosis by Bejarano and colleagues (Bejarano et al. 2018).

Considerable advancements have been made towards clearer comprehension of nanoparticle actions in biological microenvironments, but major challenges persist. Bioavailability, accumulation at target sites, and efficient payload release are just a few examples of the challenges facing nanoparticles following their administration. Limited understanding of biological barriers and drug delivery concepts, manufacturing cost-effectiveness, and regulatory issues has impacted the clinical translation of nanoparticles. Several distinct nanoparticle-based systems have been developed with the collective aim of preventing MI and restoring cardiac function. However, for promising nanoparticle strategies to achieve their purpose, general guidelines describing material types, drug classes, and targeting strategies must be deduced. Liposomes have obtained successful clinical outcomes in various applications, which give them an advantage in terms of being suitable candidates for CVD treatment. However, it cannot be denied that polymeric materials possess preferential tunable properties and superior accommodation for payloads.

The nanoparticle technologies discussed in this chapter have emerged as the most promising strategies for MI prevention and myocardial repair. To further establish heart targeted nanoparticles, improved efficacy and lowered adverse tissue accumulation following intravenous route are key requirements. The range of preclinical *in vivo* studies described here has demonstrated the potential of nanoparticle systems in the protection of cardiac function and in the induction of tissue regeneration of the infarcted myocardium. Further improvement in targeted delivery and therapeutic effects will continue to be developed as our understanding of the MI disease process and nanoparticle technology continues to improve. Therefore, it is expected that forthcoming innovations in nanomedicines, cell-based therapies, and material sciences will provide robust practical enhancements to nanoparticle-based treatments for MI therapy.

REFERENCES

Abbate, A., R. Bussani, M. S. Amin, G. W. Vetrovec, and A. Baldi. *Int J Biochem Cell Biol.* 2006; *38* (11):1834–1840. doi:10.1016/j.biocel.2006.04.010.

Acharya, S., and S. K. Sahoo. *Adv Drug Deliv Rev.* 2011; *63* (3):170–183. doi:10.1016/j.addr.2010.10.008.

Ahmadian, M., J. M. Suh, N. Hah, C. Liddle, A. R. Atkins, M. Downes, and R. M. Evans. *Nat Med.* 2013; *19* (5):557–566. doi:10.1038/nm.3159.

Alijani-Ghazyani, Z., A. Mohammadi Roushandeh, R. Sabzevari, A. Salari, M. T. Razavi Toosi, A. Jahanian-Najafabadi, and M. Habibi Roudkenar. *Int J Biochem Cell Biol.* 2020: 105897. doi:10.1016/j.biocel.2020.105897.

Aliotta, J. M., M. Pereira, S. Wen, M. S. Dooner, M. DelTatto, E. Papa, L. R. Goldberg, G. L. Baird, C. E. Ventetuolo, P. J. Quesenberry, and J. R. Klinger. *Cardiovasc Res.* 2016; *110* (3):319–330. doi:10.1093/cvr/cvw054.

An, Z., J. Yan, Y. Zhang, and R. Pei. *J Mater Chem B.* 2020; *8* (38):8748–8767. doi:10.1039/d0tb01380c.

Azarmi, S., Y. Huang, H. Chen, S. McQuarrie, D. Abrams, W. Roa, W. H. Finlay, G. G. Miller, and R. Lobenberg. *J Pharm Pharm Sci.* 2006; *9* (1):124–132.

Bae, S., M. Park, C. Kang, S. Dilmen, T. H. Kang, D. G. Kang, Q. Ke, S. U. Lee, D. Lee, and P. M. Kang. *J Am Heart Assoc.* 2016; *5* (11). doi:10.1161/JAHA.116.003697.

Bajpai, A. K., and J. Choubey. *J Mater Sci Mater Med.* 2006; *17* (4):345–358. doi:10.1007/s10856-006-8235-9.

Bei, W., L. Jing, and N. Chen. *Colloids Surf. B Biointerfaces.* 2020; *185*: 110635. doi:10.1016/j.colsurfb.2019.110635.

Bejarano, J., M. Navarro-Marquez, F. Morales-Zavala, J. O. Morales, I. Garcia-Carvajal, E. Araya-Fuentes, Y. Flores, H. E. Verdejo, P. F. Castro, S. Lavandero, and M. J. Kogan. *Theranostics.* 2018; *8* (17):4710–4732. doi:10.7150/thno.26284.

Bejerano, T., S. Etzion, S. Elyagon, Y. Etzion, and S. Cohen. *Nano Lett.* 2018; *18* (9):5885–5891. doi:10.1021/acs.nanolett.8b02578.

Benjamin, E. J., P. Muntner, A. Alonso, M. S. Bittencourt, C. W. Callaway, A. P. Carson, A. M. Chamberlain, A. R. Chang, S. Cheng, S. R. Das, F. N. Delling, L. Djousse, M. S. V. Elkind, J. F. Ferguson, M. Fornage, L. C. Jordan, S. S. Khan, B. M. Kissela, K. L. Knutson, T. W. Kwan, D. T. Lackland, T. T. Lewis, J. H. Lichtman, C. T. Longenecker, M. S. Loop, P. L. Lutsey, S. S. Martin, K. Matsushita, A. E. Moran, M. E. Mussolino, M. O'Flaherty, A. Pandey, A. M. Perak, W. D. Rosamond, G. A. Roth, U. K. A. Sampson, G. M. Satou, E. B. Schroeder, S. H. Shah, N. L. Spartano, A. Stokes, D. L. Tirschwell, C. W. Tsao, M. P. Turakhia, L. B. VanWagner, J. T. Wilkins, S. S. Wong, S. S. Virani. Heart Disease and Stroke Statistics-2019 Update: A Report From the American Heart Association. *Circulation.* 2019; *139* (10):e56–e528. doi:10.1161/CIR.0000000000000659.

Boverhof, D. R., C. M. Bramante, J. H. Butala, S. F. Clancy, M. Lafranconi, J. West, and S. C. Gordon. *Regul Toxicol Pharmacol.* 2015; *73* (1):137–150. doi:10.1016/j.yrtph.2015.06.001.

Chen, Y., Y. Zhao, W. Chen, L. Xie, Z. A. Zhao, J. Yang, Y. Chen, W. Lei, and Z. Shen. *MicroRNA-133 Stem Cell Res Ther.* 2017; *8* (1):268. doi:10.1186/s13287-017-0722-z.

Chenthamara, D., S. Subramaniam, S. G. Ramakrishnan, S. Krishnaswamy, M. M. Essa, F. H. Lin, and M. W. Qoronfleh. *Biomater Res.* 2019; *23*: 20. doi:10.1186/s40824-019-0166-x.

Cheraghi, M., B. Negahdari, H. Daraee, and A. Eatemadi. *Biomed Pharmacother.* 2017; *86*: 316–323. doi:10.1016/j.biopha.2016.12.009.

Chin, K. Y., C. Qin, L. May, R. H. Ritchie, and O. L. Woodman. *Curr Drug Targets.* 2017; *18* (15):1689–1711. doi:10.2174/1389450116666151001112020.

Chow, A., D. J. Stuckey, E. Kidher, M. Rocco, R. J. Jabbour, C. A. Mansfield, A. Darzi, S. E. Harding, M. M. Stevens, and T. Athanasiou. *Stem Cell Reports.* 2017; *9* (5):1415–1422. doi:10.1016/j.stemcr.2017.09.003.

Clemons, T. D., H. M. Viola, M. J. House, K. S. Iyer, and L. C. Hool. *ACS Nano*. 2013; *7* (3):2212–2220. doi:10.1021/nn305211f.

Clippinger, S. R., P. E. Cloonan, L. Greenberg, M. Ernst, W. T. Stump, and M. J. Greenberg. *Proc Natl Acad Sci U S A*. 2019; *116* (36):17831–17840. doi:10.1073/pnas.1910962116.

Cui, Y. *Pathophysiology*. 2010; *17* (4):307–314. doi:10.1016/j.pathophys.2009.07.006.

Dai, S., D. Wei, Z. Wu, X. Zhou, X. Wei, H. Huang, and G. Li. *Mol Ther.* 2008; *16* (4): 782–790. doi:10.1038/mt.2008.1.

Davis, M. E., P. C. Hsieh, A. J. Grodzinsky, and R. T. Lee. *Circ Res*. 2005; *97* (1):8–15. doi:10.1161/01.RES.0000173376.39447.01.

Din, F. U., W. Aman, I. Ullah, O. S. Qureshi, O. Mustapha, S. Shafique, and A. Zeb. *Int J Nanomedicine*. 2017; *12*: 7291–7309. doi:10.2147/IJN.S146315.

Dong, S., Y. Cheng, J. Yang, J. Li, X. Liu, X. Wang, D. Wang, T. J. Krall, E. S. Delphin, and C. Zhang. *J Biol Chem*. 2009; *284* (43):29514–29525. doi:10.1074/jbc.M109.027896.

Dong, Y., M. Hong, R. Dai, H. Wu, and P. Zhu. *Sci Total Environ*. 2020; *707*: 135976. doi:10.1016/j.scitotenv.2019.135976.

Dong, Z., J. Guo, X. Xing, X. Zhang, Y. Du, and Q. Lu. *Biomed Pharmacother*. 2017; *89*: 297–304. doi:10.1016/j.biopha.2017.02.029.

Duro-Castano, A., E. Gallon, C. Decker, and M. J. Vicent. *Adv Drug Deliv Rev.* 2017; *119*: 101–119. doi:10.1016/j.addr.2017.05.008.

Dvir, T., M. Bauer, A. Schroeder, J. H. Tsui, D. G. Anderson, R. Langer, R. Liao, and D. S. Kohane. *Nano Lett.* 2011; *11* (10):4411–4414. doi:10.1021/nl2025882.

Egwim, C., B. Dixon, A. P. Ambrosy, and R. J. Mentz. *Curr Heart Fail Rep*. 2017; *14* (1):30–39. doi:10.1007/s11897-017-0316-1.

Elzoghby, A. O. *J Control Release*. 2013; *172* (3):1075–1091. doi:10.1016/j.jconrel.2013.09.019.

Ernens, I., M. Bousquenaud, B. Lenoir, Y. Devaux, and D. R. Wagner. *J Leukoc Biol*. 2015; *97* (1):9–18. doi:10.1189/jlb.3HI0514-249RR.

Eshun, D., R. Saraf, S. Bae, et al. *J Appl Physiol*. 2017; *122* (6):1388–1397. doi:10.1152/japplphysiol.00467.2016.

Fam, S. Y., C. F. Chee, C. Y. Yong, K. L. Ho, A. R. Mariatulqabtiah, and W. S. Tan. *Nanomaterials (Basel)*. 2020; *10* (4). doi:10.3390/nano10040787.

Fan, C., J. Joshi, F. Li, B. Xu, M. Khan, J. Yang, and W. Zhu. *Front Bioeng Biotechnol*. 2020a; *8*: 687. doi:10.3389/fbioe.2020.00687.

Fan, C., Y. Oduk, M. Zhao, X. Lou, Y. Tang, D. Pretorius, M. T. Valarmathi, G. P. Walcott, J. Yang, P. Menasche, P. Krishnamurthy, W. Zhu, and J. Zhang. *JCI Insight*. 2020b; *5* (12). doi:10.1172/jci.insight.132796.

Fang, R., S. Qiao, Y. Liu, Q. Meng, X. Chen, B. Song, X. Hou, and W. Tian. *Int J Nanomedicine*. 2015; *10*: 4691–4703. doi:10.2147/IJN.S81451.

Ferreira, M. P. A., S. Ranjan, S. Kinnunen, et al. *Small*. 2017; *13* (33). doi:10.1002/smll.201701276.

Formiga, F. R., B. Pelacho, E. Garbayo, et al. *J Control Release*. 2010; *147* (1):30–37. doi:10.1016/j.jconrel.2010.07.097.

Frangogiannis, N. G. *Circ Res*. 2012; *110* (1):159–173. doi:10.1161/CIRCRESAHA.111.243162.

Fujiwara, M., T. Matoba, J. I. Koga, et al. *Cardiovasc Res*. 2019; *115* (7):1244–1255. doi:10.1093/cvr/cvz066.

Gao, L., L. Wang, Y. Wei, et al. *Sci Transl Med*. 2020; *12* (561):eaay1318. doi:10.1126/scitranslmed.aay1318.

Garcia-Dorado, D., M. Andres-Villarreal, M. Ruiz-Meana, J. Inserte, and I. Barba. *J Mol Cell Cardiol*. 2012; *52* (5):931–939. doi:10.1016/j.yjmcc.2012.01.010.

Gaudin, A, M. Yemisci, H. Eroglu, et al. *Nat Nanotechnol*. 2014; *9* (12):1054–1062. doi:10.1038/nnano.2014.274.

Golpanian, S., A. Wolf, K.E. Hatzistergos, J. M. Hare. *Physiol Rev.* 2016; *96* (3):1127–1168. doi:10.1152/physrev.00019.2015.

Gref, R., M. Lück, P. Quellec, et al. *Colloids Surf. B Biointerfaces* 2000; *18* (3–4):301–313. doi:10.1016/s0927-7765(99)00156-3.

Guo, J., X. Xing, N. Lv, et al. *Biomed Pharmacother.* 2019; *120*: 109480. doi:10.1016/j.biopha.2019.109480.

Gustafson, H. H., D. Holt-Casper, D. W. Grainger, and H. Ghandehari. *Nano Today*. 2015; *10* (4):487–510. doi:10.1016/j.nantod.2015.06.006.

Han, J., Q. Wang, Z. Zhang, T. Gong, X. Sun. *Small*. 2014; *10* (3):524–535. doi:10.1002/smll.201301992.

Hao, T., J. Li, F. Yao, et al. *ACS Nano*. 2017; *11* (6):5474–5488. doi:10.1021/acsnano.7b00221.

He, L., and X. Chen. *Adv Healthc Mater*. 2020; *9* (22):e2001175. doi:10.1002/adhm.202001175.

He, Y., G. Ye, C. Song, C. Li, W. Xiong, L. Yu, X. Qiu, and L. Wang. *Theranostics*. 2018; *8* (18):5159–5177. doi:10.7150/thno.27760.

Henri, O., C. Pouehe, M. Houssari, et al. *Circulation*. 2016; *133* (15):1484–1497. doi:10.1161/CIRCULATIONAHA.115.020143.

Henry, T.D., B.H. Annex, G.R. McKendall, et al. *Circulation*. 2003; *107* (10):1359–1365. doi:10.1161/01.cir.0000061911.47710.8a.

Heusch, G., B.J. Gersh. *Eur Heart J*. 2017; *38* (11):774–784. doi:10.1093/eurheartj/ehw224.

Honda, T., K. Kaikita, K. Tsujita, et al. *J Mol Cell Cardiol*. 2008; *44* (5):915–926. doi:10.1016/j.yjmcc.2008.03.004.

Hong, J., S. H. Ku, M.S. Lee, et al. *Biomaterials*. 2014; *35* (26):7562–7573. doi:10.1016/j.biomaterials.2014.05.025.

Huang, P., L. Wang, Q. Li, et al. *Stem Cell Res Ther*. 2019; *10* (1):300. doi:10.1186/s13287-019-1353-3.

Ichimura, K., T. Matoba, K. Nakano, et al. *PLoS One*. 2016; *11* (9):e0162425. doi:10.1371/journal.pone.0162425.

Inoue, T., T. Inoguchi, N. Sonoda, et al. *Atherosclerosis*. 2015; *240* (1):250–259. doi:10.1016/j.atherosclerosis.2015.03.026.

Ishikita, A., T. Matoba, G. Ikeda, et al. *J Am Heart Assoc*. 2016; *5* (7): e003872. Published 2016 Jul 22. doi:10.1161/JAHA.116.003872.

Itkin, T., S. Gur-Cohen, J.A. Spencer, et al. *Nature*. 2016; *532* (7599):323–328. doi:10.1038/nature17624.

Jinek, M., K. Chylinski, I. Fonfara, M. Hauer, J.A. Doudna, E. Charpentier. *Science*. 2012; *337* (6096):816–821. doi:10.1126/science.1225829.

Jung, J.H., X. Fu, P. C. Yang. *Circ Res*. 2017; *120* (2):407–417. doi:10.1161/CIRCRESAHA.116.309307.

Kalepu, S., V. Nekkanti. *Acta Pharm Sin B*. 2015; *5* (5):442–453. doi:10.1016/j.apsb.2015.07.003.

Karkkainen, M. J., P. Haiko, K. Sainio, J. Partanen, J. Taipale, T. V. Petrova, M. Jeltsch, D. G. Jackson, *Nat Immunol*. 2004; *5* (1):74–80. doi:10.1038/ni1013.

Katsuda, T., R. Tsuchiya, N. Kosaka, et al. *Sci Rep*. 2013; *3*: 1197. doi:10.1038/srep01197.

Katsuki, S., T. Matoba, S. Nakashiro, et al. *Circulation*. 2014; *129* (8):896–906. doi:10.1161/CIRCULATIONAHA.113.002870.

Kazemzadeh-Narbat, M., M. Reid, M. S. Brooks, A. Ghanem. *J Microencapsul*. 2015; *32* (5):460–466. doi:10.3109/02652048.2015.1046517.

Ke, Y., C. Castro, J. H. Choi. *Annu Rev Biomed Eng*. 2018; *20*: 375–401. doi:10.1146/annurev-bioeng-062117–120904.

Khan, I., A. Ali, M. A. Akhter, et al. *Life Sci*. 2016; *162*: 60–69. doi:10.1016/j.lfs.2016.08.014.

Kishore, R., M. Khan. *Circ Res*. 2016; *118* (2):330–343. doi:10.1161/CIRCRESAHA.115.307654.

Konstantinidis, K., R. S. Whelan, R. N. Kitsis. *Arterioscler Thromb Vasc Biol*. 2012; *32* (7):1552–1562. doi:10.1161/ATVBAHA.111.224915.

Krock, B. L., N. Skuli, M. C. Simon. *Genes Cancer*. 2011; *2* (12):1117–1133. doi:10.1177/1947601911423654.

Krohn-Grimberghe, M., M. J. Mitchell, M. J. Schloss, et al. *Nat Biomed Eng*. 2020; *4* (11):1076–1089. doi:10.1038/s41551-020-00623-7.

Kusumanto, Y. H., V. van Weel, N. H. Mulder, et al. *Hum Gene Ther*. 2006; *17* (6):683–691. doi:10.1089/hum.2006.17.683.

Lanahan, A. A., K. Hermans, F. Claes, J. S. Kerley-Hamilton, Z. W. Zhuang, F. J. Giordano, P. Carmeliet, and M. Simons. VEGF receptor 2 endocytic trafficking regulates arterial morphogenesis. *DevCell*. 2010; *18* (5):713–724. doi:10.1016/j.devcel.2010.02.016.

Laschke, M. W., M. D. Menger. *Trends Biotechnol*. 2017; *35* (2):133–144. doi:10.1016/j.tibtech.2016.08.004.

Lee, D. S., H. Qian, C. Y. Tay, and D. T. Leong. *Chem Soc Rev*. 2016a; *45* (15):4199–4225. doi:10.1039/c5cs00700c.

Lee, J. R., B. W. Park, J. Kim, et al. *Sci Adv*. 2020; *6* (18):eaaz0952. doi:10.1126/sciadv.aaz0952.

Lee, M. S., H. S. Park, B. C. Lee, J. H. Jung, J. S. Yoo, S. E. Kim. *Sci Rep*. 2016b; *6*: 27520. doi:10.1038/srep27520.

Leri, A., M. Rota, T. Hosoda, P. Goichberg, P. Anversa. *Stem Cell Res*. 2014; *13* (3 Pt B):631–646. doi:10.1016/j.scr.2014.09.001.

Lesizza, P., G. Prosdocimo, V. Martinelli, G. Sinagra, S. Zacchigna, M. Giacca. *Circ Res*. 2017; *120* (8):1298–1304. doi:10.1161/CIRCRESAHA.116.309589.

Li, J., Y. Guo, Y. Kuang, S. An, H. Ma, C. Jiang. *Biomaterials.* 2013; *34* (36):9142–9148. doi:10.1016/j.biomaterials.2013.08.030.

Li, R. Q., Y. Wu, Y. Zhi, et al. *Adv Mater.* 2016; *28* (43):9452. doi:10.1002/adma.201605206.

Liang, H., K. Huang, T. Su, et al. *ACS Nano.* 2018; *12* (7):6536–6544. doi:10.1021/acsnano.8b00553.

Libby, P., J. Plutzky. *Am J Cardiol.* 2007; *99* (4A):27B–40B. doi:10.1016/j.amjcard.2006.11.004.

Lin, Y. D., C. Y. Luo, Y. N. Hu, et al. *Sci Transl Med.* 2012; *4* (146):146ra109. doi:10.1126/scitranslmed.3003841.

Liu, M., M. Li, S. Sun, et al. *Biomaterials.* 2014; *35* (11):3697–3707. doi:10.1016/j.biomaterials.2013.12.099.

Liu, S., X. Chen, L. Bao, et al. *Nat Biomed Eng.* 2020a; *4* (11):1063–1075. doi:10.1038/s41551-020-00637-1.

Liu, S., Y. Li, X. Zeng, et al. *JAMA Cardiol.* 2019a; *4* (4):342–352. doi:10.1001/jamacardio.2019.0295.

Liu, W., Y. Sun, X. Dong, et al. *Regen Biomater.* 2020b; *7* (4):403–412. doi:10.1093/rb/rbaa011.

Liu, W., C. Yang, Y. Liu, G. Jiang. CRISPR/Cas9 system and its research progress in gene therapy. *Anticancer Agents Med Chem.* 2019b; *19* (16):1912–1919. doi:10.2174/1871520619666191014103711.

Lloyd-Jones, D. M., Y. Hong, D. Labarthe, et al. *Circulation.* 2010; *121* (4):586–613. doi:10.1161/CIRCULATIONAHA.109.192703.

Lu, D., J. Zhou, S. Hou, et al. *Adv Mater.* 2019; *31* (44):e1902733. doi:10.1002/adma.201902733.

Luo, L., J. Tang, K. Nishi, et al. *Circ Res.* 2017; *120* (11):1768–1775. doi:10.1161/CIRCRESAHA.116.310374.

Ma, Q., J. Yang, X. Huang, et al. *Stem Cells.* 2018; *36* (4):540–550. doi:10.1002/stem.2777.

Mahmood, E., S. Bae, O. Chaudhary, et al. *Eur J Pharmacol.* 2020; *882*: 173261. doi:10.1016/j.ejphar.2020.173261.

Majid, Q. A., A. T. R. Fricker, D. A. Gregory, et al. *Front Cardiovasc Med.* 2020; *7*: 554597. Published 2020 Oct 23. doi:10.3389/fcvm.2020.554597.

Mancuso, A., A. Barone, M. C. Cristiano, E. Cianflone, M. Fresta, D. Paolino. *Int J Mol Sci.* 2020; *21* (20):7701. Published 2020 Oct 18. doi:10.3390/ijms21207701.

Mann, A., G. Thakur, V. Shukla, M. Ganguli. *Drug Discov Today.* 2008; *13* (3–4):152–160. doi:10.1016/j.drudis.2007.11.008.

Mao, Y., J. I. Koga, M. Tokutome, et al. *Int Heart J.* 2017; *58* (4):615–623. doi:10.1536/ihj.16-457.

Marre, M., J. Shaw, M. Brändle, et al. *Diabet Med.* 2009; *26* (3):268–278. doi:10.1111/j.1464-5491.2009.02666.x.

Mendelson, A., P. S. Frenette. *Nat Med.* 2014; *20* (8):833–846. doi:10.1038/nm.3647.

Mensah, G. A., D. W. Brown. *Health Aff (Millwood).* 2007; *26* (1):38–48. doi:10.1377/hlthaff.26.1.38.

Midgley, A. C., Y. Wei, Z. Li, D. Kong, Q. Zhao. *Adv Mater.* 2020a; *32* (3):e1805818. doi:10.1002/adma.201805818.

Midgley, A. C., Y. Wei, D. Zhu, et al. *J Am Soc Nephrol.* 2020b; *31* (10):2292–2311. doi:10.1681/ASN.2019111160.

Minet, E., D. Mottet, G. Michel, et al. *FEBS Lett.* 1999; *460* (2):251–256. doi:10.1016/s0014-5793(99)01359-9.

Miragoli, M., P. Ceriotti, M. Iafisco, et al. *Sci Transl Med.* 2018;*10* (424):eaan6205. doi:10.1126/scitranslmed.aan6205.

Miyagi, Y., F. Zeng, X. P. Huang, et al. *Biomaterials.* 2010; *31* (30):7684–7694. doi:10.1016/j.biomaterials.2010.06.048.

Molavi, B., J. Chen, J. L. Mehta. *Am J Physiol Heart Circ Physiol.* 2006; *291* (2):H687–H693. doi:10.1152/ajpheart.00926.2005.

Morrison, S. J., D. T. Scadden. *Nature.* 2014; *505* (7483):327–334. doi:10.1038/nature12984.

Murphy, W. L., T. C. McDevitt, A. J. Engler. *Nat Mater.* 2014; *13* (6):547–557. doi:10.1038/nmat3937.

Nabofa, W. E. E., O. O. Alashe, O. T. Oyeyemi, et al. *Sci Rep.* 2018; *8* (1):16649. doi:10.1038/s41598-018-35145-5.

Nagaoka, K., T. Matoba, Y. Mao, et al. *PLoS One.* 2015; *10* (7):e0132451. doi:10.1371/journal.pone.0132451.

Nair, R. S., J. M. Ameer, M. R. Alison, T. V. Anilkumar. *Colloids Surf B Biointerfaces.* 2017; *157*: 130–137. doi:10.1016/j.colsurfb.2017.05.056.

Nakano, Y., T. Matoba, M. Tokutome, et al. *Sci Rep.* 2016; *6*: 29601. doi:10.1038/srep29601.

Nguyen, A. H., Y. Wang, D. E. White, M. O. Platt, T. C. McDevitt. *Biomaterials.* 2016; *76*: 66–75. doi:10.1016/j.biomaterials.2015.10.043.

Nguyen, J., R. Sievers, J. P. Motion, S. Kivimäe, Q. Fang, R. J. Lee. *Mol Pharm.* 2015; *12* (4):1150–1157. doi:10.1021/mp500653y.

Nie, J. J., B. Qiao, S. Duan, et al. *Adv Mater.* 2018; *30* (31):e1801570. doi:10.1002/adma.201801570.

Odegaard, J. I., R. R. Ricardo-Gonzalez, M. H. Goforth, et al. *Nature.* 2007; *447* (7148):1116–1120. doi:10.1038/nature05894.

Oduk, Y., W. Zhu, R. Kannappan, et al. *Am J Physiol Heart Circ Physiol.* 2018; *314* (2):H278–H284. doi:10.1152/ajpheart.00471.2017.

Oggu, G. S., S. Sasikumar, N. Reddy, K. K. R. Ella, C. M. Rao, K. K. Bokara. *Stem Cell Rev Rep.* 2017; *13* (6):725–740. doi:10.1007/s12015-017-9760-2.

Pan, Z, X. Sun, J. Ren, et al. *PLoS One.* 2012; *7* (11):e50515. doi:10.1371/journal.pone.0050515.

Park, K. M., J. P. Teoh, Y. Wang, et al. *Am J Physiol Heart Circ Physiol.* 2016; *311* (2):H371–H383. doi:10.1152/ajpheart.00807.2015.

Pascual-Gil, S., T. Simón-Yarza, E. Garbayo, F. Prósper, M. J. Blanco-Prieto. *Int J Pharm.* 2017; *523* (2):531–533. doi:10.1016/j.ijpharm.2016.11.022.

Pelaz, B., C. Alexiou, R. A. Alvarez-Puebla, et al. *ACS Nano.* 2017; *11* (3):2313–2381. doi:10.1021/acsnano.6b06040.

Peng, Y., X. Zhu, L. Qiu. *Biomaterials.* 2016; *106*: 1–12. doi:10.1016/j.biomaterials.2016.08.001.

Phatharajaree, W., A. Phrommintikul, N. Chattipakorn. *Can J Cardiol.* 2007. *23* (9):727–733. doi:10.1016/s0828-282x(07)70818-8.

Qi, Q., L. Lu, H. Li, et al. *Int J Nanomedicine.* 2017; *12*: 4835–4848. doi:10.2147/IJN.S132064.

Qiao, B., J. J. Nie, Y. Shao, et al. *Small.* 2020; *16* (4):e1905925. doi:10.1002/smll.201905925.

Qiu, J., G. Cai, X. Liu, D. Ma. *Biomed Pharmacother.* 2017; *96*: 1418–1426. doi:10.1016/j.biopha.2017.10.086.

Rawat, M., D. Singh, S. Saraf, S. Saraf. *Biol Pharm Bull.* 2006; *29* (9):1790–1798. doi:10.1248/bpb.29.1790.

Rebouças, J. S., N. S. Santos-Magalhães, F. R. Formiga. *Arq Bras Cardiol.* 2016; *107* (3):271–275. doi:10.5935/abc.20160097.

Reiter, R. J., J. C. Mayo, D. X. Tan, R. M. Sainz, M. Alatorre-Jimenez, L. Qin. *J Pineal Res.* 2016; *61* (3):253–278. doi:10.1111/jpi.12360.

Richards, P., H. E. Parker, A. E. Adriaenssens, et al. *Diabetes.* 2014; *63* (4):1224–1233. doi:10.2337/db13-1440.

Ricote, M., C. K. Glass. *Biochim Biophys Acta.* 2007; *1771* (8):926–935. doi:10.1016/j.bbalip.2007.02.013.

Robbins, C. S., A. Chudnovskiy, P. J. Rauch, et al. *Circulation.* 2012; *125* (2):364–374. doi:10.1161/CIRCULATIONAHA.111.061986.

Robinson, S. R. *J Neurosci Res.* 2001; *66* (5):972–980. doi:10.1002/jnr.10057.

Rosso, F., G. Marino, A. Giordano, M. Barbarisi, D. Parmeggiani, A. Barbarisi. *J Cell Physiol.* 2005; *203* (3):465–470. doi:10.1002/jcp.20270.

Sacco, R. L., G. A. Roth, K. S. Reddy, et al. *Circulation.* 2016; *133* (23):e674–e690. doi:10.1161/CIR.0000000000000395.

Saludas, L., S. Pascual-Gil, F. Roli, E. Garbayo, M. J. Blanco-Prieto. *Maturitas.* 2018; *110*: 1–9. doi:10.1016/j.maturitas.2018.01.011.

Santoso, M. R., G. Ikeda, Y. Tada, et al. *J Am Heart Assoc.* 2020; *9* (6):e014345. doi:10.1161/JAHA.119.014345.

Savary, G., E. Dewaeles, S. Diazzi, et al. *Am J Respir Crit Care Med.* 2019; *200* (2):184–198. doi:10.1164/rccm.201807-1237OC.

Saxena, A., K. Sachin, H. B. Bohidar, A. K. Verma. *Colloids Surf B Biointerfaces.* 2005; *45*(1):42–48. doi:10.1016/j.colsurfb.2005.07.005.

Setyawati, M. I., V. N. Mochalin, D. T. Leong. *ACS Nano.* 2016; *10* (1):1170–1181. doi:10.1021/acsnano.5b06487.

Shafiq, M., Y. Zhang, D. Zhu, et al. *Regen Biomater.* 2018; *5* (5):303–316. doi:10.1093/rb/rby021.

Shan, D., J. Li, P. Cai, et al. *Drug Deliv Transl Res.* 2015; *5* (1):15–26. doi:10.1007/s13346-014-0210-2.

Shao, M., W. Yang, G. Han. *Int J Nanomedicine.* 2017a; *12*: 7121–7130. doi:10.2147/IJN.S141549.

Shao, S., Q. Zhou, J. Si, et al. A non-cytotoxic dendrimer with innate and potent anticancer and anti-metastatic activities. *Nat Biomed Eng.* 2017b; *1* (9):745–757. doi:10.1038/s41551-017-0130-9.

Shatseva, T., D. Y. Lee, Z. Deng, B. B. Yang. *J Cell Sci.* 2011; *124* (Pt 16):2826–2836. doi:10.1242/jcs.077529.

Sluijter, J. P., V. Verhage, J. C. Deddens, F. van den Akker, P. A. Doevendans. *Cardiovasc Res.* 2014; *102* (2):302–311. doi:10.1093/cvr/cvu022.

Somasuntharam, I., K. Yehl, S. L. Carroll, et al. *Biomaterials.* 2016; *83*: 12–22. doi:10.1016/j.biomaterials.2015.12.022.

Song, Y., C. Zhang, J. Zhang, et al. *Theranostics.* 2019; *9* (8):2346–2360. doi:10.7150/thno.29945.

Sun, B., S. Liu, R. Hao, X. Dong, L. Fu, B. Han. *Pharmaceutics.* 2020a; *12* (6):575. Published 2020 Jun 21. doi:10.3390/pharmaceutics12060575.

Sun, L., Y. Hu, A. Mishra, et al. *Biofactors.* 2020b; *46* (3):421–431. doi:10.1002/biof.1611.

Swirski, F. K., M. Nahrendorf, M. Etzrodt, et al. *Science.* 2009; *325* (5940):612–616. doi:10.1126/science.1175202.
Takagawa, J., Y. Zhang, M. L. Wong, et al. *J Appl Physiol (1985).* 2007; *102* (6):2104–2111. doi:10.1152/japplphysiol.00033.2007.
Talman, V., H. Ruskoaho. *Cell Tissue Res.* 2016; *365* (3):563–581. doi:10.1007/s00441-016-2431-9.
Tang, J., T. Su, K. Huang, et al. *Nat Biomed Eng.* 2018; *2*: 17–26. doi:10.1038/s41551-017-0182-x.
Tartuce, L. P., F. Pacheco Brandt, G. Dos Santos Pedroso, et al. *Colloids Surf B Biointerfaces.* 2020; *192*: 111012. doi:10.1016/j.colsurfb.2020.111012.
Tee, J. K., L. X. Yip, E. S. Tan, et al. *Chem Soc Rev.* 2019; *48* (21):5381–5407. doi:10.1039/c9cs00309f.
Terashvili, M., Z. J. Bosnjak. *J Cardiothorac Vasc Anesth.* 2019; *33* (1):209–222. doi:10.1053/j.jvca.2018.04.048.
Tian, A., C. Yang, B. Zhu, et al. *Can J Physiol Pharmacol.* 2018; *96* (12):1318–1327. doi:10.1139/cjpp-2018-0227.
Timmis, A., N. Townsend, C. Gale, et al. *Eur Heart J.* 2018; *39* (7):508–579. doi:10.1093/eurheartj/ehx628.
Tokutome, M., T. Matoba, Y. Nakano, et al. *Cardiovasc Res.* 2019; *115* (2):419–431. doi:10.1093/cvr/cvy200.
Torella, D., M. Rota, D. Nurzynska, et al. *Circ Res.* 2004; *94* (4):514–524. doi:10.1161/01.RES.0000117306.10142.50.
Torrini, C., R. J. Cubero, E. Dirkx, et al. *Cell Rep.* 2019; *27* (9):2759–2771. doi:10.1016/j.celrep.2019.05.005.
Trappmann, B., B. M. Baker, W. J. Polacheck, C. K. Choi, J. A. Burdick, C. S. Chen. *Nat Commun.*2017;*8*(1):371. doi:10.1038/s41467-017-00418-6.
Tseng, A. S., F. B. Engel, M. T. Keating. *Chem Biol.* 2006; *13* (9):957–963. doi:10.1016/j.chembiol.2006.08.004.
Ugdutt, B. I., A. Jelani, A. Palaniyappan, et al. *Circulation.* 2010; *122* (4):341–351. doi:10.1161/CIRCULATIONAHA.110.948190.
Vander Heide, R. S., C. Steenbergen. *Circ Res.* 2013; *113* (4):464–477. doi:10.1161/CIRCRESAHA.113.300765.
Vandergriff, A., K. Huang, D. Shen, et al. *Theranostics.* 2018; *8* (7):1869–1878. doi:10.7150/thno.20524.
Vestad, B., A. Llorente, A. Neurauter, et al. *J Extracell Vesicles.* 2017; *6* (1):1344087. doi:10.1080/20013078.2017.1344087.
Viola, H. M., A. A. Shah, J. A. Kretzmann, et al. *Nanomedicine.* 2020; *29*: 102264. doi:10.1016/j.nano.2020.102264.
Wang, J. X., J. Q. Jiao, Q. Li, et al. *miR-499 Nat Med.* 2011; *17* (1):71–78. doi:10.1038/nm.2282.
Wang, K., Z. Jiang, K. A. Webster, et al. *Stem Cells Transl Med.* 2017a; *6* (1):209–222. doi:10.5966/sctm.2015-0386.
Wang, M., Y. Guo, M. Yu, P. X. Ma, C. Mao, and B. Lei. 2017b. Photoluminescent and biodegradable polycitrate-polyethylene glycol-polyethyleneimine polymers as highly biocompatible and efficient vectors for bioimaging-guided siRNA and miRNA delivery. *Acta Biomater 54*: 69–80. doi:10.1016/j.actbio.2017.02.034.
Wang, S., J. Jiang, Y. Wang, et al. *Biochem Biophys Res Commun.* 2018; *498* (1):240–245. doi:10.1016/j.bbrc.2018.02.021.
Wattendorf, U., H. P. Merkle. *J Pharm Sci.* 2008; *97* (11):4655–4669. doi:10.1002/jps.21350.
WHO (World Health Organization). 2017. *Cardiovascular Disease Fact Sheet.* Last accessed 28th December 2020. Available online. https://www.who.int/en/news-room/fact-sheets/detail/cardiovascular-diseases-(cvds).
Whyte, W., E. T. Roche, C. E. Varela, et al. *Nat Biomed Eng.* 2018; *2* (6): 416–428. doi:10.1038/s41551-018-0247-5.
Wu, Q., J. Wang, W. L. W. Tan, et al. *Cell Death Dis.* 2020a; *11* (5):354. doi:10.1038/s41419-020-2508-y.
Wu, Y., T. Chang, W. Chen, et al. *Bioact Mater.* 2020b; *6* (2):520–528. doi:10.1016/j.bioactmat.2020.08.031.
Xiao, C., K. Wang, Y. Xu, et al. *Circ Res.* 2018; *123* (5):564–578. doi:10.1161/CIRCRESAHA.118.312758.
Xiao, J., Y. Pan, X. H. Li, et al. *Cell Death Dis.* 2016; *7* (6):e2277. doi:10.1038/cddis.2016.181.
Xue, X., X. Shi, H. Dong, S. You, H. Cao, K. Wang, Y. Wen, D. Shi, B. He, and Y. Li. *Nanomedicine.* 2018; *14* (2):619–631. doi:10.1016/j.nano.2017.12.004.
Yang, B., H. Lin, J. Xiao, Y. Lu, X. Luo, B. Li, Y. Zhang, C. Xu, Y. Bai, H. Wang, G. Chen, and Z. Wang. *Nat Med.* 2007; *13* (4):486–491. doi:10.1038/nm1569.
Yang, H., X. Qin, H. Wang, X. Zhao, Y. Liu, H. T. Wo, C. Liu, M. Nishiga, H. Chen, J. Ge, N. Sayed, O. J. Abilez, D. Ding, S. C. Heilshorn, and K. Li. *ACS Nano.* 2019; *13* (9):9880–9894. doi:10.1021/acsnano.9b03343.
Yang, Y. J., H. Y. Qian, J. Huang, Y. J. Geng, R. L. Gao, K. F. Dou, G. S. Yang, J. J. Li, R. Shen, Z. X. He, M. J. Lu, and S. H. Zhao. *Eur Heart J.* 2008; *29* (12):1578–1590. doi:10.1093/eurheartj/ehn167.

Yao, C., W. Wu, H. Tang, X. Jia, J. Tang, X. Ruan, F. Li, D. T. Leong, D. Luo, and D. Yang. *Biomaterials*. 2020; *257*: 120256. doi:10.1016/j.biomaterials.2020.120256.

Ye, Y., Y. Lin, S. Manickavasagam, J. R. Perez-Polo, B. C. Tieu, and Y. Birnbaum. *Am J Physiol Heart Circ Physiol.* 2008; *295* (6):H2436–H2446. doi:10.1152/ajpheart.00690.2008.

Yokoyama, R., M. Ii, M. Masuda, Y. Tabata, M. Hoshiga, N. Ishizaka, and M. Asahi. *Cardiac Stem Cells Transl Med.* 2019; *8* (10):1055–1067. doi:10.1002/sctm.18-0244.

Yu, B. T., N. Yu, Y. Wang, H. Zhang, K. Wan, X. Sun, and C. S. Zhang. *Eur Rev Med Pharmacol Sci*. 2019; *23* (19):8588–8597. doi:10.26355/eurrev_201910_19175.

Yu, J., W. Li, and D. Yu. *Drug Des Devel Ther*. 2018; *12*: 1697–1706. doi:10.2147/DDDT.S166749.

Zhang, M., J. Zhu, X. Qin, M. Zhou, X. Zhang, Y. Gao, T. Zhang, D. Xiao, W. Cui, and X. Cai. *ACS Appl Mater Interfaces*. 2019; *11* (34):30631–30639. doi:10.1021/acsami.9b10645.

Zhang, R., W. Luo, Y. Zhang, D. Zhu, A. C. Midgley, H. Song, A. Khalique, H. Zhang, J. Zhuang, D. Kong, and X. Huang. *Sci Adv*. 2020; *6* (19):eaaz8011. doi:10.1126/sciadv.aaz8011.

Zhang, S., S. J. Chuah, R. C. Lai, J. H. P. Hui, S. K. Lim, and W. S. Toh. *Biomaterials*. 2018; *156*: 16–27. doi:10.1016/j.biomaterials.2017.11.028.

Zhang, S., J. Wang, and J. Pan. *Drug Deliv*. 2016; *23* (9):3696–3703. doi:10.1080/10717544.2016.1223218.

Zhang, Y., Z. Wang, and R. A. Gemeinhart. *J Control Release*. 2013; *172* (3):962–974. doi:10.1016/j.jconrel.2013.09.015.

Zhao, J., X. Li, J. Hu, F. Chen, S. Qiao, X. Sun, L. Gao, J. Xie, and B. Xu. *Cardiovasc Res*. 2019; *115* (7):1205–1216. doi:10.1093/cvr/cvz040.

Zhou, D., L. Cutlar, Y. Gao, W. Wang, J. O'Keeffe-Ahern, S. McMahon, B. Duarte, F. Larcher, B. J. Rodriguez, U. Greiser, and W. Wang. *Sci Adv*. 2016; *2* (6):e1600102. doi:10.1126/sciadv.1600102.

Zhou, J., X. Yang, W. Liu, C. Wang, Y. Shen, F. Zhang, H. Zhu, H. Sun, J. Chen, J. Lam, A. G. Mikos, and C. Wang. *Theranostics*. 2018; *8* (12):3317–3330. doi:10.7150/thno.25504.

Zhou, L., X. Chen, T. Liu, Y. Gong, S. Chen, G. Pan, W. Cui, Z. P. Luo, M. Pei, H. Yang, and F. He. *J Pineal Res*. 2015; *59* (2):190–205. doi:10.1111/jpi.12250.

Zou, Y., M. Zheng, W. Yang, F. Meng, K. Miyata, H. J. Kim, K. Kataoka, and Z. Zhong. *Adv Mater*. 2017; *29*(42). doi:10.1002/adma.201703285.

18

Three-Dimensional Printing: Future of Pharmaceutical Industry

Manju Bala, Anju Dhiman, Harish Dureja, and Munish Garg
Maharshi Dayanand University, Rohtak, Haryana, India

Pooja A Chawla
ISF College of Pharmacy, Moga, Punjab, India

Viney Chawla
Baba Farid University of Health Sciences, Faridkot, Punjab, India

CONTENTS

DOI: 10.1201/9781003153504-18

Introduction

Three-dimensional printing is a novel additive manufacturing technique by which three-dimensional solid objects can be produced in layer by layer sequence by using computer-aided drug design software (Lamichhane et al. 2019). The use of 3D printing in the pharmaceutical industry was once a golden dream. However, in last 15 years, we can see the growth of using 3D printing in the healthcare industry. 3D printing is playing a major role from R&D to patient treatment. It can be also known as AM: additive manufacturing, RP: rapid prototyping, and SFF: solid freeform technology. 3D printing develops a medicine which is much closer to the patients (Liu et al. 2017).

Brief History about 3D Printing

The evolution of 3D printing was originated in 1980 and named as 'Rapid prototyping' due to having economic and quick process for the development of product. In May1980, Dr. Hideo Kodama filed a first patent application for it in Japan. However, he failed to meet all the requirements for patent within one year deadline (Su and Al Aref 2018). In 1986, the first patent was issued to Charles Hull for discovery of SLA (stereolithography apparatus). So he is known as 'father of 3D printing' (Whitaker 2014). Later in 1988, they delivered 'SLA-250' as first commercial available 3D printer.

In 1988, another technique of 3D printing, i.e., selective laser sintering (SLS) was discovered and the patent was filed by Carl Deckard. Meanwhile, Scott crump filed a patent for another technique of 3D printing, i.e., fused deposition modelling (FDM) and finally in 1992, patent was issued to Scott crump for it (Cheng et al. 2017).

3D printing, i.e. fused deposition modelling and finally in 1992, patent was issued to Scott crump for it (Konta et al. 2017).

Technique Involved in 3D Printing

There are different 3D printing techniques as shown in Figure 18.1.

Laser-Based Technique

Stereolithography

It is an advanced fabrication technology in which a laser beam under the computer control is applied which cause the solidification of resin/liquid polymers in a controlled predetermined way to create 3D structure. This technique is also known as photo-polymerisation (Figure 18.2; Melchels et al. 2010).

It is of two types based on their setup:

(i) **Bottom up**: it has source of light (laser) at bottom and a moving plate form at the top of a printer.
(ii) **Top down**: it have source of light on the top and moving plate form at the bottom (Wang et al. 2017).

Advantages

The advantages are as follows: single and simple process (Carrozza et al. 1995). High resolution and high speed of printing avoid the need of high temperature (Wang et al. 2015).

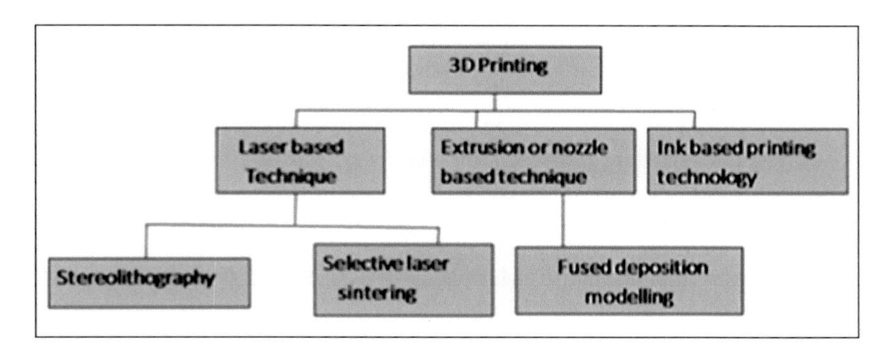

FIGURE 18.1 Different techniques of 3D printing.

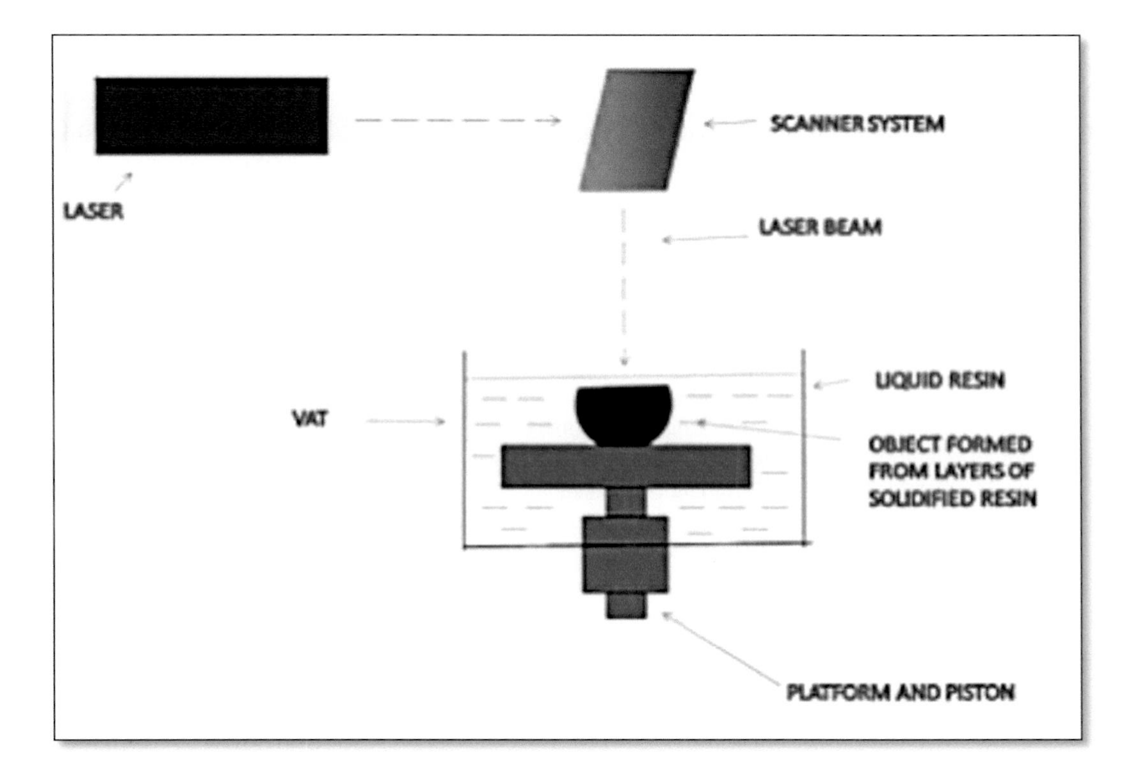

FIGURE 18.2 Stereolithography printer.

Disadvantages

The availability of photopolymerised polymers is limited. After polymerisation, residual resins and final product are different and cause genotoxicity (Pravin and Sudhir 2018).

Pharmaceutical Applications of Stereolithography

Valentin *et al* studied that hydrogels can be patterned by using sodium alginate, photoacid generator, and divalent ions via stereolithography 3D printing. Hydrogels can be used to study degradation kinetics which further verified to study toxicity on epithelial cells (Valentin et al. 2017).

Wang *et al* printed 4-aminosalicylate and paracetamol-loaded tablet via stereolithography by polymerising monomers, which is a manufacturing tool to elaborate modified drug release profile of oral dosage form (Wang et al. 2016).

Healy *et al* prepared novel photopolymerisable resin formulation of aspirin and paracetamol by stereolithography, which can be used as a manufacturing tool for personalised medicine (Healy et al. 2019).

Goyanes *et al* formulated a flexible anti acne patch of salicylic acid by stereolithography 3D printing, which has high resolution and high loading capacity than the conventional method (Goyanes et al. 2016).

Xu *et al* formulated a multidrug oral dosage form of atenolol, hydrochlorthiazide, amlodipine and irbesartan via the vat polymerisation technique, which is used as antihypertensive (Xu et al. 2020).

Selective Laser Sintering

A laser beam is applied on powder bed, which sinters the drug and polymers together by strong coherence to fabricate a 3D shape (Kumar 2003).

Advantages

The advantages are as follows: one step method, chemical resistance, speed and high strength and resolution, and solvent free and filament free method for production (Williams et al. 2005).

Pharmaceutical Applications

1. Fina *et al* formulated immediate release oral medicine of paracetamol by using polymers kollicoat IR and eudragit via sinter laser sintering technique (Fina et al. 2017).
2. Salmoria *et al* formulated implant of polycaprolactone/fluorouracil by using the SLS technique. The hydrophilic nature of the drug showed maximum drug release (Salmoria et al. 2017).
3. Allahham *et al* fabricated orodispersible printlets of ondansetron by using sinter laser sintering. The printlets showed maximum drug release (more than 90%) without depending on concentration of mannitol (Allahham et al. 2020).
4. Awad *et al* fabricated miniprintlets of paracetamol by using sinter laser sintering technique, which showed higher drug content, higher flexibility (Awad et al. 2019).

Extrusion or Nozzle-Based Technique

Fused Deposition Modelling

It includes the hot melt extrusion process for preparation of thermoplastic filament. Polymers of choice are fed from a heated movable nozzle for melting on a substrate in per determined xyz axis, which fused together or solidify to create a 3D object as shown in Figure 18.3. The shape of the 3D object is determined computer-aided design (Fuenmayor et al. 2018).

Advantages

The advantages are as follows: high resolution, better accuracy of dosage form, and good mechanical strength.

Disadvantages

The disadvantage is that it is only meant for thermal-resistant active pharmaceutical ingredients due to the use of high temperature in FDM (Tan et al. 2018).

Goyanes *et al* explored the use of fused deposition modelling for the fabrication of modified drug release tablet of 5-aminosalicylic acid and 4-aminosalicyclic acid. It proved as an efficient method for the development of a personalised oral drug delivery system for the treatment of inflammatory bowel disorder (Goyanes et al. 2015c).

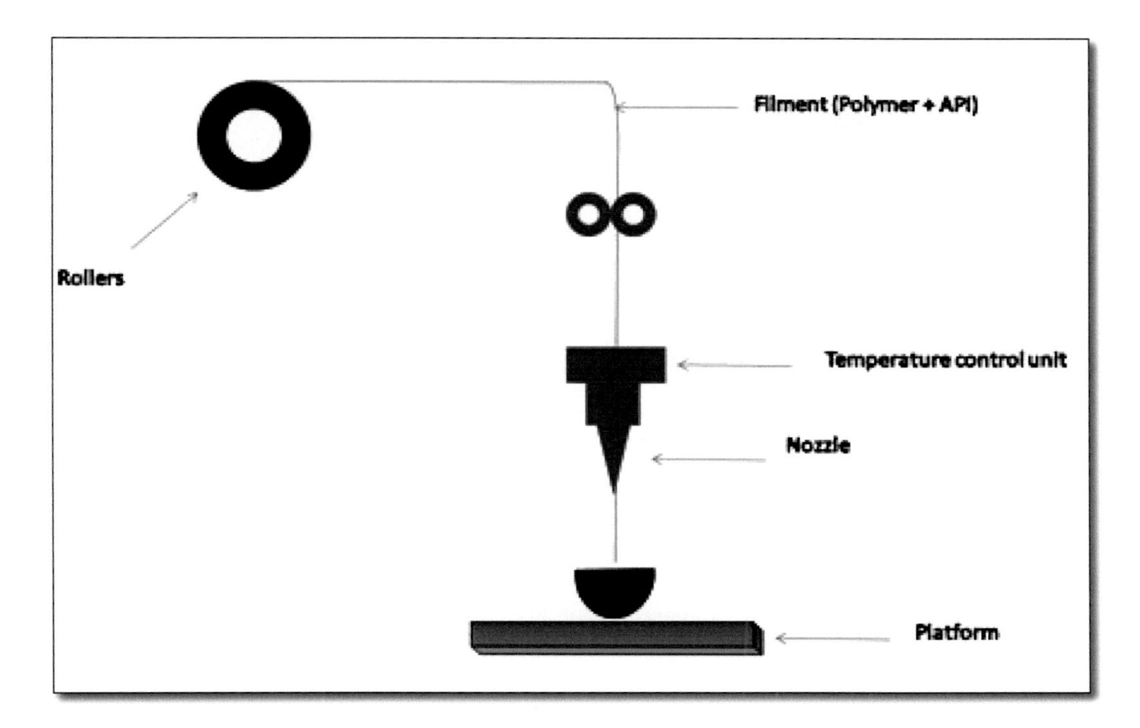

FIGURE 18.3 Fused deposition modelling (FDM) printing.

Khaled *et al* developed polypill of five in one by 3D hot melt extrusion printing. The pharmaceutical active principles used are aspirin with hydrochlorthiazide (immediate release compartment) and atenolol with remipril and pravastatin (sustained release compartment). The evaluation showed that polypill has the ability to deliver each drug in separate manner for the treatment of cardiovascular disorder (Khaled et al. 2015).

Melocchi *et al* formulated oral pulsatile capsule of acetaminophen with polymer hydroxypropyl cellulose by using fused deposition modelling and examined the difference between capsules designed by the injection molding technique and 3D printing technique. Then, it was observed that 3D technique can be utilised as an alternate to the injection molding technique (Melocchi et al. 2015).

Goyanes *et al* explored the potential of loading budesonide into polyvinyl alcohol filaments via fused deposition modelling along with film coating and hot melt extrusion. This was further evaluated for dissolution kinetics. The release characteristics showed the ability to treat inflammatory bowel disorder (Goyanes et al. 2015a).

Sadiaa Muzna *et al* employed fused deposition modelling for fabrication of immediate release tablet of 5-ASA, theophylline, prednisolone, and captopril. This developed a low-cost method for the personalised dosage form (Sadia et al. 2016).

Chai xuyu *et al* study employed the use of fused deposition modelling for fabrication of intragastric floating tablets of domperidone, which exhibited increase in drug release and floating time both *in vivo* and *in vitro* (Chai et al. 2017).

Fused deposition modelling is used for the fabrication of fixed dose combination of two antihypertensive drugs in a single tablet. This helps the physician to prescribe a single dose to the patient for maintaining patient compliance (Sadia et al. 2018).

Okwuosa designed a gastric resistant tablet of theophylline by engaging shell core structure utilising polyvinyl pyrrolidone and methacrylic acid via fused deposition modelling. Thus 3D printing overcomes the problems of hardness associated with conventional methods (Okwuosa et al. 2017).

Goyanes *et al* prepared a tablet by loading fluorescein into polyvinyl alcohol filament. It proved that the tablet of any size and density can be printed by 3D printing (Goyanes et al. 2014).

Weisman *et al* delivered antibiotics gemtamicin/methotrexate drug loading constructs within polylactic acid filaments by 3D printing, which further inhibit the growth of cancer cells (Weisman et al. 2015).

Sandler *et al* utilise nitrofurantoin (poorly soluble drug) having activity against microbes and urinary tract infections. Poly lactic acid was used as a biodegradable polymer. This further resulted in inhibition of biofilm colonisation (Sandler et al. 2014).

Genina *et al* investigated the fabrication of subcutaneous rods and the intrauterine system of indomethacine loaded in ethylene vinyl acetate by using fused deposition modelling (Genina et al. 2016).

Yan *et al* prepared an internal scaffold structure by loading drug into ethyl cellulose filaments (Yang et al. 2018).

Goyanes *et al* prepared capsule containing paracetamol and caffeine in filaments of polyvinyl alcohol via fused deposition modelling. The drug release found the same for both the drugs (Goyanes et al. 2015b).

Skowyra *et al* demonstrated the feasibility of fused deposition modelling for the preparation of tablet of prednisolone loaded in polyvinyl alcohol (Justyna et al. 2015).

Inkjet Printing (IJP) or Ink-Based Printing Technology

The IJP is a technique in which material of interest is dissolved in liquid so that an ink is formed, as shown in Figure 18.5 (Boehm et al. 2014). Accurate formation of the desired 3D structure depends on the movement of the substrate and the movement of an inkjet nozzle so it is also named as 'tool less' and 'mask less' technique. Deposition of ink takes place on the substrate in the form of drop on demand or continuous IJP. If ink is deposited in the form of drops on other drops to produce a solid object, it is known as 'drop on drop inkjet printing' and if ink is deposited in the form of a drop on a solid, then it is called 'drop on solid inkjet printing' or 'theriform process'. So it has the ability to give high-resolution printing. It includes a liquid binder and powder bed. When the process starts, powder bed is formed in the form of a thin layer on a plateform. After that, a liquid binder in the form of droplet ejects on powder bed and fused together. This process occurs in layering manner. The ejection of droplet is based on the following mechanisms: thermal, piezoelectric, and electrostatic. Working of all types of IJPs is based on the same principle (Vithani et al. 2019). The different types of IJP are shown in Figure 18.4.

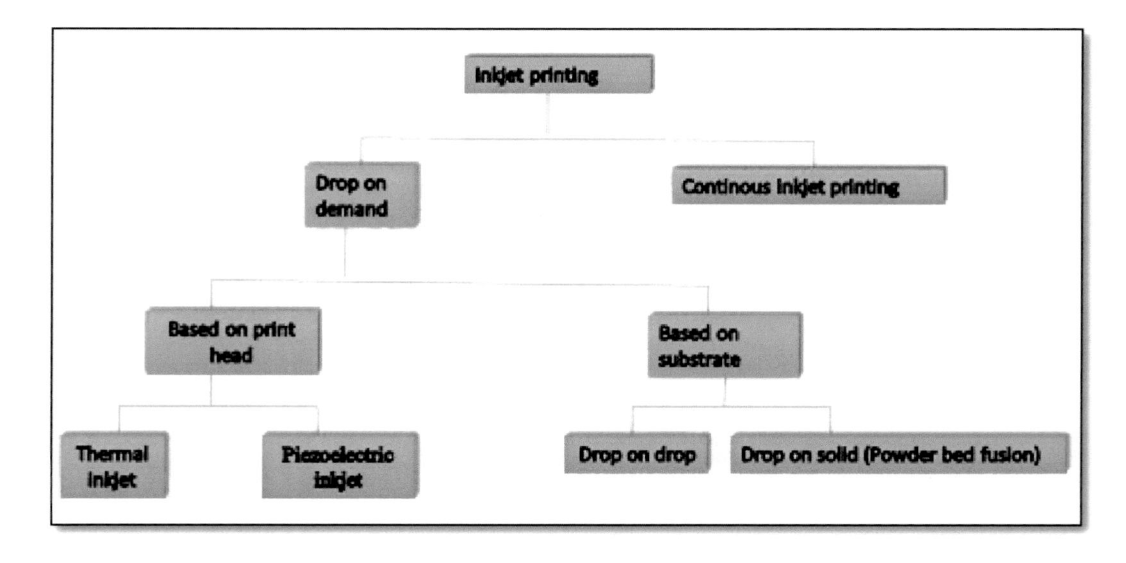

FIGURE 18.4 Types of inkjet printing.

Inkjet Printing

Continuous Inkjet Printing

When a continuous stream of fluid is ejected and deposited on the substrate of interest at high speed, this leads to more wastage of ink. The resolution is low and maintenance is costly. However, the advantage of this printing is that no clogging of the nozzle takes place due to high speed (Azizi et al. 2019).

Drop On Demand (DOD)

As the name clear, ink is ejected in the form of droplets and deposited on the substrate of interest. Economical and ease of preparation are its main features.

On the basis of its mechanism of droplet formation, it can be divided into four groups:

Thermal Inkjet Printing

It utilises a thermal actuator for the formation of droplets. The microresistor (heating source) is in direct contact with ink. Heating result in bubble formation in ink, which finally ejected and deposited in the form of a droplet (volume 10–150 pL) (Cui et al. 2012). This can be used for biological like DNA, proteins (Cui and Boland 2009). However, the main disadvantages are cell encapsulation and clogging of nozzle.

Piezoelectric Inkjet Printing

These utilise piezoelectric material (nonconducting crystals) which undergoes deformation by application of (15–25 kHz) electric field and ejected as droplets (Saunders et al. 2008). As it requires no heat, so it can be used in biomedical field. The main disadvantages associated with it are cell death and clogging of nozzle. Second, it is difficult to print over large and flat gaps (Kullmann et al. 2012).

Electrostatic Inkjet Printing

It utilises electrostatic field for the formation of droplets. It avoids heating so it can be used for printing cell and tissues (Choi et al. 2008).

Electrohydrodynamic Inkjet Printing

It utilises electric forces (high voltage 0.5–20 kV) for generation of droplet. It can be utilised in processing concentrated suspensions having size in micrometres shown in Figure 18.5 (Jayasinghe et al. 2006).

The differences between drop on drop and drop on solid are listed in Table 18.1.

Pharmaceutical and Medical Applications of Inkjet Printing

IJP is a latest established printing technology which is growing in medical and pharmaceutical field in recent years.

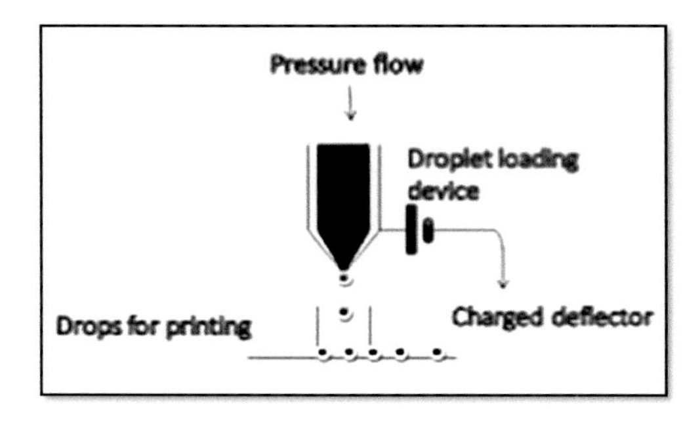

FIGURE 18.5 Inkjet printing.

TABLE 18.1

Difference Between Drop on Drop and Drop on Solid (Afsana et al. 2018)

Drop on Drop	Drop on Solid
The binder-based solidified layer produced on droplet deposition on each other	The low-melting binder along with higher-melting point particles binded together and deposited in the form of a droplet on solid powder material
The method is affected by the binding, cooling, size of droplets, solidifying property, and their interaction	The process is affected by the properties of powder like density, size of particle, wettability, and reactivity of powder with binder
No license application is required	The license granted is Theri Form™
Used for higher drug loading	Used for controlled drug delivery involving various sizes
Difficult to implement	Low drug loading limit its use

Personalised Drug Delivery System

The concept of personalised drug delivery system is not new. However, inkjet printing as a new concept in personalised drug delivery system has achieved a great success for the treatment of various diseases like cancer, diabetes, etc., by conveying dose of drug to particular site (Kumar 2000). The personalised drug delivery system can be used for the formulation of effective and suitable formulations having fewer side effects. A first 3D printed drug, i.e., levetiracetam (antiepileptic drug) prepared by the powder bed fusion method was approved by FDA and marketed by Aprecia Pharmaceuticals as Spritam having similar pharmacological activity (50).

Microparticles

These are defined as small particles with spherical shape having diameters in range 1–1000 micrometre (Kohane 2007).

Desai *et al* formulated microcapsules of calcium alginate with rhodamine 6G dye by the direct write inkjet method. This concluded that microcapsule with small size have better drug release profile (Desai et al. 2010).

Dohnal *et al* fabricated microparticles of calcium alginate containing TiO2 photocatalyst having droplet size 45–105 μm by the drop on demand inkjet printing method. He concluded that liquid having high viscosity and high solid content can be processed by this technique (Dohnal and Stepanek 2011).

Lee BK *et al* created microparticles of paclitaxel by using the piezoelectric inkjet printing technique, which showed cytotoxic effect on Hela cell lines (Lee et al. 2012).

Scoutaris *et al* studied the printed micro dot array of felodipine and hydrochlorthiazide with polymers polyvinyl pyrrolidone and polylactic-co-glycolic acid. The heterogeneous distribution of drug and polymer can be achieved in printed microdots made with polylactic-co-glycolic acid (Scoutaris et al. 2012).

Ten Cate demonstrated that the latest inkjet technique can be used for the formulation of core shell-based microcapsules comprising citric acid and calcium chloride, which showed high loading capacity as compared to conventional methods (Ten Cate et al. 2014).

Gao *et al* formulated different shaped microparticles of alginate by combining the gelation technique with drop on demand inkjet printing technique and found that microparticles with tail shape were better than spherical shape for printing (Gao et al. 2016).

Clara Lopez-Iglesias prepared microparticles of alginate aerogel loaded with a bronchodialator, i.e., salbutamol sulphate by the thermal inkjet printing technique. The obtained narrow droplet size has the potential to treat acute asthma (Lopez-Iglesias et al. 2019).

Nanoparticles

It may be defined as tiny or ultrafine particle with size 1–100 nm. (Khan et al. 2019)

Gu Y *et al* explored the inkjet printing of rifampicin and calcium phosphate micropatterns. The nanoparticles of rifampicin allow their constant release of drug for several weeks to kill bacteria (Gu et al. 2012).

Iwanaga *et al* formulated dried alginate microparticles with controlled release rate via inkjet printing (Iwanaga et al. 2013).

Wickstrom *et al* improved the delivery of poorly soluble drug, i.e., BCS class II and IV by formulation of their nanoparticles (Wickstrom et al. 2017).

Liposomes

Hanus *et al* formulated liposomes magnetic alginate beads embedded in microparticles by the drop on demand piezoelectric inkjet printing technique. Controlled diffusion from encapsulated liposomes was achieved by radio-frequency magnetic field (Hanus et al. 2013).

Pharmaceutical Dosage Forms

Sandler *et al* studied the potential of inkjet printing to produce pharmaceutical dosage form, i.e., tablet of acetaminophene+caffeine +theophylline on a porous paper substrate. This approach was used to prevent crystallisation of drug (Sandler et al. 2011).

Kogermann *et al* used IR spectroscopy for quantitative analysis of formulation of loperamide + caffeine printed by a piezoelectric inkjet printer on an impermeable substrate (Palo et al. 2016).

Montenegro-Nicolini *et al* exploited the latest method, i.e., thermal inkjet printer to deliver protein (lysozyme and ribonuclease-A) through buccal films by using sodium deoxycholate as a permeation enhancer (Montenegro-Nicolini et al. 2017).

Palo *et al* prepared the inkjet-printed oromucosal dosage form of piroxicam and lidocaine hydrochloride by using an electrospun gelatin substrate due to which physiochemical property and the release rate can be altered (Palo et al. 2017).

Balard *et al* demonstrated that the inkjet printing can be used to impregnate the antibiotics like levofloxacin, kanamycin, tetracycline, polymixin B, vancomycin, rifampicin, tobramycin along with polymer polylactic acid into 3D-printable implant devices (Ballard et al. 2019).

Acosta-Velez *et al* introduced the concept of using polyethylene glycol photocurable bioink for inkjet printing of tablets of naproxen (hydrophobic drug) (Acosta-Velez et al. 2018).

Acosta-Velez *et al* designed a pharmaceutical tablet of ropinirole HCL (hydrophilic drug) by using the photocurable polymer solution via piezo inkjet printing (Acosta-Velez et al. 2017).

Cader *et al* utilised water-based inkjet printing for printing tablets of thiamine HCL, which can be used as a universal pharmaceutical formulation for any water soluble drug (Cader et al. 2019).

Microneedle

This is a latest approach for the delivery of drugs transdermally, which get degraded in gastrointestinal tract and have low solubility in water. The mixture of drug and excipient are coated over needle and delivered into skin by piercing (Prausnitz 2004).

Uddin examined the use of inkjet printing for coating of microneedle with a combination of three drugs, i.e., 5-fluoroacil+curcumin+cisplatin. The anticancer agents delivered through the skin showed high drug release than the conventional method (Uddin et al. 2015).

Boehm used piezoelectric inkjet printing for delivery of amphotericin B through microneedle for the treatment of cutaneous leishmaniasis and fungal infections (Boehm et al. 2013).

Ross worked on the preparation of microneedles for transdermal delivery of insulin by inkjet printing and found that the solid state of insulin was delivered through microneedles for rapid drug release (Ross et al. 2015).

Boehm used polyglycolic acid microneedles for the application of voriconazole by piezoelectric inkjet printing for the treatment of fungal infections (Boehm et al. 2015).

Boehm again used polyglycolic acid microneedles for the application of itraconazole by piezoelectric inkjet printing for fungal infections (Boehm et al. 2016).

Other Applications of 3D Printing

Wound Dressing

Muwaffak reported the bactericidal potential of wound dressing prepared from polymer polycaprolactone with antimicrobial metals like zinc, copper, and silver (Muwaffak et al. 2017).

Bracaglia developed the formulation of a 3D-printed hybrid hydrogel with polyethylene glycol acrylate and homogenised pericardium matrix, which increase wound healing (Bracaglia et al. 2017).

Implant and Protheses

Herbert *et al* fabricated prosthetic foot by using 3D technology, which have the same geometry as that of residual limb of the patient (Herbert et al. 2005).

Zuniga fabricated 3D-printed prosthetic hand for children who have reduced upper limb (Zuniga et al. 2015).

Chen *et al* fabricated 3D orthopaedic cast that offer an advantage of repairing bone fracture (Chen et al. 2017).

Lee *et al* fabricated a structure of ear by using sacrificial material, i.e., polyethylene glycol, which is applied in 3D cell printing technology. Result demonstrated that 3D printing help in the regeneration of ear (Lee et al. 2014).

Young *et al* demonstrated the fabrication of soft prostheses, i.e., ear by using a desktop 3D printer. By using the scanning printing polishing casting method, artificial ear having a complicated and smooth structure can be made. The prostheses showed desirable result when fitted into defect of patient (He et al. 2014).

Zopf *et al* fabricated a bioresorbable tracheal splint by using combination of laser-based 3D printing, computed tomography, and computer-aided design and then implanted into the patient to treat tracheobronchomalacia. During estimation within three years, full resorption found in the patient (Bracci et al. 2013).

Ma *et al* designed titanium plates resembling bone segment having tumour by using 3D printing, magnetic resonance imaging, and computed tomography. After that they are implanted after removal tumour of bone. Results concluded that using titanium plates by replacing tumour bone have better clinical outcomes (Ma et al. 2017).

Dzian designed titanium sternum by 3D printing technology and implanted into the patient with a sternum tumour. Short-time estimation found better results with minimum pain (Dzian et al. 2018).

Surgical Model or Medical Phantoms

Zein *et al* developed a 3D-printed liver having network of biliary and vascular structure for the patients undergoing liver transplantation. It minimises complications associated with surgery and allow successful preoperative planning for liver transplantation (Zein et al. 2013).

Witowski *et al* developed a cost-effective 3D printed model of the liver having visible hepatic vessels, which allow performing an emergency surgery for liver malignancy (Witowski et al. 2017).

Jonthan *et al* created a 3D-printed model of renal units having visible renal vasculature by using the sterolithography technique. These models help the patients in knowing renal malignancy (Silberstein et al. 2014).

Cheung *et al* designed a cost-effective and reusable 3D-printed model of paediatric pyleoplasty and evaluated by experts. This model can be used for radical and partial nephrectomy (Cheung et al. 2014).

Sodian *et al* fabricated an anatomic structure by rapid prototyping and implanted in the patient who has required replacement of aortic valve. The results demonstrated that it provides better results in the treatment of complexity in adult cardiac surgery (Sodian et al. 2008).

Bioprinting and Organs-On-Chip

Markstedt *et al* prepared bioink consists of nanofibrillated cellulose and used by a 3D bioprinter to print human chondrocytes. The results concluded that living organs can be 3D bioprinted by using prepared bioink (Markstedt et al. 2015).

Skardal *et al* prepared a liver specific hydrogel bioink which is bioprinted by the 3D bioprinting technique for biofabrication of *invitro* liver (Skardal et al. 2015).

Lee *et al* fabricated a liver organ on a chip by 3D bio-printing technology, which showed that liver function has been increased by using it (Lee and Cho 2016).

The 3D-printed chip of an organ is an important tool for personalised healthcare and drug design. Johnson developed a 3D-printed nervous system on a chip by using the extrusion-based 3D-printed technique; by this chip, viral infections can be diagnosed (Johnson et al. 2016).

Challenges in 3D Printing

Besides having so many advantages of 3D printing, it is facing so many challenges. These limitations should be overcome before its involvement as a manufacturing technique for the development of a personalised drug delivery system.

Selection of Raw Material

The raw material should be scrutinised carefully depending on their physiochemical characteristics, viscoelastic property, print fluid characteristics, and thermal conductivity before their use in human beings.

Mechanism of a Nozzle

It includes distance between the nozzle and the powder layer, clogging of nozzles in the printer head, improper powder feeding, scraping, and migration of a binder.

Spillage of Powder

During 3D printing spillage of powder is also critical so a special area is required to perform this operation.

There are more **Chances of Surface Imperfection and Friability** in the finished product if the product is prepared by powder or the extrusion based technique because in this drying time is required. Printing rate, line velocity of a printed head, and interval time between two printing layers are different parameters are required to be optimised. There are limited number of choices for colours and material for 3D printing as compared to the conventional tablet compression technique. If 3D printing occurs at a high temperature, then it not suitable for thermolabile drugs.

Selection of Excipient

The requirement, selection of suitable excipients, and explore the use of excipients in novel drug delivery system are the major limitations. Depending upon their solubility, various hydrophilic and hydrophobic drugs require suitable excipients for their printing and these excipients should be compatible, nontoxic, and easily available. Others are how to avoid drug–drug interactions, performance of 3D-printed objects, avoid biomolecules from degradation during fabrication, posttreatment method, process of optimisation, fabrication of multiple dosage form, clinical study to assess stability, etc., are various challenges. However, one of major challenge is ensuring either safety or efficacy of printed medicines is the same as that of medicines prepared by conventional methods in the pharmaceutical sector. The laws, guidelines, quality control of 3D-printed medicines, and safety of their use are also a great challenge.

Future Prospects

The use of 3D printing in medicine is now close to become a reality. 3D printing consists of three main pillars: one is to utilise less time, second is to treat a large number of people, and third is to receive more outcome. In brief, 3D printing is made up of one sentence 'enable less physicians to treat more patients without any scarification of results'. 3D printing has opened new ways for the pharmaceutical industry. The on-demand printing of medicines is possible by emailing their database of medicines to pharmacies, which further help the pharmaceutical sector in formulating a cost-effective method of manufacturing and distributing of medicines. The medicines made by this way will increase the patient adherence. In future, it may lead to development in garage biology. As this technique is latest, so due to no regulation, security and safety are required for 3D printing. As a latest technique, 3D printing has proposed many applications in the medical sector. A new door has opened to a newer generation of advanced system of drug delivery by combination of the conventional method of manufacturing in the pharmaceutical industry with 3D printing. We believe that perseverance and patience in 3D printing can develop a safe and effective system of pharmaceutical dosage form.

Summary

A number of methods for 3D printing has been developed and differentiated depending on their working principle. It is concluded that the additive manufacturing technique is a revolutionary force in the pharmaceutical industry as it has no limit to give solutions for our problems. It is a ground-breaking technology and maturation is still going on. The application of innovative technology is becoming a hope for implantation, use of biomaterial in repairment of tissues, loading multidrug in single device, and personalised drug delivery. 3D printing is changing our world day by day. People can live a longer life by the development of this medical technique. 3D printing is facing a burden of challenges of regulation for development, design, safety, and sterilisation. In short, 3D printing can be named as 'a solution to all problems'.

ABBREVIATIONS

3D Three dimensional
SLA Stereolithography
SLS Selective laser sintering
IJP Inkjet printing
R&D Research and development
AM Additive manufacturing
RP Rapid prototyping,
SFF Solid freeform technology
FDM Fused deposition modelling
CAD Computer-aided design
5 ASA 5-amino salicylic acid
DOD Drop on demand
DNA Deoxyribonucleic acid
BCS Biopharmaceutical classification system

REFERENCES

Acosta-Velez GF, Linsley CS, Craig MC, Wu BM et al. 2017. *Bioengineering*, *4*(1): 11.
Acosta-Velez GF, Zhu TZ, Linsley CS, Wu BM et al. 2018. *International Journal of Pharmaceutics*, *546*(1–2): 145–53.
Allahham N, Fina F, Marcuta C, Kraschew L et al. 2020. *Pharmaceutics*, *12*(2): 110.
Awad A, Fina F, Trenfield SJ, Patel P et al. 2019. *Pharmaceutics*, *11*(4): 148.
Azizi MS, Mohaved S, Narayan RJ 2019. *Expert Opinion on Drug Discovery*, *14*(2): 101–13.
Ballard DH, Tappa K, Boyer CJ, Jammalamadaka U et al. 2019. *Journal of 3D Printing in Medicine*, *3*(2): 83–93.
Boehm RD, Daniels J, Stafslien S, Nasir A et al. 2015. *Biointerphases*, *10*(1): 011004.
Boehm RD, Jaipan P, Skoog SA, Stafslien S et al. 2016. *Biointerphases*, *11*(1): 011008.
Boehm RD, Miller PR, Daniels J, Stafslien S et al. 2014. *Materials Today*, *17*(5): 247–52.
Boehm RD, Miller PR, Schell WA, Perfect JR et al. 2013. *Jom*, *65*(4): 525–33.
Bracaglia LG, Messina M, Winston S, Kuo CY et al. 2017. *Biomacromolecules*, *18*(11): 3802–11.
Bracci R, Maccaroni E, Cascinu S 2013. *New England Journal of Medicine*, *368*: 2043–5.
Cader HK, Rance GA, Alexander MR, Gonçalves AD et al. 2019. *International Journal of Pharmaceutics*, *564*: 359–68.
Carrozza MC, Croce N, Magnani B, Dario PA et al. 1995. *Journal of Micromechanics and Microengineering*, *5*(2): 177.
Chai X, Chai H, Wang X, Yang J et al. 2017. *Scientific Reports*, *7*(1): 1–9.

Chen YJ, Lin H, Zhang X, Huang W et al. 2017. *3D Printing in Medicine*, *3*(1): 11.
Cheng GZ, Folch E, Wilson A, Brik R et al. 2017. *Pulmonary Therapy*, *3*(1): 59–66.
Cheung CL, Looi T, Lendvay TS, Drake JM et al. 2014. *Journal of Surgical Education*, *71*(5): 762–7.
Choi J, Kim YJ, Lee S, Son SU et al. 2008. *Applied Physics Letters*, *93*(19): 193508.
Cui X, Boland T 2009. *Biomaterials*, *30*(31): 6221–7.
Cui X, Boland T, DD'Lima DK, Lotz M et al. 2012. *Recent Patents on Drug Delivery & Formulation*, *6*(2): 149–55.
Desai S, Perkins J, Harrison BS, Sankar J et al. 2010. *Materials Science and Engineering: B*, *168*(1–3): 127–31.
Dohnal J, Stepanek F 2011. *Chemical Engineering Science*, *66*(17): 3829–35.
Dzian A, Zivcak J, Penciak R, Hudak R et al. 2018. *World Journal of Surgical Oncology*, *16*(1): 7.
Fina F, Goyanes A, Gaisford S, Basit AW et al. 2017. Selective laser sintering (SLS) 3D printing of medicines. *International Journal of Pharmaceutics*, *529*(1–2): 285–93.
Fuenmayor E, Forde M, Healy AV, Devine DM et al. 2018. *Pharmaceutics*, *10*(2): 44.
Gao Q, He Y, Fu JZ, Qiu JJ et al. 2016. *Journal of Sol-Gel Science and Technology*, *77*(3): 610–9.
Genina N, Hollander J, Jukarainen H, Makila E et al. 2016. *European Journal of Pharmaceutical Sciences*, *90*: 53–63.
Goyanes A, Buanz AB, Basit AW, Gaisford S et al. 2014. *International Journal of Pharmaceutics*, *476*(1–2): 88–92.
Goyanes A, Buanz AB, Hatton GB, Gaisford S et al. 2015a. *European Journal of Pharmaceutics and Biopharmaceutics*, *89*: 157–62.
Goyanes A, Chang H, Sedough D, Hatton GB et al. 2015b. *International Journal of Pharmaceutics*, *496*(2): 414–20.
Goyanes A, Det-Amornrat U, Wang J, Basit AW et al. 2016. *Journal of Controlled Release*, *234*: 41–8.
Goyanes A, Wang J, Buanz A, Martinez PR et al. 2015c. *Molecular Pharmaceutics*, *12*(11): 4077–84.
Gu Y, Chen X, Lee JH, Monteiro DA et al. 2012. *Acta Biomaterialia*, *8*(1): 424–31.
Hanus J, Ullrich M, Dohnal J, Singh M et al. 2013. *Langmuir*, *29*(13): 4381–7.
He Y, Xue GH, Fu JZ 2014. *Scientific Reports*, *4*(1): 1–7.
Healy AV, Fuenmayor E, Doran P, Geever LM et al. 2019. *Pharmaceutics*, *11*(12): 645.
Herbert N, Simpson D, Spence WD, Ion W et al. 2005. *Journal of Rehabilitation Research and Development*, *42*(2): 141.
Iwanaga S, Saito N, Sanae H, Nakamura M et al. 2013. *Biointerfaces*, *109*: 301–6.
Afsana Jain V, Haider N, Jain K 2018. *Current Pharmaceutical Design*, *24*(42): 5062–71.
Jayasinghe SN, Qureshi AN, Eagles PA 2006. *Small*, *2*(2): 216–9.
Johnson BN, Lancaster KZ, Hogue IB, Meng F et al. 2016. *Lab on a Chip*, *16*(8): 1393–400.
Justyna S, Katarzyna P, Mohamed AA 2015. *European Journal of Pharmaceutical Sciences*, *68*(20): 11–17.
Khaled SA, Burley JC, Alexander MR, Yang J et al. 2015. *Journal of Controlled Release*, *217*: 308–14.
Khan I, Saeed K, Khan I 2019. *Arabian Journal of Chemistry*, *12*(7): 908–31.
Kohane DS 2007. *Biotechnology and Bioengineering*, *96*(2): 203–9.
Konta AA, Garcia PM, Serrano DR 2017. *Bioengineering*, *4*(4): 79.
Kullmann C, Schirmer NC, Lee MT, Ko SH et al. 2012. *Journal of Micromechanics and Microengineering*, *22*(5): 055022.
Kumar MN, 2000. *Journal of Pharmaceutical Sciences*, *3*(2): 234–58.
Kumar S, 2003. *Jom*, *55*(10): 43–7.
Lamichhane S, Bashyal S, Keum T, Noh G et al. 2019. *Asian Journal of Pharmaceutical Sciences*, *14*(5): 465–79.
Lee BK, Yun YH, Choi JS, Choi YC et al. 2012. *International Journal of Pharmaceutics*, *427*(2): 305–10.
Lee H, Cho DW 2016. *Lab on a Chip*, *16*(14): 2618–25.
Lee JS, Hong JM, Jung JW, Shim JH et al. 2014. *Biofabrication*, *6*(2): 024103.
Liu Z, Zhang M, Bhandari B, Wang Y et al. 2017. *Trends in Food Science & Technology*, *69*: 83–94.
Lopez-Iglesias C, Casielles AM, Altay A, Bettini R et al. 2019. *Chemical Engineering Journal*, *357*: 559–66.
Ma L, Zhou Y, Zhu Y, Lin Z et al. 2017. *Scientific Reports*, *7*(1): 7626. doi:10.1038/s41598-017-07243-3.
Markstedt K, Mantas A, Tournier I, Martínez Ávila H et al. 2015. *Biomacromolecules*, *16*(5): 1489–96.
Melchels FP, Feijen J, Grijpma DW 2010. *Biomaterials*, *31*(24): 6121–30.
Melocchi A, Parietti F, Loreti G, Maroni A et al. 2015. *Journal of Drug Delivery Science and Technology*, *30*: 360–7.

Montenegro-Nicolini M, Miranda V, Morales JO 2017. *The AAPS Journal*, *19*(1): 234–43.
Muwaffak Z, Goyanes A, Clark V, Basit AW et al. 2017. *International Journal of Pharmaceutics*, *527*(1–2): 161–70.
Okwuosa TC, Pereira BC, Arafat B, Cieszynska M et al. 2017. *Pharmaceutical Research*, *34*(2): 427–37.
Palo M, Kogermann K, Genina N, Fors D et al. 2016. Quantification of caffeine and loperamide in printed formulations by infrared spectroscopy. *Journal of Drug Delivery Science and Technology*, *34*: 60–70.
Palo M, Kogermann K, Laidmae I, Meos A et al. 2017. *Molecular Pharmaceutics*, *14*(3): 808–20.
Prausnitz MR 2004. Microneedles for transdermal drug delivery. *Advanced Drug Delivery Reviews*, *56*(5): 581–7.
Pravin S, Sudhir A 2018. *Biomedicine & Pharmacotherapy*, *107*: 146–54.
Ross S, Scoutaris N, Lamprou D, Mallinson D et al. 2015. *Drug Delivery and Translational Research*, *5*(4): 451–61.
Sadia M, Isreb A, Abbadi I, Isreb M et al. 2018. *European Journal of Pharmaceutical Sciences*, *123*: 484–94.
Sadia M, Sosnicka A, Arafat B, Isreb A et al 2016. *International Journal of Pharmaceutics*, *513*(1–2): 659–68.
Salmoria GV, Vieira FE, Ghizoni GB, Marques MS et al. 2017. *Drugs*, *4*: 6.
Sandler N, Maattanen A, Ihalainen P, Kronberg L et al. 2011. *Journal of Pharmaceutical Sciences*, *100*(8): 3386–95.
Sandler N, Salmela I, Fallarero A, Rosling A et al. 2014. *International Journal of Pharmaceutics*, *459*(1–2): 62–4.
Saunders RE, Gough JE, Derby B 2008. *Biomaterials*, *29*(2): 193–203.
Scoutaris N, Hook AL, Gellert PR, Roberts CJ et al. 2012. *Journal of Materials Science: Materials in Medicine*, *23*(2): 385–91.
Silberstein JL, Maddox MM, Dorsey P, Feibus A et al. 2014. *Urology*, *84*(2): 268–73.
Skardal A, Devarasetty M, Kang HW, Mead I et al. 2015. *Acta Biomaterialia*, *1*(25): 24–34.
Sodian R, Schmauss D, Markert M, Weber S et al. 2008. *The Annals of Thoracic Surgery*, *85*(6): 2105–8.
Su A, Al Aref SJ 2018. History of 3D printing. In:*3D Printing Applications in Cardiovascular Medicine*, pp. 1–10. Academic Press.
Tan DK, Maniruzzaman M, Nokhodchi A 2018. *Pharmaceutics*, *10*(4): 203.
Ten Cate AT, Gaspar CH, Virtanen HL, Stevens RS et al. 2014. *Journal of Materials Science*, *49*(17): 5831–7.
Uddin MJ, Scoutaris N, Klepetsanis P, Chowdhry B et al. 2015. *International Journal of Pharmaceutics*, *494*(2): 593–602.
Valentin TM, Leggett SE, Chen PY, Sodhi JK et al. 2017. *Lab on a Chip*, *17*(20): 3474–88.
Vithani K, Goyanes A, Jannin V, Basit AW et al. 2019. *Pharmaceutical Research*, *36*(1): 4.
Wang J, Goyanes A, Gaisford S, Basit AW et al. 2016. *International Journal of Pharmaceutics*, *503*(1–2): 207–12.
Wang JC, Ruilova M, Lin YH 2017. The development of an active separation method for bottom-up stereolithography system. *IEEE/SICE Int. Sym. SII*: Taipei, Taiwan on 11–14 December 2017. 108–114.
Wang Z, Abdulla R, Parker B, Samanipour R et al. 2015. *Biofabrication*, *7*(4): 045009.
Weisman JA, Nicholson JC, Tappa K, Jammalamadaka U et al. 2015. *International Journal of Nanomedicine*, *10*: 357.
Whitaker M 2014. *The Bulletin of the Royal College of Surgeons of England*, *96*(7): 228–9.
Wickstrom H, Hilgert E, Nyman JO, Desai D et al. 2017. *Molecules*, *22*(11): 2020.
Williams JM, Adewunmi A, Schek RM, Flanagan CL et al. 2005. *Biomaterials*, *26*(23): 4817–27.
Witowski JS, Pędziwiatr M, Major P, Budzyński A et al. 2017. *International Journal of Computer Assisted Radiology and Surgery*, *12*(12): 2047–54.
Xu X, Robles MP, Madla CM, Joubert F et al. 2020. *Additive Manufacturing*, *33*: 101071.
Yang Y, Wang H, Li H, Yang ZG et al. 2018. *Journal of Pharmaceutical Sciences*, *115*: 11–18.
Zein NN, Hanouneh IA, Bishop PD, Samaan M et al. 2013. *Liver Transplantation*, *19*(12): 1304–10.
Zuniga J, Katsavelis D, Peck J, Stollberg et al. 2015. *BMC Research Notes*, *8*(1): 10.

Index

O

P

Q

R

S

T

U

V

Z